How can you search for and find new minor planets and comets? How can you calculate the height of mountains on the Moon? And how do you photograph the K corona of the Sun? The questions asked by serious amateur astronomers are answered in this authoritative and wide-ranging guide.

Topics range from spectroscopy of the Sun and photoelectric photometry of comets to tracking artificial satellites and predicting their re-entry and break-up. For each topic, sound practical methods of observation and the scientific background are given to lead you to better observations. Guidelines also show you how to record and catalogue your observations using the recognised professional terminology and classification schemes.

From the simplest pencil drawings of the moon to photoelectric photometry of rotating minor planets, this guide is packed with practical tips for all types of amateur observations. It will develop the observational skills of the keen novice and satisfy the more demanding needs of the experienced amateur astronomer.

The Observer's Guide to Astronomy
Volume 1

The Practical Astronomy Handbooks are a new concept in publishing for amateur and leisure astronomy. These books are for active amateurs who want to get the very best out of their telescopes and who want to make productive observations and new discoveries. The emphasis is strongly practical: what equipment is needed, how to use it, what to observe, and how to record observations in a way that will be useful to others. Each title in the series will be devoted either to the techniques used for a particular class of object, for example observing the Moon or variable stars, or to the application of a technique, for example the use of a new detector, to amateur astronomy in general. the series will build into an indispensible library of practical information for all active observers.

Titles available in this series

The Observer's Guide to Astronomy
Volume 1

Edited by
PATRICK MARTINEZ

Translator
STORM DUNLOP

CAMBRIDGE UNIVERSITY PRESS
Cambridge, New York, Melbourne, Madrid, Cape Town, Singapore, São Paulo

Cambridge University Press
The Edinburgh Building, Cambridge CB2 2RU, UK

Published in the United States of America by Cambridge University Press, New York

www.cambridge.org
Information on this title: www.cambridge.org/9780521370684

First published as Patrick Martinez, *Astronomie, le Guide de l'Observateur* by
Edition Société d'Astronomique Populaire, and © Patrick Martinez 1987

This English language edition © Cambridge University Press 1994

English language edition first published 1994

A catalogue record for this publication is available from the British Library

Library of Congress Cataloguing in Publication data

Astronomie, le guide de l'observateur. English.
The observer's guide to astronomy/edited by Patrick Martinez:
translator Storm Dunlop. – English language ed.
p.   cm. – (Practical astronomy handbooks)
Includes bibliographical references.
ISBN 0 521 37068 X (v. 1) (hc). – ISBN 0 521 37945 8 (pbk.:v.1).
ISBN 0 521 45265 1 (v. 2) (hc). – ISBN 0 521 45898 6 (pbk.:v.2).
1. Astronomy–Amateurs' manuals. 2. Astronomy–Observers' manuals.
I. Martinez Patrick. II. Title.
III. Series: Practical astronomy handbook series.
QB63.A7813   1994
520–dc20   93-29830   CIP

ISBN-13 978-0-521-37068-4 hardback
ISBN-10 0-521-37068-X hardback

ISBN-13 978-0-521-37945-8 paperback
ISBN-10 0-521-37945-8 paperback

Transferred to digital printing 2005

The contributors

Michel Blanc

Roland Boninsegna

René Boyer

Christian Buil

Roger Chanal

Serge Chevrel

François Costard

Jean Dijon

Jean Dragesco

Michel Dumont

Pierre Durand

Robert Futaully

Claude Grégory

Alain Grycan

Jean Gunther

Serge Koutchmy

Jean Lecacheux

Michel Legrand

Jean-Louis Leroy

Jari Mäkinen

Jean-Marie Malherbe

Patrick Martinez

Marie-Josèphe Martres

Jean-Claude Merlin

Thierry Midavaine

Bruno Morando

Régis Néel

Gualtiero Olivieri

Olivier Saint-Pé

Jean Schwaenen

Jean-Pierre Tafforin

Jacques Barthès (drawings)

Summary

Contents

Contents

x

Preface

Amateur astronomers

Active amateur astronomers engage in their favourite activity as a hobby, in their spare time. They are able to devote a limited amount of time to it, being restricted by their jobs, whose hours are rarely arranged to suit people who want to spend the night looking through a telescope!

Amateurs are limited financially, because they either have to provide the money themselves or obtain it from an astronomical group, most of which do not generally have large amounts of funds.

Many instruments bought commercially by amateurs are small refractors or reflectors, and the majority of these are probably gathering dust in attics, either because their owners were disappointed in their performance, or because they were purchased in a burst of short-lived enthusiasm or, finally, because their owners became bored with looking at the Moon or M31 for hours on end, did not know what else to do, and abandoned astronomy for some other pursuit. It was thinking about the latter group of observers that the germ of the idea for this book was sown: such amateurs need to be shown the vast range of useful observations that are possible, even with modest equipment, and how they should proceed.

Amateur astronomy

Just as professional astronomers may be divided into two broad groups, observers and theorists, the same applies to amateurs. Some are purely observers, and others are happier in a library than in an observing dome. Apart from these two classes of amateur, however, there are two more categories. There are the 'Telescope Nuts' (as they are often termed in North America), who spend their lives building bigger and better telescopes, but who often have no time to use their latest equipment before they start on something new, and there are the society members, who are the backbone of all the local and national astronomical groups and keep them in existence, but who have no time to look at the sky in between organising meetings, exhibitions, or conferences.

Let us concentrate on the observers. There are three types of observations made by amateurs:

Sight-seeing This consists of looking at objects in the sky purely for their beauty, and this is where most people begin. This is often the initial reason – and for some people the only reason – for taking up astronomy. Some amateurs never progress beyond this stage, but many others feel an urge, after a certain time, which may be months or years, to take things further, and frequently they are unable to find any guidance either in their circle of acquaintances or in the literature. Even if 'sight-seeing' is, by definition, unproductive, apart from the pleasure that it brings, it should not be neglected, because it enables amateurs to gain experience and become accustomed to observing; both qualities that are of considerable value when they turn to more 'serious' work.

Educational　　Here, the astronomer carries out an experiment, which may require special equipment or measurements, or else a greater or lesser degree of sophistication in analysing the results. There are many different examples, but unfortunately the results are foregone conclusions, with the outcome having been known for decades – even for centuries – and to a degree of accuracy that is far beyond amateurs' reach.

True observation　　This often forms part of a specific programme of observation, and the results contribute – albeit sometimes in a very modest fashion – to our overall knowledge of the universe. People frequently have little idea of the range of observations within the capabilities of amateurs, who are thus able to make really useful observations.

In fact, amateur and professional astronomy truly complement one another. Although professionals have powerful methods at their disposal, amateurs are able to cover those fields that require little equipment or small instruments (such as binoculars, small, wide-field telescopes, etc.), mobile resources (such as the observation of grazing occultations), or a large number of observations (examples are variable stars, double stars, sky patrols, etc.). The result has been an increase in the collaboration between professionals and amateurs, with the latter having greater access to large telescopes such as the 600-mm reflector at the Pic du Midi Observatory, or participating in meetings such as the one organised by the Société Astronomique de France in Paris in 1987, which was supported by the International Astronomical Union (IAU Colloquium 98).

If we consider the fields for which amateurs are noted, we find that France is among the first rank on an international level when it comes to the study of planetary surfaces, double stars and variable stars. It is also relatively well placed as regards the Sun, comets and meteors.

Occultations appear to be a Belgian speciality, but in many other fields such as astrometry, deep-sky work or the discovery of new objects (comets, novae, supernovae, and minor planets), the United Kingdom, United States, Australia or Japan set an example. It is time that French astronomers woke up to these particular fields, where their forebears were extremely prominent. One gets the impression that nowadays, apart from a handful of top-notch observers, whose names appear from time to time in the International Astronomical Union's telegrams and circulars, or in specialized international journals, that French astronomy has lapsed into somnolence, and has left it to other countries to make all the discoveries. (The balance-sheet when it comes to comets is particularly eloquent.)

The observer's guide

Somehow we need to convey to amateurs who have grown bored with seeing the same few notable objects that they look at every night, that there are fascinating observational programmes, in which they can also achieve extremely useful results.

It became obvious that it would be of great value if descriptions of the useful work that may be carried out could be brought together in one publication. This is what gave birth to the idea of this '*Observer's Guide*'. To avoid overawing beginners, it does include some 'sight-seeing' and 'educational' projects, but because it aims to guide observers into making useful 'scientific' work, it does concentrate on the latter.

Any such project is highly ambitious and could not be realized by a single person. In fact, each subject needs to be described by a specialist, and as no one can hope to be expert in every field, this naturally led to the idea of a collaborative work.

This book examines all the different types of object available to amateur astronomers, and for each one it describes the types of observation that are possible (paying particular attention to their scientific value), the equipment required, the methods to be employed, where appropriate information may be found, and the organisations to which the results should be reported. Each chapter has been written by a French or Belgian expert (or experts) in the subject concerned; the authors include both professional and amateur astronomers.

The first fifteen chapters describe observational programmes, divided into individual classes of astronomical object. The last five chapters discuss techniques that relate to several of the subject-areas previously described. This arrangement avoids unnecessary repetition.

An effort has been made to standardize the text provided by the various authors, and to limit the amount of editing required; some subjects that are considered to be particularly important, however, are deliberately included in more than one place in the book.

Despite its size, the '*Observer's Guide*' cannot include every aspect of amateur astronomy. It has been assumed that the reader is familiar with all the basic techniques, such as setting up a telescope, finding objects, etc. If not, details may be found in any one of a number of books intended for beginners, so it would be pointless to repeat them here.

For similar reasons, there is no chapter devoted to astrophotography as such. The basic techniques that are required are described in a number of books, which are given in the bibliography. It therefore seemed pointless to include information about this general area in this book, and to do so would have required 100 to 200 additional pages. On the other hand specific photographic techniques required by particular types of observation are discussed in considerable detail in some of the individual chapters.

Advice to the reader

The '*Observer's Guide*' is not meant to be read straight through from cover to cover. It has been designed so that the individual chapters are relatively independent, so that the reader may concentrate on the subjects of interest, and in any desirable order. Relevant cross-references are included to other chapters and sections as required.

This book should be regarded as a catalogue, where amateurs may find out what types of observation are possible given the facilities that they have available, and from which they may choose those fields that seem most attractive to them. The text describes all the steps and equipment necessary to take the first step. If this serves to confirm the initial attraction for any particular field of interest, addresses of various organisations are given that may be contacted for further information and guidance.

So now ... good reading, and good observing!

Translator's preface

In a multi-author work, there are always problems in ensuring consistency in presentation, style, and usage. Not surprisingly, there were some disparities in the original French text. In translating this work, I have tried to ensure that the whole text is consistent. In accordance with recommended practice, for example, all wavelengths in the visible and adjacent spectral ranges are now given in nanometres (nm) rather than the older, perhaps more familiar, ångström units – 1 nm (10^{-9}m) = 10 Å (10^{-10}m). At longer wavelengths, the unit used is the micrometre μm (10^{-6}m). Similarly, the dimensions of telescopes, etc. are consistently given in millimetres or metres (not in centimetres or decimetres).

The symbols for physical variables and constants have been altered to those most commonly used in English-language works and, whenever possible, agree with the recommendations of the IAU or the various IAU Commissions. Annotations and captions to the figures have also been changed to forms more readily related to the English-language terms for the terms and items designated.

Dates and times have been expressed consistently in the accepted international scientific format: Year, month, day, hour, minute, and second, with decimal forms where appropriate. In accordance with this standard format, the name of the month is given in full or as a three-letter abbreviation, not in numerical form.

One major alteration from the layout of the original volumes concerns the positioning of notes, references, glossaries and bibliographies. The original chapters were inconsistent, some giving this material within the text, some in footnotes, and some in 'appendices' at the ends of the chapters concerned. The majority of notes, glossaries, and appendices have now been incorporated in the text. Where it has been thought advisable to insert a translator's note, this is enclosed in square brackets. Section numbering and cross-referencing have also been simplified, and do not correspond precisely with the arrangement of the original edition. A few items (including some Figures) have been moved to more appropriate positions. Two general appendices on 'Time-scales' and the interesting 'T60 Association' may be found at the end of Volume 2.

Discussion of the merit (or demerit) of specific charts, atlases, etc. for a particular branch of practical work has been retained in the text, usually within a special sub-section. References to papers, articles, etc. – i.e., items with no accompanying discussion or comments – may be found with a more general bibliography (and a few notes) for each chapter in sections at the back of each volume. Please note that where books are cited as specific references, they are not repeated in the general bibliography.

The original work contained many references to French-language material, in particular to papers and articles in French journals. Obviously, many English-speaking readers will be unable to understand the original language. It was felt, however, that some items were not adequately treated in English, so some references to major French-language publications, particularly books and the journal *l'Astronomie*, have been retained. Most references to publications issued by local groups have been removed, because the majority of readers will find the material impossible to obtain or consult. Various references to easily available English-language books and journals have been added, including some material not published when the original text was written.

There are known to be inconsistencies in the details given for some of the references, many of which were missing volume, issue, or page numbers – and even in some cases date, author or title! Regrettably, it was not possible for me to spend the extremely large amount of time that would be required to check and correct every individual reference. (Many dozens of corrections have been incorporated, and I would particularly like to thank Mr Peter Hingley, the Royal Astronomical Society's Librarian, for all the work that he carried out in this respect.) It is believed that the information given will enable the appropriate papers to be located without too much trouble, even though the information may be incomplete.

All necessary amendments to the French edition have been incorporated, and in addition, I have corrected many minor, previously undetected errors, and included the latest data wherever possible – such as revised totals for numbered minor planets, and the number of supernovae discovered by Bob Evans – without reference to the authors. Many major discrepancies and errors have been referred back to the original editor, Patrick Martinez, and the authors, and appropriately corrected.

In a few cases, the precise methods used by observational groups in Francophone countries differ from those employed by the major, English-speaking organisations. No attempt has been made to 'correct' these points, not only because the various methods may be equally valid, but also because the differences may be instructive. However, attention is sometimes drawn to any disparity in a translator's note.

It is impossible for me to mention individually the many amateur and professional astronomers in all parts of the world who have so unstintingly helped me with details of methods and organisations, or with more general advice. There is not a single chapter that has not benefited from their assistance, which is greatly appreciated. If I may single out one person in particular, whose advice has extended far beyond the field of just minor planets and comets, it is Dr Brian Marsden, of the IAU Central Bureau for Astronomical Telegrams in Cambridge, Massachusetts. His advice, often received almost instantaneously, thanks to the wizardry of modern computer networks, has been invaluable.

Storm Dunlop

1 The Sun

M. J. Martres, R. Boyer, F. Costard, J. M. Malherbe & G. Olivieri

1.1 Introduction

This chapter summarizes the most important facts about the Sun and its behaviour, and continues by describing why and how interesting observations may be made. This second aspect is covered in two sub-sections: the first concerns the observation of the Sun in integrated (white) light, while the second describes observation in monochromatic light (primarily Hα). Each sub-section includes a description of certain general features: sunspots, groups of spots, and photospheric activity in the first (Fig. 1.1); flares, prominences, and chromospheric activity in the second. Specific observational techniques, which are suitable for either individual or group programmes are then described. These may be undertaken merely for the observer's personal pleasure, or be made with scientific purposes in mind. This freedom of choice is a valuable aspect of amateur astronomy.

Why should we observe the Sun? There are three primary reasons:

- The Sun is a star;
- It is the star closest to us;
- It is a variable star, and we are totally dependent upon its behaviour.

1.1.1 The structure of the Sun

The Sun is one small star among the 100 thousand million that form the Milky Way galaxy. It is quite ordinary as far as its mass, luminosity and effective temperature are concerned. Its stellar class is G2.

The energy derived from the hydrogen fusion reactions taking place in its core (at a temperature of 15×10^6 K) is emitted by the Sun as radiation. The major part of this radiation is concentrated in the visible spectrum, which is produced by a thin layer, with an average thickness of 100–200 km and a temperature of 5000 K, known as the photosphere, the apparent solar surface. This is where a sharp rise in density causes the gas to become opaque, giving rise to the visible disk (Fig. 1.2).

The temperature increases with increasing distance from the Sun, rising to 6500 K in the chromosphere, which has a thickness of about 2000 km. Higher temperatures (10^6 K) prevail in the corona, which lies farther out and extends well beyond several solar radii. The chromosphere and the corona are transparent and less dense than the photosphere and therefore cannot be detected by direct observation against the solar disk. They may be observed only in certain spectral lines. Ground-based instruments can use 'visible' and 'radio' wavelengths, but only instruments carried above the Earth's atmosphere allow spectral analysis to be undertaken in the ultraviolet region, or for measurements to be made in X-rays or γ-rays.

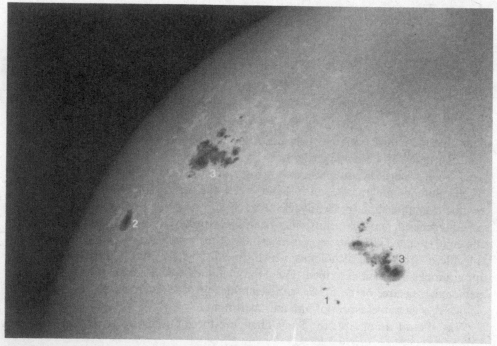

Fig. 1.1. *Several groups of sunspots at different stages of development, photographed by J. Giammertini on 1979 June 2 with a 150-mm aperture reflector.*
1: small bipolar group (early stage of formation); 2: large round spot (an advanced stage in the decay of a bipolar group: one magnetic pole has retained its spot, but the other consists of just a facula without a spot); 3: complex groups (active areas consisting of several bipolar groups).

When the Sun's disk is hidden during a solar eclipse, white-light observations of the inner corona are possible at the solar limb. The bright, irregular ring of light is the sum total of emission from all those parts of the chromosphere and corona, along the line of sight, that are sufficiently high and sufficiently bright to be seen projected against the sky, outside the Sun's luminous limb. Artificial eclipses of the Sun, as produced by coronagraphs, also allow the inner corona to be studied and for its variation with time to be followed. This is provided the experiment is carried out at a site with exceptionally clear skies, such as a mountain-top site, from the upper atmosphere, or from space.

1.1.2 Solar variability

The Sun is a star that 'varies' over the course of time, despite the somewhat limited nature and structure of the variations that are observed. Here are two examples.

Sunspots Although they were detected several centuries before our era, sunspots were identified correctly only after the invention of the astronomical telescope (in

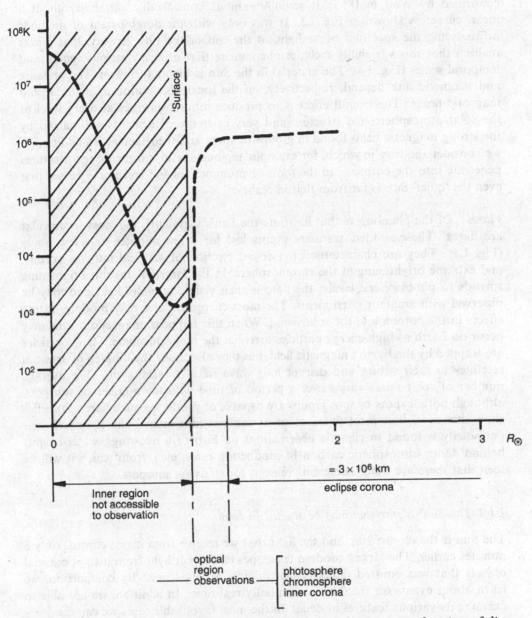

Fig. 1.2. *Curve showing the variation in temperature of the Sun as a function of distance from the centre (at 0) out to 3 solar radii.*

1610). Observers of that period, who included Galileo Galilei, Scheiner and others, realised that sunspots did not remain static on the surface of the Sun, and that they neither had a fixed lifetime nor showed any apparently orderly behaviour. They were variable. It was an amateur, Schwabe, however, who first suggested the idea of a cycle in the production of spots. The periodicity of about $10\frac{1}{2}$ years was later confirmed by Wolf in 1855. It actually exhibits considerable variability about a mean curve, as shown in Fig. 1.3. It was only with the development of methods of analysing the spectrum of sunlight, at the end of the 19th century, that it was realised that this variability includes phenomena that occur at various spatial and temporal scales (Fig. 1.4). The material in the Sun is highly turbulent. The pressure and magnetic flux depend, respectively, on the local distribution of material and magnetic fields. The overall effect is to produce inhomogeneities at every level of the solar atmosphere, the structure and very existence of which is maintained by the strong magnetic fields found in groups of spots and their faculae. In addition, if we consider the way in which, for example, regions at chromospheric temperatures penetrate into the corona – in the form of prominences and spicules – we see that even the 'quiet' Sun is far from that in reality.

Flares Of the phenomena that illustrate the Sun's variability, the most spectacular are flares. These sudden, transient events last for a few minutes or a few hours (Fig. 1.5). They are characterised by violent ejection of material into the corona, and extreme brightening of the chromosphere. In the largest flares this brightening spreads to photospheric levels; the flare is then visible in white light and may be observed with amateur instruments. The most energetic flares may produce lasting effects in the corona and the solar wind. When this happens, magnetic storms may occur on Earth as high-energy particles arrive at the magnetosphere. These particles are trapped by the Earth's magnetic field, but they also cause the braking of artificial satellites in their orbits, and disrupt long-wave radio transmissions. The average number of solar flares varies over a period of time with the number of sunspots, although not all spots or spot groups are capable of producing such flares. Statistics show, however, that most flares occur at sunspot maximum. The 10- or 11-year periodicity is found in climatic observations on Earth (in tree-rings, water-heights behind dams, atmospheric carbon-14 production rates, etc.), from which it will be seen that there are other interesting reasons for studying sunspots.

1.1.3 The Sun: a star that may be studied in detail

The Sun is the closest star, and the light that we receive from it was emitted only 8 minutes earlier. The largest modern telescopes can detect light from distant celestial objects that was emitted thousands of million of years ago. By comparison, we learn about events on the Sun in essentially real-time. In addition, we are able to examine its various features in detail. In the most favourable cases we can resolve – i.e., distinguish from one another – events that occur only 150 km apart.

The fact that the Sun is so near also provides us with a considerable amount of light, enabling us to make detailed measurements with short exposures without having to use large apertures. It is therefore particularly favourable for amateur observation. (It is useful to have a long focal length because the diameter of the primary solar image is about 1/100th of an objective's focal length.)

It should be noted that, for the Sun, it is the quality of the observing site that ultimately determines the largest aperture that may be used for collecting the sunlight. It is pointless, for example, to design an instrument that has a theoretical

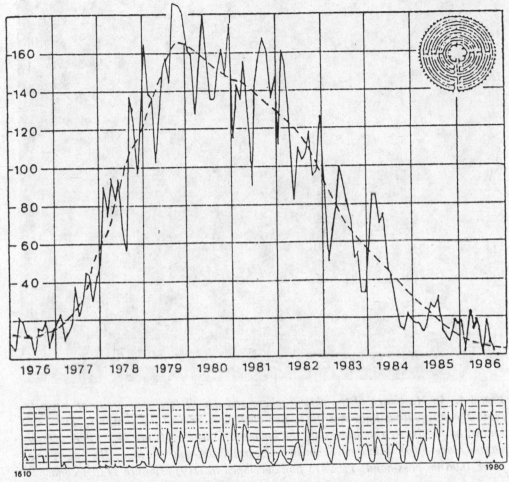

Fig. 1.3. *Solar variability.* Top: *variation in the number of sunspots over an 11-year cycle, from 1976 to 1986. This curve has been derived from the monthly average Wolf Number; it shows considerable variation around the mean (indicated by the broken line).* Bottom: *a reconstructed curve showing the secular variation in solar activity from 1610 to the present. Both curves were prepared by the Sunspot Index Data Centre in Brussels.*

angular resolution of 0.3 to 0.5 arc-seconds, if it has to be located at a site where the quality of the seeing never permits a resolution better than 2 arc-seconds. We shall see later that solar observations may be carried out with small-diameter telescopes, and that 150–200 mm is adequate. The main problem is ultimately one of rejecting the excess light, and thereby reducing the heating of the interior of the instrument and protecting the eyes of the observer (Quentel, 1980).†

† Notes, references and bibliography to volume 1 commence on p. 587

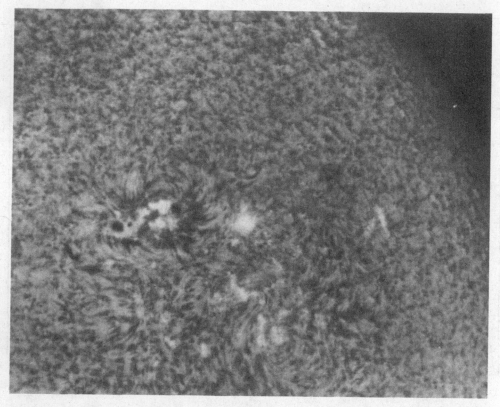

Fig. 1.4. *An Hα image of the chromosphere and a well-developed group of sunspots. Note the alternating light and dark, fine structures around the sunspots. These are known as fibrils. At this level, only the umbrae of large spots are detectable. (Observation made by D. Michaut: Schmidt-Cassegrain, 200-mm telescope; Hα interference filter, 0.06 nm pass-band; TP 2415 film developed in D19; exposure 1/30 second.)*

1.2 The changing appearance of the Sun

1.2.1 Solar rotation

The invention of the telescope at the beginning of the 17th century allowed Galileo Galilei to see spots on the Sun. We can readily imagine that in those days the quality of the image must have been very poor (given the type of optics that were used), and that observation of these, then unknown, objects must have been extremely difficult. Nowadays the Sun is much better understood, and most amateur astronomers are technically far better equipped than Galileo. They are able to learn about sunspot observation very rapidly, whereas their predecessors – the 'amateur' astronomers of their day: Schwabe, Wolf, Wolfer and others – attained the same level of knowledge only after years of patient work.

Sunspots show that the Sun's rotation on its axis is direct, like that of most of the planets (anticlockwise when looking down on the North Pole). For a terrestrial

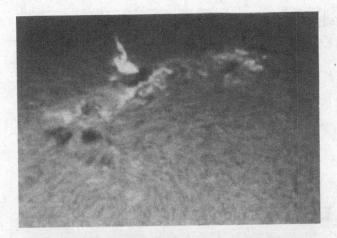

Fig. 1.5. *A weak flare observed near the limb, 1980 April 12. Both the local brightening and ejection of material are visible. Changes occurred from minute to minute (top 14:03 UT, bottom 14:04 UT). The ejection of material reached a height of more than 50 000 km above the solar 'surface' (as a comparison, the diameter of the Earth is 12 750 km).*

observer, a spot crosses the disk from left to right, i.e., from the Sun's eastern limb to the western, following a small circle that is centred on one of the Sun's poles (Fig. 1.6). A spot's lifetime may be less than half a solar rotation. Rotation may carry it round the eastern limb onto the visible disk, for example, but it may disappear before it reaches the western limb, or vice versa. On the other hand, it may live for more than half a solar rotation, or for even more than one rotation. How may it be recognised after it has disappeared for half a rotation and crossed the invisible hemisphere? Precise determinations may only be carried out by taking account of the changes in the geometry of the Earth–Sun system, which alters from day to day according to the Earth's position in its orbit and distance from the Sun throughout the year.

The relevant data are published by various institutions, such as the Bureau des Longitudes in Paris, the Nautical Almanac Office in the U.K., and the U.S. Naval Observatory in the U.S.A.

Determination of the heliographic longitude of a single spot every few days enables the angle by which the Sun has rotated in that time to be calculated. For

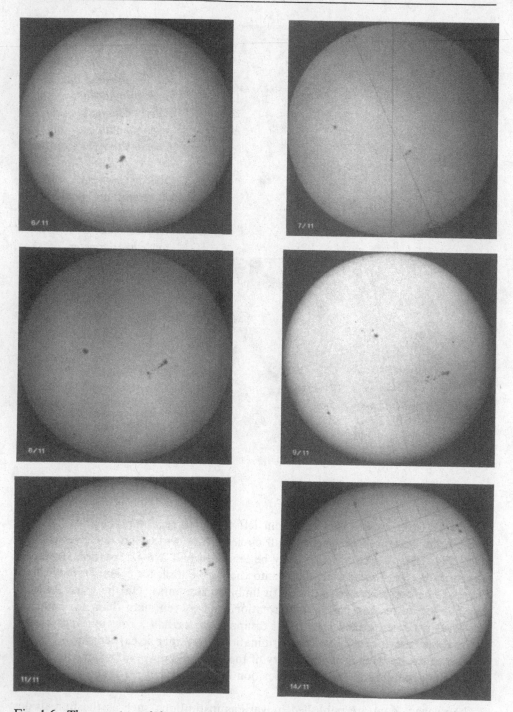

Fig. 1.6. *The rotation of the Sun, and the apparent motion and evolution of groups of spots. White-light photographs taken with instruments with apertures of between 77 and 206 mm on 1978 November 6, 7, 8, 9, 11 and 14, by M. Berthe, F. Grasse, B. Lanoë and J. Giammertini as part of the sunspot proper motion programme organised by the Société Astronomique de France.*

a spot at the equator, a complete rotation takes about 27 days. This is a synodic rotation, relative to a terrestrial observer. Because of the Earth's orbital motion, which is in the same direction as the Sun's rotation, this is about 2 days more than the true, sidereal, rotational period.

1.2.2 The Sun's differential rotation

The Sun does not rotate as a rigid body and its angular velocity decreases smoothly from the equator to the poles. A point on the equator makes a full rotation in 25 days, but a spot at 60° latitude takes more than 30 days (Fig. 1.7), and spectroscopic measurements indicate a period as long as 35 days at 80°. The differential rotation is calculated here based on sidereal rotation, i.e., the Sun's true rotation period, rather than its apparent, or synodic, period. Similar differences occur if synodic rotation is taken instead. There is also a differential rotation with depth, ranging from rigid rotation in the deep interior to a differential rotation – which is less marked than in the photosphere – in the outer layers. Apart from the statistically averaged values for the rotation, there are both large- and small-scale variations in the rotational velocity. Such differences affect spots, filaments, coronal holes, jets, etc. (giving rise to individual proper motions), and a similar variation may also occur from cycle to cycle.

Understanding the process of differential rotation is nowadays considered of vital importance in resolving many of the problems raised by modern solar physics. It is related to the basic mechanism governing the magnetic fields and their emergence from the surface and, as a result, to the source of solar activity and the associated turbulence within the outer layers. One element in investigating these phenomena is the study of sunspot proper motions, and this is where amateurs can play a part (see the collaborative programme described in Sect. 1.4.3).

1.2.3 Spörer's Law

After the discovery of the 11-year sunspot cycle, Carrington (in 1859), and Spörer (in 1874), detected the law governing the drift in latitude of the average position of spots. The first spots in a new cycle appear at about latitude 30°, whereas at the end of a cycle spots always occur near the solar equator. This effect has shown that the true length of a solar cycle is longer than 11 years, because there is always an overlap between successive cycles (Fig. 1.8).

1.2.4 The magnetic cycle

The first sign of a future group of sunspots is generally the appearance of numerous, adjacent, small pores. Two of these subsequently become spots. Groups are generally bipolar, that is they have two main spots. The westernmost spot, the preceding (or p-) spot, and the easternmost spot, the following (f-) spot, have opposite magnetic polarities. At any particular time, and for any individual cycle, all the p-spots in one hemisphere and all the f-spots in the other hemisphere have the same polarity.

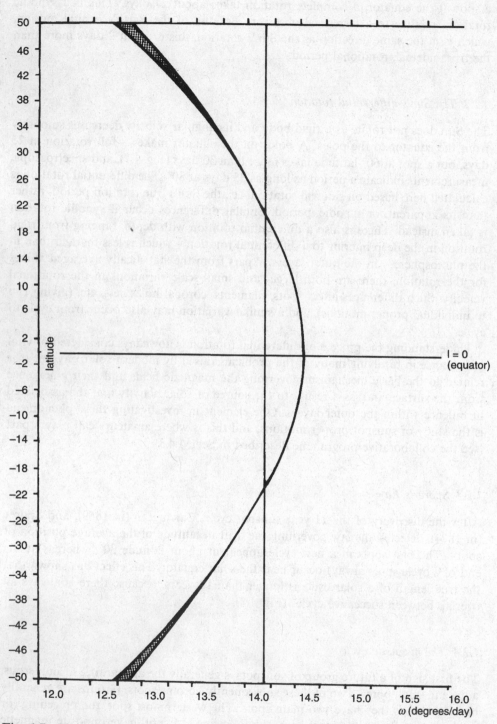

Fig. 1.7. *The differential rotation of the Sun. This graph shows the variation in the velocity of rotation in degrees per day as a function of heliographic latitude.*

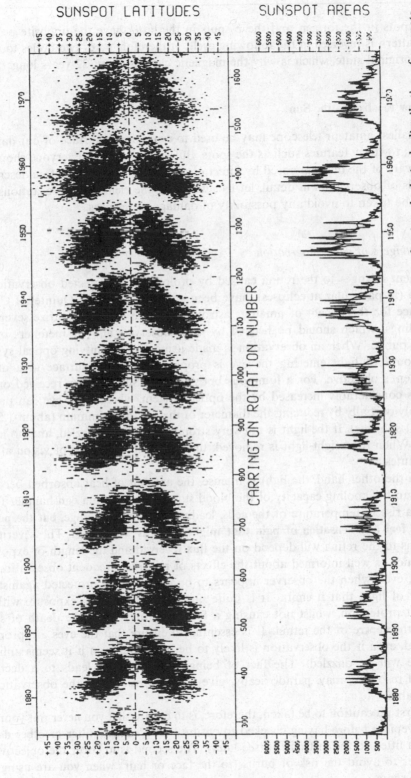

Fig. 1.8. *The variation in sunspot distribution and numbers. Top: a 'Butterfly' diagram showing the latitude-distribution of sunspots observed between 1874 and 1976 (and illustrating Spörer's Law). Bottom: the area of sunspots for the same period. (From the Royal Greenwich Observatory's Photoheliographic Results)*

The f-spots in the former, and the p-spots in the latter have the opposite polarity. This pattern is reversed in the following cycle. It therefore takes two cycles to revert to the original state, which is why the magnetic cycle is about 22 years long.

1.3 How to observe the Sun

The smallest amateur telescope may be used to examine the surface of our daytime star and to study features such as the spots. But care is needed! Everyone should be fully aware of the dangers posed by direct observation of the Sun. Before describing methods of observation in detail, let us summarize the elementary precautions that should be taken to avoid any possibility of damage to the eyes.

1.3.1 Dangers of solar observation

Permanent changes to the retina caused by prolonged, unprotected observation of the Sun (in particular at eclipses), have been analyzed by Dr G. Quintel at Créteil in France for the benefit of amateur astronomers. We shall summarize several of his findings, which should be heeded by solar observers and by members of the general public. When an observation is made using any magnifying optical system, the amount of light entering the eye is proportional to the surface area of the instrument's objective. For a luminous body like the Sun, the flux received on the retina is considerably increased by the optical system, whereas the eye can protect itself only partially by reducing the diameter of the iris to a minimum (about 1.5 mm for most subjects). If the light is not very strong, one is only dazzled, and no lesion occurs. When the bright light is removed, the eye regains its sense of vision after a few minutes.

If, on the other hand, the light is intense, the amount of heat absorbed becomes more than the cooling capacity of the blood supply to the retina can handle. There will be a rise in temperature of the cells, leading to definite damage, **but the person will not feel any sensation of pain that might serve as a warning!** The severity of the burns to the retina will depend on the flux received and the length of exposure. We are not as well informed about the effects of repeated, frequent observations of the Sun, even when the observer appears to be more or less protected against the quantity of light that it emits. It is quite probable that a weak exposure without sufficient protection, whilst not causing a full burn, may cause the death of some cells in the centre of the retina. It is essential to ensure that the eyes are properly protected, even if the observation is likely to be brief, and even if it seems unlikely that one will be dazzled. The fact of being dazzled, which leads to a decrease in visual response, may, paradoxically, give the impression that the observation is comfortable.

The first precaution to be taken, therefore, is to ensure that you **never** put your eye to the eyepiece **before** having checked to ensure that a **suitable filter or other device** has been fitted to the instrument (see later). The second is to cover the objective of the finder to avoid the risk of burns (to the face or hair) when you are using the main telescope.

In addition, you should be wary of the solar filters provided with most telescopes sold commercially. They are located at the eyepiece end, and absorb the light at the point where it is most concentrated, close to the focus of the objective. Such a filter should always be used behind a Herschel wedge or an objective filter, otherwise it may shatter in a few seconds through the intense heating to which it is subject. The whole of the luminous flux may suddenly reach the retina and cause serious damage before the observer has time to move away from the eyepiece.

Finally, an essential precaution is to ensure that the filter absorbs over a sufficiently wide spectral band. Some filters pass a significant amount of infrared or near-ultraviolet radiation, which is invisible, but may still cause burning of the retina.

1.3.2 Methods of observing the Sun

1.3.2.1 Direct visual observation

Amateurs often observe the Sun through an eyepiece, after having fitted suitable devices for reducing the amount of light (filters, Herschel wedge, reflection from a non-aluminized surface, etc.). This method is perhaps the best for obtaining a spectacular view of the surface of the Sun. We do not recommend it, however, for two specific reasons:

- Unless observers have taken every precaution they do run the risks that we have described in Sect. 1.3.1;
- The scientific value of the observations is reduced. A drawing of a group of spots made directly at the eyepiece will never have the same degree of accuracy as a photograph or even as a sketch made by projection.

1.3.2.2 Observation by projection

This method consists of directing the telescope towards the Sun by watching the shadow cast by the body of the instrument. When the shadow is reduced to a circle, the image of the Sun should be within the field of view of the telescope. It is then only necessary to hold a piece of cardboard at about 200–300 mm from the eyepiece. By gradually changing the focus of the eyepiece, a sharp image of the solar disk may be obtained on the screen. Many instrument makers provide a collar and a set of rods that can carry a screen for observation by projection. Most amateurs are also capable of making up a suitable support for themselves. If one arranges for the diameter of the image to be 114 mm, then a degree of longitude or latitude at the centre of the disk is exactly one millimetre. To determine the position of any surface detail (a sunspot, for example), the simplest method is to draw the feature in question onto a blank that has been prepared in advance (a circle with marks indicating the geographical North and South).

If the telescope has no drive, the image drifts rapidly, requiring frequent adjustment to the instrument during the observation. With a little practice an observer should be able to draw spots in no more than a few minutes (Fig. 1.9). If not, the greater the degree of perfection in the drawing, the less the actual accuracy is likely to be.

There is another procedure, which requires a somewhat larger budget. This

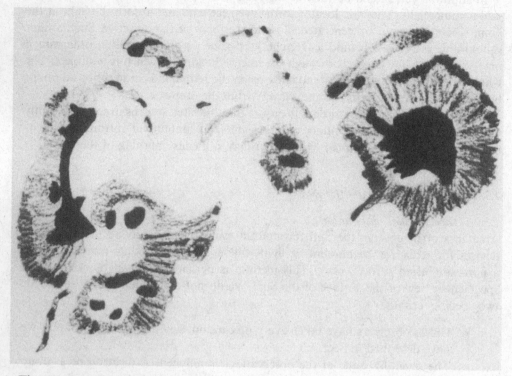

Fig. 1.9. *A sunspot group, drawn by F. Costard from a projected image obtained with the 153-mm refractor at the Sorbonne Observatory, fitted with a Herschel wedge.*

consists of photographing the solar image formed on a projection screen. The speed of this process, when the setting is carried out in advance, means that drift may be eliminated and a drive is no longer indispensable. This method does, however, require considerable care in aligning and setting the telescope properly, and in obtaining a truly circular image and checking the orientation. The method is also very sensitive to atmospheric turbulence.

1.3.2.3 Photographic observation

The best results are obtained by photographic techniques (Fig. 1.10). Direct photography of the solar image at the primary focus of the instrument is not advisable, to prevent possible damage to the camera's shutter. To obtain a proper photograph, and to ensure that observation is both comfortable and safe, it is essential to devise an arrangement that only passes a small fraction of the incident light. The first idea that comes to mind is to use a filter. For both refractors and reflectors, two positions are easily accessible: in front of the objective and in front of the focal plane.

The first solution (which is unfortunately expensive) requires a neutral-density filter with plane-parallel faces that is at least as large as the objective, which has a high degree of flatness and truly parallel faces, and is very homogeneous. Reduction of the light may be obtained either by absorption (if the filter has been dyed in the

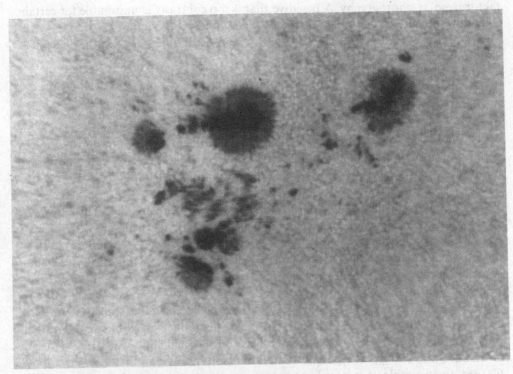

Fig. 1.10. *Photograph of a group of spots obtained by D. Cardoen with a 100-mm aperture refractor, focal length 1500 mm; Agfa Ortho 25 film, exposure 1/500 second.*

mass), or by reflection (if the sheet has a semi-transparent coating). Reduction by reflection is always preferable to reduction by absorption because it avoids heating the filter and the surrounding layers of air. Some observers use interchangeable filters for alignment and focussing and then for taking the photograph. Such filters consist of a metal-coated mylar film in a rigid holder that fits over the objective. For photographic work, a reduction factor of one thousand is generally used, in combination with a shutter speed of 1/500th to 1/1000th of a second. For visual observation, the reduction should be of the order of one million if one wishes to avoid the dangers of being dazzled that were described earlier.

The second solution (a filter near the focal plane) allows almost all the luminous flux to enter the instrument. Most of this energy is then rejected by a much smaller surface, lying in the converging cone of light at a point close to the focus. One can obtain a commercial filter that only transmits 3 % of the incident light (density 1.5), and this consists of an aluminium film deposited on a sheet of silica glass. A diameter of only 30 mm suffices for such a filter. Additional reduction is required, either by using a neutral-density filter, or by a welding glass. Another possibility is to use a Herschel wedge with one glass-air reflection, or else a pentaprism with two glass-air reflections, with either a neutral filter or one with a colour chosen specially for photography.

Contrary to the idea often put forward, high contrast is not the most important factor in solar photography. A negative that is too contrasty is unsuitable for certain scientific purposes. The solar observer wants good reproduction of the mid-tones and detail in both the 'light' and the 'dark' areas. Both the spots and the solar limb should be clearly visible.

One of the problems of direct photography of the surface of the Sun is that of over-exposure of the photographic emulsion by the large amount of light delivered by the objective. After having chosen a suitable method of reducing the amount of light from those just described, and using the fastest shutter setting (1/500th or 1/1000th second), the observer needs to experiment to find the most suitable combination of coloured filter and photographic emulsion. It is, of course, necessary to choose a slow-speed film, which has fine grain and fairly high contrast, and to develop it in a classical fine-grain developer.

Ground and polished glass colour-filters are preferable to gelatine or plastic ones, which soon degrade with the heat in the beam of sunlight. In the arrangement where most of the light is rejected just in front of the focus, these filters must be placed after the device that rejects most of the light (the semi-reflecting metal film, Herschel wedge, pentaprism or whatever), where they will not be subject to great heat and will retain their optical properties.

In choosing a filter, it is necessary to have an idea of the spectral-sensitivity curve of the emulsion being used. Current fine-grain emulsions are almost all panchromatic, but their sensitivity generally drops beyond about 650.0 nm. By using a red filter that is only transparent to radiation with a wavelength longer than about 600 nm, for example, it is possible to create a system with a passband about 45 nm wide, centred on a wavelength of about 625 nm. Figure 1.11 gives an example of a combination achieved by using Kodak Panatomic-X 5060 35-mm film and a Schott RG 610 filter (3 mm thick). Corning filters also have interesting properties.

There is an additional benefit from using this long-wavelength region of the spectrum. The Earth's atmosphere is less refractive in the red, so atmospheric effects on seeing are correspondingly reduced. This improves the quality and the resolution of the images, and gives the impression that the emulsion contrast is increased. In reality it is the image itself that is of higher contrast.

There are other excellent emulsions: Agfa Pan 25 by Agfa-Gevaert, and Kodak's AHU 1454 Infocapture microfilm may be used in the same way as Panatomic-X. Kodak also markets Technical Pan (TP) 2415 film, which is extremely sensitive out to about 700 nm. (It is the ideal emulsion for star fields, nebulae and comets.) Its use in solar photography requires attenuation by about the amount required for comfortable visual observation. To make proper subsequent use of the photographs, it is essential that the limbs of the Sun are clearly visible. This is achieved by using slow, high-contrast films, such as those just mentioned. In printing positive images on paper, the preferred choice is a matt paper in a normal to soft grade, with a polyester backing. This ensures that dimensional changes during processing are avoided, and also enables one to write on the sensitive surface with a pencil.

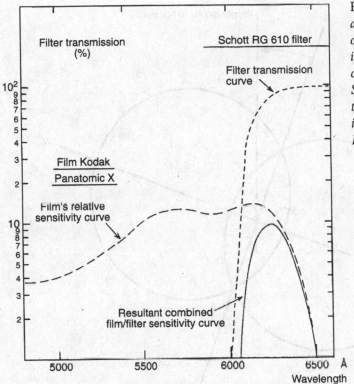

Fig. 1.11. *Example of a narrow bandwidth obtained by combining the transmission curves of the filter (a Schott RG 610) and the spectral sensitivity of the film (Kodak Panatomic-X).*

1.3.3 The orientation of drawings and photographs

1.3.3.1 Determination of geographical North and South

Let us assume that the observer is using the photographic method. It is easy to determine the required compass points by following the procedure indicated in Fig. 1.12. An initial exposure (1 in the Figure) is made with a duration t that gives a correctly exposed image. The position of the instrument is clamped and a second exposure is made 90 seconds later, on the same frame. Because of the diurnal rotation, the second image of the solar disk will have been displaced by about 1.5 times its radius from the first one. To indicate the direction of motion, the second exposure is made with a duration of $t/2$ or $t/3$; the under-exposed image will then be on the west. It only remains, after enlarging the picture, to draw the straight line NS, which passes through the points where the two circular images intersect. The precision obtained by this method is of the order of 1/2 degree.

If drawings are to be made from a solar image projected onto a screen, the East–West direction may be found rapidly by clamping the telescope and marking the position of a small spot at two different times as the image drifts across the field.

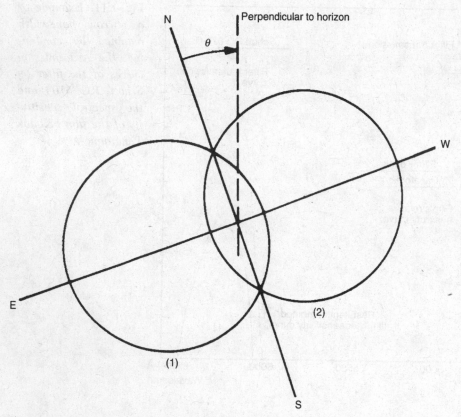

Fig. 1.12. *Determining the North–South geographical axis of an observation by experiment, using a double exposure and diurnal motion.*

1.3.3.2 *Type of mounting and orientation of the image*

If the observer uses an equatorially mounted telescope, the motion of all the objects in the sky is strictly related to a specific line in space, the North–South axis. As a result, provided the orientation of the screen carrying the projected image or that of the camera is not changed, the geographical North–South orientation of the field of the instrument remains the same. It is sufficient to carry out the orientation procedure just described once each time the mounting is altered.

With an altazimuth telescope, the local vertical serves as a reference for the diurnal motion. The direction of geographical North varies with respect to the vertical over the course of a day, and the field appears to rotate. To calculate the field rotation, consider Fig. 1.13; ϕ is the latitude of the observing site, δ is the declination of the Sun on the day of observation, and H is the Hour Angle of the Sun at the time of observation. Np indicates the North Celestial Pole, Z the zenith, and ns the projection of the meridian on the plane of the horizon. The path followed by the Sun across the sky during the day is the circular arc RMS. If θ is the angle between the geographical North–South direction (i.e., the meridian NpASp) and the vertical ZA, it may be shown that:

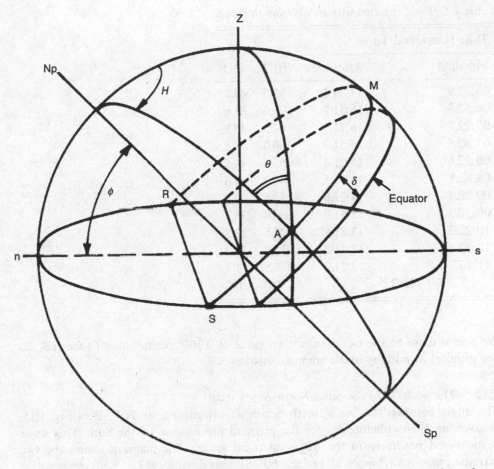

Fig. 1.13. *Definition of the elements required to determine the orientation of an observation made with an altazimuth instrument.*

$$\tan \theta = \sin H (\cos \delta \tan \phi - \sin \delta \cos H)^{-1}.$$

The angle θ depends on the observer's position; for a given point, it changes according to the hour of the day, and for a given point and time it varies according to the date. Table 1.1 gives the data calculated for Paris and 1987 June 20. Here $\phi = 48°50'$, and $\delta = 23°26'$; values for θ are given to an accuracy of a tenth of a degree. In this calculation it was assumed that the Right Ascension of the Sun was constant and equal to $5^{h}54.7^{m}$ (the interpolated value for June 20, 12:00 UT). The resulting precision is quite adequate to show how the field rotates.

Figure 1.14 shows how the North-South line rotates with respect to the vertical during a day. (The plane of the diagram is the plane tangent to the celestial sphere at the centre of the solar disk.) The rotation of the North–South line takes place in the same direction as the hands of a clock (beware of inversions or reversals in projected solar images). The rate of rotation of the field $\Delta\theta/\Delta t$ is a maximum when

19

Table 1.1. *Field rotation with an altazimuth mount*

Time (Universal Time)			
Morning	Afternoon	$H(°)$	$\theta(°)$
06:22.9	17:21.1	82.5	44.8
06:52.8	16:51.2	75.0	45.6
07:22.7	16:21.3	67.5	45.8
07:52.6	15:51.3	60.0	45.5
08:22.6	15:21.4	52.5	44.5
08.52.5	14:51.5	45.0	42.6
09:22.4	14:21.6	37.5	39.7
09:52.3	13:51.7	30.0	35.3
10:22.3	13:21.8	22.5	29.3
10:52.2	12:51.8	15.0	21.3
11:22.1	12:21.9	7.5	11.3
	11:52.0	0.0	0.0

the Sun is close to the meridian, with a value of 0.366° per minute of time (i.e., 22′ per minute) at midday at the summer solstice.

1.3.3.3 *The position of the Sun's North–South axis*
The angle between the Sun's North–South axis (indicated by NP–SP in Fig. 1.15) is given as *P* in ephemerides for the physical observation of the Sun. This angle is measured *positively* in the trigonometrical sense. The diagram shows the two extreme positions (*P* = 26.32°) of the NP–SP axis during 1987.

1.3.4 *Measurement of the positions of solar features*

Accurate measurement of positions may be obtained (to about 0.5°) by one or other of two methods: either by reading them directly from a coordinate grid (Fig. 1.16), or by calculation. In both cases the accuracy obtainable primarily depends on the quality of the images.

1.3.4.1 *Using heliographic-coordinate graticules*
There is a set of heliographic-coordinate graticules showing meridians of longitude every 10° and parallels of latitude every 5°. These graticules make allowance for the seasonal variation in the latitude of the centre of the solar disk as seen by an observer on Earth. This latitude is denoted by B on the graticules (B_0 in some ephemerides). Figure 1.16 shows the whole set of graticules used in a single year. These graticules have been calculated and drawn by the Bureau des Longitudes for values of B_0 that vary in steps of 0.5 degrees between 0° and 7°. Those who observe photographically should possess a set of graticules on film, which may be

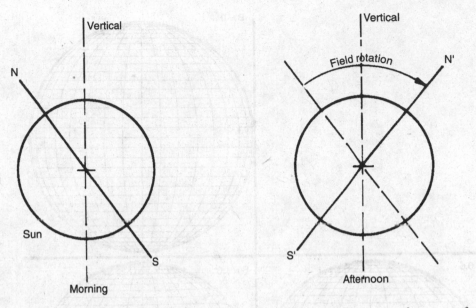

Fig. 1.14. *Variation in the geographical North–South direction as a function of time of day for an altazimuth instrument.*

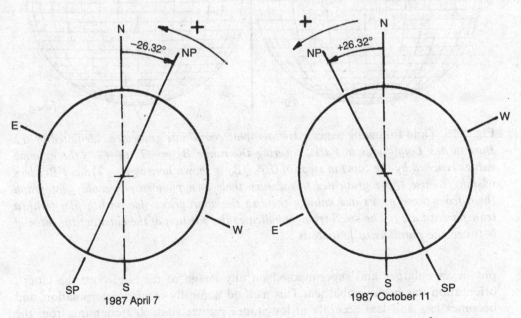

Fig. 1.15. *Determination of the direction of the Sun's axis of rotation with respect to the geographical North–South line.*

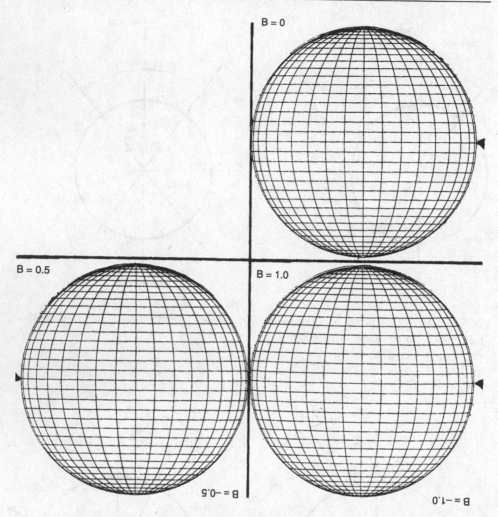

B = 0

B = 0.5

B = 1.0

B = −0.5

B = −1.0

Fig. 1.16. *(and following pages). Heliographic coordinate graticules calculated by the Bureau des Longitudes in Paris, covering the range $B_0 = -7°$ to $+7°$ (the extreme values reached by the Sun) in steps of 0.5°. [B_0 is shown here as B. – Trans.] Readers wishing to use these graticules to measure their own photographs could photograph them from these pages and enlarge them to the appropriate size on lith film (with a transparent base). The small triangle indicates the position of the equator; the interval between the parallels of latitude is 5°.*

put in an enlarger and superimposed on any image of the Sun, once the latter's orientation has been established. This method naturally requires interpolation, and becomes less and less accurate at longitudes greater than 60° (counting from the central meridian).

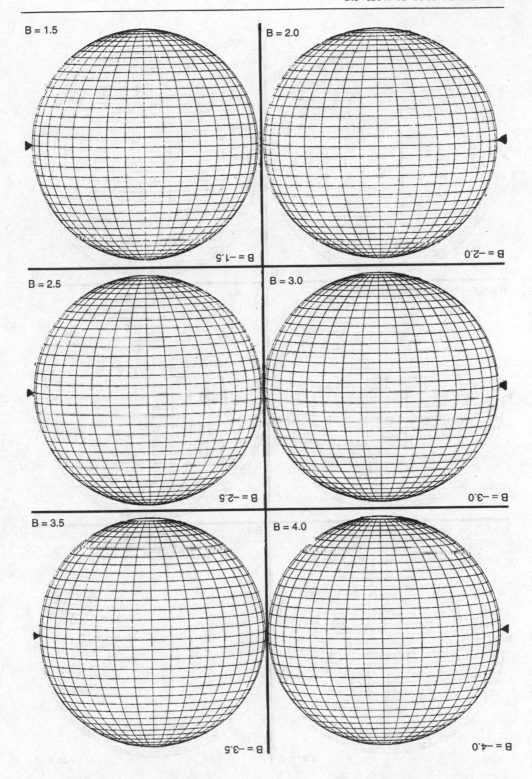

Fig. 1.16. *For caption see Page 22*

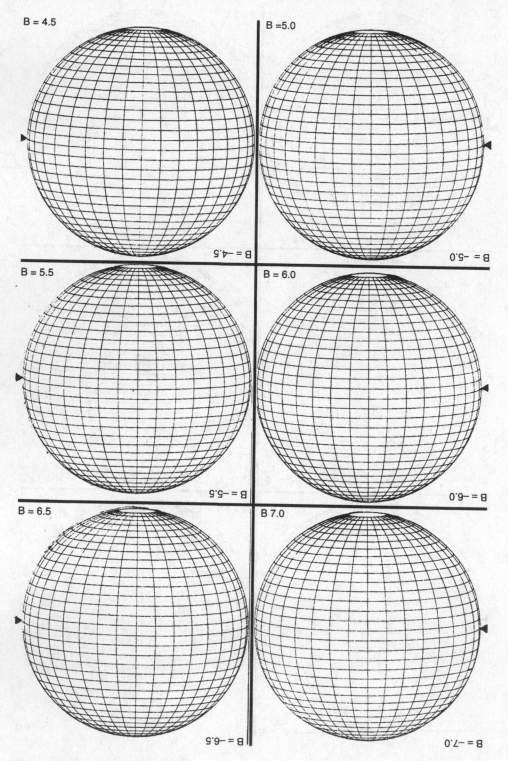

Fig. 1.16. *For caption see Page 22*

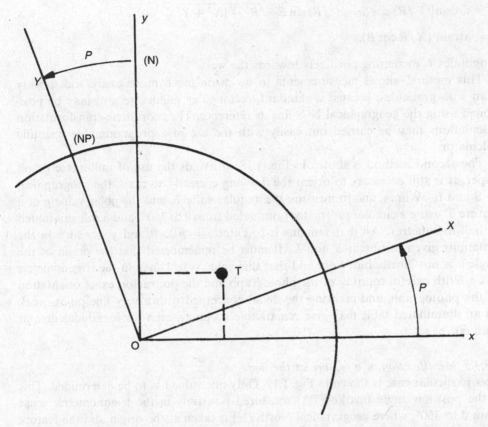

Fig. 1.17. *Determination of heliographic coordinates by calculation, using Cartesian coordinates.*

1.3.4.2 Measuring positions by subsidiary coordinates

In referring to features seen on the solar disk, we may use two methods: Cartesian coordinates, or polar coordinates. The first system is illustrated in Fig. 1.17; xOy is such that O coincides with the centre of the solar image and Oy with the geographical N–S line. OY is superimposed on the Sun's axis of rotation NP–SP; transformation from xOy to XOY is effected by rotation about O as centre through an angle P. Following measurement of the x and y values for the position of a feature T, the standard equations for the transformation of coordinates give:

$$X = x\cos P + y\sin P$$

$$Y = -x\sin P + y\cos P.$$

It only remains for X and Y to be related to latitude B and longitude L (always defined with respect to the central meridian) by equations similar to those that relate altitude and azimuth coordinates to those of the equatorial system. If R is the radius of the image of the solar disk and B_0 the latitude of the centre on the day of observation, it may be shown that:

$$B = \arcsin[(Y/R)\cos B_0 + (1/R)\sin B_0 \sqrt{R^2 - (X^2 + Y^2)}]$$

$$L = \arcsin(X/R\cos B),$$

longitudes L increasing positively towards the west.

This method allows measurements to be made much more easily and quickly than with graticules, because a standard, rectangular millimetre grid may be positioned using the geographical N–S line as reference. The coordinate-transformation calculations may be carried out easily with the use of a programmable scientific calculator.

The second method is shown in Fig. 1.18; it avoids the use of millimetre graph paper. It is still necessary to orient the drawing correctly, to mark the geographical N–S and E–W lines, and to measure the angular value θ, and the polar value p of a feature T, using a circular protractor (graduated from 0 to 360°) and a rule graduated in half-millimetres. All that remains is to enter $x = p\cos\theta$ and $y = p\sin\theta$ in the equations given to obtain B and L. It must be remembered that the origin of the angle θ is not North, but West, and that this value is reckoned in the trigonometric sense. With careful centring of the photograph and the protractor, exact orientation of the photograph, and marking the detail concerned with a very fine point, work on an illuminated table may give remarkable accuracy, even for longitudes greater than 60°.

1.3.4.3 Identification of a feature at the limb
This particular case is shown in Fig. 1.19. Only one value has to be determined. This is the position angle (marked PA) measured positively in the trigonometric sense from 0 to 360°, where geographical North (N) is taken as the origin and the feature in question (F) as the extent.

1.4 Amateur observing programmes

Once the technical problems have been overcome, curiosity leads observers into watching sunspots. Without really being aware of it, they soon start to count the spots each day, try to recognize them, and follow their changes. This naturally gives rise to two specific programmes: a programme for deriving the Wolf Number, and one for obtaining the active area.

1.4.1 Counting sunspots: the Wolf Number

On an image of the whole disk, obtained either by projection onto a white screen or by a photographic enlargement, both isolated spots and collections of large and small spots – known as sunspot groups – may be identified (Fig. 1.20). The relative Wolf Number is an arbitrary number R, calculated by the equation:

$$R = k(10g + f),$$

where k is a coefficient taking account of the observer, the quality of the image, and of the instrument; g is the number of spot-groups; f is the number of spots.

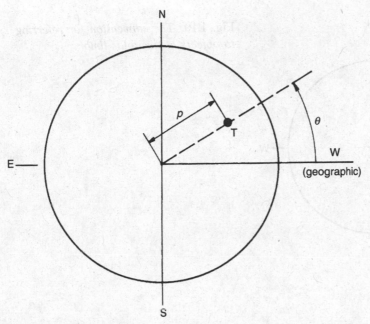

Fig. 1.18. *Determination of heliographic coordinates by calculation, using polar coordinates.*

It should be noted that such a formula gives greatest weighting to the number of spot-groups, because g is multiplied by 10. The value of k is a constant, required in standardizing observations in any collective programme, but which may be ignored in individual work.

Despite its apparent simplicity, this method of calculation must be used with care. It is not only frequently difficult to define what is a spot, in particular if there are several umbral regions within one penumbra (as with the large spot in Fig. 1.20*a*), or in the case of pores (like the middle spots in Fig. 1.20*b*), but – and this has much greater significance – there is also the problem of deciding whether a large area of activity (as in Fig. 1.20*c*) consists of one or more spot-groups.

It is also necessary to be on one's guard against an over-enthusiastic interpretation of plots showing the value of R. It must be realised that the variability of R from day to day is very large, and that it remains considerable if monthly or even yearly means are taken. It is only over long periods of observation, and with sophisticated smoothing methods that publishable curves like those from Zurich or Brussels may be obtained. These curves are often, and quite improperly, used as an indication of solar activity. It is essential to emphasize that if sunspots are a sign of this activity, they are only one of many factors. For example, the eruptive activity of a single spot (Wolf Number equal to 11) is sometimes more significant than half-a-dozen spot-groups scattered across the disk (with a Wolf Number of more than 70).

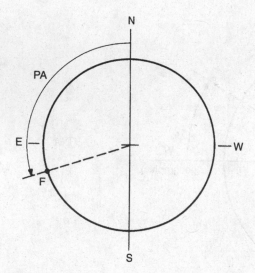

Fig. 1.19. *The convention for referring to objects at the solar limb.*

1.4.2 Calculation of the active area

This type of programme also examines the variations in one of the factors related to sunspots. The identification of individual sunspots and groups of spots is not required, which removes one of the problems with the previous method. Although the measurement of the apparent surface area of spots is relatively easy, measurement of the actual surface area requires determination of the heliographic coordinates of each spot, so that a correction may be applied to take account of the foreshortening caused by the Sun being spherical. If θ is the angle subtended at the centre of the Sun by the spot and the centre of the disk (θ varies from $0°$ for a spot at the centre of the disk, to $90°$ for a spot at the limb), then the correction may be written as: actual surface area = apparent surface area/$\cos\theta$. It is immediately obvious that errors in measuring the position or apparent area of spots close to the limb may cause major inaccuracies.

The area of sunspots is frequently expressed in millionths of the Sun's visible hemisphere (10^{-6}), or in millionths of the total solar surface area.

1.4.3 Measurement of the proper motion of spots

During the lifetime of active areas (groups of spots), sunspots move in both longitude and latitude across the surface of the Sun. The motion in longitude is generally greater than the motion in latitude. The apparent motion of a spot results from the combination of the Sun's differential rotation at the spot's mean latitude and its own proper motion. However, this proper motion is not the same for the eastern (p-) spot and for the western (f-) spot in a simple group (Fig. 1.21). It always tends

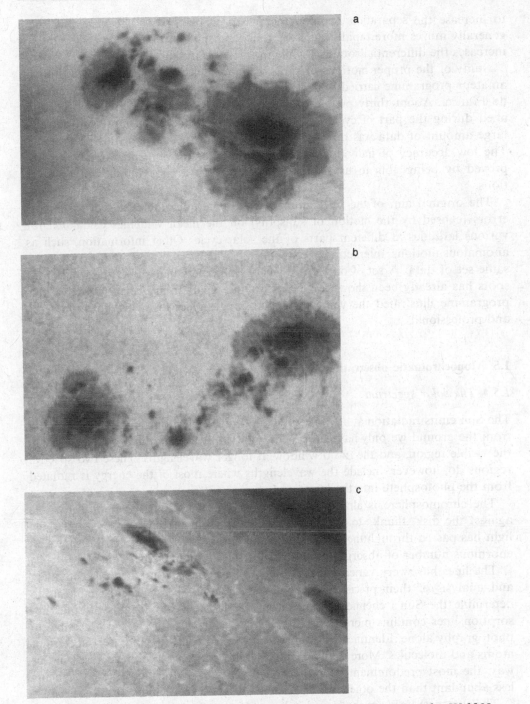

Fig. 1.20. *Examples showing the problems encountered in determining the Wolf Number. In a, there is ambiguity as to the large spots, which have more than one umbra; among the middle spots in b, there are many pores and examples of penumbrae without umbrae; in b and c, which show the same group, ambiguity arises from the different positions: at the centre of the disk and near the limb.*

29

to increase the separation between the two principal spots. The preceding spot generally moves more rapidly than the following spot. In the first case the motion increases the differential rotation, while in the latter it reduces it.

Study of the proper motions of spots was the subject of a special, collaborative, amateur programme carried out by the Solar Section of the Société Astronomique de France. About thirty people took part, being specifically alerted and encouraged during the part of cycle 21 (1974–82) that produced the most sunspots. A large amount of data was received and analyzed in the course of this programme. The low accuracy of individual determinations (about 0.3° heliographic) was improved by being able to use statistical methods on a large number of observations.

The original aim of the programme was to determine the effects of systematic errors (caused by the motion of sunspots) on the mean velocities calculated for various latitudes at different parts of the solar cycle. Other information, such as anomalous motions, interactions between spots, etc., may also be obtained from the same set of data. A set of data from various observers that shows the rotation of spots has already been shown in Fig. 1.6. Quite apart from being instructive, this programme illustrated the value of similar programmes involving both amateurs and professionals.

1.5 Monochromatic observation of the Sun

1.5.1 The solar spectrum

The Sun emits radiation at all wavelengths (λ) of the electromagnetic spectrum, but from the ground we only have access through two windows: the optical window in the visible region, and the radio window at longer wavelengths. These two spectral regions do, however, include the wavelengths where most of the energy is radiated from the photosphere into the lower corona.

The chromosphere is almost completely transparent, and is therefore invisible against the disk, thanks to the intense radiation from the photosphere. After the light has passed through the photosphere and chromosphere, its spectrum shows an enormous number of absorption lines: this is the Fraunhofer spectrum (1).

The lines have very varied widths, relative intensities and profiles. Measurement and analysis of their precise wavelengths throughout the spectrum enable us to determine the Sun's chemical composition. A table of wavelengths of solar absorption lines contains more than 25 000 different lines in the region accessible to photography alone. Identification is made by comparison with laboratory spectra of atoms and molecules. More than 80 different elements have been recognised in this way, the most predominant being hydrogen, calcium and helium (the latter being less abundant than the other two by a factor of between 10 and 50.)

The strongest lines are the H and K lines of ionized calcium (396.8 and 393.3 nm). Their widths amount to a considerable fraction of a nanometre. Ionized calcium also has infrared spectral lines at 849.8, 854.2 and 866.2 nm, but their intensity is about one-hundredth of that of the H and K lines. A narrow line of neutral calcium is visible at 422.7 nm.

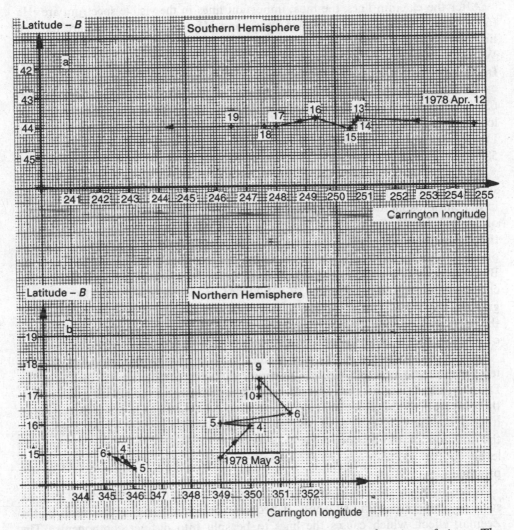

Fig. 1.21. *Graph showing the displacement of sunspots as a function of time. The abscissae give the longitudes, and the ordinates the latitudes. Dates of observation are indicated alongside the points plotted: + indicates leading spots, • following spots. a: measurements of an old spot at a very high latitude; the drift towards the east is primarily caused by differential rotation. b: measurements of the two principal spots in a young bipolar group. The positions are accurate to approximately 0.5° heliographic. These graphs are from data obtained during the sunspot proper motion programme organised by the Société Astronomique de France.*

After the calcium lines, the most important lines in the visible spectrum are: Hα (656.3 nm), Hβ (486.1 nm), Hγ (434.0 nm), and Hδ (411.0 nm). These are the first lines in the Balmer series of hydrogen. The width of the Hα line is about 0.1 nm. We shall return to the importance of the Hα line later. It is currently the one most frequently used, particularly for the study of prominences and flares.

Then there are the numerous lines of iron, many of which (for example, those at 630.2, 525.0 and 371.9 nm) are used for measurements of the magnetic field in the lower atmosphere. There is also the magnesium triplet (518.4, 517.3 and 516.7 nm). Finally, helium is represented in this region of the spectrum by the D_3 line at 587.6 nm.

Monochromatic observation is observation carried out at the wavelength of a single, strong line in the Fraunhofer spectrum. The image obtained shows features in the chromosphere that are normally invisible, such as facular plages at the centre of the disk (in the K line of ionized calcium), or filaments seen in projection against the sky at the limb, when they are known as prominences (in K_3 or Hα). For this one uses an adjustable slit, which selects the desired wavelength from the spectrum. By shifting this slit within the width of these strong lines, it is possible to observe the chromosphere at different depths (Fig. 1.22). At the centre of the line, the opacity is greatest (for the K_3 line), and the light comes primarily from the highest region of the solar atmosphere in which it may be formed. As one moves out into the wings of the line, one is generally observing lower and lower regions.

From measurements of the line-profiles we are able to derive values for the temperature, gas pressure, electron pressure, and microturbulence. Furthermore, the Doppler effect acts on the different lines in proportion to the velocity, relative to the observer, of the point in the solar atmosphere from which the lines originate: it is a measure of the radial velocity. Such measurements may be applied to the rotation of the Sun itself, as well as to the motion of material at different levels in the photosphere or chromosphere, and to determining the motion of prominences.

When light crosses regions of strong magnetic fields (from 10 to 4000 Gauss), such as active areas and sunspots, it becomes polarized. This is the result of the Zeeman effect, which causes the lines to be split into distinct components. The separation between the individual components is proportional to the strength of the component of the magnetic-field vector that lies along the line of sight for each individual point being examined. In this way it is possible to draw maps of the magnetic field and to follow the evolution of spots and active areas.

It is these magnetic fields that govern changes on the Sun. They are a sign of the specific characteristics that distinguish sunspots from one another, and cause particular configurations of faculae or prominences. They are the cause of flares and of the solar magnetic cycle (with a period of about 22 years). They control all the features in the solar atmosphere, and sustain them, sometimes for as long as 10 or more rotations. This is true for some filaments, which are long streamers of cool material (at about 4000 K) that is surrounded by coronal material, which is a plasma that may be at several hundred thousand K, or even a million K.

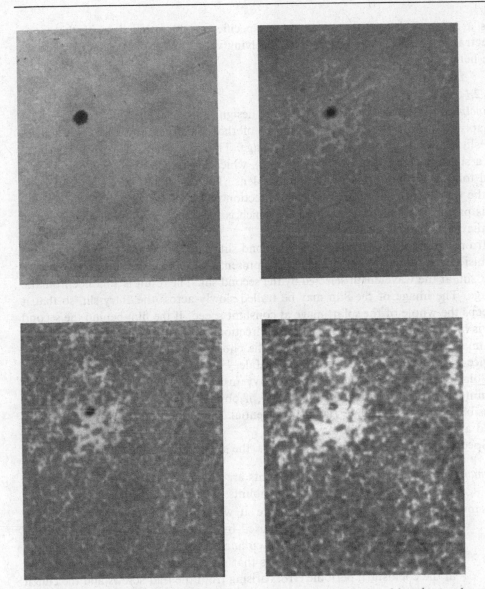

Fig. 1.22. *The appearance of a sunspot region as a function of height in the solar atmosphere.* Left to right, top to bottom: *at the surface (the spot alone is observed); at the photospheric level (spot and facula); then, in the chromosphere, the spot gradually is lost in the brilliance of the facula. Photos by L. and M. d'Azambujar, Meudon Observatory.*

1.5.2 Specific instrumentation

Various types of instrument have been developed to obtain monochromatic images of the Sun, or of specific features, such as spots, faculae, or prominences. Two sorts of equipment may be distinguished. The first, capable of covering a wide spectral range and giving high resolution, consists of spectroheliographs. The second type has a range restricted to one or more, specific, neighbouring lines. It has good spectral resolution, and is capable of resolving very rapid events: such instruments are heliographs.

1.5.2.1 Spectroheliographs

Principles of spectroheliographs In the design of a classical spectrograph (*see* Chap. 18), an objective forms an image of the object being studied, and an entry slit selects a narrow region of this image. The light from this strip is dispersed by a suitable element (a prism or grating), which is preceded by a collimating lens and followed by an appropriate 'camera' lens. The dispersion varies in proportion to the wavelength and takes place in a direction perpendicular to the axis of the slit. This produces a spectrum, each line of which is the image, at a specific wavelength, of the area covered by the entry slit.

If one spectral line is isolated by a second slit, placed in front of a photographic emulsion, the image falling on the latter represents the area of the Sun selected by the first slit, at the wavelength selected by the second slit. The result is a monochromatic image. The image of the Sun may be trailed slowly across the entry slit, so that it sweeps the whole of the solar image at constant speed. If the film behind the second slit is moved simultaneously in the same direction and at the same speed, it captures an image of the whole solar surface at the chosen wavelength (Fig. 1.23). Such a device is known as a spectroheliograph (Hale, 1935).

Solar spectroheliographs are very heavy instruments and cannot therefore be mounted on a refractor. On the contrary, if observations are to be of the greatest possible accuracy, absolute rigidity is essential. Generally the spectroheliograph is fixed and light is fed to it via a coelostat.

Spectroheliographs are subject to errors, the sources of which may be:

Permanent If the displacements of the slits are not precisely parallel, or if the first and second slits have unequal amounts of residual curvature;

Random If there is variation in the rate at which the slits move relative to the fixed parts of the equipment, caused by wear in any part of the variable-speed drive train (producing non-circular images of the Sun, which must be distinguished from similar images that may be produced by the same fault in the coelostat); periodic errors arising from the lead-screws and the clutch for the worm and nut, or other periodic errors arising from inaccuracies in any part of the whole drive train.

The coelostat The classical form of coelostat has two mirrors (Fig. 1.24). The first mirror, known as the primary, rotates at half the Sun's diurnal rate around an axis parallel to the Earth's axis and located in the same plane as the reflecting surface. As a result, it reflects light into a fixed direction. A secondary mirror subsequently reflects the beam towards the optical axis of the objective at the entrance to the spectroscope.

To allow for the Sun's changes in declination, the secondary mirror may be shifted in a north–south direction so that it intercepts the beam from the primary mirror. In winter, when the Sun is low, the secondary is close to the primary and may

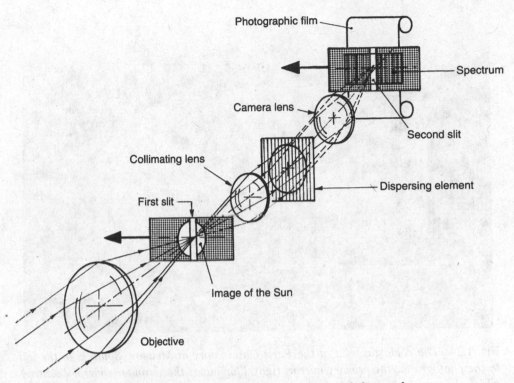

Fig. 1.23. *The principles underlying the design of a spectroheliograph.*

even cast a shadow on it. If this occurs, the primary is shifted towards the east in the morning, and towards the west in the afternoon. The field rotation that results requires a correction to the Sun's position angle to be made.

When undertaken on an amateur scale, a portable coelostat may allow the less robust part of the instrument to be permanently mounted under cover. But alignment has to be carried out with the greatest care. The coelostat may introduce errors, especially if observation includes long exposures. These may be:

Permanent   Errors caused by the setting of the hour angle, which may become evident if observations are not always carried out at the same time of day; errors caused by the poor mechanical state of the tangent screw in the drive train, which it is difficult to take into account in short exposures, because it is impossible to know at which point in the screw's rotation the exposure was made. Another error whose effects are impossible to estimate is that of variations in the speed of rotation of the hour axis caused by variations in the frequency of the electricity supply.

Random   An error in alignment is not of great significance in summer, when the primary mirror coincides with the optical axis of the instrument. The same does not apply in the winter, when the coelostat is east or west of the optical axis. Any error causes field rotation and thus alters the position angle.

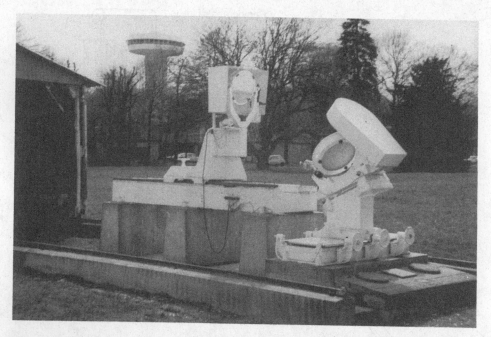

Fig. 1.24. *The coelostat used at the Paris Observatory at Meudon. South is to the left in the photograph; the primary mirror,* right, *illuminates the secondary mirror (*centre*), which directs the beam of light horizontally towards the solar instrument located to the right, outside the picture.*

The spectroheliograph at Meudon Observatory   Details of this instrument are given here by way of example.   The objective, a two-element lens, 250 mm in diameter with a focal length of 4m, gives an image of the Sun about 37 mm across on the entrance slit.   The collimator's objective has a diameter of 140 mm and a focal length of 1.3m.   The camera objective is a simple lens with a focus of 3m.   The instrument enlarges the primary image by $3/1.3 = 2.3$.   The final image therefore has an average diameter of about 85 mm.

The image is obtained by synchronized displacement of the objective (thus traversing the image across the entrance slit), and of the photographic plate, which lies behind the second slit.   (There are actually two secondary slits: the one for K_3 work being set to give a bandwidth of 5 nm, and that for $H\alpha$ a bandwidth of 2 nm.)   The translational motion of the objective and the plate-holder is provided by a screw with a pitch of 1 mm.   Its speed of rotation may be varied by suitable gearing.   This provides 12 different speeds, giving a range of time to sweep across the diameter of the Sun of between 0.4 and 6.4 minutes.   The drives to each of the two plate-holders have continuously variable speeds, which enables the ratio between the speeds of the objective and the plate-holders to be adjusted to suit the overall magnification produced by the spectroscope system.

1.5.2.2 Heliographs

When the wavelength to be studied is predefined or, even better, when just one specific line is to be studied (in monochromatic observation), it is possible to do away with the cumbersome spectroheliograph by using suitable filters. Whether commercial interference filters or the more sophisticated professional filters such as the Lyot (Paul, 1953; Petit, 1953; Kitchin, 1984), Halle or other types are used, the method may be adapted to a simple refractor fitted with small-aperture, long-focus optics.

For amateurs this is the simplest method of studying features in the chromosphere, such as filaments/prominences, spicules and flares at both the limb and on the disk, for example in Hα light. Interference filters remain expensive, however, and they are fragile. They have a limited lifetime, but the results obtained with them are generally exciting.

At Meudon, two monochromatic heliographs are used for automatic cine-photography of chromospheric activity (flares and Doppler effects). The twin telescopes are carried by a single mounting. They are fitted with two Lyot (i.e., polarizing) filters.

1.5.2.3 Amateur Instruments

Attempts to build spectroheliographs or heliographs, which are required for mono-chromatic observation of the solar disk, are very rare amongst amateurs (Fig. 1.25). The former require relatively complicated mechanical arrangements, but we may expect more such instruments to become operational in future. Those who are undaunted by the potential problems should take inspiration from professional designs that have fully proved their worth – which is why the details have been described above. As for the heliographs, they require an interference filter with a very narrow bandwidth (less than 0.1 nm), which requires care in handling and is also very demanding in use.

Monochromatic observation of prominences is easiest to undertake. Amateurs make this form of observation using one of three techniques:

(i) A monochromatic interference filter with a narrow bandwidth of less than 0.5 nm (or 0.1 nm for filaments on the disk), located at the focus of a reflector or refractor (Fig. 1.26).

(ii) A prominence telescope, the optical design of which resembles that of the coronagraph invented by B. Lyot (Roques, 1961). Here, use of an interference filter is essential to avoid any risk of burning the retina if any error in setting occurs. The image-resolution largely depends on the quality of the filtration used. This type of instrument is described in Chap. 2.

(iii) A prominence spectroscope, located at the focus of either a refractor or reflector, and using a widened slit. This method, although little-known, is very effective, and has the advantage of being adaptable to a reflector or refractor that is not specifically designed for solar observation. This method is described in the following section.

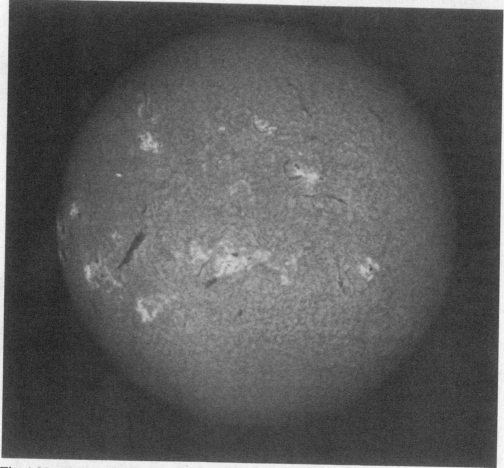

Fig. 1.25. *The solar chromosphere at a time of maximum activity on 1980 May 20. A monochromatic image obtained by D. Cardoen; aperture 50 mm, focal length 1500 mm, Hα interference filter with 0.06 nm bandwidth; TP 2415 film developed for 10 minutes in D19; exposure 1/30 second.*

1.5.2.4 The 'wide slit' method

This method's equipment resembles both a coronagraph and a spectroscope. As in the coronagraph, a disk occults the surface of the Sun; but this disk is, in effect, the internal edge of an annular slit, which is the entry slit of the spectrograph. The lower corona, which is the region observed, produces emission lines; its spectrum consists of adjacent images of the region selected by the slit at each of the wavelengths emitted by the corona. The one chosen, Hα, is isolated from other lines, so one may use a wide slit, which gives an image of a relatively large region of the Sun that is not affected by neighbouring lines.

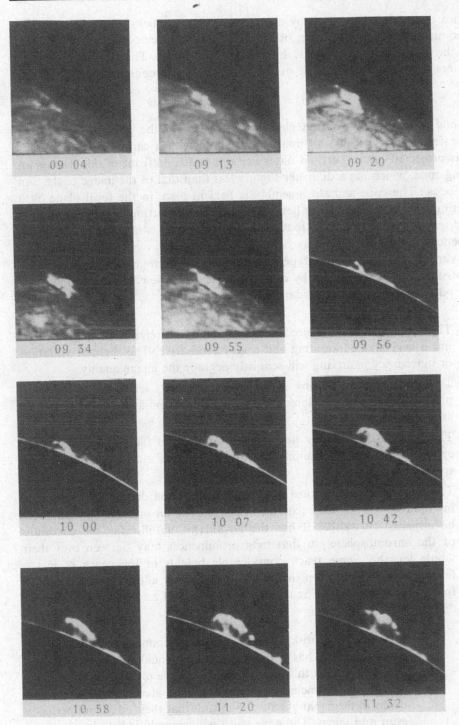

Fig. 1.26. *A sequence of observations showing a chromospheric eruption close to the solar limb, obtained by D. Michaut; 200-mm Schmidt–Cassegrain, Hα interference filter with 0.06 nm bandwidth. The black disk masking the Sun was added at the printing stage.*

Such a device has been used by amateur astronomers belonging to the Société Astronomique de France (among others), and has been employed on the refractor at the Sorbonne Observatory, in the heart of Paris. The following description refers to this equipment and may inspire readers to construct similar instruments.

Optics and mounting The object glass of a refractor or the primary mirror of a reflector forms, at its focus, an image of the Sun with a certain diameter. This image is projected onto a slit that has a variable width. In front of this there is an occulting cone, which has a diameter slightly less than that of the image of the Sun at the focus. The cone and slit assembly is held in place in the focal plane by a collimating lens. Behind the latter lies a 600-lines/mm, reflecting, diffraction grating, which directs the solar spectrum towards a small telescope, where the Hα line from prominences at the solar limb may be observed in emission.

The various elements in this type of prominence spectrograph (Fig. 1.27) may be constructed in different ways by different amateurs. A few comments about the various parts may be given as guidance:

(i) The whole spectroscope needs to be mounted on a refractor or reflector that has a very stable mounting and a very even drive. The slightest vibration or inaccuracy in driving will seriously degrade the image quality.

(ii) To avoid excessive heating of the spectroscope, its refracting or reflecting primary should have a long focal length and, if possible, a focal ratio of around f/15.

(iii) The occulting cone may be made by applying a file, held at 45°, to a cylindrical bar of aluminium, held in the chuck of a drill, running at high speed.

(iv) The slit, which is the most important part of the design, may be made in different ways. We recommend the use of an annular slit that may be adjusted in width. It has the advantage of following the curvature of the chromosphere, so that long prominences may be seen over their whole length. Some reach considerable heights in the space of just a few minutes, so variation in the width of the slit allows the events to be followed continuously. The annular slit consists of an occulting disk and a diaphragm:

- The occulting disk, which is located immediately behind the occulting cone, has a diameter exactly equal to that of the Sun at the focus. As this varies over the course of a year, 3 or 4 disks of different diameters are required. Considerable care should be taken in their manufacture to ensure that their edges are perfectly even and sharp. The cone and disk assembly is held in place and centred by a tube that may be adjusted in length and which is rigidly connected to the collimating lens, which is drilled in the centre;

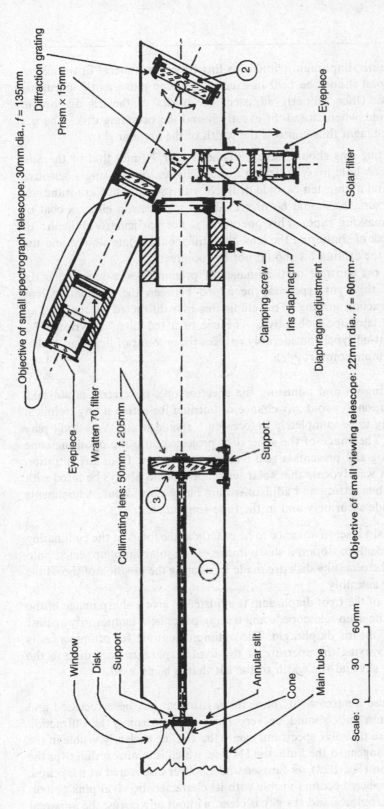

Fig. 1.27. *Schematic diagram of a prominence spectroscope made by François Costard (after l'Astronomie, 1985 June).*

Diffraction grating

Prism × 15mm

Objective of small spectrograph telescope: 30mm dia., f = 135mm

Eyepiece

Solar filter

Clamping screw

Iris diaphragm

Diaphragm adjustment

Objective of small viewing telescope: 22mm dia., f = 60mm

Eyepiece

Wratten 70 filter

Collimating lens: 50mm. f. 205mm

Support

Window

Disk

Support

Main tube

Annular slit

Cone

Scale: 0   30   60mm

- The iris diaphragm (similar to those used in cameras or on microscopes) should be held in exactly the same plane as the occulting disk. Once properly adjusted, the blades of the iris diaphragm should, when closed, fit exactly round the occulting disk. The iris diaphragm thus controls the width of the annular slit.

(v) The collimating lens should have a diameter 2 or 3 times that of the Sun at the focus. A hole may be drilled through its centre by using a sensitive drill-stand and a tungsten-carbide drill fed with 180-minute Carborundum. The optical surfaces should be protected beforehand with either a coat of varnish or masking tape. This operation is not particularly difficult: to avoid any risk of shattering the lens, the drill should rotate slowly, and the pressure on the drilling bit should not be too great.

(vi) To improve the contrast of the images of prominences, a second iris diaphragm of the Lyot type may be added between the collimating lens and the diffraction grating. This eliminates rays diffracted by the edge of the primary telescope's objective. The use of a red filter, a Wratten 70, which is a cut-off type, considerably reduces the amount of light and makes observation more comfortable.

Adjustments   In setting up and adjusting the spectroscope, observers should take every possible precaution to avoid any chance of burning the retina if any technical hitch – which is likely to be completely unforeseen – should occur. You don't play about with the Sun! The image of the solar disk projected onto the occulting cone is particularly blinding and presents a real danger. Quite apart from any filtration within the instrument we advocate that solar telescopes should always be fitted with a Wratten 70 filter when setting and adjustment are being carried out. Adjustments should always be made accurately and in the following order:

(i) The annular slit is set in advance to lie exactly at the focus of the collimating lens. Adjustments to obtain a sharp image of the solar limb, projected onto the edge of the occulting disk are made by altering the position of the whole spectroscope assembly.

(ii) The position of the Lyot diaphragm is adjusted to give a sharp image of the objective of the main telescope, using tracing paper held temporarily against the blades of the iris diaphragm. The optimum contrast for prominences is obtained by varying the aperture of the Lyot diaphragm according to the clarity of the sky and the width of the slit that is being used.   –

The wide slit   First use a narrow slit; once the instrument has been focussed and adjusted the absorption lines should be very narrow. By turning the diffraction grating we may explore the solar spectrum, where the Hα line is clearly visible in the red. When the slit is tangent to the limb, the Hα line, which is in absorption over the disk, shows in emission (Fig. 1.28), as Janssen and Lockyer discovered at the eclipse of 1868. The chromosphere becomes visible with its characteristic vivid pink colour. When there is little turbulence and the sky is clear, without any cirrus, the apparent top of the chromosphere becomes finely structured and resembles an enormous field

Fig. 1.28. *The wide-slit method (after l'Astronomie, 1985 September). Top: the slit is over the photosphere and the Hα-line appears in absorption. Bottom: the slit is placed tangent to the limb of the Sun, the Hα-line appears in emission, enabling the prominences to be seen.*

of wheat; the individual 'stalks' are spicules (Fig. 1.29), which are small tubular structures, about 9000 km high and 1000 km wide, which were discovered by Secchi in 1872, using this method of observation.

Some prominences attain exceptional heights, and may often exceed 100 000 km. When this happens they may be seen in sections by moving the slit. To see them as

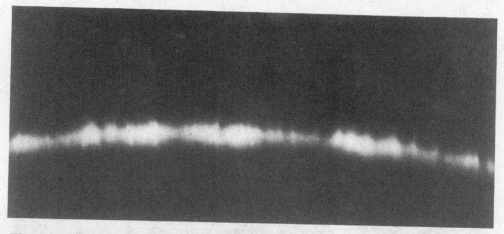

Fig. 1.29. *Spicules (chromospheric fine structure) photographed on 1985 September 12, by S. Costard, at the Sorbonne Observatory.*

a whole, all that is required is to widen the slit of the spectroscope by a considerable amount. This is the 'wide slit' method. At the Sorbonne Observatory of the Société Astronomique de France, slits as wide as 0.8 mm are commonly used. When the sky is particularly clean, the slit-width may be more than 2 mm.

Advantages of this method This method has a number of advantages. For example, although the Parisian sky is not at all favourable, observation of prominences from the Sorbonne is often excellent, and may even exceed 1 arc-second resolution! Turbulence reduces the resolution of photographs, but it may be minimized in a spectrograph, thanks to the narrow bandwidth centred on the Hα line. The seeing is less affected at these wavelengths, and the image appears almost completely steady.

Scattering in the atmosphere and within the equipment decreases the image-contrast. The simultaneous use of a narrow bandwidth, a Lyot diaphragm, and the second-order spectrum (which doubles the spectral resolution), reduces this scattering and simultaneously improves the image-quality.

Another advantage of the spectroscope is that it gives a real image, although this may be distorted by variations in the Hα line's radial velocity that produce the Doppler effect. The value of the velocity (v) measured along the observer's line of sight may be determined by measuring the shift ($\Delta\lambda$) with respect to the 'base' of the Hα line (λ) by the equation:

$$\Delta\lambda/\lambda = v/c$$

where c is the velocity of light and v is the radial velocity. It may reach several hundred kilometres per second in certain eruptive prominences. To the observer, the Doppler shift appears as a shift in the feature observed; for example, a prominence may appear superimposed on the dark image of the occulting disk. Although this deformation renders the interpretation of the images of prominences more difficult, it is, on the other hand, very useful for predicting eruptive events. In fact, these

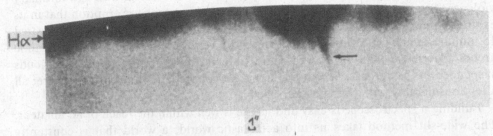

Fig. 1.30. *High-resolution photograph of a jet of chromospheric material seen in projection against the disk (by Doppler shifting), during the growth phase of an eruption at the solar limb (arrowed). The Doppler shifts at the Hα line indicate the large variations in radial velocity found over very short distances. In this case the differences reach more than 100 km/s between points that are only 1800 km apart. Observation by F. Costard, prominence spectrograph, Sorbonne Observatory.*

are generally *preceded* by the occurrence of high radial velocities (Fig. 1.30) and the observer may have sufficient time to film the whole event.

The use of a wide slit has the advantage of showing the whole of an eruptive event, whereas an interference filter with a 5-nm bandwidth only shows a part. Before and during a chromospheric eruption, features develop that have very high radial velocities, producing high Doppler shifts that lie outside the narrow bandwidth of a monochromatic filter.

Prominence photography In general, excellent images of prominences may be obtained, particularly about two hours after sunrise on slightly misty days with anticyclonic conditions. In Paris and most other cities, the quality of photographs declines at the beginning of the afternoon when a thin veil of cirrus forms, a primary cause being aircraft. When taking a sequence of photographs with a coronagraph it is essential to point the instrument away from the Sun about every 15 minutes, to prevent air currents, caused by excessive heating of the occulting cone and the annular slit, from building up inside the equipment.

The camera body, without a lens, should be fixed behind the eyepiece, after having replaced the normal focussing screen with one having a clear central spot with double cross-hairs. The focussing at the slit and at the eyepiece must be carried out with great care. At present the best film is Kodak Technical Pan 2415. When developed in D19 for 7 minutes at a temperature of 20°C, it gives a high-contrast negative with very fine grain.

Regular monitoring of the surface of the Sun shows the position of groups of spots and enables the location of possible active prominences to be predicted. Even a small 'spyglass' type of refractor with a Herschel wedge and a solar filter would suffice to see the position of the main active areas.

On good days, prominences have a characteristic fibrous structure and are utterly fascinating to watch as they change continuously. Although some prominences, the quiescent prominences (Fig. 1.31), change only very slowly with time, others, known as eruptive prominences, suddenly erupt into the corona with extraordinary violence. Figure 1.32 shows one of these events: a chromospheric eruption that in its explosive phase attained a velocity of about 500 km/s, and that entirely disappeared 15 minutes later! The forms of prominences are very varied; some resemble gigantic arches like those in a Gothic cathedral, others, more diffuse, look like clouds suspended in the corona, without any apparent connection to the chromosphere; all alter considerably with changing perspective.

Building a spectroscope is easy and its use is well within the reach of an amateur. The wide-slit method takes us into a fantastic world; a world that is constantly growing, changing and vanishing. This should inspire more amateur astronomers to construct a similar instrument in order to monitor solar prominences regularly.

1.5.3 Amateur prominence observation programmes

Curiosity alone may be sufficient motive for beginning a specific programme of observation of prominences. Watching them, discovering their variety, their slow and rapid changes, and their sizes is one thing. Measuring the coordinates at which they arise, so that they are easier to follow from one day to another, or even from one limb to the other, is rather more difficult, but is not really a forbidding task. The information given on p. 26 will enable this to be undertaken.

Taking notes, making drawings, and obtaining 'artistic' photographs already constitute an observational programme. It is also possible to take part in collective, scientific programmes, like those that were organised by the Solar Section of the Société Astronomique de France to cover the period of maximum of solar cycle 22 (1987–8 to 1992–3 and beyond), and the programmes mounted by other national groups.

The three main areas of research proposed by the SAF were:

- Study of connections and the exchange of material over large distances;
- Monitoring of eruptive prominences:
  - *a* What may be termed 'true' eruptions (i.e., no features visible beforehand, and none afterwards);
  - *b* Eruptions associated with previously existing prominences (instabilities or detachment from the surface);
- Sudden disappearance of prominences.

Those undertaking such programmes are strongly recommended to use observational techniques that employ either an interference filter or a wide-slit spectroscope. If such observations are properly made and if a fair number are available, it is possible to obtain valuable statistical information about the behaviour of the magnetic-flux tubes in the disturbed areas of the Sun (by enabling one to obtain the radial components of their motion, which are generally indeterminate in events observed on the disk itself).

The study of prominences is of considerable current interest to professional solar workers, because these features present problems for solar magnetohydrodynamics, and are among the most difficult effects to explain. Every effort needs to be made to

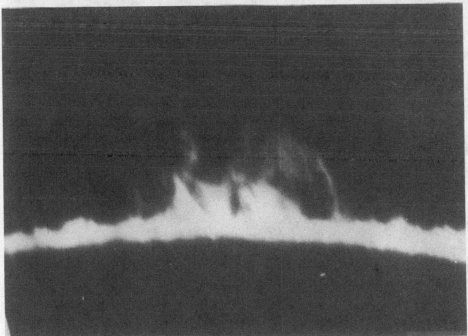

Fig. 1.31. *Quiescent prominences (apparent height 40 000 km), 1986 October 1 (top: 9:52 UT; bottom: 10:06 UT). Exposure: 1/30th second; film: TP 2415; observation by F. Costard, prominence spectrograph, Sorbonne Observatory.*

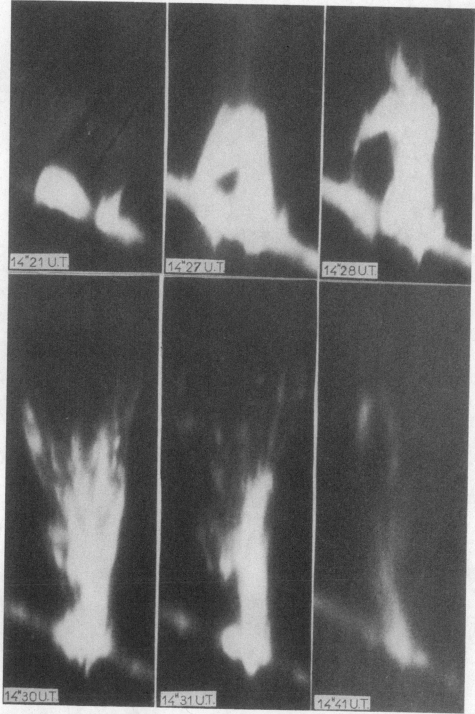

Fig. 1.32. *A chromospheric eruption, 1985 September 12, photographed by F. Costard with the prominence spectrograph, Sorbonne Observatory (from* l'Astronomie, *1986 February).*

understand them, because they now appear to hold the key to internal phenomena (such as convection) that cannot be observed directly.

1.6 Conclusion

Experience shows that people soon become disheartened when they think that they have exhausted all the possibilities offered by their equipment. They then think of increased performance, which often leads to the decision to obtain a new, larger telescope. But then what should be done with the old refractor, of perhaps 100 to 150-mm aperture, with which one began, and to which one is sentimentally attached? After reading this chapter – which does not pretend to be a complete introduction to solar physics, far less an exhaustive discussion of all the techniques that are available to amateurs (no mention has been made of photoelectric work, polarimetry, radio observations, etc.) – no one should ever be disheartened again. What could be simpler, after having observed throughout the evening with the new telescope, or a larger instrument at the nearest observatory, than to set up the small refractor as a solar telescope for the next day? One can be sure of never being disappointed, because the Sun presents a new aspect every day.

2   Observing the Sun with a coronagraph

J. L. Leroy

2.1 Why create artificial eclipses?

As most people realize, total eclipses of the Sun are magnificent spectacles. The Moon has essentially the same apparent diameter as the Sun, and when it passes between an observer on Earth and the Sun, it just masks the dazzling solar disk. The corona becomes visible, and appears as a form of halo, the brightness of which declines with distance from the eclipsed Sun. Some bright rays may be followed out as far as 10 solar radii. The corona emits approximately as much light as the Full Moon, so it is not completely dark during the few minutes that a total solar eclipse may last. Nevertheless, the corona is only about one millionth of the brightness of the Sun, so its observation with an ordinary telescope, outside eclipses, is quite impossible.

Despite this, when it was realised about a century ago that the corona is the Sun's outer atmosphere, numerous astronomers tried to study it outside eclipses. Total eclipses, which will be described in more detail in the next chapter, are rare, of short duration, and visible from very restricted areas of the Earth. Initially, astronomers only succeeded in observing prominences, which are specific regions – denser and cooler than the surrounding material – at the very base of the corona. They are also much brighter, especially if they are observed at wavelengths corresponding to their emission lines, using the spectrograph described in the previous chapter. The instrument that allows observation of the corona outside eclipses, the coronagraph, was invented by the French astronomer Bernard Lyot, who, in 1930, succeeded in eliminating, one after the other, all the sources of scattered, stray light, which is considerably stronger than the corona if no specific precautions are taken.

Making observations of the white-light corona, like those made at eclipses, using a coronagraph remains an extraordinarily difficult operation, which requires both a perfect instrument and an ideal site. It is still possible to observe monochromatic emission from the corona under slightly less perfect observing conditions, provided one has a suitable monochromatic filter. This last requirement suddenly became less costly with the development, around 1950, of interference filters. Subsequently, small-sized coronagraphs suitable for amateurs could be devised, and interesting designs that allowed proper study of prominences were described at the beginning of the 1960s.

Twenty-five years later, study of the corona has taken on a new lease of life for professional astronomers, with the development of radio methods, and through ultraviolet observations from artificial satellites. The observational networks set up by professional astronomers are not always able to follow transient events, and observations made with amateur coronagraphs may make a very important contribution. This will be discussed more thoroughly later.

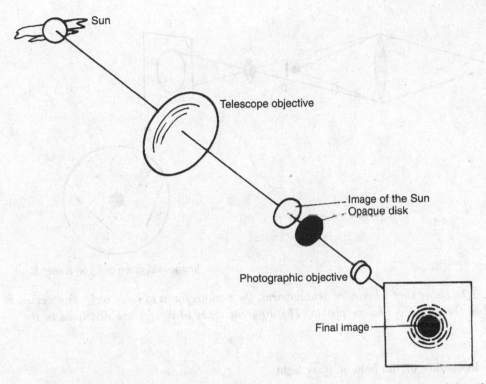

Fig. 2.1. *A diagram of the arrangement that is often initially considered by those wishing to observe the Sun's outer atmosphere, and which attempts to hide the image of the solar disk at the telescope's focal plane. Unfortunately, such an arrangement is quite impractical, because an intense halo of scattered light renders the final image quite useless.*

First, it is necessary to discuss in some detail the principles behind coronagraphs. These instruments give the false impression of being simple: the mental picture that generally springs to mind is like that shown in Fig. 2.1. An object glass forms an image of the Sun and its surrounding corona at a focus, where an opaque disk is located so that it just hides the brilliant disk of the Sun.

At first sight it would appear that if we photographed the resulting image of the Sun, then we ought to obtain an image of the corona like that seen during an eclipse. The reality is completely different. The basic instrument shown in Fig. 2.1 will generally reveal nothing more than a bright halo of stray light that has been scattered by the Earth's atmosphere and by the instrument itself. The key to the problem is to eliminate the various sources of scattered light. If, like Lyot, we are able to eliminate them one by one, we shall obtain an instrument that has only a small, residual halo. Many interesting observations then become possible, given good, clear weather. It is essential to note, however, that failure to consider any particular source of scattered light could ruin all attempts at observation, so a systematic approach to the instrumental problems is essential.

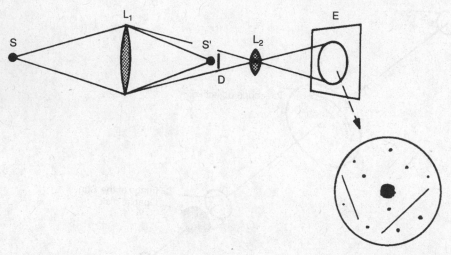

Image of L₁ given by L₂ on screen E

Fig. 2.2. *Using this laboratory arrangement, the various forms of stray light, that occur within the lens* L₁, *become visible. The different types of defects are discussed in the text.*

2.2 Reduction of the halo of stray light

2.2.1 Faults in the objective

The arrangement shown in Fig. 2.2 is required: the lens to be examined is illuminated by an intense source of light S, an image of which it forms at S′. We place a small opaque disk D at S′, and produce an image of L₁ on a screen E with another lens L₂. The disk D should block all the light from S at S′, so we might expect to see no light at all on the screen E. No such thing happens. If the source S is sufficiently intense, a very detailed image of the lens L₁, like that shown *bottom right* in Fig. 2.2 is visible at E. It shows:

- Various small bright spots and streaks scattered across the image of L₁;
- A brilliant spot of light at the very centre of the image; and
- A bright ring surrounding the whole circumference of L₁.

In other words, defects in the lens L₁ that we are examining produce scattered light even when the geometrical image of a luminous source is hidden by an obstacle. The same thing will apply if the source of light is the Sun: even after the disk of the Sun is blocked at S′ there will still be various sources of stray light within the instrument that will create a luminous halo, which will seriously interfere with observations.

Let us examine the various effects seen on the screen E, in the order in which they have been mentioned. The bright spots and lines observed on the image of L₁ are caused by defects in the lens: the material may be imperfect and contain bubbles or striae ('threads' of different refractive index); the surfaces may not be perfectly polished and may have small pits or scratches; and the surfaces may not be perfectly clean and have greasy spots or specks of dust, etc. on it. To minimize the halo of scattered light, it is essential, first, that the lens should be made from a piece of glass that is completely homogeneous, and second, that it should be ground and polished with extreme care. These two conditions can be met, but this assumes that the optical worker making the coronagraph's lens has powerful inspection methods available (and is able to use phase-contrast testing, for example). The difficulties in making the objective may be reduced by using just a simple lens (which has only two surfaces to be optically worked), made of a fairly hard glass that is not prone to scratching when cleaned. (In general a borosilicate-crown glass is used.)

Making a coronagraph objective is likely to be beyond the facilities available in most amateurs' workshops, so this optical component will have to be purchased. Under these circumstances, the arrangement shown in Fig. 2.2 is ideal for testing any lens offered by a manufacturer, and for deciding whether to accept or reject it.

Now let us turn to the brilliant spot of light in the centre of the image of the lens L_1 that is observed on the screen E, because has a completely different cause. Figure 2.3 shows a cross-section of the objective L_1, which is just a simple lens, normally chosen to be plano-convex. Most of the light incident on the lens will form an image S′ of the light-source, and this will be hidden behind the disk D. But transmission of a beam of light through a lens is never perfect, and partial reflections arise at the entry and exit surfaces. (These secondary beams of light are easily seen when a lens is illuminated with a laser.) In our objective, a small fraction (about 5 %) of the light will be reflected by the outermost surface and will be sent back towards the Sun. This will obviously not cause any problems. An approximately equal amount of light will be reflected at the glass-air exit surface (the plane surface in Fig. 2.3). This portion will suffer a second partial reflection at the entry surface, which will send a small fraction of the beam back to form a secondary image of the Sun, somewhere near S″. Naturally, this image is very faint because it is formed by a double reflection, and its intensity will be approximately $5/100 \times 5/100 = 25/10\,000$ of that of the principal beam. But it is particularly inconvenient that it should be formed at a point where it cannot be hidden by the opaque disk D, as may be seen from Fig. 2.3. The bright spot observed in the centre of the image of L_1, shown in the diagram in Fig. 2.2, is caused by this light that has undergone a double reflection from the surfaces of the objective.

One solution to this difficulty is often suggested: surely all that is required is to treat the surfaces of the objective lens L_1 with anti-reflection coatings, thus reducing the partial reflections and the stray light? In fact, the cure is likely to be worse than the complaint, because it would be difficult to obtain anti-reflection coatings that are sufficiently free from defects themselves to meet our stringent requirements. Moreover, it is most unlikely that they could be maintained in that state, given that the lens of a coronagraph has to be cleaned repeatedly.

A simple, effective solution to the problem was devised by Lyot: an image of the objective L_1 is actually formed at a point within the coronagraph assembly (which will be described later). A tiny disk, placed at that point, will block out the troublesome bright spot caused by the double reflection within the lens.

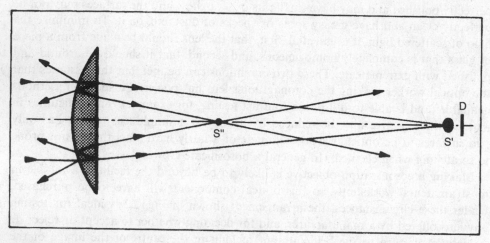

Fig. 2.3. *The geometrical paths of rays doubly reflected by the surfaces of a simple lens.*

2.2.2 The role of diffraction

We come to the third source of stray light that we identified: the brilliant ring surrounding the image of the objective L_1 on the screen E (*see* Fig. 2.2). The somewhat obscure explanation is sometimes offered that this consists of light diffracted by the circumference of the lens, but this description is not completely satisfactory. As is well-known, the image of a star given by a perfect astronomical objective is not an infinitely small spot of light, but appears instead as the Airy disk: a small central disk surrounded by rings of light. This is indicated (*top*) in Fig. 2.4, where the diagram on the right shows the variation in the intensity of light that is measured if the image is scanned in the direction indicated by the arrow.

Let us assume that we are observing, not a star, but a very narrow, straight line. This time (Fig. 2.4, *centre*) the image given by the objective will be a slightly widened, bright line, bordered by weaker fringes with minima where the intensity does not completely drop to zero.

Finally, consider the image of a luminous area that is sharply bounded at its right-hand edge (Fig. 2.4, *bottom*). It is not difficult to see that we may consider the luminous area as consisting of numerous, narrow, luminous lines side by side. The image given by the objective cannot be properly represented in a drawing, but it will gradually diminish, which means that a certain amount of light will reach areas of the image that ought to be completely black.

The disk of the Sun resembles a luminous spot, the brightness of which falls to zero (or almost to zero) as soon as we pass the limb. Although the Sun in the sky may be like that, because of the diffraction effects that we have just described, the Sun's image will not be sharply defined, but will fall off gradually instead. This stray light will not be blocked by the opaque disk inside the coronagraph but will be mingled with the weak light of the corona that we want to examine.

The problem of suppressing stray light caused by the diffraction fringes is well-

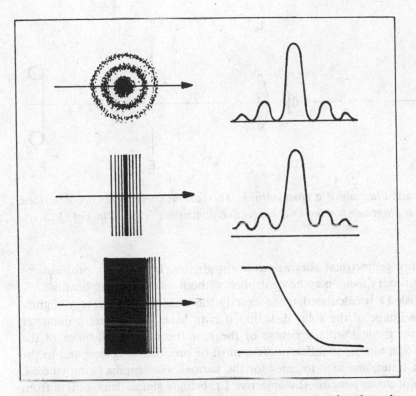

Fig. 2.4. *Because of diffraction phenomena, even a perfect lens does not give an in-finitely small luminous point, but instead the well-known pattern (top) of a disk sur-rounded by rings. Similarly, the image of an infinitely fine line is enlarged and bordered by fringes (centre). Finally, the image of a bright area with a sharp edge will appear as a more gradual decline (bottom). This causes considerable degradation of obser-vations at the solar limb.*

known to optical workers, who speak of 'apodizing' optics. One of Lyot's strokes of genius was to show that in a coronagraph, apodizing could be carried out simply using the arrangement shown in Fig. 2.2. The stray light that concerns us may be suppressed by masking the brilliant fringe around the image of L_1 by a slightly smaller diaphragm. In practice, a distinctly smaller diaphragm (about 2/3 or 3/4 of the size of L_1's image) is used. It is better to lose a little light rather than to run the risk that any slight misalignment will accidentally pass part of the bright fringe, which would be particularly deleterious, given the level of sensitivity that we want to attain.

2.2.3 The design of a Lyot-type coronagraph

Figure 2.5 illustrates schematically the final design of the coronagraph as devised by Lyot. The objective L_1, the most important part of the instrument, is a simple plano-convex lens, perfectly polished, with a focal ratio of about 1/20. Thanks

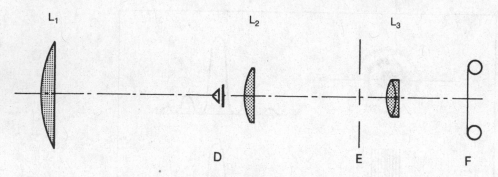

Fig. 2.5. *Schematic diagram of a coronagraph. The various components are described in the text and a diagram showing true respective dimensions is given in Fig. 2.25.*

to this long ratio, geometrical aberrations are negligible. In addition, problems of heating at the primary focus may be controlled without excessive complications.

The metal disk D is calculated to be exactly the same size as, or very slightly larger than, the image of the solar disk that it is to hide. The apparent diameter of the Sun varies throughout the course of the year (because the distance of the Earth from the Sun alters), so different disks must be provided to correspond to the size at different times, and also to cater for the various wavelengths being studied. This second point arises because the objective L_1, being a simple lens, suffers from chromatic aberration. This has important consequences, which will be described later.

Figure 2.6 shows a method of mounting the reflecting cone and occulting disk D that has proved successful in practice. The two are held by a central rod several centimetres in front of the lens L_2, which has been drilled through the centre. An insulating sleeve protects the glass from the heat transmitted by the cone. In order to avoid thermal problems, the cone should be highly polished, or even aluminized. (Only one cone is required, provided it is slightly smaller than the various occulting disks. This why it is wise to separate the two functions: the precise occultation required is given by the disk, and the cone rejects the heat.)

As we have seen, lens L_2 forms an image of L_1 at the diaphragm E. Figure 2.7 shows schematically the image that should be visible at E when the coronagraph is properly adjusted. We have already said that the diaphragm should have a diameter somewhat less than the full aperture of L_1. This does mean that some light is lost, but more stray light will be lost than light from the corona, which is the main thing. The resulting arrangement is also reasonably insensitive to small errors in alignment (caused by flexure of the coronagraph, for example). The central disk, which may become hot, should not be in contact with any optical component: in particular, it should not be cemented to L_2.

Behind diaphragm E in Fig. 2.5, all the instrumental sources of scattered light have been eliminated. It only remains to form an image of the 'eclipsed' Sun onto a recording medium, for example onto a film F, by means of a photographic objective L_3. To avoid increasing the length of the overall instrument more than absolutely

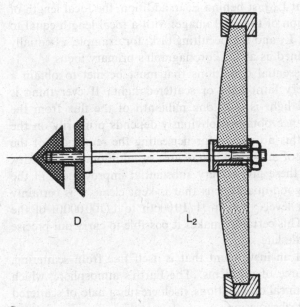

Fig. 2.6. *A method of mounting the cone and occulting disk D in front of lens L$_2$ in a coronagraph.*

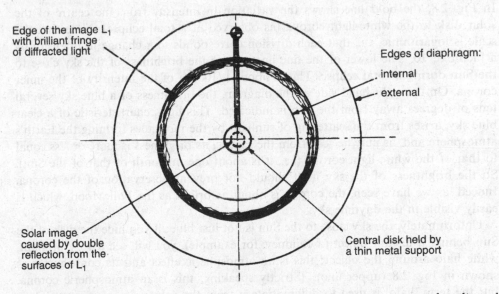

Edge of the image L$_1$
with brilliant fringe
of diffracted light

Edges of diaphragm:
internal
external

Solar image
caused by double
reflection from the
surfaces of L$_1$

Central disk held by
a thin metal support

Fig. 2.7. *The appearance at the plane of the Lyot diaphragm in a properly adjusted coronagraph.*

necessary, it is advisable to mount L_3 just behind E. In addition, the focal length of L_3 will determine the magnification of the final image: with a focal length equal to half that of the distance between L_3 and the occulting disk, for example, essentially the same image size will be obtained as at the coronagraph's primary focus.

We have just described the essential conditions that must be met to obtain a coronagraph that produces a very faint halo of scattered light. If everything is favourable the level of scattered light is only one millionth of the flux from the non-occulted Sun. The performance obtained obviously depends primarily on the quality of the objective L_1; but this no reason for neglecting the remainder of the optical elements. At the expense of an acceptable degree of additional complexity, and of difficulty in collimation, these give a very substantial improvement in the final image. With a reasonably good-quality lens that is kept clean, it is certainly possible to obtain scattered-light levels that are 1/10 000th to 1/100 000th of the non-occulted flux from the Sun. This certainly makes it possible to carry out precise observations, as we shall see a little later.

Although we have constructed an instrument that is itself free from scattering, we have eliminated only one source of problems. The Earth's atmosphere, which severely limits all sorts of astronomical observations, itself creates a halo of scattered light that may sometimes prevent observation.

2.2.4 The atmospheric halo

In Fig. 2.8, the bold line shows the variation in intensity from the centre of the solar disk to the white-light corona (as observed at a total eclipse). Note that the scale is logarithmic, i.e., that each division corresponds to a change in intensity by a factor of 10. The lower of the fine lines shows the brightness of the sky close to the Sun during a total eclipse. This is about 1/1000th of the intensity of the inner corona. On the right-hand side of the diagram, the brightness of a blue sky several tens of degrees away from the Sun is indicated. This light, characteristic of a clear blue sky, arises from the scattering of sunlight by the molecules forming the Earth's atmosphere and, as may be seen from the Figure, its brightness is more or less equal to that of the white-light corona (i.e., it is about one millionth of that of the Sun). So the brightness of the sky itself should not prevent observation of the corona. Indeed, as we have seen, the corona is about as bright as the Full Moon, which is easily visible in the daytime sky.

Unfortunately, the sky close to the Sun is not just blue; if you hide the disk of the Sun behind a distant object (a chimney, for example), you will see a very brilliant, white halo around the object: this is an atmospheric effect and its contribution is shown in Fig. 2.8 (upper line). [Strictly speaking, this is an atmospheric corona, but the term 'halo' is used to differentiate it from the solar corona. – Trans.] It is caused by dust, crystals, droplets, etc. that are suspended in the Earth's atmosphere. These aerosols are less numerous than the molecules of air, but, as they are far larger, they have the unwelcome property of scattering sunlight almost exclusively in the forward direction. Their role is therefore negligible when one looks at the sky well away from the Sun, but it becomes 100 to 1000 times more significant than molecular scattering when one wants to observe very close to the Sun, as we do.

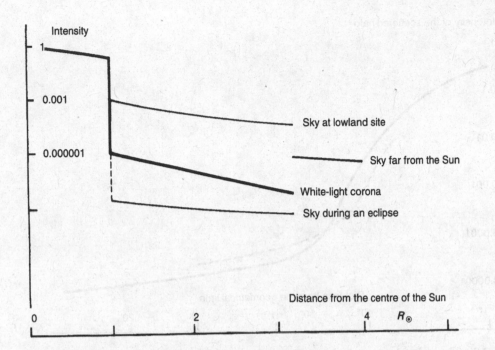

Fig. 2.8. *The ordinate gives the luminous intensity expressed on a logarithmic scale with respect to the intensity at the centre of the solar disk. The abscissa gives distances from the centre of the Sun, expressed in solar radii. The curves show the variation in the intensity of the Sun's outer atmosphere (the white-light corona) as well as that of the sky through which it is observed, both during and outside an eclipse.*

Is it possible to avoid this halo caused by aerosols? It is, of course, possible to avoid places (such as large towns) where the air has a particularly large concentration of particles. Equally, there is little point in observing under certain weather conditions: in hot fine weather, masses of tropical air are well-known for bringing a heat-haze. Fine days when thin cirrus clouds cover the sky around the Sun and create a small, but very brilliant halo, should also be avoided. Finally, the most certain remedy is to try to gain altitude: aerosols are mainly concentrated in the lower layers of the atmosphere and their density decreases by a factor of 3 for each increase in altitude of 1000 m. If one is prepared to go up 3000 m (the altitude of the Pic du Midi Observatory), the atmospheric halo would be weaker than that at sea-level by a factor of approximately $3 \times 3 \times 3 = 27$. Under certain meteorological conditions when there are 'temperature inversions', a very obvious gain may be obtained by going only a few hundred metres higher. Very fine skies may be found under these conditions in the winter, when the wind is in the north.

The halo caused by scattering by aerosols is not, unfortunately, the only detrimental effect arising in the Earth's atmosphere. Figure 2.9 shows (solid line) the brightness of the atmospheric halo observed at a high-altitude site with a good-quality coronagraph that does not add any scattered light (note that in this Figure both the brightness and the distance from the geometrical limb of the Sun are

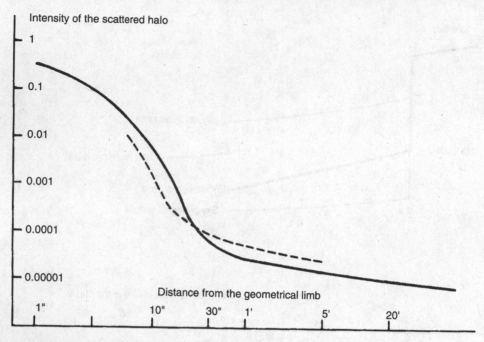

Fig. 2.9. *The ordinate gives the luminosity expressed on a logarithmic scale with respect to the centre of the solar disk. The abscissa gives the distance from the Sun's limb in seconds and minutes of arc, also on a logarithmic scale. The solid line shows the variation in the scattered corona at the final focus in a properly adjusted coronagraph. The dashed line shows how this scattering varies between good seeing conditions, and those with a sky that causes a slight degree of scattering.*

expressed on logarithmic scales). It will be seen that there is a change in the slope of the curve at about 30 arc-seconds from the limb of the Sun. Beyond that distance scattering by aerosols causes most of the halo. Closer to the Sun, the light encountered is the result of refraction caused by fluctuations in atmospheric density, which act to increase the apparent diameter of the solar limb. This is no more than the usual effect of bad seeing on telescopic images, but here it assumes overriding importance because the phenomena that we want to examine are so faint. (No one observing the Moon would be worried by a halo of scattered light, 1/10 000th of the Moon's brightness, about 30 arc-seconds away from the limb!)

The solid line in Fig. 2.9 is an average curve, which varies according to atmospheric conditions, just as it does for aerosols. At the risk of complicating the picture, we should mention that the two sources contributing to atmospheric coronae do not obey the same rules: fine, warm, anticyclonic weather, which increases the halo caused by aerosols, often gives rise to good images, i.e., to a decrease in the inner part of the atmospheric halo (as shown by the dotted line in the Figure), whereas the opposite effect on the curve is often observed when the wind is from the north. As a result, conditions are never perfect, and the observer should, as always, try to come to terms with the situation as it exists. It may be possible to adapt one's

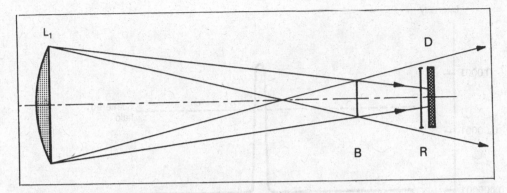

Fig. 2.10. *The objective of a coronagraph is generally a simple lens, so it is impossible to block out the whole solar disk at all wavelengths. Here the occulting disk is placed at the focal plane for red light (R), and it can be seen how part of the blue image (B) is not obstructed.*

programme according to the prevailing conditions: for example, in order to observe low prominences or even a flare on the limb, it is generally necessary to have good seeing; but for photography of monochromatic emission from the corona it is better to have less scattering by the sky.

We have discussed the characteristics of the atmospheric halo in detail, because the properties that have just been described are rarely found in the literature. The principal reason, however, is because any experienced amateur who has constructed a good coronagraph will find that observations are limited, not by the instrument, but by atmospheric scattering. This is readily seen from Fig. 2.9, bearing in mind that the halo of light scattered by a good coronagraph is 1/100 000th to 1/1 000 000th that of the Sun, and is therefore less than the atmospheric halo, even under the relatively favourable conditions at the top of a mountain!

2.2.5 Removal of stray light caused by chromatic aberration

As discussed earlier, a coronagraph's objective should be a simple lens: a doublet has a number of sources of additional scattering in the four glass surfaces and in the cement used to assemble the two lenses. Unfortunately, the simple lens that we have chosen is affected by chromatic aberration: the image of the Sun in red light (R in Fig. 2.10) is larger and farther from the lens than it is in blue light (B in the Figure). The distance between these two points, which is greatly exaggerated in Fig. 2.10, is about 20 mm for a simple lens with a focal length of 1m. This implies that provision should be made for the latter part of the optical train (cf. Fig. 2.5) to be focussed according to the wavelength at which observations will be carried out.

The occulting disk D should itself be located in the primary-image plane corresponding to the working wavelength, and this complicates the design slightly. But there are more important consequences: as will be seen from Fig. 2.10, if the occulting disk is placed over the red image of the Sun, some of the blue light will

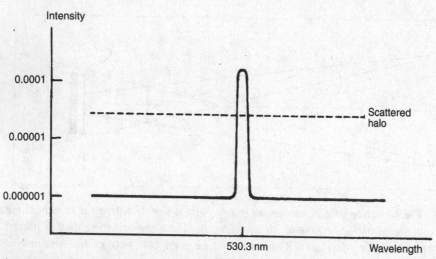

Fig. 2.11. *The solid line is a schematic representation of part of the coronal spectrum. In the presence of scattered light, it is much easier to observe the corona if it is possible to isolate wavelengths close to an emission line by using a filter.*

not be intercepted. In other words, the artificial eclipse will only be total for a small range of wavelengths and the level of scattered light in the coronagraph will increase sharply at other colours. When the occulting disk is adjusted to fit the red-light image of the Sun, there will be a very considerable flux of non-occulted blue light within the instrument.

Is this fault prohibitive? No, providing the coronagraph is used with a good-quality monochromatic filter, which is why we stress the importance of this particular optical element.

Recall that in the best case (Fig. 2.9) the scattered halo at 1 arc-minute from the solar limb has an intensity 1/100 000th of the solar disk's. Because the white-light corona is about 1/1 000 000th of the brightness of the Sun, it will always be difficult to detect without employing special methods (such as analyzing the degree of polarization for example), and the conditions never resemble those in a natural eclipse.

Under these circumstances, a coronagraph may seem of doubtful advantage; luckily, specific features of the spectrum of the corona and of prominences allow the situation to be very considerably improved.

Consider Fig. 2.11, which shows a tracing of a spectrum of the corona: apart from the continuum (which, at eclipses, appears as the 'white-light' corona), there are narrow, but strong, emission lines in the spectrum. The principal one is in the green at 530.3 nm; another is visible in the red at 637.5 nm, etc. Exactly at the wavelength of an emission line, the corona is as much as 100 times stronger than the continuum. The diffraction halo arising in the Earth's atmosphere, on the other hand, typically shows a continuum with narrow absorption lines (because it is basically light from the solar disk that is scattered in the Earth's atmosphere).

Although, taking the spectrum overall, the corona is much weaker than the scattered halo, it predominates if it is possible to make observations at the exact wavelength of emission lines.

It only remains for us to find a good filter capable of selecting a suitable coronal-emission wavelength. The gelatine or glass filters that are used in photography, for example, are far from being sufficiently selective (curve G in Fig. 2.12). On the other hand, interference filters, which are now available at reasonably affordable prices, give bandwidths that may be very narrow, giving maximum reduction of the scattered halo (curve I in Fig. 2.12). It will readily be appreciated that the visibility of coronal emissions is increased the narrower the filter's bandwidth. All considerations of cost aside, however, it should be noted that it is more difficult to use filters having bandwidths of less than 1 nm, because very narrow-pass filters only give the desired bandwidth at a specific temperature and for a truly parallel beam of light. To detect coronal emission it is absolutely essential to reduce the scattered light as much as possible, so a filter with a bandwidth of 0.2 nm must be used. This will require the use of a temperature-controlled enclosure (generally sold by the filter manufacturers). The filter should be installed in a beam that converges or diverges as little as possible: a good solution (Fig. 2.13) is to use two doublets instead of the lens L_3 (cf. Fig. 2.5), and to mount the interference filter between them.

This is not the place to discuss in detail the subject of filters, which is covered in other publications. It will suffice to mention two conditions that must be satisfied if good observations are to be made with a coronagraph. First, the filter's transmission should be very low outside the bandwidth (in optical jargon, the transmission curve should not show 'toes'). Otherwise, light from the Sun that is not blocked by the disk (Fig. 2.10) will succeed in getting through, and cause considerable degradation of the image; typically the red image of the sky and prominences given by the Hα-interference filter will show a blue fringe. Second, the filter should be of good optical quality, and in particular, it should be reflection-free; this is not always easy to ensure, and may alter as the filter ages.

The preceding discussion has concerned observation of the corona. The situation is similar with respect to prominences, i.e., observation is greatly facilitated by using a filter that can isolate the emission lines. Figure 2.14 shows a schematic representation of the most intense emission lines in the visible region. It will be seen that observations may be made most easily if the Hα hydrogen-line can be isolated by using a filter centred on 656.3 nm. Thanks to the strength of this emission, a bandwidth of 5–10 nm suffices to show prominences clearly, although a narrower filter obviously allows observation when there is greater scattering from the sky.

This is perhaps an opportune moment to remind readers that the use of monochromatic filters obviously means that it is pointless to use colour films. With a filter centred on the Hα line, only red light is transmitted and it is cheaper to obtain black-and-white images, and subsequently project them through a red filter, than it is to use colour film, which is more expensive to develop. (This is not falsifying the image. Far from it: when projected, a colour transparency does not show the true, Hα colour, but merely the red dye that the film manufacturer has chosen to use, and which is deposited during development.) It is only during a total eclipse, when one

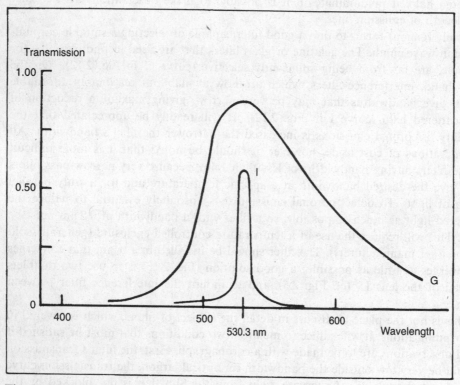

Fig. 2.12. *The general shape of the bandwidth of a filter made from coloured gelatine (G), and of an interference filter (I).*

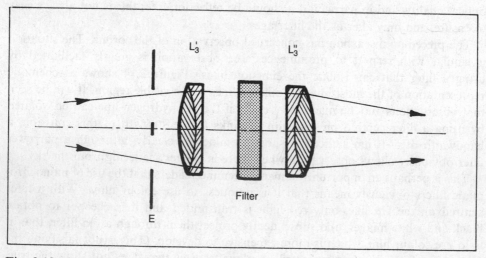

Fig. 2.13. *Interference filters should be located where the beam of light is essentially parallel. This may be achieved by splitting lens* L_3, *for example, as indicated here.*

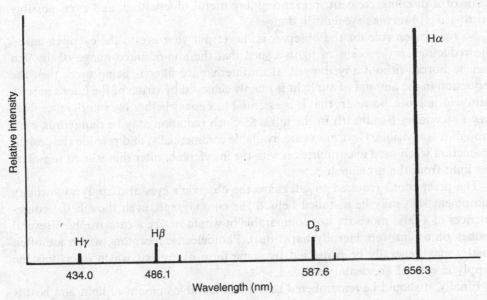

Fig. 2.14. *In the visible region the spectra from prominences mainly consists of hydrogen lines (Hα, Hβ, Hγ, etc.) and a helium line (D₃). The red Hα line is by far the strongest, and thus the easiest to observe.*

can dispense with monochromatic filters, that photography in colour comes into its own (*see* Chap. 3).

2.3 Observation of prominences at the limb

2.3.1 Essential precautions

In all solar investigations, observers must never forget that precautions are always essential if the intense flux is not to damage the instrument, or worse, the observer's eye. These warnings should be couched in even more draconian terms when it comes to observations with a coronagraph. Because the blinding solar image is 'normally' hidden by the occulting disk, it is tempting to look at the prominences without any additional attenuation of the beam of light. But any accidental decentring of the coronagraph (by an involuntary movement, a gust of wind, or failure of the drive motor, etc.) would uncover the Sun and could immediately cause a burn, which might be very severe.

This only serves to emphasize that particular attention must be paid to the equatorial drive of a coronagraph: in any case, a good drive is essential if good photographs are to be taken. Here the potential problem is far more serious. If someone is observing a planet through a telescope that is tracking badly, there will simply be a slow shift of the object in the eyepiece, and this is not very inconvenient for visual observation or short exposures. With a coronagraph, however, any fault in tracking will cause the Sun, originally hidden by the occulting disk, to appear in the form of a dazzling crescent, preventing any useful observation, and even possibly putting the observer's eyesight in danger.

So the golden rule for the observer is: never put your eye to the eyepiece unless the reduction in the beam of light is such that the non-occulted image of the Sun can be borne without any danger. If an interference filter is being used, then the reduction in the amount of sunlight is mainly achieved by virtue of the filter's narrow bandwidth. Note, however, that it is essential to know whether the interference filter has a secondary bandwidth in the infrared. Such radiation may be dangerous, even though it is invisible. Heat filters are available commercially, and provide the desired protection when used in conjunction with the interference filter that is used to isolate the light from the prominences.

The precautions required to safeguard the observer's eyes also apply to ancillary equipment that may be installed behind the coronagraph, even though the consequences of a false move are not comparable (it would not be a catastrophe if several frames on a film were literally burnt out). Photoelectric receptors, which are often costly, may generally be protected by some form of cut-out which interrupts the supply in case of accidental overload.

Finally, it should be remembered that a considerable amount of light and heat is rejected by the occulting disk at the coronagraph's primary focus, and that adequate free space must be maintained around the disk and the occulting cone.

2.3.2 *The types of emission that may be observed*

The curve given in Fig. 2.9, showing the variation in the brightness of the halo of stray light observed at the final focus of a good coronagraph, is repeated in Fig. 2.15, with the addition of curves showing the variation in brightness of prominences observed in Hα (at a wavelength of 656.3 nm), and in Hβ (wavelength 486.1 nm) and D_3 (587.6 nm) – both of which are weaker – and the curve for the corona at 530.3 nm (radiation characteristic of highly ionized iron, Fe XIV). In Fig. 2.15, the relative intensities of the stray light and the emissions correspond to those observed through narrow-pass interference filters: the bandwidths are 0.1 nm for the corona and 1 nm for the prominences.

It is obvious that observation of the corona in monochromatic light is very difficult, and should only be attempted by an experienced observer having the benefit of an excellent site. We have already seen that observation of the white-light corona, when the atmospheric corona cannot be reduced by suitable filtration, is scarcely feasible. Note that the images seen in monochromatic emission only show coronal material that is at a very specific temperature, depending on the wavelength studied, so the visible features entirely depend on the filter employed. (The white-light corona shows material over a wide range of temperatures simultaneously, and therefore gives a broader-based picture.)

It is generally agreed that the best way of becoming familiar with a coronagraph is to observe prominences in Hα light. Figure 2.16 is a superb amateur photograph, taken by J. M. Roques, which shows a particularly fine, spectacular set of prominences. Observers new to this type of observation should be warned that such beautiful prominences are both rare and very short-lived (especially large

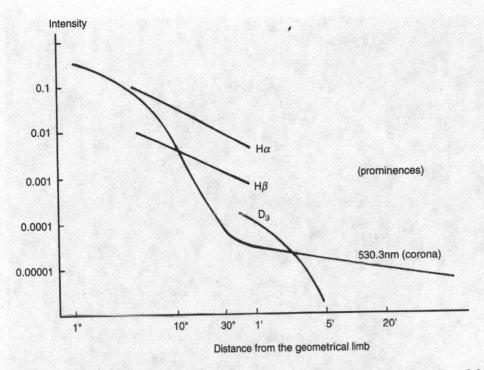

Fig. 2.15. *The variation in the intensity of scattered light, as shown in Fig. 2.9, is here compared with the intensities of prominence and coronal emission lines observed through an interference filter.*

prominences like that visible on the left of the photograph). On some days the Sun is practically featureless, without the coronagraph or the sky being to blame. Prominences are constantly changing features in the solar atmosphere, and it is time to describe some of their characteristics.

2.3.3 The telescopic appearance of prominences

Figure 2.16 shows that prominences vary considerably in shape and brightness: some are very compact, bright objects (which are generally low), and others have very complex filamentary structures and are distinctly fainter. Naturally, the darker the sky background, the greater the amount of detail that can be seen in the prominences. The differences in brightness that may be encountered pose a slight problem for photographers, who should avoid using high-contrast films.

One interesting characteristic of prominences is their extremely fine structure: the best photographs taken at observatories show details that are only 0.5 arc-second across, and even these are probably enlarged by turbulence. This means that the detail accessible with an amateur coronagraph, under the most favourable circumstances, is likely to be limited by the resolving power of the objective (i.e., between 1 and 2 arc-seconds, depending on the aperture used), or in most cases, by

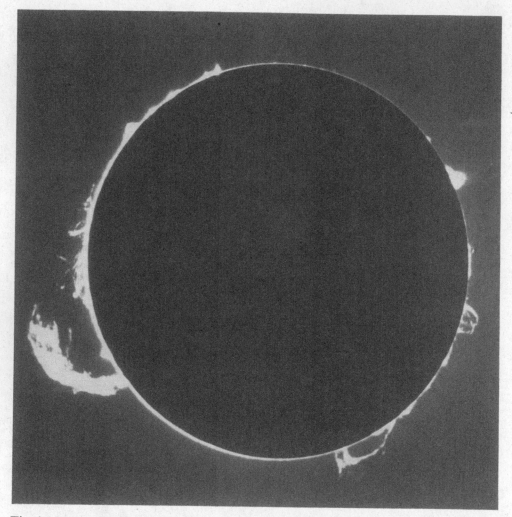

Fig. 2.16. *Prominences photographed by J.M. Roques, 1968 January 16, at 11:54 UT. 100-mm diameter coronagraph; Hα filter with a 0.38 nm bandwidth; exposure: 5 seconds on 45C62 film.*

turbulence. It should be remembered that average daytime turbulence amounts to several seconds of arc, and that it is rare to have both a very clear sky and stable images at the same time. In any case, the most favourable conditions occur in the early morning.

Photography of prominences in Hα light naturally requires films sensitive to red light, which were not easy to obtain at one time. Thanks to the arrival of Kodak 2415 film, which is well-known to amateur astronomers, this is no longer a problem. Photoelectric detectors such as charge-coupled devices (CCDs) are being rapidly developed, however, and in a few years they will doubtless be available at affordable prices. Such devices would allow images of prominences to be captured onto a magnetic medium, and it would then be feasible to make long series of observations, only retaining those images that show features of specific interest.

The observer who decides to take up photography with Kodak TP 2415, 35-mm film can still experiment with the degree of enlargement given by lens L_3. If the final diameter of the solar image does not exceed 20 mm, the whole image can be recorded on a 24×36-mm frame at once, which offers many advantages, although focussing is very critical, given the extremely small size of the images of prominences. Photographing just a small part of the solar limb on 24×36-mm format certainly allows finer detail to be captured; but it has the disadvantage of requiring decentring of either the camera, which is undesirable for both mechanical and optical reasons, or of the occulting disk, which makes surveying the whole solar disk a very laborious task.

Exposure times should be altered depending on the clarity of the sky. If the sky is bright, it is useless to take long exposures: only bright prominences will be seen (but useful observations may still be made, especially if there is a limb flare). With a very clear sky, on the other hand, exposures may be increased to capture fainter prominences, which are generally higher ones. Typical exposure times will lie between 1 and 1/30 second. The film should be developed to a moderate level of contrast to avoid losing both faint and bright details in the prominences.

When seen through a telescope, prominences give the observer the impression of being located on the plane of the sky. The reality is completely different, of course: prominences are generally regions of relatively cool gas; these have thicknesses of about 10 000 km, and a length easily 10 times that amount. When viewed in Hα light against the solar disk, prominences appear as dark filaments which characteristically look rather like a long, thin viaduct. Generally, they extend farther in longitude than in latitude. When they cross the limb they are therefore seen in profile, and their appearance is greatly altered because of perspective. Reconstructing their true shape in space would entail having a thorough knowledge of their appearance against the disk, and also knowing that they are evolving very slowly with time. These two conditions are very rarely met (Fig. 2.17).

Figure 2.18 shows some fine images of (different) prominences obtained by R. Bibault with a coronagraph. It is obvious that there is a wide variety to be observed, and that they are interesting just from the aesthetic point of view!

Before discussing how prominences change with time, we will just mention – although it is fairly obvious – that the slight fluctuations in the images that may be observed visually with a coronagraph are caused by variations in the Earth's atmosphere, and have nothing whatsoever to do with any changes on the Sun. These fluctuations, which make prominences appear 'alive' are, of course, of considerable nuisance when trying to photograph fine detail.

2.3.4 *Changes in the appearance of prominences*

The appearance of prominences alters over three typical time-scales:

Under very stable seeing conditions, the finest detail in prominences is seen to change over a few minutes or tens of minutes, sometimes reproducing an earlier state almost exactly. The cause of these changes is not fully understood. Contrary to

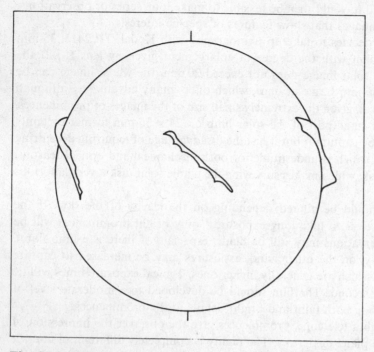

Fig. 2.17. *Because of the rotation of the Sun, prominences may be seen with a coronagraph at the eastern limb, then observed against the disk with a spectroheliograph, and finally become visible at the western limb before disappearing behind the Sun. The same object often appears very different because of changes in perspective.*

appearances, however, prominences are not hot 'flames' extending from the Sun into cold, empty interstellar space. Instead, they are relatively dense, cool regions (8000 K) immersed in the corona, which is far too hot (2 000 000 K) to emit any radiation at the Hα wavelength. Prominences are therefore the site of extremely complex thermal-instability phenomena, which are doubtless the cause of the changes that occur on time-scales of a few minutes.

Over longer times, i.e., from one day to another, prominences change their appearance. But here, as we have seen, it is difficult to separate real changes from variations caused by the alteration in perspective. The Sun rotates by about 13° per day, and even a very high prominence does not remain visible for more than about 2–3 days each time it crosses the limb. The life-times of prominences may exceed a month, so it is frequently possible to see a prominence disappear over the western limb and then recover it at the same latitude at the eastern limb 14 days later (Fig. 2.17), after the Sun has completed half a rotation. Despite this, the appearance is generally very different after a period of two weeks, and not just because of differences in perspective.

The third time-scale of changes in prominences arises from the 11-year solar cycle. Prominences are much less frequent around maximum, when the disk has most spots. In addition, the latitudes at which prominences appear depends on the

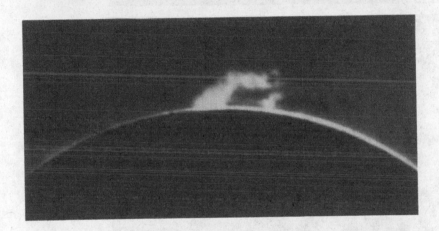

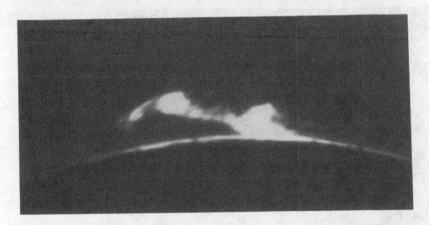

Fig. 2.18. *These photographs of prominences obtained by R. Bibault with his 100-mm coronagraph show the variety of forms that may be encountered. Taken with a filter having a 0.5-nm bandwidth. Exposures: 1/40 second on TP 2415 film.*

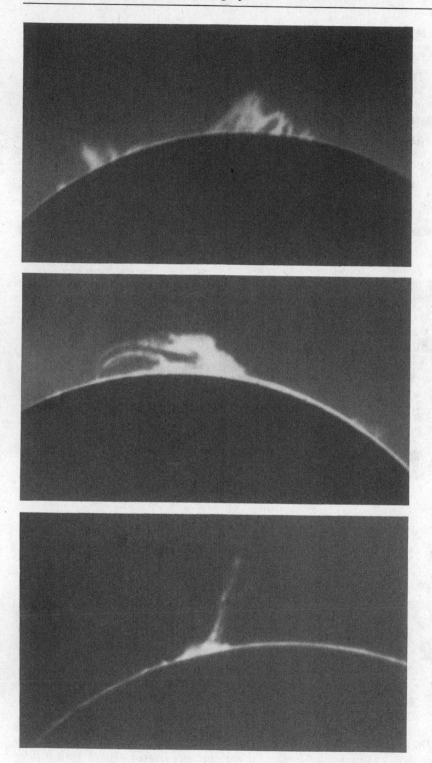

Fig. 2.18. (cont.) *For caption see Page 71*

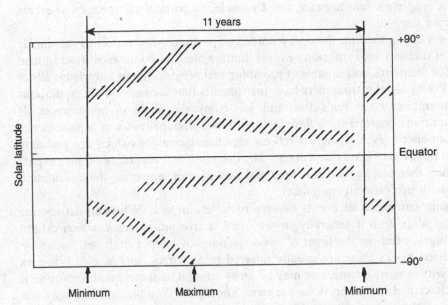

Fig. 2.19. *The bands of latitude where prominences most frequently appear (shown by the shaded zones in this diagram) vary over the course of the 11-year cycle.*

phase in the cycle. If the latitudes are plotted throughout the cycle, we obtain a diagram like that in Fig. 2.19. At minimum there are just two bands of prominences on the Sun, at about 45° latitude in each hemisphere. Prominences may therefore be seen crossing the limb in the northeast, northwest, southeast and southwest. As the cycle progresses, some of the prominences shift towards higher and higher latitudes, forming what is known as the polar cap, finally more or less reaching the poles at the time of maximum. It is then possible to watch prominences rotating in one spot over a considerable period of time. A second zone where prominences appear, more or less corresponds to the latitudes at which spots are seen, i.e., starting at about 40° at the beginning of the cycle and ending at about 10° just before minimum. Naturally there are exceptions and variations to this scheme, not just in each solar cycle, but also from one cycle to another. Overall, however, the pattern shown in Fig. 2.19 allows one to predict reasonably well which regions of the Sun are likely to show prominences.

2.3.5 Rapid events

Many amateur astronomers will have seen the marvellous film *Flames on the Sun* prepared by B. Lyot and J. Leclerc. They will certainly have retained the impression that solar prominences display extremely violent movements (explosions, jets of material, etc.), which appear frequently in the film in question. In fact, most prominences do not behave in this way. They are known as quiescent: in other words they vary less drastically and more slowly than those just described. True

73

large-scale eruptive events are rare, and Lyot selected particularly spectacular events in preparing the film.

Anyone viewing the Sun through a coronagraph should not expect to see similar events. Yet it is just such transitory events that are likely to be the most fruitful field of study for amateurs, and capable of providing real progress in our knowledge about the Sun. Professional astronomers have instruments that take essentially continuous cine-photographs of the Sun's disk, and, less comprehensively, of prominences. It is therefore quite possible for a flare, or the sudden disappearance of a prominence to pass unnoticed, even though it reflects significant changes (which are probably magnetic) in the active area concerned. The detection and recording of such events by amateurs therefore represents a very real way of complementing the information gathered with professional equipment.

Two main categories of events deserve to be monitored. When an active area crosses the solar limb it generally shows small, active prominences, which exhibit rapid changes, often in the form of mass-ejections of greater or lesser frequency. In specialized works these are usually referred to as surges, sprays, etc. When an area's activity is most intense, one may be lucky enough to observe a chromospheric flare, as described in the previous chapter. Through a coronagraph one may see, very close to the Sun, arches, and large and small clumps of gas that appear very bright. Some hours later, these may have turned into spectacular loops, which look like fountains of light, with material apparently appearing in the corona and raining down onto the surface of the Sun.

Another category of event is the sudden disappearance of what had previously been a quiescent prominence. This may happen to prominences at any latitude, but most frequently affects high prominences that are fairly close to active areas. Through the coronagraph, the prominence may be seen rising to considerable heights (several minutes of arc) and progressively dissipating, while some of the material appears to fall back on each side. The very large prominence visible on the left of Fig. 2.16 was just going through such a stage. Fig. 2.20 (taken by J. M. Roques) shows its appearance at 12:22 UT; i.e., only half-an-hour later!

Disappearances are not absolutely final, and it is not unknown for a very similar prominence to reappear in the same place a day or two later. This somewhat paradoxical behaviour may be explained in general terms as follows: the material in a prominence, which is relatively cold and therefore heavier than the surrounding corona, is sustained by magnetic forces. The prominence material is confined to a sort of magnetic trap (Fig. 2.21), which also helps to isolate it thermally from the surrounding coronal material. If an increase in temperature does nevertheless arise (for example at one end of the prominence), the gas, which was initially at a temperature of 8000 K, will become hotter and will soon cease to emit Hα radiation. The prominence will therefore disappear. But if the magnetic trap that favoured the condensation of cold material continues to exist, obviously conditions may later become suitable for a prominence to reform in the same place.

The disappearance is final, however, if the magnetic field itself is disrupted. This may happen during the evolution of an active area: if a new group of spots emerges, for example. In such cases, the alteration in the local magnetic field may cause a chromospheric flare and destabilization of prominences in the surrounding area.

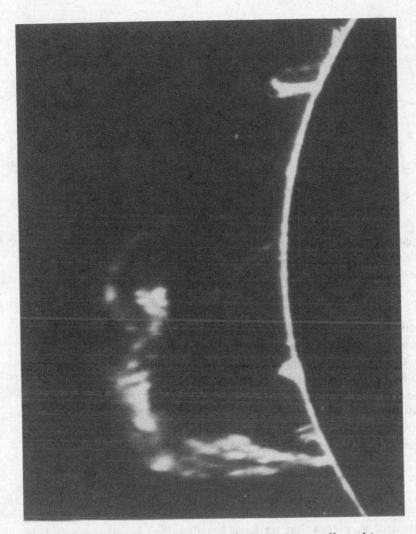

Fig. 2.20. *Eruptive prominences frequently evolve rapidly and in a spectacular fashion. This photograph by J. M. Roques shows the appearance at 12:22 UT of the large prominence that is clearly seen on the left-hand side of Fig. 2.16. The considerable alterations in half an hour are quite apparent.*

What is the best method of observing these different events? Obviously changes with time are of most interest, so photographs should be taken at regular intervals throughout the duration of the event (which may be one or two hours), beginning as soon as it is detected. A frequency of one image per minute would appear to be appropriate. The exact times of the photographs must be recorded, of course.

Flares and disruptive events are generally accompanied by fairly bright emission which is not too difficult to record. (Care should be taken not to over-expose limb flares.) It should be borne in mind that the velocities encountered are quite

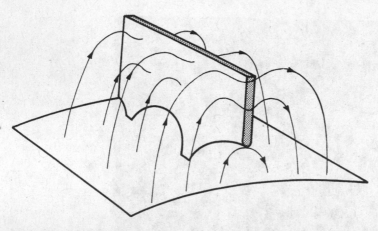

Fig. 2.21. *It is generally accepted that the relatively cold, dense material in prominences is held in a sort of magnetic trap by the lines of force. However, the model shown here is not unique, and in any case probably does not apply to all prominences.*

considerable, and that the shift in the Hα emission, caused by the Doppler effect, may be quite considerable. Photographs taken through a filter with a very narrow band-pass may therefore only capture part of the eruptive prominence, so it is advisable to use filters with bandwidths of at least 0.5–1 nm.

Finally, it should go without saying that the photographs obtained should not remain in the bottom of a drawer! For proper use to be made of them, the observer should belong to a group of amateurs who specialize in solar observations. This was also discussed in the previous chapter.

2.4 Conclusions

Study of the catastrophic events that we have just described is not the only field of research open to an amateur with a coronagraph. With an instrument that is particularly free from scattered light, one could, for example, increase the exposure times and search for faint extensions of prominences out into the corona. The problem of the origin of prominences is still controversial, and it is quite possible that different phenomena are being lumped under the general heading of 'prominences'. Some prominences may result from condensation of coronal material, whereas others do seem to be extensions of the chromosphere.

Very useful studies may be carried out if a group of amateurs has several solar instruments. A spectroheliograph (*see* Chap. 1) or a very narrow-band filter allow prominences on the disk to be seen as dark filaments against the chromosphere. There is a close relationship between filaments and prominences seen at the limb, but this sometimes differs from what one might expect – the filaments that are most visible on the disk are not necessarily the highest ones – and it is always interesting to gain an overall picture of the different structures at various levels of the solar atmosphere. The precise location of the prominences that are associated with active regions with respect to the spots is always unique, and merits proper study.

Figures 2.22, 2.23 and 2.24 show other interesting records obtained by amateurs (J. M. Roques, A. Doucet and D. Cardoen) and prove beyond all doubt that the work that we have described may be carried out with coronagraphs constructed with modest means. Plenty of valuable advice may be obtained from amateurs who have successfully constructed their own coronagraphs. Naturally, professional astronomers who use coronagraphs, particularly those at the Pic du Midi Observatory, are prepared to help with those points that this chapter has not covered.

An indication of some of the books and articles that will give further information about the subject may be useful. As regards the coronagraph itself, the articles that Lyot wrote in *l'Astronomie* may be reread with profit (Lyot, 1931, 1932, 1937). On the fiftieth anniversary of the coronagraph's invention, Lyot's work was described by A. Dollfus in issues of *l'Astronomie* (Dollfus, 1983a and 1983b), as well as in several articles in *Journal of Optics* (Dollfus, 1983c), published in Paris under the editorship of Prof. Françon. The latter author (who developed the method of polishing coronagraph lenses with Lyot) has also published a book on optical filters that is indispensable for amateurs who are confronted with the problem of spectral filters (Françon, 1984).

The latest book (Mazereau and Bourge, 1985) contains a lot of details about techniques that will certainly help with the practical problems encountered in building a small coronagraph. But it should be noted that the optical layout suggested (their Figure 4) is not correct, and does not allow the light scattered by L_1 to be eliminated. Other amateur designs have been described and should also be consulted (see, for example, the description of the coronagraph by Roques, 1961). There is a more extensive bibliography available in English.

2.5 A typical coronagraph

This section gives details of one of the coronagraphs used at the Pic du Midi for photographing monochromatic emission from the solar corona. Figure 2.25 shows a schematic diagram of the various elements. An amateur-sized instrument could be obtained by scaling-down the various elements to 1/2 or 1/3 of the dimensions given.

The identification of the various elements is, of course, exactly the same as that given in Figs. 2.5, 2.6, and 2.13. (Objective: L_1, field lens: L_2, transfer lenses: L_3' and L_3'', occulting disk: D, and annular diaphragm: E.)

Primary lens L_1 Plano-convex, borosilicate-crown lens; diameter: 200 mm; focal length: 4007 mm at 637.4 nm; 3968 mm at 530.3 nm.

Occulting disk D One of a set of disks in 1-mm thick brass, with diameters: 36.0, 36.2, 36.4, ..., 39.0 mm. It is located at the focus of lens L_1 for the working wavelength, and has ahead of it a cone of polished dural, 34 mm in diameter. (The distance between the cone and the occulting disk is approximately 10 mm.)

Fig. 2.22. Left, *a portable corona-graph built by J. M. Roques: OG diameter = 50 mm; F = 715 mm; effective focal length = 1225 mm; Mizar mount. Below, a photograph obtained with this instrument using an Hα filter with a 0.5-nm bandwidth.*

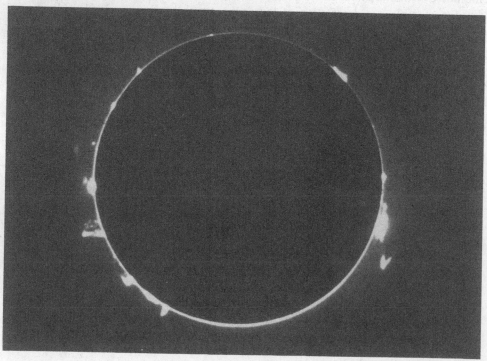

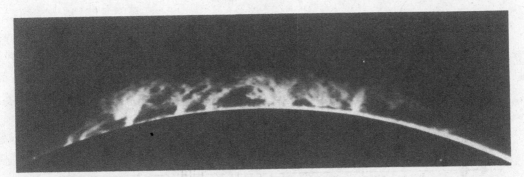

Fig. 2.23. *A fine example of a hedgerow prominence, photographed on 1979 August 5 by A. Doucet with a 110-mm coronagraph with an eccentric occulting disk and an interference filter with a 0.4-nm bandwidth. Exposure 1/60 second on SO 115.*

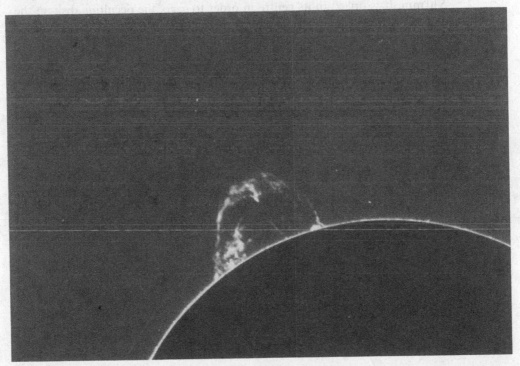

Fig. 2.24. *A prominence photographed on 1980 October 31 by D. Cardoen with a 100-mm coronagraph. Exposure 1/8 second on TP 2415 film.*

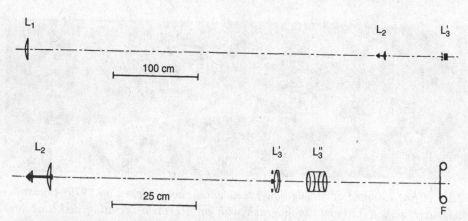

Fig. 2.25. *Optical diagram of a coronagraph used at the Pic du Midi for photographing the solar corona.*

Field lens L_2 Plano-convex lens in ordinary crown; diameter: 80 mm; focal length: 530 mm. It is drilled in its centre in order to carry the occulting disk on the end of a metal rod. The distance between the disk D and the lens L_2 is 40 mm.

Annular diaphragm E Situated 650 mm behind L_2. The image of L_1 given at E by L_2 is 33 mm in diameter. The aperture of the diaphragm has a diameter of 25 mm. The central disk, 6 mm in diameter, hides the image of the Sun caused by double reflections from the surfaces of L_1, which has a diameter of about 2 mm.

Lens L_3' Situated 10 mm behind E. A Clavé doublet: diameter 60 mm; focal length 720 mm.

Lens L_3'' Situated 100 mm behind L_3'. The interference filter, within its oven, is located in the space between these two lenses L_3' and L_3''. The lens is a Benoist–Berthiot triplet: diameter 55 mm; focal length 350 mm.

Final image of the Sun This is formed about 350 mm behind L_3'' on the film, which is carried in an Exakta 24 ×36 mm body. The diameter of the (occulted) solar disk is 18 mm. Focussing is carried out by longitudinal adjustment of L_3''.

3 Solar eclipses

S. Koutchmy

3.1 Introduction

Total solar eclipses are certainly the most impressive astronomical events that may be observed, so much so that in the past they evoked fear and powerful emotions, even among the more scientifically minded. Nowadays, with our greater understanding of science, thousands of people travel to see the magnificent natural spectacle that suddenly reveals the solar corona. It is disconcerting, even if not menacing, to be plunged into heavy darkness that allows stars to be seen even at midday. The effect is heightened because the phenomenon happens so suddenly, in the space of a few seconds, following the gradual disappearance of the solar disk over a period of about an hour, and with an appreciable drop in temperature. During totality, the black disk of the Moon is ringed by bright pink light and surrounded by a silvery halo. The analysis of records of historical eclipses is doubtless very worthwhile, but the phenomenon enables extremely valuable astrophysical observations to be made. We shall therefore just summarize the circumstances giving rise to the event, and concentrate on the detailed study of the corona – undoubtedly the finest 'nebula' ever visible in the sky – that is revealed during those few precious minutes of totality. The solar corona is the source of the solar wind and strongly scattered light within our Solar System. Total eclipses of the Sun allow thorough investigations of the corona to be made, subject to the limitation imposed by the need to transport dedicated experiments (particularly those relating to faint nebulosity) to suitable sites on the central line of totality.

3.2 Circumstances of eclipses

Figure 3.1 shows the Earth and the Moon in space, illuminated by the rays of the Sun. Note that the Moon, which completes one rotation around the Earth in 27.3 days (a sidereal month), has an orbit that is inclined at an appreciable angle (5°09′) to the plane of the Earth's orbit around the Sun (the ecliptic). Note also that the Earth–Moon distance varies around the lunar orbit and that the apparent size of the lunar disk is, on average, similar to that of the Sun (about 32 minutes of arc). This happy coincidence arises from the fact that the ratio of the distance from the Earth to the actual diameter has a very similar value for each of the two bodies. As a result, the Earth sometimes intercepts the Moon's shadow-cone under reasonably favourable conditions. When the apparent size of the Moon is greater than that of the Sun there is a total eclipse: when the apparent size is smaller, the eclipse will be annular. The duration of totality is also governed by the ratio of the apparent diameters of the two bodies, which therefore determine the magnitude of the eclipse. When the Earth's rotation is taken into account, this duration never exceeds 8

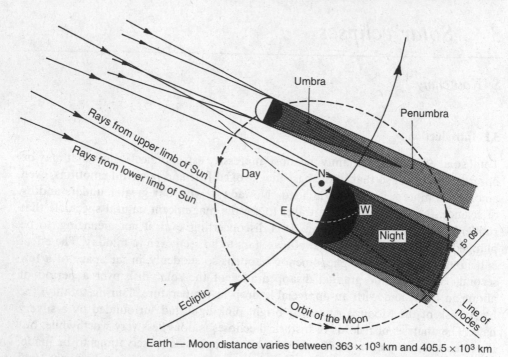

Earth — Moon distance varies between 363×10^3 km and 405.5×10^3 km

Fig. 3.1. *Schematic diagram showing the conditions under which the Moon's shadow may be intercepted by the Earth, producing a total eclipse of the Sun.*

minutes. These few minutes are extremely precious for observers, and for research, because they offer the finest view of the corona. Anyone within the shadow cone (which races across the surface of the Earth at a speed of more than 2000 km/h), sees the brightness of the daytime sky fall to about 1/10 000th of its normal value. Prominences, which are extensions of the chromosphere, may sometimes be seen around the lunar disk. Surrounding everything is the magnificent halo of light scattered by the plasma in the solar corona and by particles in the interplanetary medium.

Around 1870, T. R. Oppolzer investigated past eclipses and calculated new ephemerides. More recently, J. Meeus, C.E. Grosjean and W. Vanderleen have refined these calculations, publishing tables and maps with details of all eclipses up to the year 2510 (Meeus *et al.*, 1966). These detailed studies have, of course, confirmed the ancient eclipse (or Saros) cycle, where eclipses repeat with a period of 18 years 11.3 days. This cycle may be considered as a commensurable (or 'beat') period arising from of the relationship between the draconitic month of 27.21 days – the Moon's period crossing the ascending node of its orbit – the synodic period of 29.53 days, and the draconitic year of 346.62 days. This may be expressed as: $223 \times 29.53 + 242 \times 27.21 = 19 \times 346.62 = 18$ years 11.3 days. To summarize, the conditions required for any eclipse to be total are (taking the factors shown in Fig. 3.1 into account):

- It must be New Moon;
- The Moon must be near the line of nodes;
- The apparent size of the lunar disk must be larger than the apparent size of the Sun.

These conditions are met 14 times during a Saros period. Figure 3.2 shows the central paths of totality for events until the end of this century. The finest is undoubtedly that of 1991 July 11, which provided as many as 7 minutes of totality from various sites in Hawaii and Mexico. A total eclipse will also be visible in 1999 over part of western Europe. This is quite a rare event, because on average a total eclipse is visible only once every 370 years from any one point of the surface of the Earth.

3.3 Observation of the event

The diagram below shows the various phases of a total eclipse, assuming the apparent disks of the Sun (filled) and of the Moon (open) to be perfectly circular.

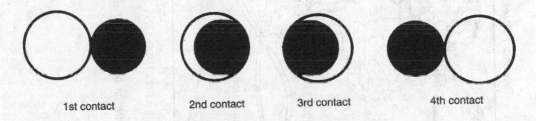

| 1st contact | 2nd contact | 3rd contact | 4th contact |

The disk of the Moon, naturally, crosses in front of the disk of the Sun, so the latter is completely covered between the 2nd and 3rd contacts, which actually last several seconds because the lunar limb is very irregular (Fig. 3.3). Variations in the profile of the maria and highlands may allow a certain amount of the solar photosphere to be seen, giving rise to Baily's Beads. The difference between the apparent motion of the two bodies across the sky gives a typical relative velocity of 1 second of arc in 2 seconds of time. (The Earth's diurnal rotation gives a 'velocity' of 15 seconds of arc per second of time.) When variations in the profile are taken into account, it is obvious that the duration of the visibility of Baily's Beads may be considerable. This is one of the difficulties facing experiments designed to measure the duration of eclipses (or 'occultations' as they should properly be called).

3.3.1 *Measurements of the solar diameter*

Published ephemerides for the Sun and the Moon allow the average apparent diameters of the two disks as seen from the Earth to be determined very accurately from the topocentric Earth–Moon and Earth–Sun distances. By making the reasonable assumption that the diameter of the Moon remains constant over a time-scale of several centuries, and by measuring the precise interval between 2nd and 3rd contacts, it is possible to determine if the diameter of the Sun is truly constant – at

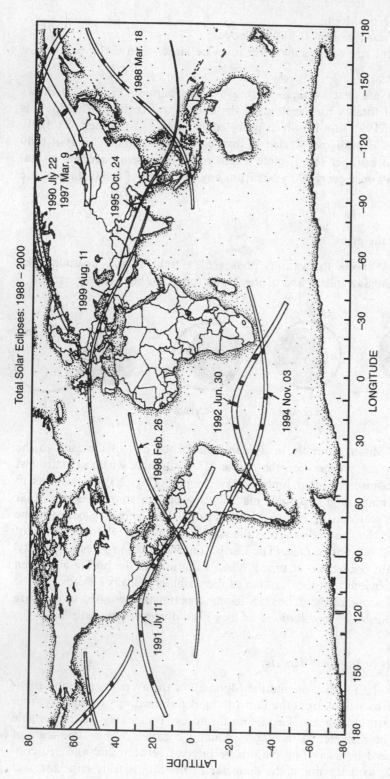

Fig. 3.2. A map of the paths of certain total solar eclipses, with an indication of the width of totality, after F. Espenak (NASA).

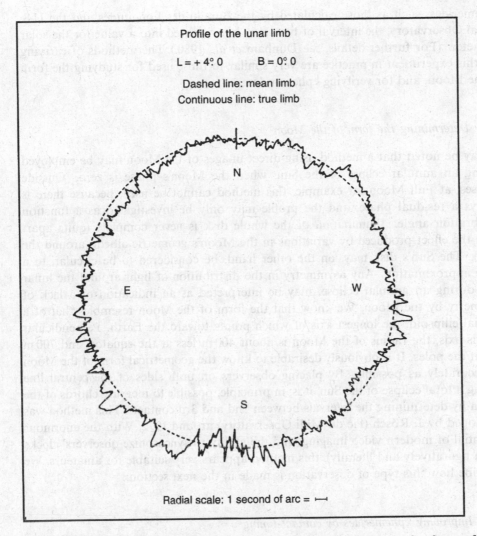

Profile of the lunar limb

L = + 4°.0   B = 0°.0

Dashed line: mean limb
Continuous line: true limb

N

E   W

S

Radial scale: 1 second of arc = ⊢⊣

Fig. 3.3. *Typical profile of the lunar limb as calculated for the total eclipse of 1984 November 23 (from a publication of the U.S. Naval Observatory). Note that the radial scale is greatly exaggerated (about 60 times). After M. Sarazin.*

least from one eclipse to the next. Recently, there have been some doubts about the constancy of the solar diameter. Provided allowance can be made for the profile of the lunar limb, a very precise measurement is, in principle, possible. The intervals to be measured are generally of the order of a few minutes, and an accuracy of the order of 0.02 seconds of time may be expected. This corresponds to measuring the Sun's diameter to 0.01 arc-seconds (or the actual size of the Sun to about 10 km). If the observer is not on the central line (and depending upon the eclipse site's position within the path of totality), it may or may not be necessary to introduce a correction to allow for the relative positions of the disks. Using the calculated ephemerides, such as those calculated by the Bureau des Longitudes and the U.S. Naval Observatory, the interval of time may be converted into a value for the solar diameter. (For further details, *see* Dunham *et al.*, 1980.) The methods of carrying out this experiment in practice are very similar to those used for studying the form of the Moon, and for verifying ephemerides.

3.3.2 *Determining the form of the Moon*

It may be noted that a method using direct images of the Moon may be employed during an annular eclipse of the Sun, when the Moon's phase is zero. Outside eclipses, at Full Moon for example, this method cannot be used, because there is always a residual phase, and the profile may only be investigated as a function of libration angle. Illumination of the whole disk is never complete (quite apart from the effect produced by variations in the Moon's geometric albedo around the limb). The Sun's disk may, on the other hand, be considered to be circular, to a close approximation. Any asymmetry in the distribution of light around the lunar disk during an annular eclipse, may be interpreted as an indication of a lack of symmetry by the Moon. We know that the form of the Moon resembles that of a triaxial ellipsoid, the longest axis of which points toward the Earth. Perpendicular to this axis, the radius of the Moon is about 400 m less at the equator, and 700 m less at the poles. It is obviously desirable to know the geometrical form of the Moon as accurately as possible. By placing observers on both sides of the central line during a total eclipse of the Sun, it is, in principle, possible to measure chords of the Moon by determining the intervals between 2nd and 3rd contacts. This method was developed by J. Rösch (Pic du Midi Observatory) around 1972. With the enormous potential of modern video imaging, and the ability to synchronize observers' clocks (both figuratively and literally), this method appears very suitable for amateurs. We describe how this type of observation is made in the next section.

3.3.3 *Improving ephemerides by contact-timing*

The precise times of contact for a specific place may be calculated from ephemerides, which obviously have a limited accuracy. This generally produces errors in the times of contact that are of the order of a few seconds of time. One of the sources of error, of course, lies in our insufficient knowledge of the form of the Moon, and even of the diameter of the Sun, but various other factors affect the calculations. As a result, the times of contact calculated from the two sets of ephemerides prepared by the Bureau des Longitudes in Paris and by the U.S. Naval Observatory may differ by 2 seconds of time. (See the article about this in the special edition of *l'Astronomie* entitled 'Total solar eclipses' published in November 1986.)

Contact times may be measured to a high degree of accuracy in quite a simple experiment. In the past, a photometer was required to measure the solar luminosity during the few seconds before and after totality. Time signals must be recorded simultaneously, and in general this presents no problems, even at a remote site, provided a good short-wave receiver is available. It is also frequently possible to

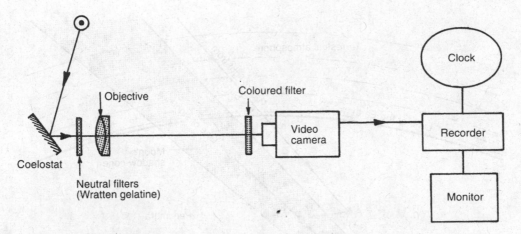

Fig. 3.4. *Schematic diagram of an experiment designed to measure the 2nd and 3rd contacts of an eclipse.*

use a microcomputer to provide an adequate time-base. Nowadays, it is possible to do far better, either by using a cine camera, exactly synchronized with the clock, or a video camera. Both may be analyzed later, but the latter may even be examined line by line. This method has a very considerable advantage over the photometer method, where the total flux (or a poorly-known fraction of the flux) from the solar crescent is measured. In fact, it allows the problem posed by the irregular edge of the Moon (Fig. 3.3) to be tackled, by providing a visual record of the Baily's Beads that are produced by gaps in the Moon's profile. For such a record to be usable, it is important to make measurements of the variation in the amount of flux received during the last few seconds before totality (and the first few after 3rd contact), which is why it is essential for the response of the camera to be precisely calibrated. To obtain a good-quality image, the equipment should be installed on an equatorial mounting, or else fed by a coelostat, and provided with filters (both coloured and neutral-density). The video camera's sound channel may be used to record the time signals, or even better, these may be superimposed on the image. An example of a possible form of mounting is shown in Fig. 3.4.

3.4 Inner coronal photography and photometry

Before beginning this section, which concerns imagery and photometric measurements, it is just as well to recall that given the short duration, not to say suddenness, of a total eclipse, there is no question of observing it a second time! Preparations for the observation of the corona during totality should be made with the greatest care. Professional astronomers often do this several years in advance; visiting the site several times before the fateful day. Experiments must function without fail; there must be no question of improvising or counting on luck. It is preferable to undertake reliable, but less accurate, experiments than to attempt some special feat. Even if the resulting photograph is unique, you will not be able to make proper use

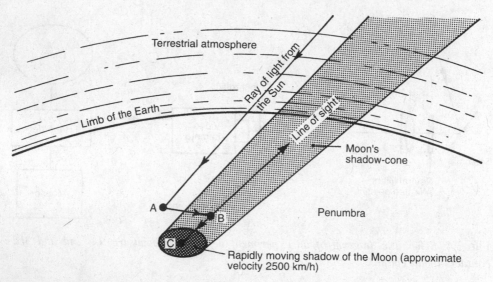

Fig. 3.5. *This diagram shows the origin of the sky illumination during an eclipse. A: primary scattering by the ground and the lower atmosphere of sunlight which then illuminates B, the line of sight in the Earth's atmosphere, which further scatters some of this radiation towards the observer at C.*

of it unless it is properly calibrated and all the instrumental factors are well-known. Subject only to limitations of cost, there are still many extremely interesting experiments that should be attempted, because the full glory of the corona is visible only during a total eclipse.

We may see why from Figs. 3.5 and 3.6. The latter diagram shows the brightness distribution, measured at the ground, of the Sun, the Moon and the sky (the Earth's atmosphere). It shows that even in the mountains, the sky close to the solar disk is not dark enough to allow the full corona to be observed outside an eclipse. (The inner portions may be investigated using the special methods described in the previous chapter.) During totality, however, the luminance of the sky drops by a factor of at least 10 000, because the illumination of the sky background is then primarily produced by multiple scattering in the atmosphere and from the ground. With a completely clear sky, the atmosphere is illuminated by light originating outside the Moon's shadow cone. As a result, the illumination is slightly variable, because the shadow moves rapidly with respect to the observer (Fig. 3.5). As yet, no theory has fully explained the phenomena that are observed, especially with respect to the polarization of the sky background during an eclipse, or its colour, so work still remains to be carried out in this area. Nevertheless, Fig. 3.6 shows clearly why the solar corona may be observed during totality: the sky becomes sufficiently dark for the contrast between the various features to be perceptible. Even the faint Earthshine on the face of the Moon may be measured.

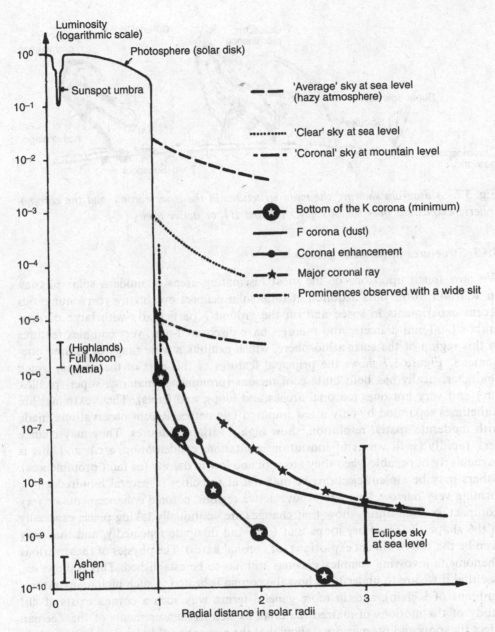

Fig. 3.6. *The radial distribution of intensity close to the Sun of the daytime sky and of various objects observed during an eclipse. Note the great dynamic range* (10^{10}) *covered by the intensity scale: a moonless night sky would have an intensity between* 10^{-13} *and* 10^{-14} *of that of the disk of the Sun.*

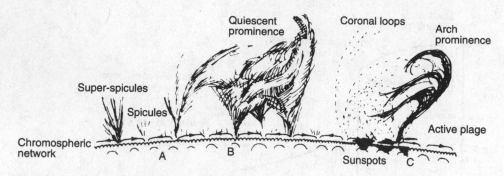

Fig. 3.7. *A diagram showing the main structures in the inner corona, and the chromospheric extensions over various quiet (A and B), or active areas (C).*

3.4.1 Structures in the inner corona

We now touch upon one of the most fascinating areas of modern solar physics. In addition to the data acquired at total solar eclipses over many years, numerous recent experiments in space and on the ground – particularly with large coronagraphs (200-mm diameter and more) – have discovered new, very complex features in this region of the solar atmosphere, which exhibits a wide range of plasma phenomena. Figure 3.7 shows the principal features of this part of the corona, which characteristically has both quite cool regions (prominence material, super-spicules, etc.) and very hot ones (coronal arches and loops, and flares). These extremes are sometimes separated by only a few hundred kilometres. Recent observations, made with moderate spatial resolution, show highly variable features. They may change very rapidly (with waves of ionization, excitation, condensation, etc.), and this is particularly noticeable when they are cool and low in density (as faint prominences). Others may be violent ejections of material at velocities of several hundred km/s, forming very narrow, long jets. Above active regions, coronal enhancements or very 'compact' condensations show that changes are continually taking place, especially in the shape of temporary loops that form and dissipate repeatedly, and that may even be the site of violent explosions (or coronal flares). The physics of these various phenomena involving complex plasmas still has to be established. This is, however, essential if we are to understand how the corona is heated to such high temperatures (millions of Kelvin), and, in more general terms, why such a corona exists at all! Study of the motions of ionized material, as well as measurements of the Zeeman effect (in spots and prominences) show that the magnetic field plays a primary role in determining most of the phenomena that are observed. Thermodynamic properties that vary with time are also important factors to be taken into consideration. It is important to obtain as much hard information as possible by observations made during totality. The key element in the various experiments that may be carried out is high spatial resolution.

3.4.2 Observations at the extreme limb

This type of observation is similar to those made at the contacts, except that selective filters should be used (or preferably a spectrograph – see later). The exact details depend on whether the desired spectra are chromospheric or photospheric ones – the latter are, of course, emission spectra. For studies of the extreme outer limit of the photosphere it is preferable to work in the red and the near infrared, where there are spectral windows that are almost completely, or totally free from lines. These allow the 'true' continuum to be measured and cover fairly wide spectral regions, thus ensuring that there is a good signal-to-noise ratio for photometric measurements. Because allowance must also be made for the irregularities at the lunar limb, a high magnification should be employed (with an effective focal length of at least 10 m).

If the photographic method is used, exposure times should be chosen to give correct rendition of the solar limb (over-exposure is to be avoided). To ensure that measurements may be made at the extreme limb, where the intensity is less than 1 % of that at the centre of the disk, contrast should be kept low ($\gamma < 1$). Finally, very careful calibration should be carried out, so that the profile of the brightness at the extreme edge of the solar photosphere may be deduced. This is poorly known within 2 arc-seconds of the limb.

The narrow layer the chromosphere and the spicules, up to about 10 arc-seconds from the limb, may also be studied with an appropriate filter (for example, an Hα filter with a bandwidth of at least 1 nm). Investigation of the super-spicules that extend out to 20–30 arc-seconds may also be attempted, but this is very difficult to do photographically. Visual methods may be tried with a telescope, using the optimum magnification to show these low-contrast, elongated features, which have average widths of 1–2 arc-seconds. It would be extremely interesting to count these and to determine their orientation and length. This type of observation could also be made with a telescope fitted with a spectrohelioscope, by placing a fairly wide slit parallel to the (hidden) limb, but the resolution is likely to be less. The best method might be to use a very broad-pass filter – perhaps even a Wratten gelatine one – that transmits the Hα line, and to concentrate on studying extremely fine detail.

3.4.3 Observation of the inner corona

There are many different features that may be studied, provided good resolution is attainable. For our purposes, photography gives the best results. To distinguish cool material, whose radiation is dominated by the orange and red emission lines of D_1, D_2, D_3 and (above all), Hα, from ionized material at 2×10^6K, with 'white' emissions, colour emulsions, i.e., colour sheet-film, should be used. (It does, in addition, give excellent results.) With a refractor designed for visual use, radiation in the blue is generally more 'diffuse' because of chromatic aberration. Cool material emits also some strong lines in the blue (Hβ; H and K of Ca II, etc.). If these wavelengths are removed by filtration, better discrimination between coronal and cooler material is obtained. Using this method, the green layer of the emulsion will be dominated mainly by coronal radiation, and the red layer by radiation from cooler material. The cool material is in ceaseless motion, and it is easy to detect changes during totality. The same does not hold for the ionized corona, which is 'fixed' by the magnetic field. In general, it is not likely to show any variations during totality, except perhaps the extremely low-amplitude oscillations that are predicted by certain theories.

To obtain the greatest possible coverage, observations should begin several seconds before 2nd contact, when most of the corona is visible on the opposite side of the Sun, to a few seconds after 3rd contact. If the shutter release mechanism does not allow repeated exposures to be made at a sufficiently high rate (let us say one exposure every 1/2 second), it is preferable to wait until the disappearance of the last Baily's Bead at 2nd contact, before making the first exposure.

The success of this type of work largely depends upon the care with which the photographic equipment has been set up. It is important to ensure that the image is of high quality and has an adequate scale, so that the resolution will not be too dependent on the film's grain size, which is generally the limiting factor for amateurs. A long focal length is therefore required, preferably at least 5m. (The 6 × 6 film format may then be used). A focal length of 10m with 13 × 18 sheet film will obviously give even better results. Because the focal ratio is so small (around f/100), a simple achromatic doublet will give excellent results. With fine-grain (and therefore slow) film, exposure times between 1/4 and 1 second are required, which means that guiding must be accurate if second-of-arc resolution is to be preserved. With such a long focal length, the best solution is to use an eclipse coelostat to feed the camera. If the objective is 100 mm in diameter, then a flat coelostat mirror at least 150 mm in diameter is required, allowing the beam to be directed horizontally, which is the most convenient direction for observation and photography. The camera should be held on a very stable mounting to prevent any vibration from the shutter. If the telescopic field does not cover the whole corona, one specific region will have to be chosen. Generally, the area near the point of 2nd contact is given priority, and then, towards the end of totality, that around 3rd contact. Such a choice does not really allow monitoring of changes in various features with time. In any case, the most interesting features sometimes occur at the poles, which are areas that are never close to the points of contact. It is worth stressing again, however, that above everything else, priority should be given to spatial resolution.

3.5 Photographing the K, or plasma, corona

3.5.1 Scientific interest

We now come to another field of research, involving study of the outermost region of the solar atmosphere: the K, or plasma, corona (i.e., the region that consists of completely ionized material), which contains mainly protons and electrons trapped by the Sun's overall magnetic-field lines. This also entails investigation of mass-loss from the Sun, and of its counterpart in the interplanetary medium, the solar wind. Figure 3.8 shows the principal features that are involved. Very narrow features, like the exceptionally long coronal streamers, form part of this corona, which fades (at several solar radii) into another, tenuous corona that we will discuss later. During a solar eclipse, attempts are made to photograph these streamers out to as great a

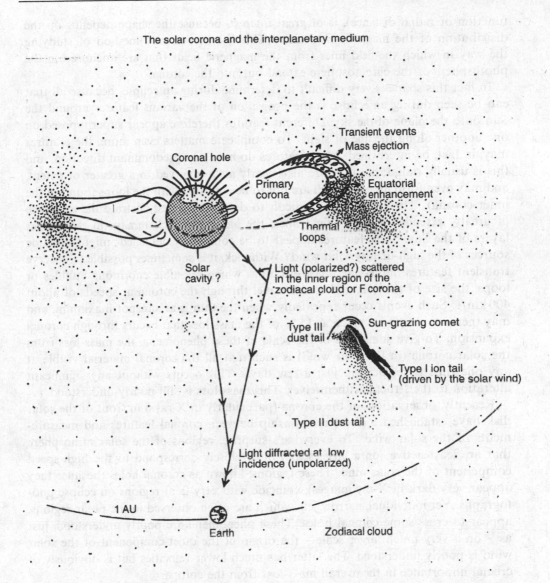

The solar corona and the interplanetary medium

Fig. 3.8. *A very schematic representation of the solar corona and its extensions into the interplanetary medium.*

distance as possible. The base of these features is also important, so there is a very considerable range of intensities to be considered. The light arises from scattering of solar radiation by free electrons in the ionized gas (Thomson scattering). This radiation is therefore polarized, and measurements of the degree of polarization enable the location of the features in space to be determined, and enable them to be 'seen' out to greater distances. In fact, at their streamers' greatest extent from the Sun, the radiation shows almost 100 % linear polarization. The shape of the corona (which may be derived from measurements of the degree of flattening as a function of radial distance), is of great interest, because the shape depends on the distribution of the magnetic-field lines. This is still our only method of studying the way in which the field lines from the magnetic fields that are anchored in the photosphere or the chromosphere extend out into the corona.

In fact this shape is very difficult to determine during an eclipse, because all that can be seen during an eclipse is the projection of the various features around the Sun onto the plane of the sky. Different features therefore appear superimposed on one another along the line of sight. To complicate matters even more, the features may, in fact, be intertwined. Most features do have one predominant direction, and this is usually radial; but they are also nearly always curved to a greater or extent, and may even form arches (which are not to be confused with the loops found in the inner corona). These arches are difficult to detect; they demonstrate the existence of greatly extended magnetic-field lines, the bases of which remain in the deeper layers of the Sun. Some features appear to be completely detached; might these be sources of the high-density solar wind? With luck, it is sometimes possible to observe transient features, large-scale mass-ejections, which resemble enormous bubbles or loops, the size of the Sun, and which travel through the corona at speeds of about 400 km/s. Such events occur repeatedly when the Sun is at sunspot maximum, and may then account for as much as 15 % of the mass-loss that occurs through coronal expansion. To give some idea of the scale of these phenomena, the mass-loss from the solar corona (in the solar wind) is such that all the coronal material visible at any one time is replaced in just a few days. This occurs without any significant alteration to the structures themselves. This mass-loss is still poorly understood.

Recently, observations of the corona (particularly in X-rays) in front of the solar disk have established a direct relationship between coronal features and measurements of the solar wind. To everyone's surprise, regions of the solar atmosphere that are least active (on a large scale) most closely correspond to the high-speed component of the solar wind. These regions, known as coronal holes (because they appear very dark in X-ray images), coincide with very faint regions on eclipse photographs. Yet individual, narrow jets, which are often observed in the polar regions, appear to cross some coronal holes. These phenomena are poorly understood, just as – on a very much larger scale – the origin of the quiet component of the solar wind is poorly understood. The latter has much lower velocities but is obviously of crucial importance in the overall mass-loss from the corona.

We see that new problems have recently arisen regarding the physics of the corona and solar wind: observations of eclipses should offer a relatively easy way of solving these questions.

3.5.2 Methods of photographing the plasma corona

Interesting studies may be carried out with modest equipment, provided the film is correctly calibrated. This work involves a form of absolute photometry, which consists of measuring the overall distribution of brightness in the corona. By making certain assumptions, we may deduce the total mass of the corona at the time of observation. In addition, the records may be used to create a series of isophotes, from which the degree of flattening of the plasma corona may be established. Along with drawings of specific structures – made visually during totality, or else from photographs examined at a later date – such analysis has formed a significant part of the scientific work carried out at eclipses for a very long time. It is essential that it should be continued in future. Nevertheless, more elaborate methods have been developed to carry out precise photometry of coronal structures, notably in compensating for the extreme radial luminosity-gradient – there is a difference of a factor of 1000 in the brightness at the solar surface and at one solar radius from the limb, as may be seen from Fig. 3.6 – and also in increasing contrast sufficiently to permit features to be readily identified.

3.5.2.1 Wide-field photographs

Thanks to modern high-speed emulsions, and also thanks to excellent cameras equipped with reliable shutters, it is quite possible to carry out this wide-field photography with just a fixed mounting, without having to compensate for diurnal rotation. With a 1000-ISO film, or faster (such as Tri-X or even better, T-MAX 3200), and an aperture of f/5.6, for example, typical exposure times are between 1/1000 and 1/15 second (*see* Fig. 3.9). Care must be taken to frame the corona properly and to use a focal length of at least 300 mm to avoid the damaging effects of vignetting. To allow for any possible errors in the shutter speed, individual frames should be repeated, and a range of exposures used. Using a longer-focus lens and a 6×6 camera will obviously give results with even better resolution, but not necessarily with greater photometric accuracy. Sensitometry calibration must be carried out on the same film, if possible the same day, before and after the eclipse. Photographs should be taken of a grey-scale, uniformly illuminated by sunlight reduced by a factor of about 50 000. To obtain a proper calibration, however, it is preferable to construct an 'absolute' sensitometer, i.e., one without any optics. This type of measurement is difficult for amateurs to make, so we recommend careful, relative calibration instead. From such a calibration the response curve of the film (the characteristic, or Hurter–Driffield curve, HD) may be drawn, enabling numerical values to be obtained for each of the isophote levels in the image. This takes exposure times into account, and thereby allows the isophotes in different images to be related in a common system. A complete set of isophotes is obtained in this way (Fig. 3.10).

To calculate the integrated luminosity of the corona, i.e., its stellar magnitude, the scale of intensities measured (for a specific, arbitrary exposure-time, as just mentioned) has to be related to an absolute scale. The measured luminosities may then be converted into solar magnitudes, or to the average brightness of the Sun's disk – which comes to the same thing because the angular diameter of the Sun is known extremely accurately for the date of the eclipse (from published ephemerides). In making the calculations, care must be taken to relate the distribution of coronal brightness to that of the chromosphere (i.e., the base of the corona), and not to that of the lunar limb at some arbitrary time during totality. The scales may be correlated by using the same equipment to photograph the solar disk, attenuated one million times, but this is difficult to achieve accurately. Another method consists of using the same equipment to photograph a bright planet or even a bright star, defocussing the image sufficiently to ensure that accurate photometry of the object is

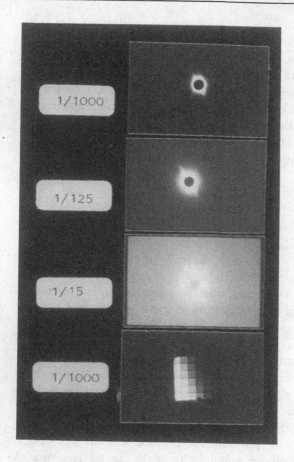

Fig. 3.9. *A set of photographs obtained by R. Verseau (SAF) with different exposure times, but the same 24 × 36-mm camera fitted with a 300-mm focal length lens, operating at f/5.6, during the eclipse of 1984 November 22–3. The lower photograph was obtained by the same equipment, fitted with a special sensitometer and looking directly at the Sun. This allows the intensities measured in the corona to be related to the average intensity of the solar disk.*

possible. From the known absolute magnitude of the object photographed, it is then possible to derive an absolute scale of intensities for the overall system. With a wide bandwidth (which would apply to observations made without filters, for example), allowance must also be made for the colour index of the object that is used as a calibration standard.

To guarantee success, it is obviously advisable for either of the two methods (absolute or relative calibration) to be developed and tested several months in advance, by carrying out observations of the twilight sky with the eclipse equipment. In addition, some knowledge of photographic photometry is required.

The degree of flattening of the corona as a function of its radial distance (Fig. 3.11) may be deduced from a set of isophote contours that have been obtained by one of the analogue methods used in photography. To do this, the average radial distance of each isophote is estimated (taking the mean over a 45-degree sector) near the equator (giving R_{eq}) and then near the poles (giving R_{pol}). The graph showing consecutive values of $[(R_{eq}/R_{pol}) - 1]$ as a function of R_{eq} gives the value of the factor $(a+b)$, which is a statistical property of the corona at the time of observation. This factor is linked in a fairly complicated manner to the stage of the sunspot cycle.

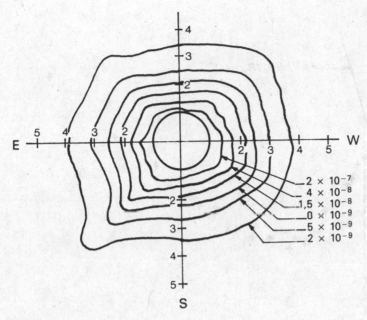

Fig. 3.10. *An example of a set of isophote contours, calibrated in absolute units (the average luminosity of the solar disk). Eclipse of 1983 June 11, after C. Nitschelm.*

3.5.2.2 High-resolution photography

In the section dealing with observation of the inner corona, we have already discussed the difficulties in obtaining a good-quality image, i.e., one where the spatial resolution is limited only by the Earth's atmosphere (image motion, blurring, scattering, etc.). The problem posed by obtaining images of the white-light, plasma corona is more complicated, because the field to be covered is much greater (which implies a much greater range in the luminosities involved). In addition, the exposure times required to photograph the outermost, and therefore faintest, features are 100 to 1000 times as long – for a given aperture and film speed, of course. Such observations require particularly accurate guiding. To obtain the necessary rigidity, we would again advise the use of an eclipse coelostat – i.e., one where a single mirror reflects the beam of light horizontally into a fixed azimuth, either east or west of the meridian (according to whether one is observing in the morning or in the afternoon) – together with a special photographic camera (Fig. 3.12). This camera, which is essentially an eclipse coronagraph, is fed by an astrographic-type, photographic objective of fairly large diameter and with a focal length considerably longer than one metre. This camera could be supported on an equatorial mounting, provided this were solid enough, to track the solar corona.

The enormous dynamic range of luminosity in the corona gives rise to a strong, radial intensity gradient. To avoid gross over-exposure of the inner corona, a radial mask, or better, a radial neutral-density filter is required. The structures within the corona exhibit a very low contrast, because several are seen (and photographed) superimposed one on another along any particular line of sight. To obtain the

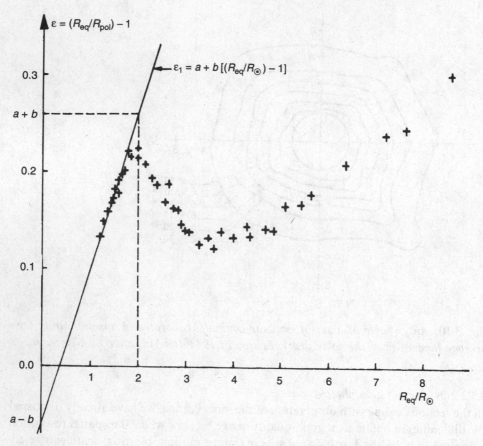

Fig. 3.11. *Variation in the flattening factor ϵ in the coronal isophotes on 1983 June 11, as a function of radial distance. The parameter $(a + b)$ represents an average value extrapolated to $1R_\odot$ from the limb, and which depends on the solar cycle. After C. Nitschelm.*

best response from the film and reproduce the variations as different densities, it is essential for the exposure to be correct over the whole photograph. A radial-density neutral filter to compensate for the radial gradient may be obtained by evaporating a thin film of metal of varying thickness onto a glass support (which may be of tinted glass). The support must be rotated rapidly during the evaporation process to ensure that the thickness varies correctly from centre to edge over the whole area of the filter. The thickness of the layer is calculated so that it attenuates the inner part of the corona, and largely compensates for the average radial gradient, which may be predicted from models of the solar corona. This filter must be located close to the focal plane, immediately in front of the film. It must be perfectly uniform and centred on the image of the corona, i.e., on the Sun before the eclipse. Focussing of the objective, which is very critical for good resolution, should be made with this filter in place. Because the central portion coincides with the centre of the Moon

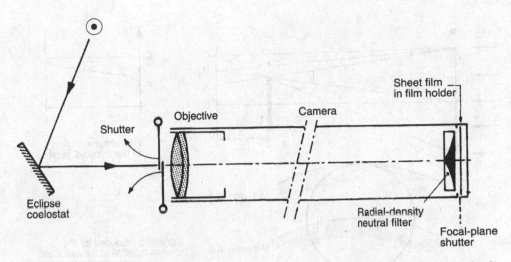

Fig. 3.12. *Schematic diagram for an eclipse experiment designed to photograph the features in the solar corona with a high spatial resolution.*

during totality, it does not need to be very dense, and it is very convenient if an area equivalent to about 0.8 lunar diameters is left clear. This will allow the disk of the Sun to be observed before totality, ensuring that the equipment is properly centred. To avoid excessive heating of the filter while setting up the equipment, it is advisable to fit an easily operated, secondary shutter in front of the objective. If the focus is chosen to coincide with the rear face of the filter (which is very convenient), it should be possible to apply the sheet film to this face, using a special holder. The dimensions of the filter and sheet film should be chosen to suit the focal length of the objective and the required field (for example: with a focal length of 1500 mm and a field of about 6 solar radii, 10 × 12.5 sheet film is suitable).

Such a filter is obviously an expensive optical item and difficult to make. Another method may be used to obtain a suitable filter from Schott neutral density glass. This method consists of grinding away the glass in a radially symmetrical way. The thickness is chosen to give an attenuation of at least 1000 close to the centre, and the filter is made thinner and thinner towards the edge. This component is then cemented to a second piece of clear glass that has been worked to the complementary shape. The two then form a plane-parallel assembly, with the same thickness throughout, thus avoiding any aberrations. It is also worth mentioning here that it is possible to correct the slight field curvature often present in the objective, and which gives rise to images at the edge of the field that are often elongated (producing an effect like classical coma). This may be done by cementing an appropriately calculated, thin, plano-convex lens to the front face of the filter. [Or, better, by working an appropriate curvature on the face of the clear glass component of the composite filter. – Trans.]

If the equipment has a long focal length – which ensures good photographic resolution even with fast, and therefore somewhat grainy film – an old, but very

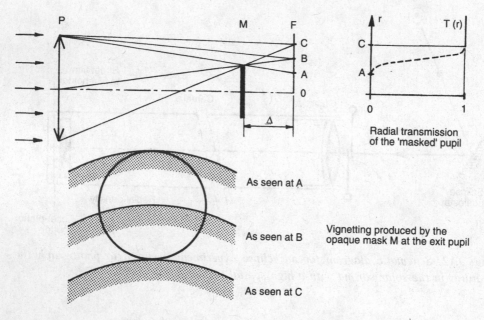

Radial transmission
of the 'masked' pupil

As seen at A

As seen at B   Vignetting produced by the
opaque mask M at the exit pupil

As seen at C

Fig. 3.13. *The principle behind using a defocussed mask to attenuate the great intensity gradient to photograph the inner corona.*

effective technique may be used. The radial attenuation required may be obtained very approximately by using a mask, rotating at several revolutions per second. The profile of this mask must be calculated to give the required effect. This arrangement has been described in an excellent article (Laffineur, 1969).

Finally, let us mention a cheap and fairly simple method that may be used with small equipment (with a focal length of less than 1m). This involves using an opaque circular mask. (A circular 'spot' of blackened metal, cemented onto a support of optical quality glass is very suitable.) If this central obstruction is placed at an appropriate distance from the focus it causes vignetting across the field that may be made to compensate for the radial gradient in the coronal luminosity. Figure 3.13 illustrates the principle on which this method works.

The performance of this last method (*see* Fig. 3.15) is, of course, inferior to the others mentioned. The resolution is considerably decreased in the innermost part of the corona, because the pupil is almost completely masked. Such a mask, however, may prevent gross over-exposure of the inner corona and may therefore prove quite effective.

If any photograph is to be used for photometry, it should be calibrated by adding a sensitometry grey-scale (on the edge of the film, for example).

Various methods of spatial filtration of the final photographic image may be used to carry out a detailed analysis of the structures in the corona. A duplicate negative and a positive may be superimposed: with a slight shift in azimuth, weak variations

Fig. 3.14. *A high-resolution photograph of the plasma corona taken during the total eclipse of 1980 February 16 in India, by J. Dürst (Zurich); the original has been photographically enhanced to bring out lateral gradients and to show features that otherwise are superimposed.*

in azimuth are amplified, for example. A digitized image may also be manipulated by numerical methods, but this requires considerable computing capacity for example Fig. 3.14.

3.6 Photography of the outer corona

3.6.1 Scientific interest

From the diagram in Fig. 3.8 and also the graph in Fig. 3.6 we can see that the outermost region of the solar corona, beginning 2 solar radii from the limb and also known as the F-corona, is dominated by the dust component. This is demonstrated by the fact that the Fraunhofer absorption lines in the solar spectrum are very prominent in that region. Unlike the process occurring in the plasma corona, sunlight is reflected by particles with quite low velocities, and these have been identified as belonging to the inner part of the interplanetary dust cloud. Study of this component can give us valuable information about the evolution, the nature and the origin of interplanetary dust. The most disconcerting, and perhaps the most exciting, aspect concerns theoretical predictions about this part of the solar corona.

Fairly simple calculations, using a model for the distribution of dust where the density increases towards the Sun, show that there is a 'forbidden' region around the Sun that is free from dust, and also one or more other regions, at about 4 solar radii from the centre, and farther out, where dust accumulates. Some eclipse observations appear to confirm these predictions, but no definitive answer has been obtained, especially as other observations, made both during eclipses and at other times, have given negative results. A point to be made at the outset is that observations should be carried out in the red or even the infrared, because the dust particles close to the Sun acquire high temperatures through absorption of solar radiation, and re-emit the energy in those regions of the spectrum. Theoretical predictions are incapable of saying with any certainty whether the relatively enhanced distribution of dust particles near the Sun should be in the form of a shell or a ring. Study of this amorphous, unpolarized and predominantly 'red' component of the solar corona is of considerable interest. Sensitive measurements are required, which cannot really be undertaken at any time other than at solar eclipses.

3.6.2 Methods of observing the F-corona

Several methods of observing the F-corona have been tried in the past. It is not really possible to say which is the best. One thing does seem to be essential in obtaining a successful observation: the presence of a sky that is sufficiently 'dark' in the near infrared, and over a wide field. Observations need to be undertaken from a high-altitude site, at 2500m or more. Any faint atmospheric haze, even fine cirrus, will prevent good measurements during totality. The sky background should be both low and constant throughout the eclipse. Moreover, because the outer part of the corona is being studied, the equipment itself should scatter very little light. It is advisable for the light from the brightest part of the corona (the bright ring of the inner corona) to be masked out before the light strikes any optical surface. The ideal arrangement would therefore be to place a mask at a considerable distance in front of the objective (to avoid too large a penumbral shadow). This obviously poses some problems.

Taking photographs in the infrared also requires certain precautions and even a considerable degree of practice in handling emulsions and filters. A few important details need to be mentioned. Focussing of the objective should be carried out photographically. Infrared films have to be handled in absolute darkness – even when taking a cassette from its container! In addition, these films do not tolerate high temperatures (see the manufacturers' recommendations). Wratten gelatine filters should be distrusted; all are more or less transparent in the infrared, and they are very fragile.

In taking the actual photographs, there are problems in determining appropriate exposure times and then in calibration. The simplest method consists of bracketing on either side of the estimated exposure time, trying to obtain a correct rendering of the reflected Earthshine, for example (Fig. 3.6). The exposure times will be long, given the filtration required and the generally slow speeds of available emulsions.

Fig. 3.15. *A photograph of the solar corona taken with a defocussed radial mask and a telephoto lens of 400 mm focal length on Ektachrome 400 at 1/30 second. SAF eclipse team, New-Caledonia, 1984 November 22–3.*

Use of a cooled CCD camera with integration times of a few seconds would, of course, be very productive. Such equipment is not particularly robust, and we refer the reader to the chapter on modern detectors (Chap. 20) for more information.

3.7 Spectroscopic study of the corona

3.7.1 The scientific value

In this final section, we come to some of the most sensitive methods of analysing stellar atmospheres. In the solar corona, however, we only encounter the 'optically thin' case, which, in practice, means that we are only dealing with emission lines.

Study of these lines, which are now quite well documented, should enable us to measure the temperature of the corona, or even its distribution, as well as the velocity of the plasma.

Observation of the variation in width of the strongest coronal lines as a function of distance from the surface, or even better, determination of any Doppler shifts, would give exceptionally valuable information about the heating process, and about expansion velocities in the corona. It would be interesting to compare spectra obtained from a dense region of the corona with those from an 'empty' region or coronal hole, where the escape velocities are very high. Finally, an experiment that is very difficult to conduct, but which is within our reach nowadays thanks to modern detectors, might, if observations sufficiently closely spaced in time can be made, detect the waves in the corona that are predicted theoretically, and which might partly explain its very high temperature.

On a rather more modest scale, other still-unsolved problems might be tackled by spectrophotometry around the times of contacts. Observation of the narrow crescent, and then of the lower chromosphere (once known as the reversing layer because the lines suddenly appear in emission) would investigate problems concerning the contrast of faculae in the continuum at the extreme limb, or those of emission from cool features in the transition region between the chromosphere and corona.

3.7.2 Some suggested experiments

Most of the time spectrophotometric, or even just spectroscopic, analysis requires the use of the latest techniques, at least as far as detectors are concerned. Attempting such observations at an eclipse is really taking a gamble. It is, however, interesting to occasionally make use of the specific circumstances accompanying the event, notably its short duration. As the light is considerably dispersed in such experiments, the use of components that have a high degree of luminous efficiency, such as Fabry-Pérot (etalon) interferometers and interference filters, are desirable.

3.7.2.1 Monochromatic and Fabry-Pérot imaging

By using narrow bandwidth interference filters that are available commercially (with bandwidths of about 0.2–0.3 nm), it is easy to photograph the corona in one of the emission lines in the visible or even in the near infrared – at least it would appear to be so. In practice, it is not so easy, because these filters have very limited angular fields and thus small apertures; their temperature control is very critical (having to be better than 1°C); their position and any possible tilt are also critical; and they are expensive. The combination of a Fabry-Pérot etalon and an interference filter with a moderate bandwidth (about 2–3 nm) is perhaps of more interest. A Fabry-Pérot etalon may also be used to obtain interference fringes covering the whole field. Analysis of these gives the profile of the emission line used (generally this is the green line at 530.3 nm) and, by comparison with the system of fringes produced by an artificial calibration line, the Doppler shift of the line. It should be noted that recording these images requires the use of a very fast emulsion, or even better, an image tube.

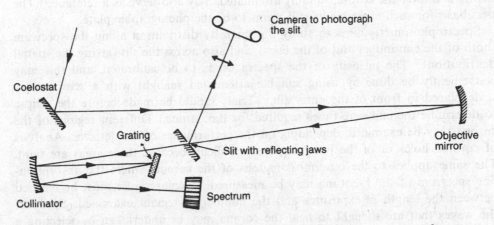

Fig. 3.16. *An example of the arrangement of the various elements in an eclipse experiment designed to carry out spectroscopic analysis of the solar corona.*

3.7.2.2 Spectroscopy and spectrophotometry

Many different experiments are possible according to whether one is most interested in the corona, the transition region (with its 'flash' spectrum), the dust corona, in searching for new lines, in simply identifying them, or even in measuring specific thermodynamic properties. Figure 3.16 shows an example of the arrangement of the various components in an experiment designed to observe part of the coronal spectrum. To capture the flash spectrum of the chromosphere, the widened entry slit should be placed parallel to the limb of the Moon, first at 2nd contact, and then towards the end of totality, at the position of 3rd contact. Care should also be taken to keep exposures fairly short (about one second). They should also be taken in succession to obtain good spatial resolution. (But be warned that the analysis of such observations is not simple.) Photographs of the area being studied may be obtained at the same time as the spectrograms, by capturing the image reflected from the front side of the slit. To simplify the spectrograph, the plane diffraction grating may be replaced by a concave grating, but it must be borne in mind that the focal ratio is slow (f/15 or more). It is also possible to replace the mirrors (objective and collimator) with lenses, and to add components to correct aberrations to make the final part of the system more efficient – to the detriment of resolution, of course.

Spectroscopic analysis of the solar corona should preferably be made of a bright area of the inner corona. This will generally be close to the equatorial zone, over tive regions that are crossing the limb (observe these areas before the eclipse in order to place the slit over the appropriate area). This is not always easy unless one has a field rotator, or has a spectroscope that may rotate about the optical axis of the entry slit. To obtain maximum coverage of the spectrum during totality, it should be possible to make slight adjustments to the grating. It is also possible to record the whole spectrum at once, but with a lower spectral resolution. To identify the lines and even perhaps accurately determine their positions, it is necessary to use the same system to photograph lines from a reference source. (The solar spectrum before and after the eclipse, suitably attenuated, may also serve as a reference.) The best basis for such experiments still seems to be the photographic plate.

Spectrophotometry aims to study the intensity distribution along the spectrum (both of the continuum and of the lines), and also across the slit (giving the spatial distribution). The intensity of the spectra needs to be calibrated, and this may conveniently be done by using suitable attenuated sunlight with a sensitometry scale placed in front of the entry slit. (Trials should be made before the eclipse to determine the exposure-times required for the corona.) Different regions of the corona may be examined, depending on the performance of the detector. Analysis of coronal holes or of the polar regions is difficult because these areas are faint. The same applies to the outermost regions of the corona, where the absorption-line spectrum of the F-corona may be measured. A compromise must be reached between the length of exposures and the number of regions examined. Study of the waves that are thought to heat the corona may be undertaken by selecting a specific, well-defined region and obtaining as many spectra possible. A statistical analysis may then be undertaken, aiming at discovering periodic phenomena, which may have periods between a fraction of a second and about one hundred seconds. In this particular experiment it appears to be difficult to analyse more than one line at a time.

3.8 Conclusions

We have discussed a wide range of experiments that may be carried out during a total eclipse, with numerous and very varied scientific aims. The results already accumulated during past eclipses fill many volumes of publications. New methods of observation, and particularly new technologies – especially in the field of detectors – justifies considerable investment of time, energy and finance in future eclipses. That of 1991 (Fig. 3.2) was a major astronomical event for various reasons (excellent observing sites, favourable time, long duration, etc.). To succeed in obtaining good observations, several years of preparation may be required. Besides, such observations are always very difficult, so team-work is required. International collaboration is a natural consequence of work on eclipses. The preparations themselves are a fascinating part of the work, and there is always the reward of a nice trip to some far-away place …

4 The Moon

S. Chevrel & M. Legrand

4.1 Introduction

The Moon is the closest astronomical object to the Earth and it is so easily visible and its phases are so regular that, early in human history, it helped to define and mark the passage of time. Our modern calendar, where the month is based on the Moon's motion around the Earth, reflects those early observations. Astronomy certainly began with observations of the Moon, which, along with the Sun, was the most familiar heavenly body.

The constantly changing aspect of the Moon, with its phases and the numerous surface features that may be seen even with the naked eye, mean that it remains the most notable object in the sky. Because it is often present in the sky, and is easily accessible with even modest instruments, it is still the object most frequently and easily observed by amateur astronomers. Generally, however, they only 'do some sight-seeing'. Believing that everything is now known about the Moon, they rapidly move on to more distant, difficult, and also more fashionable, objects. But since the first observations of the Moon by Galileo Galilei around 1610, an impressive amount of data has been accumulated.

The Moon has been mapped in detail since the end of the 18th century, and photographed since the beginning of this century. It was visited by space-probes in the 1960s, and finally parts of it were explored in the Apollo programme between 1969 and 1972. The Moon has revealed many of its secrets, and it might seem that there is little remaining to be done, at least from the surface of the Earth.

Yet many people (few of them, it is true, in France) still observe the Moon. Many of the answers to problems of the formation of the Solar System in general and of the planets in particular, are relevant to observation of the Moon, especially as the other, more distant, planetary bodies cannot be observed regularly or in great detail.

Active research is still being carried out on the data and the materials (the lunar rocks) obtained by the Apollo programme. Taking planetology, for example – just one of the disciplines that are being followed – analysis of the lunar soil has been carried out for several years from Earth by spectrophotometry.

Faced with the considerable resources that have been brought to bear on lunar studies in the past, and with those that are still being developed, it is hardly surprising that amateurs should feel rather discouraged and excluded from active research.

The lunar surface is not subject to any changes: it appears the same to us now as it did to primitive man, with exactly the same features. Research is restricted to formations that have already been extensively studied, or to obtaining finer and finer detail. Under such circumstances it is difficult for amateurs to undertake serious research work. But perhaps that is not the true goal. For amateurs, the main thing is

107

to make discoveries for themselves and to understand the reasons for the phenomena they observe. In this respect, the Moon is an ideal body to study, because effectively we have what amounts to a whole planet that is easy to examine. The work may be done at several levels, and several fields may be suggested, which are more than just simple 'sight-seeing', and which even border on areas of research (such as studies of lunar composition). Others may offer new insights into certain phenomena about which there are still a number of unanswered questions (such as transient lunar events). Some phenomena, such as luminescence, have not been included because they are too complex.

4.2 The motion of the Moon

Before starting to observe the Moon, it helps to be familiar with its orbital motion around the Earth. The fundamental properties of this determine its apparent movement and the various aspects that it presents to terrestrial observers.

The Moon's behaviour is more complicated than that of any other body in the sky, when its irregular apparent motion, its phases, libration, etc., are taken into account. It seems logical to start with phenomena visible to the naked eye, because this serves as a basis for understanding the mechanisms underlying the behaviour.

4.2.1 The Earth–Moon system

When the planets in the Solar System are considered – and if the Pluto–Charon system of two icy bodies is omitted – it is noticeable that the terrestrial planets either have few (or no) satellites, but that the gas giants have large numbers (as many as 16, 18, and 15 for Jupiter, Saturn, and Uranus, respectively). In all these cases the satellites are small when compared with the planet that they orbit. (Jupiter and Saturn are 13.6 and 11.6 times as large as Ganymede and Titan, for example.) The satellites of Mars are minute.

The Moon is the exception to this rule: not only is it the Earth's sole satellite, but it is far from negligible in terms of size. The Earth is only 3.7 times its diameter. In addition, it orbits relatively close to the Earth, whereas satellites in general are fairly distant from their planet. The two bodies are closely related and no other satellite in the Solar System influences its primary to such an extent.

Our satellite therefore occupies a unique place in our Solar System. The term 'double planet' is often applied to the Earth–Moon system, and this term should be restricted solely to the physical sense.

4.2.2 Orbit

The Moon's orbit around the Earth is elliptical, with the considerable average eccentricity of 0.054 90. This means that perigee is at 363 296 km, apogee at 405 504 km, and that the average distance is 384 400 km (0.002 58 AU). The inclination of the orbit to the plane of the ecliptic is 5°08′ (Fig. 4.1).

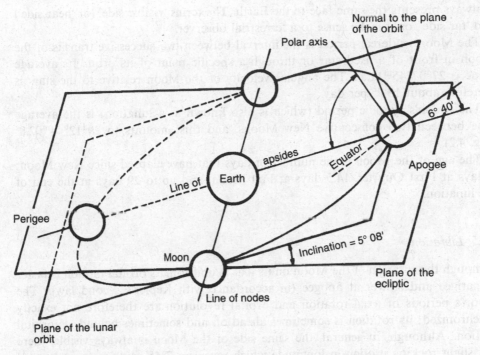

Fig. 4.1. *The relative positions of the orbits of the Earth and Moon.*

4.2.3 Phases of the Moon

Everyone must have noticed the phases of the Moon, although many people apparently remain confused about the mechanism that causes them. Apart from the Sun itself, the Moon and all the other bodies in the Solar System are visible only because they reflect sunlight. During the course of an orbit (or revolution) of the Earth, anyone on Earth sees variations in the amount of the Moon that is illuminated (the phase) and this varies with its position relative to the Sun. A specific phase has no significance except to an observer on Earth.

The phase is known as waxing between New and Full Moon, and waning between Full and New. The Moon is a crescent when less than half of the disk appears illuminated (between New Moon and First Quarter, and between Third Quarter and New Moon), and gibbous when more than half appears illuminated (between First and Third Quarters).

4.2.4 Rotation and revolution

An observer on the Moon would see the Earth exhibit phases, and these would be complementary to the Moon's as seen from Earth. Of course, the lunar observer would have to be on the so-called 'visible' side of the Moon. Because the latter completes one orbit of the Earth in the same time as it rotates once on its own axis, it always presents the same face to the Earth. The terms 'visible side' (or 'near side') and 'far side', only make sense to a terrestrial observer.

The Moon's sidereal period is the interval between two successive transits of the Moon in front of a fixed star or through a specific point of its orbit. Its average value is $27^d07^h45^m11.5^s$. The angular velocity of the Moon relative to the stars is therefore about 13.2° per day.

The Moon's synodic period (which is also known as a lunation) is the average time between two consecutive New Moons, and this amounts to $29^d12^h44^m02.8^s$ (Fig. 4.2).

The age of the Moon is the number of days that have elapsed since New Moon: 7 days at First Quarter, 14.5 days at Full Moon, and up to 29 days at the end of the lunation.

4.2.5 Libration

Although the rotation of the Moon on its axis is uniform, its orbital motion is faster at perigee and slower at apogee (in accordance with Kepler's second law). The Moon's periods of axial rotation and orbital revolution are therefore not exactly synchronized: its rotation is sometimes ahead of, and sometimes behind, its orbital motion. Although, in general, the same side of the Moon is always visible, there is a slight rocking motion in longitude which reveals a 7°45'-wide segment beyond the eastern and western limbs. In other words, the mean centre of the Moon's disk (the origin of the selenographic coordinate system) shifts from east to west (and back again), during one orbital revolution. The displacement reaches 7°45' (in selenographic coordinates) with respect to the apparent centre of the disk as seen from Earth. This is libration in longitude (Fig. 4.3). In addition, because the Moon's axis of rotation is not perpendicular to the plane of its orbit (the inclination varies between 83°11' and 83°29'), a similar phenomenon in latitude results, so that the mean centre shifts in a north–south direction by a value of ±6°44' with respect to the apparent centre of the disk. This is libration in latitude (Fig. 4.4).

A third type of lunar libration, known as diurnal libration, arises from the rotation of the Earth. It has a value of about 1° (Fig. 4.5). The combined effects of these librations enable 59% of the surface to be seen from Earth. However, practically all of the surface is known, thanks to the numerous lunar probes (Luna, Lunar Orbiter, etc.) and manned flights (Apollo) that have orbited the Moon.

4.2.6 Irregularities in the motion of the Moon

Because of the combined gravitational perturbations of the Earth, the Sun, and (to a lesser extent), the other planets, the motion of the Moon around the Earth is very irregular – the deviations from uniform motion amount to several per cent – and very complex. These perturbations are all periodic (with longer or shorter overall periods), and if this were not the case, the Earth–Moon system would not have remained stable. Because of these irregularities, the orbital properties given at the beginning of this section are no more than average values.

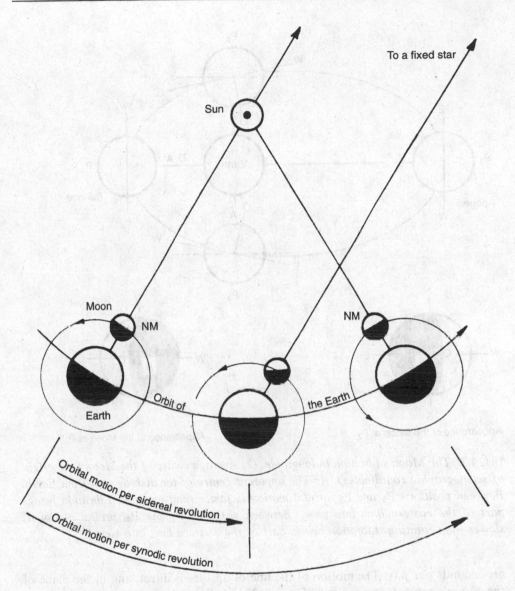

Fig. 4.2. *The Moon's sidereal and synodic rotation. After the Moon has finished one revolution around the Earth, it still has to cover an additional part of its orbit before it returns to the same position NM, relative to the Earth and Sun.*

The two most significant irregularities affecting the apparent motion of the Moon are the retrograde motion of the line of nodes (the line of intersection between the Moon's orbit and the ecliptic), and the advance in the line of apsides (the line joining perigee and apogee in the plane of the orbit). The first of these is a retrograde motion in the plane of the ecliptic (i.e., opposite to the Moon's orbital motion) with a period of 18.6 years (6793.5 days). It is by no means negligible, amounting to 191

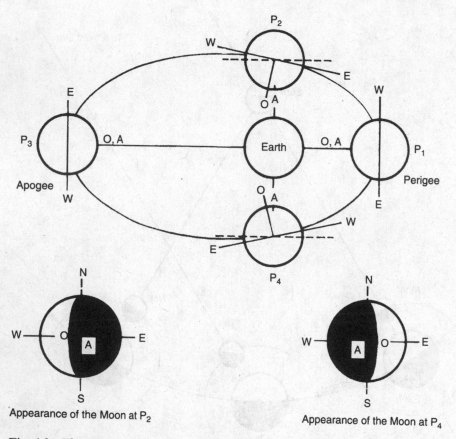

Appearance of the Moon at P_2

Appearance of the Moon at P_4

Fig. 4.3. *The Moon's libration in longitude. O: Average centre of the Moon (the origin of selenographic coordinates). A: The apparent centre of the disk as seen from Earth. Between positions P_1 and P_2, orbital motion is faster than rotation. Libration brings part of the eastern limb into view. Between positions P_3 and P_4, orbital motion is slower than rotation. Libration brings part of the western limb into view.*

arc-seconds per day. The motion of the line of apsides is direct, and in the plane of the Moon's orbit. Its period is 8.85 years (3232.6 days).

Another perturbation, caused by the Sun, is known as evection. Perigee and apogee may occur at any particular phase of the Moon. When they coincide with New or Full Moon, the line of apsides must obviously be pointing towards the Sun. When this occurs (every 206 days) the eccentricity of the Moon's orbit reaches a maximum: the distance of the Moon at perigee is reduced, whilst that at apogee is increased. When, on the other hand, the line of apsides is perpendicular to the direction of the Sun – either perigee or apogee coincides with First or Third Quarter – the eccentricity reaches a minimum. The distance at perigee increases and that at apogee decreases. The combined result of these changes is that the distances at perigee and apogee vary considerably from 363 296 km and 405 504 km, the average values quoted earlier.

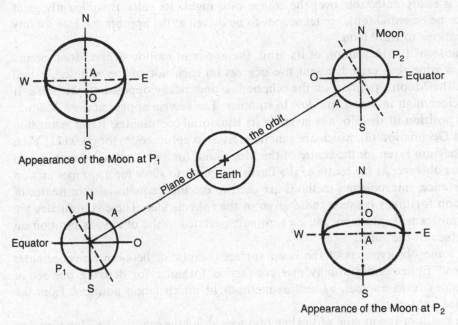

Fig. 4.4. *The Moon's libration in latitude. O: Average centre of the Moon (the origin of selenographic coordinates). A: The apparent centre of the disk as seen from Earth. At P_1 and P_2, the Moon alternately reveals parts of the northern and southern limb.*

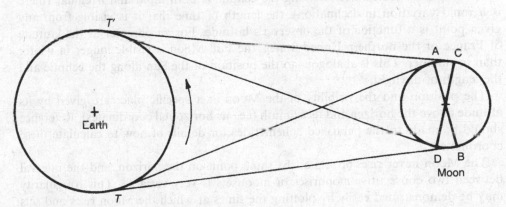

Fig. 4.5. *Diurnal libration. Because the Earth is rotating, an observer sees the hemisphere AB at the beginning of the night (at time T), and hemisphere CD at the end of the night (at T').*

4.3 Data for the observation of the Moon

4.3.1 The Moon's apparent motion

The orbital motion of the Moon is direct, so it moves from west to east relative to the stars. This movement in right ascension (13.2° per day, or about 0.5° per hour), is easily detectable over the course of a night. Its value is sufficiently great for it to be essential for large telescopes to be driven at the appropriate rate for any observations of the Moon.

Because of the inclination of its orbit, the apparent motion of the Moon occurs within a zone that extends about five degrees on each side of the ecliptic. At Full Moon, the Moon's position on the ecliptic is diametrically opposite to the Sun's. It is therefore high in winter and low in summer. The reverse applies at New Moon.

The position of the Moon is given by its equatorial coordinates: Right Ascension (α) and Declination (δ), which are generally given in ephemerides for 00:00 UT each day. They are given for the centre of the Moon, and for a geocentric observer (i.e., a fictitious observer at the centre of the Earth). In order to allow for its proper motion and distance, interpolation methods are used to calculate equatorial coordinates of the Moon for times between those given in the ephemerides. These coordinates are then transformed into topocentric coordinates, related to the observer's position on the surface of the Earth.

For some observations of the lunar surface, ecliptic or heliocentric coordinates are used. Ephemerides usually give conversion formulae for deriving one set of coordinates from another, as well as methods of interpolating positions from the tabulated values.

As a result of the motion of the line of nodes along the ecliptic, the Moon reaches its maximum declination ($\pm 28.5°$) when the ascending node coincides with the vernal equinox (or First Point of Aries ♈. The opposite holds when the descending node is at the vernal equinox.

Because the Moon's motion along the ecliptic is both rapid and irregular (there is a rapid variation in declination), the length of time that it is visible from any given point is a function of the observer's latitude. For an observer at the latitude of France or the northern United States, the Full Moon is visible longer in winter than in summer. This is analogous to the position of the Sun along the ecliptic and the length of daylight.

The position and the visibility of the Moon at a specific place are given by its altitude above the horizon and its azimuth (i.e., its horizontal coordinates). Reference should be made to the published ephemerides for details of how to calculate these coordinates.

The Moon never rises or sets at the same point on the horizon, and the interval between two consecutive moonrises or moonsets is very variable. This irregularity may be demonstrated easily by plotting the times at which the Moon rises and sets for every day in a lunation. If the times of meridian transit are also plotted, it will be seen that these fall more or less on a straight line, which is the line of symmetry between the curves for the times of moonrise and moonset; each transit occurring, on average, about 50 minutes later than the previous one.

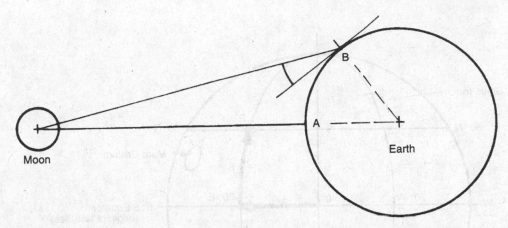

Fig. 4.6. *Because the Earth is spherical, an observer at A, where the Moon is at the zenith, is closer to the Moon than an observer at B, where the Moon is low on the horizon. As a result, the diameter of the Moon for A is larger than that for B. The difference may amount to 1.7 % of the Moon's diameter.*

4.3.2 The appearance of the lunar disk

4.3.2.1 Apparent size

The apparent diameter D_A of the Moon, i.e., the angle that it subtends in the sky, varies slightly as a function of its distance from the Earth around its orbit. It is given by the equation:

$$D_A = D_M/d,$$

where D_M is the actual diameter of the Moon (3476 km) and d is the Earth–Moon distance; D_A is expressed in radians. At the average distance of 384 400 km, D_A is 31.08 arc-minutes.

The diameter also depends on the height of the Moon above the horizon (Fig. 4.6). The Moon ought to appear smaller at the horizon than at the zenith. This is actually the opposite of what is observed, because an optical illusion makes the Moon seem larger when it is close to terrestrial objects on the horizon than when it is isolated high in the sky.

4.3.2.2 Selenographic coordinates

Like the Earth, the Moon has a system of latitude and longitude that defines a grid of selenographic coordinates. Any point on the surface may be referred to this system of coordinates, the reference plane of which is the lunar equator, perpendicular to the Moon's axis of rotation, which runs through the North and South Poles. The zero, or central, meridian is the meridian that passes through the mean centre of the visible disk (allowance being made for libration).

As the Moon appears in the sky, characteristic formations such as Mare Crisium, Grimaldi, Sinus Iridum and Tycho are close to the western, eastern, northern and southern lunar limbs, respectively. These are the directions taken in the

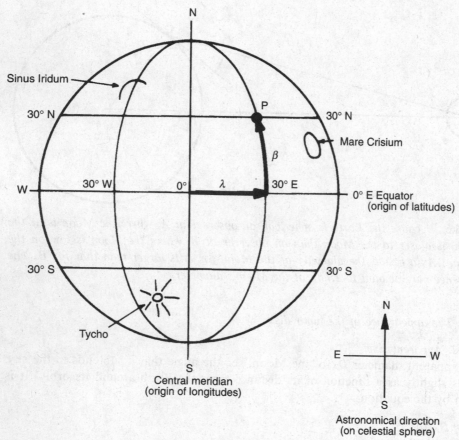

Fig. 4.7. *Selenographic coordinates. Longitude (λ) and latitude (β) are reckoned positively towards the east and north, respectively. The selenographic coordinates of a point P are $\lambda = +30°$ and $\beta = +30°$, or 30° E, 30° N.*

astronomical sense (on the celestial sphere), which applied until 1961, when the International Astronomical Union (IAU) decided to adopt the astronautical east–west convention. According to the new orientation, Mare Crisium is on the east and Grimaldi on the west. An observer on the Moon now sees the Sun rise in the east and set in the west, whereas the converse applied previously. Some peculiarities have resulted: for example Mare Orientale is now in the west!

Selenographic longitude, generally denoted λ, is the angle at the centre of the Moon subtended by the point under consideration and the central meridian (Fig. 4.7). It is reckoned from 0° to 180°, positively towards the east, and negatively towards the west. The notation is thus $\lambda = -45°$ or $\lambda = +45°$, although this form is often replaced by 45°W (west) and 45°E (east).

Selenographic latitude, denoted β, is the angle between the point under consideration and the lunar equator. It varies between 0° and 90°, positive towards north, and negative towards south. Again, 45°N and 45°S are often used instead of $\beta = +45°$ and $\beta = -45°$.

Any point on the surface of the Moon therefore may be referred to by its selenographic coordinates λ and β. The notation is given as (λ, β), e.g., $(-44°, -04°)$ or, more frequently, $(44°W, 04°S)$.

4.3.2.3 Phase

The line separating the illuminated portion of the surface from that in darkness is known as the terminator. Its shape is a portion of an ellipse, except at the times of First or Third Quarter when it is a straight line, dividing the lunar disk into two equal, illuminated and dark halves.

Phase is merely the position of the terminator on the visible disk of the Moon. This depends on the positions of the Sun and Moon relative to a terrestrial observer. In other words, it is the difference in longitude of the two celestial bodies. The definition of phase does not take libration into account. At First or Third Quarter, the position of the terminator in selenographic coordinates is not, therefore, necessarily that of the central meridian: it may lie anywhere between 7°45' on either side of that point.

Phase and libration are two factors that must be taken into account in studying the lunar surface. They determine the visibility of any specific formation on the Moon and the way in which it is illuminated. At the terminator, sunlight is at grazing incidence. Shadows then have their maximum extent, and this is when lunar relief appears most pronounced. (In fact, to ensure that the shadows are not too confusing, observations are made of areas relatively close to the terminator.) The terminator sweeps over the surface of the Moon from east to west (in selenographic coordinates), at a rate of about 12.2° per day or about 0.5° per hour. Its motion is therefore quite considerable and it is possible to watch the Sun rise (or set) on any particular formation.

A distinction is made between the morning terminator, the points on the surface where the Sun is rising, and the evening terminator, where it is setting. (Both move from east to west.) The first sweeps across the Moon between New Moon and Full, and the second from Full Moon to New.

Because the lunar day is equal to 29.5 terrestrial days, any point on the surface of the Moon is illuminated for about 15 successive days and then plunged into darkness for the same amount of time. If one wants to observe a particular formation, it is therefore essential to check that the phase enables the formation to be seen, and also that the position of the terminator enables it to be observed under conditions that are appropriate for the type of observation envisaged.

The crater Copernicus, for example, is visible only between the 9th day of a lunation (when the Sun rises over the formation) and the 24th day (when the Sun sets), i.e., roughly between two days after First Quarter and two days after Third Quarter – which includes Full Moon, of course. Around those two specific dates the terminator will be close to the crater and the lighting conditions will be most favourable for studying its relief. Between those two dates the change in lighting will show different aspects of the crater.

For any formation close to the terminator, shadows will fall towards the west if it is the morning terminator, and towards the east if it is the evening one. At Full

117

Table 4.1. *Times of recurrence of a specific phase in successive lunations*

Lunation	Recurrence of specific phase	
0	00:00 UT	midnight
1	12:44 UT	in daylight
2	01:28 UT	about 1.5 hours later than lunation 0
3	14:12 UT	in daylight
4	02:56 UT	about 1.5 hours later than lunation 2
5	15:40 UT	in daylight
6	04:24 UT	about 1.5 hours later than lunation 4
7	17:08 UT	in daylight
8	05:52 UT	about 1.5 hours later than lunation 6
9	18:36 UT	in daylight
10	07:20 UT	about 1.5 hours later than lunation 8
11	20:04 UT	in daylight
12	08:48 UT	about 1.5 hours later than lunation 10
13	21:32 UT	in daylight
14	10:16 UT	about 1.5 hours later than lunation 12
15	23:00 UT	roughly the same time as lunation 0

Moon there are practically no shadows visible on the Moon; because the Earth is between the Sun and the Moon, no relief is apparent.

Phase may be defined as the phase-angle (α_{ph}), which is the Sun–Moon–Earth angle. A quantitative way of expressing the phase at a specific time is by giving the fraction of the lunar disk that is illuminated. This value is usually given in ephemerides, and corresponds to the ratio of the illuminated area to the total area of the disk. It therefore varies from 0 (New Moon) to 1 (Full Moon).

If a particular phase occurs during the night in one lunation (which we will call lunation 0), the same phase recurs during the day in the next lunation (lunation 1). It will occur during the day in odd lunations, and during the night in even ones, becoming about $1^h 30^m$ later each time (Table 4.1). After 15 lunations, or about 433 days, the phase will occur at about the same time as in lunation 0.

4.3.2.4 The degree of libration

Libration is an important factor in studying the surface, because it changes the orientation of the lunar disk with respect to the observer. It is important to take it into consideration when observing regions close to the lunar limb.

Ephemerides give the selenographic longitude (λ) and latitude (β) of the Earth for each day at 00:00 UT. These coordinates are those of the point on the Moon at which the Earth is at the zenith. They therefore correspond to the selenographic coordinates of the point on the Moon that is at the centre of the apparent lunar disk, as seen from the Earth (for a geocentric observer), at the time concerned. The values λ and β thus define the amount of libration in longitude and latitude.

When libration is zero, the mean centre of the Moon's disk – the origin of selenographic coordinates – coincides with the apparent centre of the Moon – the centre of the lunar disk as seen from Earth.

If the selenographic longitude of the Earth (λ) is positive, the mean centre is shifted towards the west relative to the apparent centre, and libration then brings areas at the eastern limb of the Moon into view (Fig. 4.3, P_2).

If the selenographic latitude of the Earth (β) is positive, the mean centre is shifted towards the south with respect to the apparent centre, libration then bringing northern areas into view (Fig. 4.4, P_1).

When λ and β are negative, the western and southern limbs of the Moon, respectively, are visible (Fig. 4.3, P_4; Fig. 4.4, P_2).

The combined effects of libration in longitude and latitude, the maximum values of which are $7°45'$ and $6°44'$, may displace the mean centre of the disk by as much as $10°16'$ from the apparent centre. Figure 4.8 will enable the region of the limb most favourably placed for observation to be determined from the values of λ and β for any given date.

It should be noted that at perigee and apogee, libration in longitude is zero (Fig. 4.3).

4.3.2.5 Position of the terminator

The position of the terminator on the surface of the Moon is determined by the selenographic longitude of the Sun ($L'_\odot$) at the time concerned. ($L'_\odot$ corresponds to the lunar meridian vertically beneath the Sun.) It will be seen that the selenographic longitude of the terminator is then given by $L'_\odot \pm 90°$.

The value of $L'_\odot$ is given by the equation:

$$L'_\odot = L_\odot + 180° - L_m,$$

where $L_\odot$ is the longitude of the Sun (in ecliptic coordinates), and L_m is the mean longitude of the Moon.

The position of the morning terminator is defined by:

$$S = 90° - L'_\odot \quad \text{or} \quad S = L_m - L_\odot - 90°,$$

where S is the selenographic colongitude of the Sun, measured westwards from the central meridian, and from $0°$ to $360°$. (It is given in certain ephemerides.)

The quantity L_m is given by:

$$L_m = L - i\cos(L - \Omega)\tan U - \lambda,$$

where L and U are the longitude and latitude of the Moon (in ecliptic coordinates); i is the inclination of the lunar equator relative to the ecliptic ($i = +1.534°$); λ is the selenographic longitude of the Earth; and Ω the longitude of the Moon's ascending node:

$$\Omega = 259°08' - 69\,629''(t - 1900.0),$$

where t is expressed in years.

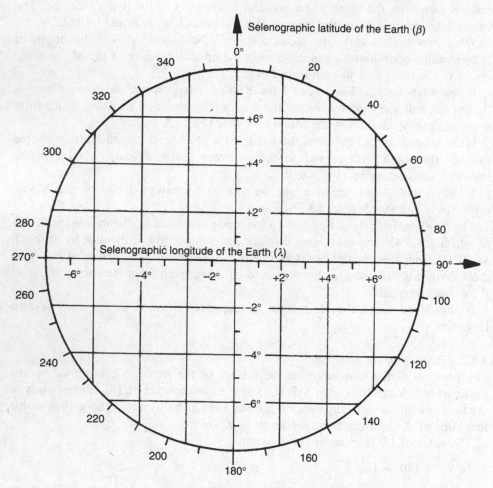

Fig. 4.8. *Libration chart. To find the area of the limb that is uncovered by libration, locate the point (λ, β) corresponding to the Earth's selenographic coordinates. Extend a line from the centre through the point to the circumference. Its intersection indicates the position angle of the area of the limb that is most favourably placed for observation. (The position angle is reckoned from 0° to 360° from North through East.)*

4.3.3 The surface of the Moon

4.3.3.1 Moonlight

The light that reaches us from the Moon is reflected sunlight. A spectrum of moonlight is therefore the same as the solar spectrum (with slight, additional components caused by luminescence and absorption, both produced by the interaction of solar radiation with the lunar soil). The spectrum of the Moon also shows a continuum that is stronger towards the red, showing that the Moon is redder than the Sun.

The reflectance of a surface depends on the composition or structure (or both) of the material of which it consists. In the case of planetary objects, it is measured by albedo. This represents the ratio of the total scattered flux to the total incident flux from the Sun. (It is therefore independent of distance from the Sun, and is equal to 1 for a perfectly reflecting surface and 0 for a perfectly absorbing one.)

Measured in this way, the average albedo of the Moon in the visible region is 0.07, indicating that the lunar soil absorbs 93 % of sunlight. Although it appears very bright, the Moon is, in fact, a very dark body.

Similarly, although the highlands appear relatively bright when compared with the maria, the albedo differences are quite low over the whole lunar surface. The albedoes average 0.05 for the maria, and 0.10 for the highlands, so the latter are only twice as bright as the mare areas. They may seem much brighter, but this is only a contrast effect (Fig. 4.9). This relatively uniform, low albedo is explained by the pulverized nature of the lunar soil. Any material in powdered form has a lower albedo than it does in the mass. The older a soil – i.e., the more it has been subjected to solar radiation (the suntan effect) and meteoritic bombardment – the lower its albedo.

At any given point on the Moon, the albedo that we have just defined (which is known as the spherical, or Bond, albedo) varies slightly as a function of phase.

When the value of albedo is required in a specific investigation, the so-called geometrical albedo is used. This takes account of the level of sunlight at the Earth and the Moon, the Earth–Moon distance, and the phase. It has a constant value whatever the phase. It is usually measured for a wavelength of 560 nm, and its value increases monotonically as a function of wavelength between 400 and 800 nm. The individual curves determined for different lunar formations may show slightly different slopes (Mikhail & Koval, 1974).

4.3.3.2 Illumination of the lunar surface

The albedo at any particular point on the Moon should not be confused with its apparent brightness (sometimes called 'apparent albedo'), which depends on its degree of illumination by the Sun, and thus on the phase. (A surface, even if only weakly illuminated, may have an albedo of unity.) For the Moon as a whole, the curve of luminosity as a function of phase-angle shows a rapid rise to maximum illumination at Full Moon, when the Moon is 12 times as bright as it is at quadrature. (In fact the Moon would be 19 times as bright, if one were to take the theoretical Full Moon, which actually occurs at the time of a lunar eclipse.) The curve also shows that the lunar crescent is invisible at angular distances of less than 7 degrees from the Sun, around the time of New Moon (Kopal, 1966).

This behaviour indicates that the lunar surface has considerable large-scale roughness. The rapid rise at Full Moon is caused by the sudden disappearance of shadows from the surface of the Moon.

If individual points on the Moon are considered, it will be found that wherever they may be on the lunar disk, their light-curve is identical to that of the Moon as a whole; i.e., they all reach their maximum brightness close to Full Moon. The reason for this behaviour is exactly the same as just described, but this time the phenomenon acts on a different scale, which is that of the roughness of the lunar soil itself.

Fig. 4.9. *The Moon at First Quarter, photographed by C. Arsidi with a 102-mm, f/24 fluorite refractor. Exposure 1 s on Kodak Infocapture film. The differences between the albedoes of the various maria may be readily seen.*

4.3.3.3 *The colour of lunar soil*

Studies of differences in colour of the lunar soil have shown that, in general, any variations are minor, and are related to differences in composition and age of the surface materials, and thus to the albedo. (Materials that have been darkened in the manner discussed earlier appear correspondingly redder.)

Differences of colour are more significant at longer wavelengths (670–790 nm, for example) than at shorter wavelengths (400–550 nm). For a single type of formation (highlands or maria, for example) the differences are much weaker (Asaad & Mikhail, 1974).

The materials forming the highlands appear to be the reddest, whilst those of the youngest craters, rays and ejecta, are bluest. The difference between the two extremes does not exceed 20 %, however. The lunar maria are intermediate in colour between these two types of formation, and the youngest are bluer than the older ones. There are, however, certain exceptions, such at Mare Tranquillitatis (an old mare), which appears blue, reflecting a difference in composition.

4.3.4 *Maps and atlases of the Moon*

Lunar maps, photographic atlases and specialized books and articles are absolutely essential to anyone wanting to study the Moon. Unfortunately, it is often a problem to obtain this literature, because it is relatively rare and difficult to locate. Most people will soon exhaust the possibilities of material normally intended for amateurs, and will turn to the professional literature. Consulting this generally requires access to libraries at a university, observatory, or other research institution. This is not necessarily impossible for any amateur who is sufficiently motivated. The references that follow are divided into amateur (o) and professional (•) material.

4.3.4.1 *Moon maps*

An overall picture of lunar features is given by:

o *Philips Moon Map*, 1982; George Philips and Son Ltd, London. Orthogonal projection (the Moon is shown as it appears from Earth). Scale: 1:7 700 000 (lunar diameter: 45 cm).

o *The Moon*, Hallwag A. G., 1980; Berne, Switzerland. Orthogonal projection. Scale: 1:5 000 000 (lunar diameter: 69 cm).

o *Moon, Mars and Venus*, Rükl, A., 1976; Hamlyn, London. Orthogonal projection, 76 plates (11.5 × 8.5 cm). Scale 1:3 700 000 (lunar diameter 94 cm). Data on lunar formations (diameters of craters, heights, origin of names, accurate coordinate grid). An excellent work, now regrettably out-of-print in English. [French and German versions are still available, but this has now been superseded by Rükl's *Atlas of the Moon*, Hamlyn, London, 1991. – Trans.]

• *Lunar Atlas Chart* (LAC) series. Aeronautical Chart and Information Center, U.S. Air Force, St Louis, Missouri, U.S.A. Mercator projection. Scale: 1:1 000 000. A series of 68 charts of the visible side of the Moon.

• *Geologic Map of the Near Side of the Moon*, Wilhelms, D. E. & MacCauley, J. F., 1971; Map I-703. U.S. Geological Survey, Washington, D.C., U.S.A. Scale

1:5 000 000 (lunar diameter: 69 cm). Geological data on lunar formations, detailed description of stratigraphic units.

4.3.4.2 Photographic atlases of the Moon

There are very few amateur-level publications available. Luckily, the recent appearance of the atlas by G. Viscardy has filled a serious gap in this field.

○ *Photographic Atlas of the Moon*, Kopal, Z., 1965; Academic Press Inc., New York & London. A work that is relatively difficult to find.

○ *Atlas-Guide Photographique de la Lune*, Viscardy, G., 1986; Editions Masson, Paris. 220 plates, 160 black-and-white photographs, description of the most important formations, date and time of the photographs, scales, appendix on certain notable formations, technical data about the way in which the photographs were taken, treatment of the film (TP-2415) and printing. A remarkable work, absolutely indispensable for any amateur undertaking lunar research.

● *Photographic Lunar Atlas*, ed. Kuiper, G., 1960; University of Chicago. Based on the best photographs obtained at Mount Wilson, Lick, Pic du Midi, MacDonald, and Yerkes Observatories. 230 plates, 50×40 cm, various lighting conditions (including at Full Moon); details for measurement of the plates (selenographic coordinates, heights, distances on the Moon). A remarkable work, quite rare, but indispensable for researchers; to be handled with great care therefore if you are lucky enough to be able to consult it.

● *Photographic Lunar Atlas*, vol.2, U.S. Air Force, 1961. A work that reproduces the plates from the preceding Atlas, with additional selenographic coordinate and orthogonal grids.

● *Consolidated Lunar Atlas*, U.S. Air Force Photo Lunar Atlas, 1967. A supplement to the atlases mentioned, which is based on photographs obtained by the Lunar Orbiter probes. Wonderful photographs, which are of the greatest value for workers studying the lunar surface in detail. To be handled with care ...

4.3.4.3 Amateur organizations studying the Moon

The lunar sections of the British Astronomical Association (BAA) in the United Kingdom, and of the Association of Lunar and Planetary Observers (ALPO) in the U.S.A., aim to encourage and coordinate amateur work about the Moon. Their publications, the British Astronomical Association's *Journal*, and *The Strolling Astronomer*, respectively, are absolute mines of information about lunar formations and the Moon as a whole.

● BAA Lunar Section, c/o British Astronomical Association, Burlington House, Piccadilly, London W1V 9AG, U.K.

● ALPO Lunar Section, c/o J. W. Westfall, Dept of Geography, San Francisco State University, 1600 Holloway Ave, San Francisco, California 94132, U.S.A.

4.4 Amateur observation of the Moon

4.4.1 Principal fields for amateur work

The surface of the Moon has been the subject of highly detailed research. Although in the past advanced amateurs were on even terms with professionals as far as practical work was concerned, and as a result, made a significant contribution to the overall understanding of the Moon, since the development of space research, professionals have had the means of carrying out very significant investigations. For each type of research, orbital data (close-up photographs, including stereoscopic ones; altimetry; gravimetry; etc.) and surface data (analysis of lunar samples; seismology; etc.) are systematically integrated with telescopic data (spectrophotometry; multi-spectral imagery; radar results; etc.). This complementary nature of the various forms of data is absolutely essential to arriving at the correct interpretation of the lunar features that are being studied.

The aim of amateurs is not to rival the professionals: the fact that they do not have access to research facilities, the fragmentary nature of the fundamental data to which they have access, and the modest instrumentation that they possess, all make this impossible. Their aim is to carry out a coherent programme, the results of which may perhaps be included in the data used in more detailed studies carried out elsewhere. It should suffice just to present the facts, and to be circumspect in their possible interpretation.

Amateurs are generally limited to two basic methods of observation: visual or photographic observation. In addition, they generally have only modest equipment with which to carry out those observations: 400-mm telescopes are still rare, even among societies and groups of amateurs. It is with this scale of optical equipment in mind that we shall discuss what useful work amateurs can do in studying the Moon.

We shall, however, discuss photometric methods, which are indispensable in obtaining more precise information about the lunar surface. The development of new detectors and the use of microcomputers have accelerated to such an extent over the last few years that these methods have become much more readily available to amateurs. One type of detector in particular, the CCD (Charge-Coupled Device) has become more and more widespread in astronomy, and amateurs have become used to seeing digitized images that it has captured in both specialized and popular journals. For this reason, and also because a later chapter (Chap. 20) is devoted to it, we shall mention the use of this type of detector in certain lunar studies.

Given that we are observing in the visible region, the only way of studying the lunar surface is to extract the maximum amount of information from variations in illumination of various lunar features. Preference may be given to studying either the topography (the surface relief) or the composition of the lunar soil. In the first case (Fig. 4.10), work is carried out under low angles of illumination, i.e., in a zone close to the terminator. This type of investigation might, for example, involve determining the height of certain features by measuring the length of their shadows.

In the second case, work is carried out under full illumination (at Full Moon). The effects of relief are then eliminated and the illumination is the same for the whole of any specific zone, so differences in albedo indicate differences in the physical nature or the chemistry (or both), of the surface rocks. These variations may give important

Fig. 4.10. *Clavius and its craterlets photographed by G. Viscardy: 520-mm reflector, exposure 1 s on Kodak TP-2415 film.*

information regarding the geology and other properties of the lunar formation under observation.

Amateurs cannot hope to add anything to lunar cartography or to the study of topography, because of the detailed photographic coverage of the Moon by various spacecraft that have orbited it, some of them at low altitude. It would be misguided to expect to discover, from Earth, some new, permanent detail on the surface of the Moon. Yet there is still a region that has not been mapped properly on the southwestern limb of the Moon. This has been called Luna Incognita.

On the other hand, detailed investigation of the topography and (above all) of the composition of certain formations would be very useful, because this type of work is part of modern research projects.

Numerous observers have also reported non-periodic variations in albedo of the lunar surface. These temporary brightenings, darkenings or coloured events are collectively known as Transient Lunar Phenomena or TLP. [Also called Lunar Transient Phenomena, LTP, in North America – Trans.]. They seem to occur in certain specific areas of the Moon. Although the existence of these events is still subject to some doubt, monitoring of these areas might provide definite information about their nature, and whether they are actually of lunar origin or not.

The Moon may also serve as an indirect means of studying our own planet. Measurements of the size of the Earth's shadow cone and of the intensity of the coloration of the lunar surface during a lunar eclipse provide information about the terrestrial atmosphere (such as the presence of volcanic or meteoritic particles within it).

These various forms of research are some of the fields in which the moderately equipped amateur can be most active. They are described in more detail in the remainder of this chapter.

4.4.2 Methods of visual observation

4.4.2.1 Limits of visual observation

In the case of a purely visual observation, the information contained in an image of the Moon is translated by the observer into a drawing or sketch, together with a related description of specific details. This drawing is the end result of the observation and serves as a basis for any later studies.

Apart from atmospheric factors (turbulence and transparency), instrumental ones (performance and optical quality of the instrument), and those related to the point on the Moon being investigated (lighting conditions and complexity), the observer is a major factor in assessing the information obtained. The way in which detail on the Moon is recognized depends on the observer's visual perception (the physiological state of the eye and the visual acuity). The observer's ability to describe the visible details or to reproduce the information as a drawing, as faithfully and as objectively as possible, is also involved.

There is therefore a certain subjective element present in all visual observation, which means that it is not appropriate on its own as a scientific analytical technique. Frequently, if at all possible, it should be supplemented by photographic observations.

Greater weight may be given to data from visual observations if the observer has had considerable experience. Lunar observation contains traps for the unwary: the extreme variety of forms that exist on the lunar surface and the variations that they exhibit under various lighting conditions (at various phases), may easily lead to false conclusions. Similarly, certain atmospheric effects, which are easily visible with an extended object such as the Moon, and the amount of light that it reflects, may give rise to optical effects that an inexperienced eye might well take for fleeting phenomena on the surface of the Moon itself. A degree of training is therefore required for lunar observation, especially if certain specific types of observation, such as monitoring transient events, are to be carried out.

Consequently, it is not a waste of time to begin by regularly observing and making detailed drawings of characteristic lunar features, even if these are so well-known as to be almost 'archetypal'. (On the contrary, with such formations, there will be a large number of references available, which may be compared with one's own findings.) Such observation will also enable experience to be gained in using instruments (the use of large magnifications, the effects that these may cause on the visibility of fine detail, etc.), while developing observational skills and the facility for reproducing visible details by means of a drawing.

4.4.2.2 Observing-instruments

Any instrument, no matter how small, may be used for observation of the Moon. It is even possible to start without any equipment at all. Naked-eye observation of the Moon reveals about as much detail as that visible on planets observed through a large telescope.

With a 50–60-mm refractor or a 115-mm reflector, one can really begin to discover the Moon. The latter would reveal enough detail for a lifetime of study by an observer who wanted to draw everything in detail.

Naturally, if the aim is to undertake a detailed study of the lunar surface for scientific purposes, larger diameters are required. A minimum would be a 110-mm refractor, or a 150–200-mm reflector with a long focal length (a large focal ratio). For the type of work that we have in mind, the ideal instrument is capable of withstanding magnifications of 200–300 times under good conditions. Diameters of 200–400 mm are therefore in order. We shall see, however, that large instruments suffer from certain limitations because of atmospheric conditions.

Even the smallest instruments may still be used for serious work. Binoculars are best for seeing the shading and tints of the eclipsed Moon, for example, and a transient event might well be visible in a 115-mm reflector.

4.4.2.3 Instruments for visual observation

Lunar formations cover a great range of sizes and, depending on the equipment used, are therefore more or less easily observed. Major features like the maria, are visible with the naked eye. Others, like the craters, have a vast range of diameters. Copernicus, for example (93 km in diameter), has an apparent diameter of 48 arc-seconds, i.e., the same as Jupiter at opposition. Both therefore appear the same size in a telescope whatever the magnification. Aristarchus (45 km in diameter) has an apparent size of 24 arc-seconds, the same as Mars at opposition. Narrow features like the Triesnecker rilles (see Fig. 4.25) are 3.5 to 0.9 km wide on the Moon, with an apparent width of 2 to 0.5 arc-seconds. A 200-mm instrument, with a theoretical resolving power of 0.6 arc-second, allows details on the Moon about 1 km wide to be distinguished. But this does depend on a feature's degree of contrast with the surrounding area and, above all, on atmospheric turbulence, to which lunar features are particularly susceptible. As a result, the limit of 1 km will very rarely be attained.

The theoretical performance of specific instruments, particularly resolving power and magnification, should therefore be treated with some caution. The resolving power is independent of the magnification, but with equal magnifications, the larger the diameter of a telescope, the greater the amount of detail that will be visible (even though the apparent diameter of the object remains the same). In practice, when turbulence is taken into account, resolutions below 0.5 arc-second are rarely attained in lunar work.

As with any astronomical body, the best observing conditions occur when the Moon is close to the meridian. The choice of an observing site free from bad seeing, and equalizing the interior and exterior temperatures if one is using any form of covered observatory, will reduce the effects of turbulence. A reflector (unless fitted with an optical window) is always more susceptible to poor seeing than a refractor and, in general, the larger the diameter, the more an instrument is affected.

The effective field in any particular instrument is a function of the ratio between the apparent field of the eyepiece (which is fixed) and the magnification. If an eyepiece has an apparent field of 50° and a magnification of 100×, the whole of the Moon will just fit in the field, because its apparent diameter is 0.5°.

Details to be studied should be located close to the centre of the field, because optical quality is best near the optical axis. At magnifications greater than 90–100×, the Moon must be observed piecemeal, so a good map or lunar atlas becomes essential to keep track of which region is being observed.

The magnification should be appropriate to the size of the feature being examined, and to the specific telescope being used. With the Moon, moderate and high magnifications – corresponding, respectively, to approximately 10 and 20 times the diameter expressed in centimetres – may be employed (although the latter will only be under good seeing conditions). A magnification of 15 times the diameter is generally ideal. Too large a degree of magnification leads to a loss of contrast.

One unusual problem with observing the Moon is that the surface is sometimes too bright, so that the eye is sometimes dazzled and unable to see fine detail. It is not advisable to obtain a fainter image by increasing the magnification or by reducing the telescope's aperture, because this reduces the field of view and lowers the resolution. It is best to use neutral density filters instead. Too bright an image rapidly leads to tiredness of the eye, with a corresponding effect on visual acuity.

4.4.2.4 Drawing the lunar surface

Over a decade, in the 1960s, a whole series of spacecraft, ranging from orbiting satellites (such as Luna and Lunar Orbiter) to soft-landing automatic probes (Surveyor, Lunakhod) and manned craft (Apollo) revealed the surface of the Moon in detail. The days when amateurs had a major, active part to play in lunar cartography are long past. Detailed mapping of individual formations on the Moon is now undertaken for the personal benefits that it brings, rather than in the hope of making new discoveries about the lunar surface.

The main advantage of making drawings of the lunar surface is an observational one: drawing what is visible through the eyepiece is the best way of training the eye to pick out fine detail (Fig. 4.11). With a simple visual observation, all that is remembered is a general impression of the particular formation, but making a drawing requires the observer to make an effort to pick out and decide upon the

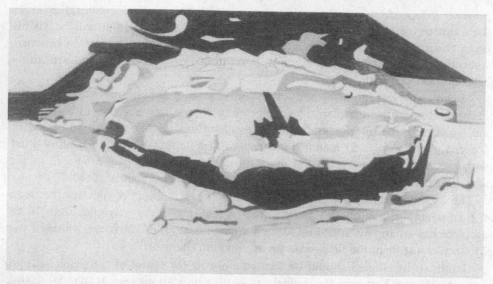

Fig. 4.11. *A drawing of the central peak of Pythagoras by P. Morel on 1982 September 30, between 20:40 and 22:16 UT: 108-mm refractor, magnification 173×, seeing 2.*

most significant detail. The act of drawing the details on paper helps to train the eye to 'see' properly.

Even though any drawing of the Moon is bound to have a particular, personal style and is, by its very nature, subjective, it may nevertheless often still reveal features that are too fine to be recorded photographically. (This is because of the limited resolution of films, and the combined effects of turbulence and poor seeing, etc.).

Drawings are therefore of considerable interest, and if sufficient care is taken in executing them, they may become valuable records. There is nothing wrong in producing beautiful drawings of the Moon, but the most important point is to record details accurately, particularly their relative positions and sizes. It is also essential to make any drawings with an unbiassed mind, completely free from preconceptions about the appearance of the feature being recorded.

Every effort should be devoted to this goal of faithfully representing lunar detail. It is not easy, and only concentration, patience, and perseverance will ensure that one's drawings do eventually meet this standard. The observer's comfort is of considerable importance here. For preference, the observer should be seated, and the drawing paper should be fixed to a suitable board and provided with adequate illumination (but this should not be too bright). Anything else that may be required should be easily accessible.

Before starting a drawing, it is essential to check that the seeing conditions are suitable for satisfactory work. Turbulence is the principal factor limiting visibility of fine detail, so its value must be recorded on the final drawing. It may be estimated 'by eye' using an appropriate scale, which may be refined with experience. This estimate should be made using a high magnification on the part of the Moon to be studied, or by examining the lunar limb. If a star is used, it should be as close to the Moon as possible.

If seeing is of class 3 on the Antoniadi scale – which runs from 1 (perfect) to 5 (bad) – observation would be difficult at high magnifications but possible with moderate ones. Class 3 is the limit for making drawings or taking photographs of the Moon. One other method of estimating the seeing is to use an atlas to check the finest detail visible in the area being studied.

Transparency should also be estimated. The presence of haze or slight mist is generally easily seen by examining the clarity of the sky around the lunar disk. Although the presence of a slight haziness might not appear to detract greatly in lunar observations, it does nevertheless cause a certain loss of contrast and limits the amount of fine detail visible.

The first point to note about making a good drawing is to use the minimum magnification necessary to see the details that you want to record, and to limit yourself to a small area of the lunar surface (perhaps a group of four or five craters, a single crater and its ejecta, or just part of a mare). Too large an area is difficult to draw in detail and takes too long. Similarly, the actual format of the drawing should be appropriate for the size of the formation and for the amount of detail that is visible. Ordinary drawing paper is generally used, and medium-to-soft pencils. Start by drawing the main outlines of the larger formations and the most important features before moving on to finer detail. The latter are sometimes visible only momentarily because of turbulence: include them only when you are perfectly sure of their existence.

Because the terminator moves relatively rapidly, details close to the shadow may appear or disappear rapidly – depending on whether it is the morning or evening terminator – during the course of an observing session. It is important, therefore, to decide upon a set of priorities for recording detail close to the terminator.

Once the drawing shows the positions of all the details, some shading may be added. It is not realistic to attempt to reproduce the whole range of tints actually visible on the surface. There are several methods of estimating the brightness of particular areas of the surface of the Moon and planets. A numbered scale is generally used, for example from 0 to 10, where the lowest number is used for the most brilliant area and the highest for the darkest (shadows in the case of the Moon). [Note that this scale is the one normally used by most European observing groups. The American Association of Lunar and Planetary Observers (ALPO), uses a scale running in the opposite direction, where 0 indicates deepest shadow or perfectly black, background sky. – Trans.]

In practice, one works by determining the brightest and darkest areas, and then derives the intensity of intermediate areas by successive approximation. The brightnesses may be indicated by using hatching or by actual numbers on the drawing. Once again, it is only experience that enables one to judge intensity easily.

Once a drawing has been finished at the eyepiece, neither it nor the accompanying notes should be altered. The drawing may be somewhat untidy, however, and thus difficult to interpret, or the paper on which it is drawn may be creased or otherwise damaged. A clean copy should therefore be prepared soon after finishing the observation and this should be as faithful a copy of the original as possible.

An additional copy may also be made using pencil shading instead of hatching. Whatever happens, the original drawing should always be kept in the record book.

4.4.3 Photographic observation

Photography of the Moon is very similar to photographing the planets (*see* Chap. 5). At the quarter or gibbous phases the region near the terminator is similar in brightness to Jupiter, so we are able to take high-resolution photographs with fairly short exposure-times.

As with planetary photography, our main enemy is atmospheric turbulence. The choice of observing site is therefore all-important, and the quality of the photographs will largely depend on the state of the atmosphere. Because of its orbit's relatively high inclination to the ecliptic, the Moon may reach declinations that are considerably higher than those of the planets, which is a favourable factor. To a first approximation, however, the path followed by the Moon may be considered to be more or less that of the ecliptic, which means that it has a high declination at New Moon in summer, at First Quarter in spring, at Full Moon in winter, and at Last Quarter in autumn.

Turbulence prevents long exposures: one second is about the maximum that is generally attainable. Because the Moon's brightness varies considerably as a function of phase, longer focal ratios may be used around Full Moon than at the quarters and, of course, even more so than at New Moon. A value of $F/D = 50$ is an average that has been found to be very suitable for Kodak TP-2415 film. Naturally, if the whole disk is to be included, then the focal length must be chosen to ensure that the image size does not exceed the film format (about 2.2 m for a 24×36 mm format). Unfortunately, such an image is very small, and even on high-resolution film the finer detail will be lost.

It is essential to remember the Moon's own considerable proper motion, which is of the order of 0.5 arc-second per second of time in Right Ascension. This is sufficient to affect the photographic resolution if no compensation is made. Ideally, the telescope should be driven, not at sidereal rate, but at lunar rate – by using a variable-frequency drive, for example. It should also be noted that the Moon's motion in Declination may reach 0.26 arc-second per second of time under the worst conditions (when the Moon is close to the celestial equator). A declination drive would then be highly desirable, but this is found on very few telescopes.

Because of the shadows, relief details are easier to record close to the terminator. This area should be photographed first. The whole surface of the Moon may be covered in this way, over the space of half a lunation.

The film generally chosen for lunar photography is Kodak TP-2415 (Figs. 4.10, 4.12 and 4.13). To be used to best effect, this film should be developed to a high contrast (with D19 or HC110 developer). This poses a problem, because different areas of the Moon have very different brightnesses, and these are largely a function of distance from the terminator. It is therefore difficult to obtain an exposure that is ideal over the whole of the field. The difficulty appears most markedly in the printing stage, because papers have a much smaller dynamic range than negatives. A standard print of a lunar photograph shows a limb that is a uniform white,

Fig. 4.12. *The area around Walter, photographed by C. Arsidi: 305-mm, f/48 reflector, exposure 1 s on Kodak TP-2415.*

Fig. 4.13. *The area around the crater Birt and Rupes Recta (the Straight Wall). Photograph by D. Cardoen: 406-mm reflector, film TP-2415, exposure 0.5 s.*

without details, and a terminator that disappears completely into the black of the background sky. Use of a soft paper is a trap to be avoided: the whole of the print then becomes a muddy grey. In any case the range of such a paper is still too small to cope with the difference between the limb and terminator. The solution is to use dodging (with the hands, cloth or pieces of card) over the clearest parts of the negative during part of the exposure. It is quite possible to give the correct exposure to as many as ten different areas on a single negative in this way, although care must be taken to ensure that each receives the proper exposure. Printing lunar photographs soon becomes an art, and it is interesting to read the comments made by one of the artists, Georges Viscardy, in the introduction to his book *Atlas-Guide Photographique de la Lune.*

Naturally, the larger the field covered by the photograph, the greater the difficulty caused by differences in the brilliance of parts of the lunar surface. A particularly difficult instance occurs when the photograph includes detail in both a mare area crossed by the terminator, and in a highland region away from the terminator. When taking a photograph of the whole Moon, one might opt to use a softer developer. For such a photograph the F/D ratio used has to be small, so there is no need to try to obtain the maximum resolution that would be given with a more energetic developer.

Colour films are not suitable for lunar photographs (except for lunar eclipses). They are inadequate for use in colorimetry (when it is better to use filters and black-and-white film), and their resolution is far less than that of TP-2415 film.

4.4.4 Photometric methods

In certain work, the aim is to compare the relative albedo of different areas of the Moon's surface or else to determine variations within a specific region. Visual comparison is feasible, because the eye can detect very minor differences, but it is often difficult to estimate these on a specific scale, or to record them on a drawing, or by any other method. It is also of interest to give precise values for the differences, with a view to detecting any possible variations from other observations. Accurate measurements of the amount of light reflected by the Moon can only be obtained by photometric methods, and we will only describe a few practical implications here.

Measurement of the albedo of a point on the Moon should be confined to a very small area, to avoid including neighbouring parts of the surface. The use of standard types of detectors, such as small-aperture photomultiplier tubes, together with a long focal ratio allows this type of selective measurement.

It is therefore essential that the equipment should include a means of selecting a specific area of the surface. A flip mirror or a beam-splitting cube in the light-path in front of the detector will send the image to a cross-hair eyepiece. (With the Moon, the cross-hairs do not need illumination.) In photometry, certain significant characteristics of the detector must be taken into account, in particular its sensitivity and linearity.

The sensitivity S is determined by the ratio between the excitation flux φ and the response of the detector (i.e., the voltage produced, V):

$$S = V/\varphi.$$

The output voltage measured is:

$$V = V_m + S\varphi,$$

where V_m is the dark-current voltage. The detector is linear if the curve of $V = f(\varphi)$ is a straight line when φ increases in a linear manner.

When the aim is to measure differences in albedo over a specific area of the Moon (such as part of a mare, for example), a large number of individual spot measurements are required. Imaging appears more appropriate in this case, because

135

the ideal would be to have a digitized image of the region concerned. This would show the topography and simultaneously contain the exact value of the luminosity of every individual point (every pixel) in the image.

With recent, vast improvements in technology, such an image may now be recorded by using a CCD detector. The image of the lunar surface may be recorded directly by a two-dimensional array (a CCD matrix), or by trailing the lunar image across a linear detector (a linear CCD), the image being built up line-by-line. This technique is particularly interesting for lunar work, both because of the very high performance of this type of detector, which has strong linearity and good spectral sensitivity (typically from 400–1100 nm), and also because there are many ways in which the images may be processed with appropriate software: luminance ratios may be derived, differences may be detected, and false-colour images produced.

Corrections must be applied to individual spot measurements or images before they can be used for photometric purposes. The raw data should first be adjusted by subtracting the dark-current reading. The latter is obtained by operating the detector in exactly the same way as when making an actual measurement, while ensuring that no light is falling on the detector itself (*see* Chap. 20).

The dark current is mainly of thermal origin. Its value increases with the length of time a detector is exposed to light. Because the luminosity of the Moon is high, this duration is minimal, which means that the detector does not have to be cooled.

If a detector has an extended surface area (as in a CCD, for example), the sensitivity varies slightly from point to point. The original image must therefore be corrected by subtracting an evenly illuminated, reference image. This evenly illuminated field should be obtained under exactly the same conditions as those under which measurements are made: with the same optical elements in front of the detector, the same filter, and at the same temperature. The dark-current voltage should also be subtracted from this measurement. Once the raw data has been modified in this manner, the image may be said to be calibrated:

$$\text{calibrated image} = \frac{\text{raw image} - \text{image noise}}{\text{uniform-field image} - \text{image noise}}.$$

Measurements of albedo at a specific point on the Moon are measurements either relative to another point that is part of the area being studied or, for spot measurements, to some standard point on the Moon. With imagery, the measurement is relative to some specific point in the image. The relative albedo of a point (after calibration) is therefore:

$$V = V_{\text{point to be measured}} / V_{\text{standard point}}.$$

The standard photometric points usually used on the Moon are:

- Mare Serenitatis 2 (MS-2): $\lambda = 21°25'\text{E}$; $\beta = 18°41'\text{N}$, and
- Plato: $\lambda = 09°00'\text{W}$; $\beta = 52°30'\text{N}$.

This procedure has the advantage of eliminating instrumental and atmospheric effects, provided the interval of time between each pair of measurements (point measured / standard point) is short (i.e., less than a few minutes). A whole series of measurements is made of each point to ensure consistency: taking the intensity ratios between two images of the same area should result in a uniform field, for example.

4.5 Lunar formations

4.5.1 Origin and appearance

The Moon has two basic forms of terrain:

- The highlands, which are light in colour and form 80 % of the lunar surface (65 % of the visible face); and
- The maria, darker in colour, forming 20 % of the Moon's surface (35 % of the visible hemisphere).

4.5.1.1 The maria

The maria are large, generally flat expanses of basaltic lavas, which were laid down 3900 to 3000 million years ago in giant basins in the lunar crust, which were enormous impact structures (impact basins). This flooding took place following a general heating of the Moon about 1000 million years after the last impact basins had been formed, not after each impact as people generally tend to believe.

The oldest basins, considerably degraded by later impacts, have poorly defined outlines, so the maria that fill them have irregular borders. (Examples are Mare Tranquillitatis and Mare Fecunditatis.) The younger basins have retained much of their original form, and the maria inside them are distinctly circular. (Examples are Mare Crisium – the only truly isolated mare – Mare Serenitatis, and Mare Nectaris.)

One of the Moon's most recent impact basins is the Imbrium basin, which contains the mare of the same name (Fig. 4.14). Its structure is that of a vast depression, 1100 km in diameter, surrounded by three concentric rings. This type of structure is known as a multi-ringed basin. These rings arise where blocks of crust are elevated by the force of the impact. The innermost is almost completely covered by the mare lavas, and only a few peaks such as La Hire and Montes Recti remain above the surface. The second ring appears on the north-eastern side only, where it forms the Montes Alpes. The third ring forms the impressive Montes Apenninus and the Montes Caucasus in the west.

Although impact basins are found all over the surface of the Moon, the maria essentially occur on the visible hemisphere only. The reason for this is that the lunar crust is thinner on the visible side, offering less resistance to the passage of lavas generated at depth. A typical example of a recent multi-ringed basin on the far side of the Moon is Mare Orientale, which shows very little lava flooding.

Terms that are used for mare areas – more or less in descending order of size and significance – are *Oceanus* (ocean in Latin), *Mare* (pl. *Maria*; sea), *Lacus* (lake), *Palus* (swamp) and *Sinus* (bay).

4.5.1.2 The highlands

The maria are relatively thin deposits (just a few kilometres thick), but the highlands (sometimes known as *Terrae*: lands) are the upper part of the lunar crust, which

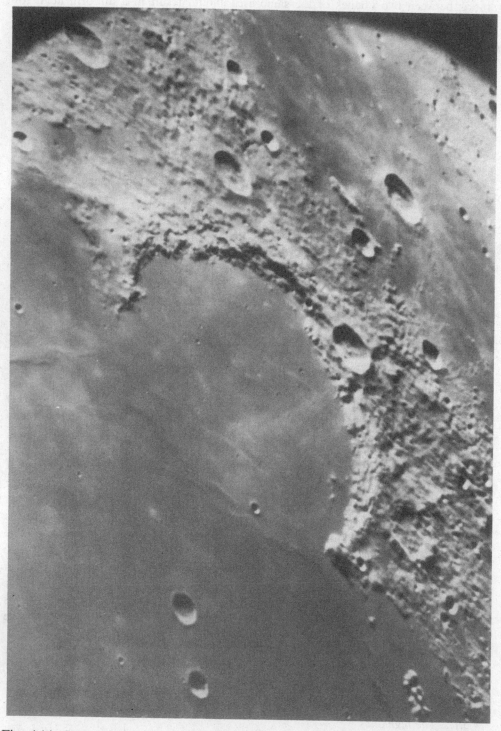

Fig. 4.14. *Sinus Iridum and the northern border of Mare Imbrium. Photograph by G. Viscardy: 520-mm reflector, TP-2415 film.*

is several tens of kilometres thick. They typically have an extremely irregular relief and a great number of craters, to such an extent that the latter frequently overlap.

The stratigraphy of the highlands consists of a series of crater and ejecta units lying on top of one another. The predominant rocks are breccias, an indication of the numerous impacts that have taken place. None is younger than 3800 million years, and the oldest go back to 4600 million years, practically the age of the Moon itself.

4.5.1.3 Craters

Craters are circular features that may be of either meteoritic-impact or volcanic origin. Craters are found on all the solid bodies in the Solar System, and most of their ages lie between 4000 and 3000 million years. They are best preserved on those bodies that have the least geological activity. This is the case on the Moon, whereas on Earth, where internal processes and erosion are very active, most craters have been erased from the surface. (Apart from a few exceptions, all that can be seen in most cases are degraded remnants, known as astroblemes.)

On the Moon, the vast majority of the craters are of meteoritic origin; only a few small craters along faults or on top of some domes are volcanic.

Lunar craters are very varied in size and appearance. The diameters of those visible from Earth vary continuously from a few hundred metres to hundreds of kilometres. There is an equally smooth gradation between the largest craters and the giant impact basins, and it is difficult to say where a distinction might be drawn. Impact basins are not included in lists of craters, however, just as terrestrial continents are not normally considered to be the largest islands on Earth. The generally accepted transition between these two types of impact structures is around 200 km in diameter.

A typical crater profile shows a floor that is slightly below the level of the surrounding surface, and an elevated rampart, which has a steeper slope on its inner face. Craters smaller than about 10 km in diameter typically have a bowl shape. The largest craters generally have flat floors but a more complex structure, with one or more central peaks, or terraced ramparts. These features arise from isostatic rebound and slumping of the rim. (Examples are Copernicus, Eratosthenes, Theophilus and Euler.) But a crater's complexity is not an inevitable consequence of a larger diameter.

The apparent diversity of crater forms arises from later modifications, which have occurred by various processes:

- Meteoritic erosion: examples of such heavily eroded craters are Deslandres (Fig. 4.15), Hipparchus, and Maginus. Significant overlapping of craters may be seen in Maurolycus and Stöfler. Other examples are easy to find, especially in the highland areas.
- Deposition: the craters may be covered by ejecta from impact basins or other craters, or through the floor being flooded by mare lavas. Examples of partially flooded craters are Doppelmeyer, Fracastorius (Fig. 4.16), Letronne, Le Monnier, and Fra Mauro, the last of these being particularly degraded. The sequence continues with craters that are practically sub-

Fig. 4.15. *Deslandres (arrowed), an example of a heavily degraded, highland crater. Photograph by G. Viscardy: 520-mm reflector, TP-2415 film.*

merged, such as Kies, Bonpland or the ring Flamsteed P, and on to ghost craters such as Stadius. Such craters may be found along the boundaries between mare areas and the highlands.

- Tectonic modification: changes caused by isostatic compensation, leading to the formation of terraces, faults in the crater floors, etc. Examples are Gassendi, Alphonsus, and Posidonius (Fig. 4.17). Such modifications affect large craters at the edges of the maria.
- Volcanic alteration: the direct outflow of lava onto the crater floor.

Secondary structures associated with impacts are ejecta deposits (solid and molten material excavated and ejected by the impact that decrease outwards from the rim of the crater), and secondary craters, sometimes arranged in radial lines, loops and chains, generally beyond the ejecta covering (e.g., Copernicus, Fig. 4.18).

When the major impact that created the Imbrium basin occurred, large quantities of material were ejected and deposited all round the basin, creating furrows and radial deposits (known as the Imbrium Sculpture). These features may be seen clearly around the north-eastern rim of the crater Ptolomaeus, for example (i.e., about 600 km from the southeastern edge of the Imbrium basin).

Bright rays of finely powdered material may extend far beyond the craters from which they originate. (Examples are: Tycho, Kepler, Messier B, Aristarchus and Proclus. The last three have asymmetrical systems of rays.) These extremely fine, ray deposits are rapidly darkened by the solar wind and eroded by micrometeoritic impacts. As a result, they are only associated with relatively young craters. (Tycho is about 100 million years old.)

Counting the craters on a specific area of the surface allows its age relative to another area to be determined, because the younger the surface, the lower the number of craters.

4.5.1.4 *Formations of tectonic origin*

Linear structures that appear like low ridges (or folds), some tens of metres to 200 or 300 m high, are visible in the maria. They are sometimes several hundred kilometres long and only a few kilometres wide. These mare wrinkle ridges cross basaltic flows that differ in nature and age, showing that the ridges are of tectonic, rather than volcanic origin. (They are compression phenomena.)

When associated with circular maria like Mare Crisium or Mare Serenitatis, the ridges occur both just inside the edges of the maria, where they form systems concentric with the centres of the maria, and in the middle of the maria themselves, where their general orientation is north–south. The latter type of ridge probably arose from shallow tectonic processes (changes in the level of the mare basalts after their eruption onto the surface). The concentric ridges indicate the existence of faults in the rocks underlying the maria. These faults arose from isostatic compensation that occurred following the impact, and also from the compression produced by the weight of the lavas. Certain mare wrinkle ridges that run into the surrounding highlands likewise appear to have a deep-seated origin.

Other tectonic structures (but this time caused by tension), are the ordinary faults (the rilles) and the grabens (where a block has subsided between two faults), which

141

Fig. 4.16. *Fracastorius, on the southern border of Mare Nectaris, is an example of a crater that has been invaded by mare material. Photograph by G. Viscardy: 520-mm reflector, TP-2415 film.*

generally appear concentric to the outer edge of basins, such as Rima Hippalus southeast of Mare Humorum (*see* Fig. 4.26), or Rima Menelaus and Rima Plinius on the inner border of Mare Serenitatis. They are also caused by isostatic recovery of the rocks underlying the basins.

The general picture is one where the central part of a basin has subsided beneath the weight of the basalts (giving rise to compression and folding of the surface), accompanied by elevation around the rim (producing expansion features in the rocks). The basic tectonic events that occurred were vertical motions of the underlying rocks.

The expansion structures are not confined to just the edges of basins; many appear on the lunar surface as rectilinear features with various orientations. It may

Fig. 4.17. *Posidonius, on the northeastern border of Mare Serenitatis. Photograph by G. Viscardy: 520-mm reflector.*

be noted, however, that a large number (30 %) of these features have either a radial or tangential orientation to the Imbrium basin, the formation of which appears to have been a major event in the Moon's history.

4.5.1.5 Volcanic features

As mentioned earlier, the lunar maria are volcanic features *par excellence*. The great extent of the flooding inside the basins is explained by the extreme fluidity of the lavas and by the way in which they arose. They originated in fissure eruptions with very high flow-rates that produced lava lakes without any obvious sources, and also without any clear differentiation between flows. Such eruptive rocks are known as flood-basalts and plains-forming basalts.

The history of the eruption of the mare basalts is complex: several types followed one another, with a general tendency for the later basalts to be more viscous than the earlier ones. This is shown by the presence of sinuous rilles, which are either channels cut by turbulent lavas or lava tubes whose roofs have later collapsed. They

Fig. 4.18. *Copernicus and its ejecta under grazing illumination. Photograph by G. Cardoen: 406-mm reflector, TP-2415 film, exposure 0.5 s.*

may attain lengths of 100 km or more, with widths of 1–2 km. Typical examples are Hadley Rille (where Apollo 15 landed) and Schröter's Valley.

Volcanic edifices such as those found on Earth are rare on the Moon. They are represented by rimless craters (calderas), or more commonly, by low domes 5–15 km in diameter, sometimes surmounted by craterlets 1–2 km in diameter. Their low relief (100–200 m maximum) means that they may be observed under low angles of illumination only. Domes generally occur in groups, such as those near Hortensius, Lansberg D and Tobias Mayer, or else as part of volcanic complexes such as the Marius Hills and the Aristarchus plateau where they are found together with sinuous rilles, calderas and lava flow-fronts.

Many structures on the Moon, such as the crater chains associated with grabens (Rima Hyginus, for example), and craters with dark haloes are suspected of being of volcanic origin.

4.5.2 Observation of lunar features

Although every form of lunar formation has been the subject of numerous studies by professionals, this should not prevent amateurs from observing the lunar surface in detail, and from undertaking observations that are somewhat different from those normally made. We shall describe a few examples of work that will allow observers to attempt certain analyses for themselves, and which enable them to move on from being just 'sight-seers' to making a more serious study of the surface of the Moon.

Even if, at the end of the day, the scientific contribution by amateurs remains small, they will still have gained a better understanding of the methods that are applied to the study of the surfaces of planetary bodies, and of the Moon in particular.

As mentioned previously, analysis of the lunar surface may be divided into topographic studies, and investigations of the composition of the surface. The former use classical, visual and photographic methods, while the latter employ the more complex methods of spectrometry and photometry. These two disciplines are, of course, complementary.

4.5.2.1 Topographic studies

Any study of lunar topography might begin by examining the largest features as individual units, involving (for example) a comparison of the morphology of the maria. For each of them the appearance of their borders (the relationship between them and the neighbouring highlands or mare areas) could be determined, together with their general albedo, their degree of cratering, and the development of tectonic structures on their surfaces.

The superimposition of features and the way in which structures intersect (easily seen on the Moon), enable the relative ages of different features to be determined, and give an idea of the relationship between the formations found in any particular area (Fig. 4.19).

This type of study, like those to be described shortly, requires long periods of observation: it is not possible to study the system of rilles in a mare area and determine its albedo at the same time (to give but one example).

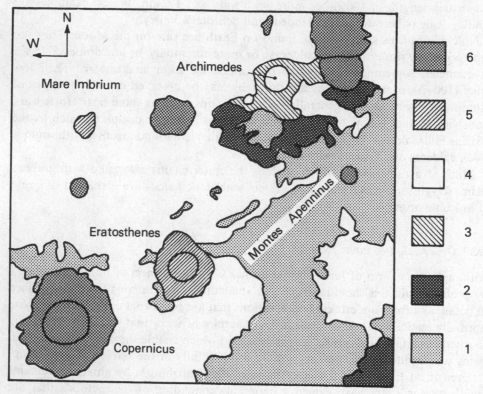

Fig. 4.19. *Simplified geological map of the south-eastern section of Mare Imbrium, as determined from the degree of superimposition and intersection of various features, running from the oldest (1) to the youngest (6).*

A survey of craters might consist of recording their different morphologies as a function of factors like their size and age. Their degree of degradation might be examined by determining for each crater the contribution from meteoritic erosion, flooding by mare materials, infilling by ejecta, and tectonic activity.

A specific, but more difficult, study would be the examination of craters around the edges of impact basins. Some are affected by the faults surrounding the basins (*see* Sect. 4.5.1.4). It is sometimes possible to determine the direction of the movement. This applies, for example, to Hippalus and Posidonius, which lie on the eastern borders of Mare Humorum and Mare Serenitatis respectively.

Examination of the floor of Posidonius shows the effect of one of the faults that divides the crater into eastern and western halves (Fig. 4.17). An initial flooding by lava, which surrounded the central peaks, was followed by a second flooding, which affected just the western half of the floor. This portion had subsided relative to the eastern half in between the two episodes of flooding. Detailed examination of such craters reveals important information about the tectonic movements that occurred in the mare–highland transition regions around the impact basins.

To conclude our discussion of craters, we may mention some typical research

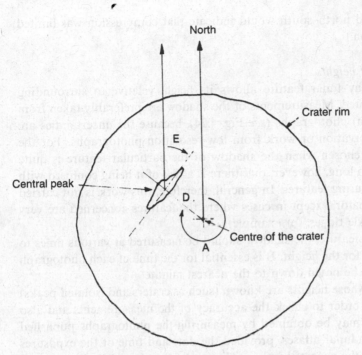

North

Crater rim

E

Central peak

D

Centre of the crater

A

Fig. 4.20. *Determining the position of a central peak. A: azimuth of the peak; E: direction in which the peak is elongated; D: distance from the centre of the crater.*

concerning the central peaks. There are still a number of theories about their origin. They may have been caused by simple isostatic rebound or by slumping of the crater rims towards the centre along parabolic faults created during the impact. Another possibility is that the shockwave from the impact was reflected from deeper layers with sufficient energy to raise the floor of the crater and form a peak. Detailed examination of the relationship between central peaks and other factors, such as the diameter of the craters, will provide data that may be used as constraints on the various models describing the peaks' formation. A way of making such a study would be to make quantitative measurements from photographs. Height of the peaks (relative to the crater floor), position (azimuth and distance from the centre relative to the radius of the crater), and direction of elongation might all be derived in this way (Fig. 4.20). Measurements should be made of a large number of craters, classified as to whether they belong to the maria or to the highlands. The directional information could be presented in the form of rose diagrams.

Study of the lunar maria might take the form of measuring the orientations of mare ridges, which, as we have seen, are compression phenomena occurring in the basalts that fill the basins. If the wrinkle ridges were caused by phenomena (such as contraction) that arose as the basalt cooled, the ridges would show random orientations. On the other hand, any tendency towards a preferential orientation, which would be shown by plotting directions on a rose diagram, would indicate that the compression that occurred was subject to certain constraints. This would enable the average direction in which the forces were applied to be determined. (For

147

example, ridges oriented north–south would indicate that compression was limited to an east–west direction.)

4.5.2.2 Determination of heights

The shadow cast by any lunar feature allows its height relative to surrounding formations to be calculated. Measurements of the shadow are preferably taken from detailed (high-resolution) photographs (*see* Fig. 4.24), because the uncertainties are too great in visual observation or work from low-resolution photographs. For the same reason, a time is chosen when the shadow of the particular feature is quite long. It should not be so long, however, that there is a risk of it being confused with the shadows of neighbouring features. In general, therefore, this work is not carried out at the actual terminator, except in cases where the features concerned are very low (as with mare wrinkle ridges, for example).

Ideally, for any given feature, the shadow should be measured at various times to obtain an average value for the height. It is essential for the time of each photograph or visual observation to be noted down to the nearest minute.

Initially, formations whose heights are known (such as craters and isolated peaks) should be measured, in order to check the accuracy of the measurements and also the procedure. Practice may be obtained by measuring the photographs published in various journals or in lunar atlases, provided the date and time of the exposures are given, together with the geographical coordinates of the observing site.

Figure 4.21 shows schematically how measurements are carried out. The altitude (a) of the Sun at the point P is equal to the difference between the longitude of this point L_p and that of the terminator L_T:

$$a = |\, L_T - L_p \,|. \tag{4.1}$$

The selenographic longitude of the terminator is given by the selenographic colongitude of the Sun (S) if it is the morning terminator, and ($S + 180°$) if it is the evening terminator. The selenographic longitude of the point P may be measured from a chart of the Moon. In the first case we have:

$$a = |\, S - L_p \,|. \tag{4.2}$$

The height (h) of the point P is then given by:

$$\tan a = h/L. \tag{4.3}$$

The length of the shadow (L) is measured on the photograph, the scale of which has been established by using a map or an atlas. If, for example, we want to determine the height of the rim of a crater, we might have:

$$L = D/l/d, \tag{4.4}$$

where D = diameter of the crater in km; d = size of the latter on the photograph, in mm, and l = length of the shadow in mm.

Such a calculation is only approximate, because it does not take account of various important factors, such as the position of the Earth relative to the point being measured (Fig. 4.22), or the selenographic latitude of the Sun. Two methods

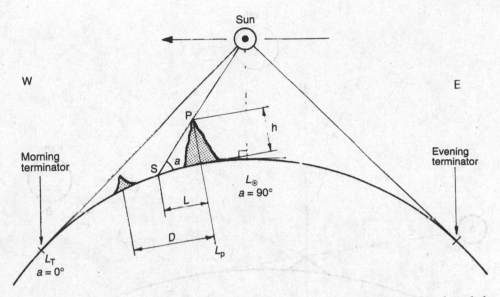

Fig. 4.21. *Determining the height of a feature on the Moon. h is the height of the point P; L is the length of the shadow; S is the tip of the shadow; a is the altitude of the Sun at point P; D is the diameter of the crater. [Note that this diagram is drawn as if seen from the south Pole of the Moon. – Trans.]*

for calculating heights follow: the first (A) is quick, while the second (B) is longer and more rigorous, giving a more accurate result.

Method A The following parameters appear in the calculation:

- λ and β, the selenographic longitude and latitude of the Earth (given in ephemerides)
- $l_\odot$ and $b_\odot$, the selenographic longitude and latitude of the Sun (the coordinates of the point on the Moon where the Sun is at the zenith). These elements are given by the equations:

$$l_\odot = L_\odot + 180 - L_m,$$

$$b_\odot = -(\pi_\odot U/\pi) + I\sin(L_\odot - \Omega), \qquad (4.5)$$

where:

L_m	=	mean longitude of the Moon (Sect. 4.3.2.5);
$L_\odot$	=	longitude of the Sun (ecliptic coordinates);
$\pi_\odot$	=	8.794 arc − seconds, the solar parallax;
π	=	3422.62 arc − seconds, the parallax of the Moon;
U	=	longitude of the Moon (ecliptic coordinates);
I	=	1.534°, the inclination of the lunar equator
		relative to the ecliptic;
Ω	=	longitude of the Moon's ascending node.

149

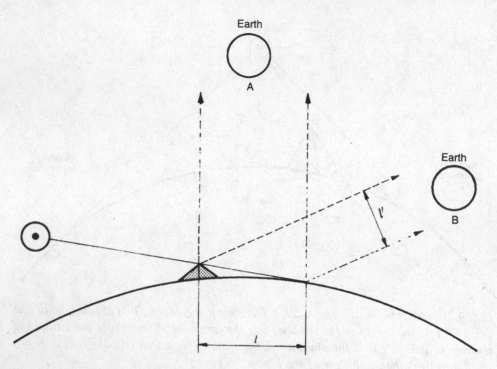

Fig. 4.22. *The effect of the Earth's position on the apparent length of a shadow. At position B, the shadow appears shorter than at A: $l' < l$.*

- λ' and β', the selenographic longitude and latitude of the point for which one wishes to calculate the height (measured from a map of the Moon)
- L', the length of the shadow expressed in lunar radii. From (4.4) we find:

$$L' = L/\rho_M,\tag{4.6}$$

where $\rho_M = 1734\,\mathrm{km}$.

The altitude of the Sun (a) above the point measured is given by:

$$\sin a = \sin \beta' \sin b_\odot + \cos \beta' \cos b_\odot \cos(\lambda' - l_\odot).\tag{4.7}$$

We may replace $l_\odot$ by S the selenographic colongitude of the Sun (frequently given in ephemerides), using the equation $S = 90° - l_\odot$. We thus have:

$$\sin a = \sin \beta' \sin b_\odot + \cos \beta' \cos b_\odot \sin(S + \lambda').\tag{4.8}$$

The height (h') expressed in lunar radii, is:

$$h' = (L' \sin a / \sin \theta) - (L'^2 \cos^2 a / 2 \sin^2 \theta),\tag{4.9}$$

where θ is approximately the angle between the Earth and the Sun as seen from the centre of the Moon:

$$\cos \theta = \sin \beta \sin b_\odot + \cos \beta \cos b_\odot \cos(\lambda - l_\odot).\tag{4.10}$$

As before, we may eliminate $l_\odot$ and incorporate S, so the term $[\cos(\lambda - l_\odot)]$ may be replaced by $[\sin(S + \lambda)]$. The final result is the height in kilometres:

$$h = \rho_M h'. \tag{4.11}$$

Method B Figure 4.23 shows the various geometrical parameters of the shadow (the position and distance of the observer with respect to the point being measured, the curvature of the Moon, etc.) that are required in this method. Again we take the calculation of the height of a point P (λ', β') on the rim of a crater:

- The geocentric angle γ is given by:

$$\cos\gamma = \sin\delta \sin\phi + \cos\delta \cos\phi \cos(\alpha - \mathrm{LST}). \tag{4.12}$$

  where: α, δ are the equatorial coordinates of the Moon; ϕ, latitude of the observing site; LST, local sidereal time; $(\alpha - \mathrm{LST})$ is the hour angle.
- The angle η is:

$$\cos\eta = \sin\beta \sin\beta_S + \cos\beta \cos\beta_S \cos(\lambda - \lambda_S). \tag{4.13}$$

  where: λ, β have been defined previously (Method A); λ_S, β_S are the selenographic longitude and latitude of the tip of the shadow S (measured from a map).
- The distance of the tip of the shadow (S) from the observer is R_S.
  If $\rho_\oplus$ is the radius of the Earth at the observer's position,

$$\rho_\oplus = R \sin\pi(1 - 0.002\,35 \sin^2\phi), \tag{4.14}$$

  whence:

$$R_S = R[1 - (\rho_\oplus \cos\gamma/R) - (\rho_M \cos\eta/R)], \tag{4.15}$$

  where:

  R = $384\,400$ km, the average distance of the Moon;
  π = 3422.62 arc-seconds, the lunar parallax;
  ρ_M = 1734 km, the Moon's radius.

- The angular length of the shadow (N) seen by the observer.
  If ϕ is the apparent diameter of the Moon in degrees; D and L are respectively the diameter of the Moon and the length of the shadow (in mm) measured from the photograph; we have:

$$N = \phi L/D. \tag{4.16}$$

- The altitude of the Sun (a) above the lunar horizon at point P (λ', β'):

$$\sin a = \sin\beta' \sin b_\odot + \cos\beta' \cos b_\odot \cos(\lambda' - l_\odot), \tag{4.17}$$

  (*see* (4.7), Method A).
- The angle between the direction of the Sun and the observer is θ, giving (4.10), previously derived:

$$\cos\theta = \sin\beta \sin b_\odot + \cos\beta \cos b_\odot \cos(\lambda - l_\odot). \tag{4.18}$$

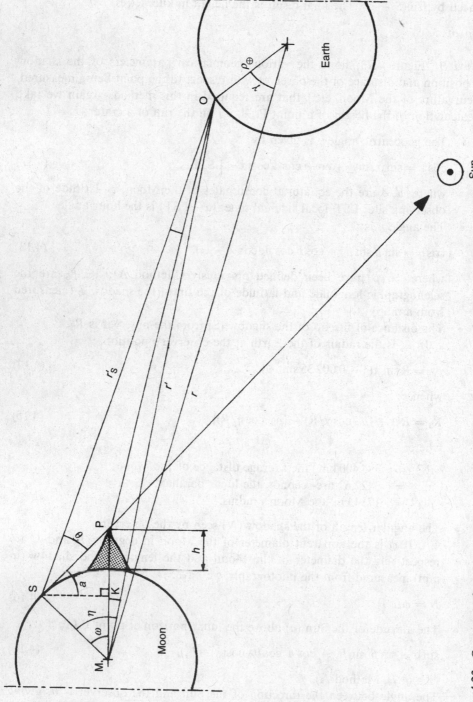

Fig. 4.23. *Geometrical construction for the accurate determination of heights.* (*See Method B.*)

- The selenographic angle of the shadow ω, is given by:

$$\sin \omega = R_S \sin N \cos a / \rho_M \sin(\theta + N)$$

or:

$$\cos \omega = \sin \beta' \sin \beta_S + \cos \beta' \cos \beta_S \cos(\lambda' - \lambda_S). \tag{4.19}$$

- The height h (or more accurately, the ratio h/ρ_M, where ρ_M = radius of the Moon), is given by:

$$h/\rho_M = [\cos(\omega - a)/ \cos a] - 1 = \sin \omega \, \tan a - 2 \sin^2 \omega/2. \tag{4.20}$$

The accuracy of the final result strongly depends on the degree of care that is taken to measure the length of the shadow on the photograph and the selenographic coordinates of the various points on the Moon. To reduce errors in the measurement, big enlargements are used. It should be possible to determine the length of the shadow to within 0.1 mm, using an accurately divided rule and a magnifying glass. It will be found, however, that it is sometimes difficult to determine where a shadow starts and ends on a photograph. Depending on the observer, there is often a tendency to over-estimate or under-estimate the length of the shadow.

Amateurs with bifilar micrometers can measure the lengths of shadows visually. If a and b are respectively the values of the diameter of the crater and the length of the shadow, from (4.4) we have: $L = D \, b/a$. Another (but far less accurate) visual method is to measure the time taken by the lunar disk (t) and the shadow (t') to cross the field of the eyepiece. To obtain an accurate reference point for this it is best to use a graticule. Again, from (4.4), the length of the shadow is given by $L = D \, t'/t$.

4.5.2.3 The composition of the surface

Studies of the morphology of lunar features may be complemented by studies of the composition of the surface rocks, only a simple approach to which will be described here.

Sunlight reflected from the lunar surface has certain characteristics that are governed by the physical, mineralogical and chemical properties of the soil. It therefore provides a means of studying these from a distance. Differences in the composition of the lunar surface may be established by several types of measurement, some of which merely give general information (albedo and colour measurements, for example). Others (spectrophotometric measurements) allow more accurate estimates – and even quantitative measurements – of the nature of some of the elements or minerals that make up the soil. This type of study has to be undertaken when the phase is close to zero, to avoid effects caused by topography (mainly shadows).

Differences in albedo and colour Study of the differences in composition of the lunar surface might begin with an investigation of differences in albedo. With relatively homogeneous areas like the maria, these differences may logically be taken to indicate differences in the composition of the volcanic material. It is

Fig. 4.24. *Tycho with its central peak, the height of which may be calculated from the length of the shadow. Photography by G. Viscardy: 520-mm reflector, exposure 1 s on Kodak TP-2415.*

therefore possible to make a rough distinction between the various basalt units that make up the maria.

It must be remembered, however, that variations in albedo are very sensitive to differences in the physical nature of the surface. In the case of the maria, such differences may indicate materials that are not of volcanic origin (such as ejecta, crater rays, etc.). In interpreting them, it is essential to pay close attention to the topographic detail of the region being examined.

Apart from a few rare cases like Mare Serenitatis, where the difference in albedo between the centre and the southern border is clearly visible, indicating two distinct basalt units, the boundaries are generally too indistinct and the contrasts too low for them to be estimated visually and reproduced in drawings, or to be assigned values on an intensity scale.

Photography seems to be the easiest way of determining differences in albedo over fairly large areas of the Moon, either by standard photographic techniques (by giving different exposure times in the enlarger or by using different grades of paper), or by densitometry measurements of the image.

Photometric measurements are the only ones that allow accurate studies of differences in albedo. This is particularly the case with CCD imagery, which may be used for examining variations in the 'continuum' over a particular area of the Moon (*see* Sect. 4.4.4). Albedo is determined somewhere within the range 550–620 nm (most generally at 560 nm).

Differences in colour give more comprehensive information about variations in the composition of the soil than do differences in albedo. They should, nevertheless, be interpreted with just the same amount of care, because they are also subject to other factors such as the surface age of the materials (i.e., the time that they have been exposed to the lunar environment). The differences are most easily detected in the maria, and in certain cases they do allow the different basalt units to be distinguished, thus enabling the different phases in the filling of the basins to be determined.

A photographic composite, consisting of a positive made at 600 nm (in the red) and a negative made at 350 or 400 nm (in the blue), for example, makes it relatively easy to distinguish between red material (which appears light) and bluer materials (which appear dark). Naturally, the film should have a spectral sensitivity appropriate to the wavelengths being used – Kodak TP-2415 is appropriate in this particular example.

The difference between the centre and the edge of Mare Serenitatis that we mentioned earlier appears very clearly on such a composite photograph. A very similar case is Sinus Iridum, which appears to be filled by a much redder material than that covering the neighbouring area of Oceanus Procellarum, where the relationship with the albedo is far less evident.

Photometry, using filters with various bandwidths, would give a lot more accurate data that could be used for this sort of work. CCD imagery would allow more rigorous analysis of the differences between various images than is possible with photography. It would also enable quantitative measurements of colour indices to be made.

Measurements of colour differences on the Moon use either the classical UBVRI

Fig. 4.25. *Rima Hyginus* (centre) *and Rima Triesnecker, farther to the north (towards the bottom of the photograph), close to the crater of the same name. Photograph by C. Arsidi: 305-mm telescope, TP-2415 film.*

Fig. 4.26. *The area around Bullialdus (left). The concentric faults around Mare Humorum may be seen near the crater Hippalus at upper right. A lunar dome is visible to the right of the ghost crater Kies (above Bullialdus). Photograph by C. Arsidi: 305-mm telescope, TP-2415 film.*

system, or else other wavelength regions, such as the bands 400–550 nm, 470–670 nm, 470–790 nm, or 670–790 nm.

Spectrophotometry measurements   Spectral analysis of sunlight reflected from the Moon, i.e., the lunar reflection spectrum, reveals various spectral signatures (such as absorption bands), which are dependent on the mineralogy and chemistry of the soil. The lunar rocks returned by the Apollo and Luna missions have been analyzed and their reflection spectra obtained enabling correlations to be established with material on the lunar surface (Pieters, 1978).

These methods have established that the absorption bands in the infrared between 900 and 2500 nm indicate the presence of certain silicate minerals, such as pyroxene, olivine, and pagioclase, which are some of the major components in the various groups of lunar rocks that have been examined.

Another spectral signature, that of the ratio between the brightness in the ultraviolet (at 400 nm) and that in the visible (at 560 nm), is indicative of the amount of titanium in the soil, which may therefore be expressed in quantitative terms. This element plays a significant part in classifying lunar mare basalts.

Values obtained for certain spectral characteristics (such as the intensity – i.e., depth – of the absorption bands at 1000 and 2000 nm, or the value of the UV/visible intensity ratio), have allowed lunar reflectance spectra to be classified into a number of types, each of which is representative of one form of geological unit on the Moon. Among these spectral types there are several that correspond to units that were sampled by the lunar missions. (In such cases, the telescopic reflectance spectra are identical with those obtained in the laboratory.) Others represent types of rocks that have not yet been sampled, but the general characteristics of which are known. This was how, before the rocks obtained by Apollo 17 were analysed in the laboratory, their general geological characteristics were known from studies of reflectance spectra of the proposed landing site. This shows how spectrophotometric data may be applied to sites on the Moon that have not been visited by astronauts or automatic probes.

At present, in the absence of any lunar exploration programme following the Apollo missions, measurements made from a distance are the only way of continuing the study of lunar geology. Numerous teams of professional scientists are currently working in this field. Although the techniques that have to be employed are fairly sophisticated, this area of research is not completely beyond the reach of amateurs, who may therefore certainly play a part.

Spectrophotometric data may be obtained with a standard photometer and a series of interference filters (filters with narrow bandwidths of the order of 15 nm). The spectral reflectance curve is built up in sections, but it generally has a very low resolution, because of the limited number of filters that are normally available.

A more appealing method is to use a standard spectrograph, with a slit and a dispersing element, obtaining the spectrum of a part of the Moon, and allowing this to fall on a linear CCD. The curve obtained by this method has a high resolution, which allows the variations in strength and the absorption bands to be studied in detail. Measurement of the width and depth of a band gives information about the relative abundance of the mineral involved. Any asymmetry indicates a composite band, and thus the presence of a second mineral.

A spectral measurement should involve only a small area of the Moon's surface (say some 10 km in diameter, or 20 km at maximum) so that data from various points are not combined. This requires the use of long focal lengths. Methods of obtaining and reducing spectrophotometric measurements are the same as those described for photometric measurements in Sect. 4.4.4 (i.e., repeated observations, short intervals between the measurements, and subtraction of the dark current).

Studies of reflectance spectra involve obtaining relative spectra with respect to the standard area MS-2 in Mare Serenitatis:

relative spectrum of area A =

   spectrum of area A/spectrum of standard area MS-2.

This procedure eliminates instrumental and atmospheric effects (particularly atmospheric lines) and accentuates the minor differences that exist between lunar spectra. These do not, in fact, vary by more than about 10–20% over the whole Moon. For some lunar formations, such as the maria, variations often do not exceed a few per cent, which is why a high degree of accuracy (around 1%) is required from the instrumentation. The interpretation of reflectance spectra is usually difficult because several factors may be involved. Taking the absorption bands, for example, their intensity may be decreased by the presence of opaque elements in the soil (causing the spectral contrast to decrease), which sometimes makes it difficult to distinguish them from the continuum or to interpret. An idea of the significance of this effect may be gained by obtaining the UV/visible ratio. Similarly, because the principal absorption bands are confined to a fairly restricted spectral region (between 900 to 2500 nm), they often appear composite and difficult to interpret. (One mineral masks another.)

4.6 Luna Incognita

There are some lunar observation programmes where amateurs can play an important part. Because of their numbers and availability, amateur observers are able to monitor the sky regularly and thoroughly. This is one way in which they may help with the cartography of the southern and southwestern limb of the Moon (Fig. 4.27).

First sight of this area through a telescope is likely to leave the observer rather perplexed, if not totally discouraged! Early selenographers called this jumble of hills, peaks, crests and valleys, the Doerfel and Leibnitz Mountains. This gives an indication of the complexity of this southern limb of the Moon. Yet it was by observing this very area that Lucien Rudeaux contradicted the generally held belief that lunar relief is abrupt and extreme. He showed that, on the contrary, the tops of lunar mountains were rounded and that it was solely the shadows that they cast that gave the impression that they were rocky peaks rising abruptly from lava plains.

If a chart of the southern limb of the Moon is examined, it will be seen that a large part of Luna Incognita lies well beyond the 90° meridian, on the hidden side.

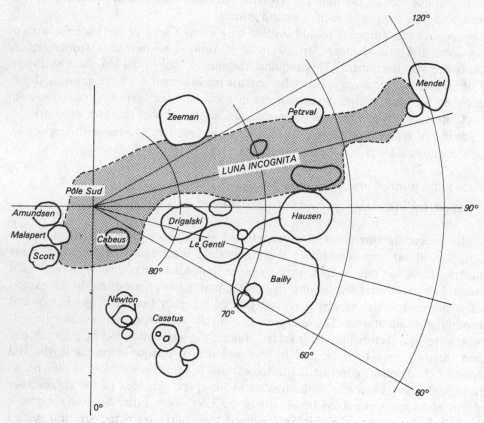

Fig. 4.27. *Map showing the region of Luna Incognita (hatched zone). Drawing by M. Legrand from* l'Atlas Universalis.

The actual area of this region is about 270 000 km² and it has yet to be covered by photography from space. Both the Lunar Orbiter and the Apollo missions were primarily concerned with studying the equatorial regions. Although the Soviet Luna and Zond probes passed over high latitudes, the lighting conditions were not suitable for them to obtain usable photographs.

This is why John Westfall, a geographer at the University of San Francisco and secretary of the American Association of Lunar and Planetary Observers (ALPO), undertook the study of this region in 1972. [A chart, derived from ALPO observations, has recently (1991) been published (Westfall, 1991). Additional checking of this region is still required. – Trans.]

4.6.1 Visibility

As we have just said, Luna Incognita lies on the southern and southwestern limbs of the Moon; this means that observations may only take place at certain specific times during a lunation. These times are when libration in both latitude and longitude are favourable. In latitude, so that the South Pole is turned towards us, and in longitude, so that the western limb (IAU) – with Mare Humorum and Oceanus Procellarum – is turned towards the Earth. Ephemerides normally give the values of libration in longitude λ, and in latitude β. These values apply to both selenographic longitude and latitude and to the geocentric coordinates of the centre of the apparent disk at the time of observation. Positive selenographic latitude is towards the North (towards Mare Serenitatis) and positive selenographic longitude towards the West (as seen from Earth!), i.e., towards Mare Crisium. (Note that here East and West do not correspond to the IAU convention, so beware of confusion!)

Anyone wanting to draw up their own ephemerides for the observation of Luna Incognita should determine the dates of maximum negative libration in latitude and longitude, and in particular check when the dates of both agree. Predictions for observation have also been published each year in *The Strolling Astronomer*, the ALPO journal.

An important factor that must not be neglected in observing Luna Incognita is precession of the lunar orbit, which causes a long-term variation in the values of libration. For example, 1987 was the last year of a three-year period that allowed observation of the northern part of Luna Incognita under the most favourable conditions of libration and lighting. It will actually be necessary to wait six years to see this particular area again.

4.6.2 Observation of Luna Incognita

For both physical reasons and for convenience, Luna Incognita is divided into three areas:

- Zone A includes the South Pole (Cabeus, Malapert);
- Zone B includes the southwestern limb from Drygalski to Hausen;
- Zone C, the northernmost, runs between Hausen and Mendel.

Before tackling Luna Incognita, it is best to have acquired plenty of practice on easier regions. We mentioned earlier how an observer is likely to be perplexed by the chaotic nature of the region around the Moon's South Pole, primarily because of the grazing-incidence lighting that always prevails. However, anyone with a 100-mm refractor or a 150-mm reflector (or larger, of course) can consider doing effective and useful work on this region. As with all astronomical observations, experience and patience are indispensable to anyone who hopes to take part in any active programme.

The conditions under which Luna Incognita is visible mean that its observation is possible about one week every month, and that week shifts through the month over the course the year. Because a high negative libration means that the Moon is above the ecliptic, and because Luna Incognita is close to the South Pole, this region is always high in the sky when observed from the northern hemisphere. For once, northern observers are favoured!

'Artistic' drawings of Luna Incognita are more or less impossible, given the complexity of the region, and also because we are trying to carry out serious cartography. The ALPO team prepared masters that help greatly in locating features correctly (Fig. 4.28). These masters show Luna Incognita under various librations in latitude and longitude. Observers have to fill in this basic outline, noting down as accurately as possible all the detail visible with their equipment under the prevailing conditions. As with every observation, the times of the beginning and end of the observation should be noted to the nearest minute, together with the seeing conditions, and details of the telescope. One additional factor requires particular attention: this is the selenographic colongitude of the Sun. We have seen that this angle may be defined, to an accuracy of about one degree, as the longitude of the morning terminator. This is visible before Full Moon, whereas the evening terminator is visible when the Moon is waning. This information is essential, as we shall see later, when reducing results.

Despite the fact that it is subjective, drawing is the ideal means of recording fine detail (Fig. 4.29). Under good observing conditions a 200-mm reflector allows fine topographical detail to be seen, because it has a resolving power of rather better than 1800 m on the Moon. (This applies in a direction perpendicular to the line of sight. Close to the limb, the resolution along the line of sight is far less, because of the low angle of incidence.)

It is, nevertheless, now possible for amateurs to obtain high-quality photographs of the Moon (Fig. 4.30). There is no need to stress that such records are valuable, both because they are objective and because they can be examined later under far more comfortable conditions than those prevailing at the eyepiece. Recording Luna Incognita can usefully combine both techniques, and it would be a good idea to take two sets of photographs, before and after making the drawings. The fine detail on the one will complement the objectivity of the others.

4.6.3 Reducing the observations

Undoubtedly many observers will be content just to collect data, but most amateurs would probably like to carry out some form of reduction of their observations. It is here that the most significant problems arise. Several methods may be used, and their reliability depends, to a large extent, on the way in which they try to represent the physical features present on the surface of the Moon by a map. Some idea of the problems facing cartographers may make this easier to appreciate.

Once the rectangular coordinates of any particular point on a drawing or photograph are known, a map is constructed by plotting that position on a reference surface that is specifically related to a coordinate system established for the body concerned (in this case, the Moon). It is well-known, however, that representation of a sphere by a plane surface is accompanied by unavoidable distortion of any region that departs, by however small an amount, from the point or line of contact between the projection plane and the reference sphere. Various methods of projection try to limit this distortion.

Orthographic projection is the form most widely used for lunar cartography. It assumes that the disk is viewed 'face on' as in a telescope. The main inconvenience is that any object away from the centre of the disk is deformed radially, whereas its size in a direction perpendicular to the radius that joins it to the centre of the disk is correctly represented. This projection is neither 'conformal' (representing the true

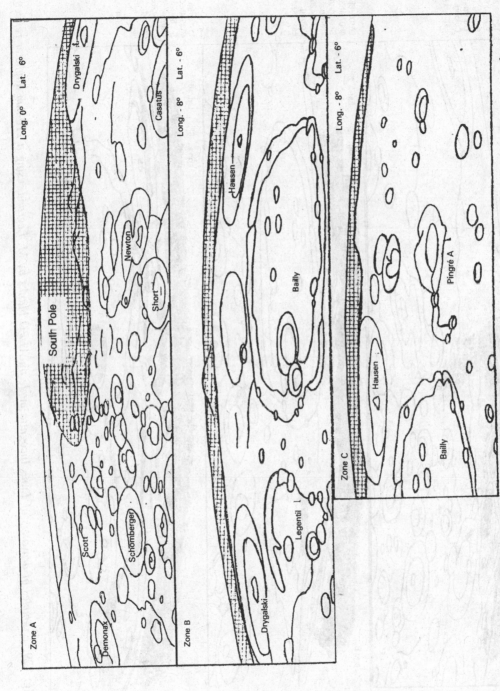

Fig. 4.28. *The area of Luna Incognita has been shaded on these examples of the ALPO blanks, so that it can be seen easily. The blanks are those showing the most favourable libration.*

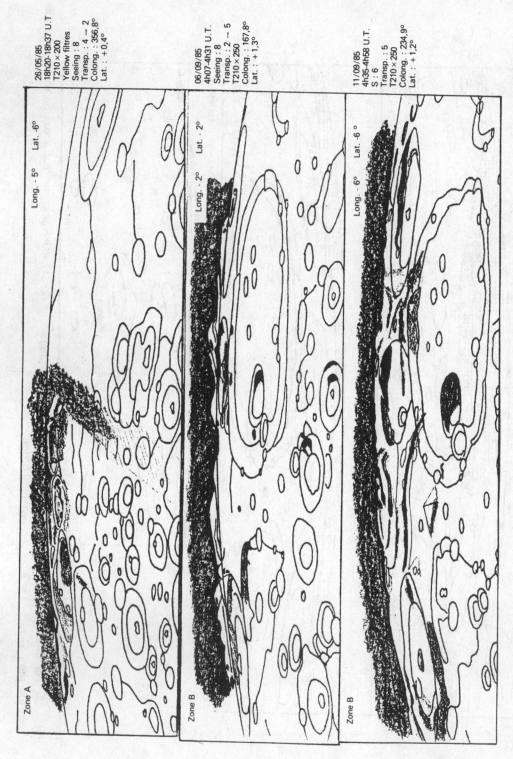

26/05/85
18h20-18h37 U.T
T210 × 200
Yellow filtres
Seeing : 8
Transp. : 4 — 2
Colong. : 356,8°
Lat. : + 0,4°

Long. - 5°   Lat. -6°

Zone A

06/09/85
4h07-4h31 U.T.
Seeing : 8
Transp. : 2 — 5
T210 × 250
Colong. : 167,8°
Lat. : + 1,3°

Long. - 2°   Lat. - 2°

Zone B

11/09/85
4h35-4h58 U.T.
S : 6
Transp. : 5
T210 × 250
Colong. : 234,9°
Lat. : + 1,2°

Long. - 6°   Lat. -6°

Zone B

Fig. 4.29. *Examples of drawings of Luna Incognita made on ALPO blanks by M. Legrand, using a 210-mm diameter reflector.*

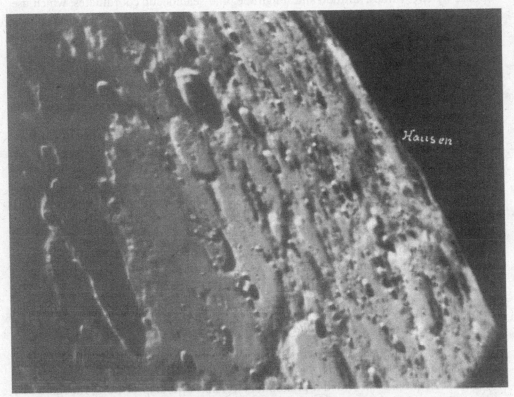

Fig. 4.30. *Behind Bailly and Pingré lies the crater Hausen, on the edge of zones B and C of Luna Incognita. G. Viscardy: 520-mm reflector.*

shape of the object), nor 'equivalent' (where the ratio of areas is the same as the ratio between the scales of the feature and map). It is easy to see that this projection is only suitable for a smooth object: any elevation above the map's reference level appears with a displacement that becomes greater, the closer the point is to the limb.

In Lambert's projection, the features are projected onto the surface of a cone, where the line of contact with the reference surface is the arc of a great circle. Distortions are far less than in the previous case, and the correct angular relationships are preserved. It is conformal.

In conformal azimuthal projection, or stereographic projection, radial and tangential deformations are equal, preserving the shape of even small areas, and all directions are correct. So a circle on the surface always appears as a circle on the map.

Finally, 'rectified' cartography allows the actual physical topography to be represented to a high degree of accuracy over a small area of the surface. This is what ALPO have attempted to obtain with their maps of Luna Incognita.

One final point about techniques: it is important to distinguish between selenographic coordinates, which define the position of a point on an ideal reference sphere by two values, latitude and longitude, and selenodetic coordinates, which use an additional coordinate, altitude relative to the reference sphere. We have seen that the choice of a particular projection may affect the values of these coordinates.

Our records (whether drawings or photographs) may now be used to project the points of interest onto a plane – or better, onto a sphere, which may then be re-photographed to restore the real appearance. (We have, however, not yet tried this method.)

Determination of coordinates may also be made by setting up a series of primary control points; the positions of these points are known to 10^{-3} degrees of arc in longitude and latitude, and 10^{-2} km in altitude. These allow the scale of the drawing or photography to be established and, from this, secondary control points may be derived. Pairs of photographs under different librations but with identical solar colongitudes would provide a stereoscopic view of the region. The greater the libration, the greater the accuracy obtained.

Digitization of high-resolution photographs, and appropriately powerful software, would undoubtedly improve the accuracy of positions, even thought these may already have been derived using standard spherical trigonometry.

4.6.4 Amateurs and Luna Incognita

The work done by the amateur observers involved in the ALPO programme follows the outline that we have just described. Current work by the team under John Westfall has already provided a preliminary version of a map of Luna Incognita. In his 1985 report, John Westfall gave the following details: 773 positions determined from 2078 measurements; 307 crater diameters obtained (850 measurements); and 994 altitude measurements of 404 points. Undoubtedly many more measurements have been made since then and the accuracy has probably improved over the 7 km mentioned in the report. But 7 km represents 2.8 mm on a map at a scale of 1:2 500 000, i.e., with a lunar diameter of 695 mm. [A more recent report (Westfall, 1991) quotes 1893 observations: 1509 photographs and 384 drawings. – Trans.]

As we have said, any amateurs with reflectors at least 150 mm in diameter or 100-mm refractors, may make a useful contribution by obtaining data, which they can then either reduce themselves – but as we have seen this is quite a complicated procedure – or forward to ALPO. It is important to finish this work within the next few years, because the USSR and France are considering a joint project to study the poles of the Moon from a satellite in polar orbit in the 1990s. In any case, it will be interesting to compare the photographs taken by such a spaceprobe with maps drawn up from ground-based observations.

One theory that reappears from time to time is that remnants of the Moon's (hypothetical) early atmosphere may be found in permanently shadowed hollows near the South Pole. It is true to say that so far not the slightest trace of any atmosphere or volcanic activity has ever been noticed in this area of the Moon.

4.7 Transient phenomena

It is always a problem to know how to deal with something that is the subject of controversy. This is particularly the case with the temporary anomalies (the apparent increase or decrease in albedo, or change of colour) that have been observed on the surface of the Moon, and which are known by the term TLP (Transient Lunar Phenomena) – or LTP (Lunar Transient Phenomena) in North America.

The fact that the transient phenomena are, with rare exceptions, purely visual, and that they are relatively diffuse and indistinct, implies that many reports are actually observations of atmospheric or instrumental effects, or have physiological causes (i.e., they are optical illusions). The fact that the phenomena are repeatedly observed at specific points on the Moon, sometimes by several independent observers, working under different conditions, means that these reports cannot be dismissed out of hand.

It has yet to be formally established that these events do originate on the surface of the Moon, and this will only be possible through more accurate, and more homogeneous, visual observations, and in particular, through confirmation by other methods such as photography, photometry or spectroscopy.

From a scientific point of view, the study of transient events is only beginning. Important data may be secured with very limited means, and these could lead to interesting results for lunar physics or for observation in general, depending on whether the phenomena are of lunar origin or not. If the former is true, then any observations made with the additional methods just mentioned would help to define constraints on the many models that have been advanced to explain the phenomena. Amateurs can undoubtedly play an important part in these investigations.

4.7.1 The history of transient events

Numerous reports of transient phenomena on the Moon are known from past observers. They are, unfortunately, generally too vague and difficult to confirm to be taken into account in modern studies. The first detailed observations date from the end of the 19th century. Some were made by professional astronomers (such as Barnard, for example). It is only from about the 1950s that this phenomenon really attracted the attention of amateurs.

On 1958 November 3, the Soviet astronomer N. A. Kozyrev, who was studying luminescence on the surface of the Moon, using the 1.25-m telescope at the Crimean Astrophysical Observatory, observed a reddish coloration affecting the central peak of Alphonsus. His observation lasted for about thirty minutes. During that time he was able to take a spectrogram, the most striking feature of which was the spectrum of gaseous carbon. Despite this observation, and studies of the luminescence of lunar soil, which have occasionally revealed unusual phenomena, some coinciding with areas known to amateurs as the location of such events, professional astronomers have never undertaken a systematic study of the sites. Kozyrev's spectrum still remains the only tangible evidence suggesting a lunar origin for these transient events.

Observation of transient phenomena attained greatest popularity among amateurs early in the 1960s. Two astronomical societies the BAA, beginning in 1965, and

ALPO, beginning in 1975, undertook the collation of all past and future observations of transient phenomena on the Moon, and created networks of observers. Their aim was to develop methods of observation that would encourage the gathering of data in a standardized form, thus facilitating their classification and analysis.

The results are regularly published in the journals published by the societies concerned: the *Journal* of the BAA, and the *Strolling Astronomer* (ALPO). Lists of the reported events are regularly revised by these groups. For each observation they give the date, location on the Moon, description, and observers.

The methods suggested are primarily based on visual methods, which allow a large number of observers to participate in the programmes, and minimize the limitations imposed by equipment.

4.7.2 *Characteristic features of transient phenomena*

The transient phenomena observed to date have various characteristics and these may be classified as follows:

- Brightening of an area on the Moon with respect to its immediate surroundings. This apparent increase in albedo may be gradual or rapid, generally taking 0.25–5 seconds. This category also includes starlike points of light: the exact timings of these are only rarely obtained when they occur on the sunlit portion of the Moon.
- Darkening of an area with respect to its surroundings. Again this may be more or less gradual.
- Local, coloured glows. These are often described as being reddish or bluish, more rarely yellow, orange, greenish or brown. The tints are generally weak and unsaturated.
- Obscurations, where an area that normally shows fine detail becomes indistinct, compared with its surroundings, which remain sharply defined.

Some events may combine several of these effects. The average size of a transient on the Moon is 5–15 km. Their duration is about 10–20 minutes, but a phenomenon may persist intermittently for several hours. However, no permanent changes have ever been detected at the site of an event.

4.7.3 *The sites of transient events*

From the 1400 observations reported, about 100 sites appear to be particularly prone to these anomalous events. Their distribution over the face of the Moon is not random: they show a marked preference for the edges of maria (this is particularly striking on the northern edge of Mare Imbrium, the southern border of Mare Serenitatis, and around Mare Crisium), and also certain craters. Very few transient phenomena are reported in the highlands. Their appearance may therefore be linked to zones of weakness in the lunar crust.

Among the 100 most likely sites, 12 make up 80% of the total number of observations, and just 6 contribute 60 %. The crater Aristarchus alone amounts to 33 % of the observations.

The best sites to monitor include Aristarchus, Alphonsus, Bullialdus, Copernicus, Gassendi, Plato, Proclus, Theophilus (Fig. 4.31), Menelaus, Herodotus, and Eratosthenes. Pages 447 and 448 of Georges Viscardy's *Atlas-Guide Photographique de la Lune* give a fairly complete list of sites on the Moon that should be monitored, as well as a description of the principal events that have been observed.

4.7.4 Possible causes

In the absence of sufficiently numerous and accurate observational data that would serve to limit the possible models, there are numerous possible physical mechanisms that are capable of accounting for the observed features of these events. We shall mention just a few of the proposed theories, including some of the most interesting, which invoke an internal origin.

A process that would account for some of the events that occur on the illuminated portion of the Moon was proposed by Mills (Mills, 1980). The emission of gas from the surface would levitate fine particles of the soil, and put microscopic grains of the regolith into ballistic trajectories. This would produce a very tenuous cloud of dust that would scatter sunlight. Its appearance (brightness, opacity, and colour) would be a function of the size, density and nature of the particles involved. Scattering from microscopic dust particles amounting to between 1 and $20 \, kg/km^2$ would suffice to account for the observed phenomena.

Processes that produce and emit light are, however, required to account for the observations of events seen on the dark side of the Moon, as well as some of the brightenings observed on the illuminated side. As far as the latter are concerned, luminescence or thermoluminescent phenomena caused by solar radiation must be discounted, because there is a lack of materials that are capable of producing them, and also because of the low flux involved.

The same applies to incandescence from the possible emission of lava over an area too small to be detectable from Earth. Certainly, no emission of fresh lava has ever been detected on the Moon, and in any case, if such an event were to occur, an infrared source would persist at the site of the transient event. No such occurrence has ever been detected.

One possible process that could account for the emission of light is a phenomenon similar to that described earlier. The friction within a heterogeneous cloud of dust (i.e., one containing particles of different sizes), that is levitated by a sudden expansion of gas may, under certain circumstances, produce electrical discharges (by what is known as the triboelectric effect).

In the absence of spectral data, the nature of the gases required by these theories must remain speculative. Cosmic abundances and the spectrum obtained by Kozyrev suggest that hydrogen is a significant component, along with helium. The reddish colour associated with the events where emission has been thought to occur would support this hypothesis.

The 'periodic' release of gas could take place from deep fractures, which are indeed found at the junction of maria and the highlands. Transient phenomena

Fig. 4.31. *Theophilus, a crater that is one of the most active sites of transient phenomena. Photograph by C. Arsidi: 305-mm telescope, TP-2415 film.*

might therefore have a deep origin, and would mainly be associated with tectonic structures that were involved in volcanism. The mechanism behind the emission of gas remains obscure. For a long time it has been known that transient phenomena occur preferentially around perigee and apogee. Emissions might therefore be related to the tidal effects between the Earth and the Moon. In fact it seems that the agreement is poor. There is an equally poor correlation with the zones of seismic activity found by the Apollo programme with just three coincidences out of the 100 transient sites. Even those sites are of minor importance as far as the number of events is concerned.

As for external causes, it may be noted that certain correlations have been observed between changes in brightness on the Moon and specific events. These include sunrise at the site concerned; passage through the Earth's magnetotail (at Full Moon), or through the shockwave and the magnetopause; and even terrestrial magnetic storms. All these might produce luminescent effects in the lunar soil. These correlations have been observed in very few cases, however, so no definite conclusions can be drawn.

Impact of meteorites does not appear to be an adequate explanation. A small meteorite would not be capable of causing luminous effects as significant as those observed, and a large one would leave visible traces.

It would therefore seem that at present no mechanism, either internal or external, can be suggested to explain these transient phenomena.

4.7.5 Observation of transient events

Potential observers should realize that the observation of transient phenomena is demanding, and even difficult. It is easy for any observer, even experienced, to be misled by various types of optical effect. A certain amount of experience in lunar observation is absolutely essential before undertaking this work.

It is impossible to study transient events on one's own, because an event can only be regarded as definite if it is observed by at least two independent observers. It is essential to work in a team and to follow what may appear to be a somewhat rigorous procedure that requires recording atmospheric and instrumental effects as well as details of the event itself. This procedure is required to obtain a large number of compatible results that can be readily compared.

The two main groups who have specialized in this study are the Lunar Section of the BAA, and ALPO (*see* Sect. 4.3.4.3).

4.7.5.1 Observing within a monitoring programme

Observing transient phenomena is not as spectacular as some might think. The majority of observations are routine monitoring, when measurements of relative albedo are made at various points suspected of activity. Reduction of the measurements made over a period – generally one lunation – enables the detection of anomalies even though they may not have been noticed by the observer because of their low amplitude. This type of observation forms the basis of the programmes run by the BAA and ALPO.

Table 4.2. *Lunar albedo standards*

Value	Shade	Standard points
0	black	Shadows, non-illuminated part of the Moon
1		Darkest parts of the surface, Grimaldi
3	grey	Central region of Mare Serenitatis, Aristillus
5		Floor of Tycho
7	light grey	Anaxagoras, central peak of Alphonsus
8	white	Pico, central peak of Tycho
9	brilliant white	Rim of Proclus
10	dazzling	Central peak of Aristarchus

Albedo measurements In the ALPO programme, four transient sites and two additional sites, one for comparison purposes (a non-LTP area) and the other associated with seismic activity, are allocated to each observer joining the programme. One or two additional comparison points are also used, relatively far from, and directly north or south of each area covered. These are chosen so that they are under the same lighting conditions (i.e., they are the same distance from the terminator). At each site, measurements of apparent albedo are made of various fixed points (on the rim, the floor, or the central peak of a crater, for example) by comparison with a reference scale. Each observer establishes this scale from their own measurements at Full Moon. To carry out the reduction it is sufficient to correlate the observer's scale with a standard scale used by everyone. This, known as Elger's scale, has 21 levels running from 0 to 10 in steps of 0.5. The reference points given in Table 4.2 may be used.

Another method, used by the BAA observers in particular, consists of inserting a series of neutral density filters (set in a filter wheel) in the light-path until the point to be measured is extinguished. The density required for this to occur then gives a numerical value for the apparent albedo at the point being observed. (Photographic negatives of identical density may be superimposed instead of using a series of graduated neutral density filters.)

Observation of a transient site should be made at least twice a night, with about 15 minutes between observations, in accordance with the average duration of an event. Each site should be observed throughout its period of visibility in any lunation.

Observing conditions The observing conditions, which are usually given in subjective terms in other types of astronomical observations, should, in this case, be expressed in a more quantitative fashion, because they may be one of the factors involved in these transient events. They should be determined before and after each observing session by observing the disk shown by an out-of-focus star. (The defocussing should always be in the same direction and by the same amount, and with the eyepiece used for the actual observations.) The star should be as close as possible to the Moon.

The measurements to be made are:

(i) The length of time required for the disk to cross the field of the eyepiece, and the time it requires to disappear when brought to the edge of the field. The drive must be stopped. These times enable the diameter of the stellar disk to be expressed in seconds of arc, which is useful for subsequent reduction of the actual estimates.

(ii) The drive should now be switched on. An estimate should be made of the amount of lateral movement of the centre of the disk (or of one of the edges if that is more convenient) on either side of its average position. The shift should be expressed as a fraction of the field of the eyepiece. The average time (in seconds) between each extreme should also be given.

(iii) Finally, the pulsation (expansion and contraction) of the star's disk should be observed, giving an estimate of the ratio between the extreme sizes of the disk, and also the average time between pulsations. This last measurement may be very different from that found for the lateral displacement of the disk, because the two types of measurement made in (ii) and (iii) involve different atmospheric layers.

The measurements of the displacement and amount of pulsation of the stellar disk give the amplitude of the atmospheric waves that give rise to turbulence, whereas the times give the wavelengths. The larger the intervals, the better the seeing conditions. These measurements are important, because they are available for comparison with the times of any possible variations in brightness (or other factors) that may be recorded during an event.

Data reduction An albedo map for each site is prepared, giving the apparent albedo values determined for each point as a function of the Moon's age. For any given point, any deviation of 4 steps (or more) on Elger's scale from the mean value measured over a lunation, is suspected of being a possible anomaly.

The albedo measurements for various points at a specific site are generally plotted on a diagram as a function of the Moon's age. Plotting the various times of sunrise, of the Sun's meridian passage (local noon), of Full Moon, of apogee and of perigee, and passage through the magnetopause, etc., on such a diagram allows possible correlations with the various measurements and the observed anomalies to be investigated.

Results of observational programmes, showing examples of the reduction and the interpretation of 'albedo maps' may be found in the journals of the BAA and ALPO.

4.7.5.2 Visual observation of a transient event

If any anomaly is detected during the course of a regular observation as part of a monitoring programme, or at any other time, the observer should provide the maximum amount of detail about the phenomenon observed, such as albedo, colour, and any other details that may seem of interest, together with any variations that are observed. (The period of any such changes should be timed). The observer should avoid bias as far as possible, and should also carefully examine the surrounding area, and particularly similar areas north or south of the suspected transient that are under similar lighting conditions, as well as similar formations nearby. Atmospheric conditions should be determined immediately after the observation in the manner described earlier.

One approach to the problem of determining the physical nature of transient phenomena is to employ colour filters. L. E. Fitton, of the BAA, suggested a method of observation using red (Wratten 25) and blue (Wratten 44a) filters, whereby it is possible to differentiate between emission and absorption mechanisms occurring at the site of any event that shows a reddish or bluish coloration (Fitton, 1975). If, at the point in question, a red transient event is identified – red is the colour most frequently observed – there are two possibilities: either the point is selectively reflecting red light, other colours being eliminated by absorption, or else the point is emitting an excess of red light.

In the first case, observation of the event through a blue filter will show a dark area (there is a lack of blue, which is why red is seen), whereas the surroundings will appear blue. In the second case, the point will appear bright when seen through a red filter – and the stronger the emission, the brighter it will be – while the surroundings also appear red. The effects of the filter are the opposite when a blue event is observed.

More precise observations of the coloured area (how sharply defined it is, and the degree of darkening) should also, according to Fitton, allow the determination of whether the absorption or emission are taking place in a gas or on a solid medium (i.e., directly on the surface of the Moon).

There are certain limitations to this method, however, because it only takes two colours into account, and in both cases these have to be relatively strong. To interpret the filter effects correctly, it is essential to observe any coloration that may be apparent in integrated light. The degree of magnification also plays a part in the ease with which these effects can be seen.

4.7.5.3 Other methods of observing transient events

The study of transient events on the Moon suffers greatly from the almost complete lack of data about their chemistry. In our current state of knowledge, work should be directed towards solving this problem if at all possible. Spectroscopic measurements would be particularly valuable, and would allow considerable steps to be made towards understanding the phenomenon. Luminescence would be established, for example, by observation of the H line of ionized calcium at 397 nm. This is relatively wide (0.9 nm at mid-height) and normally weak, which means that if present it should be easy to detect. The group of iron lines (at about 545 nm) and, of course, the hydrogen lines (both in the ultraviolet and the Hα line at 656 nm) may also be used.

In fact, the whole spectral range from 400 to 1000 nm ought to be studied, with a spectral resolution of about 1 nm, in order to detect emission lines. Methods that could be used to study these various wavelengths, comprise, as we have already seen, photometry (using interference filters for either spot measurements or imagery); photography (imagery through filters, and spectroscopy); and spectrophotometry.

These types of observation could be made systematically during patrol observations. When some abnormality is observed, measurements should be made in rapid succession (every 1 or 2 minutes) throughout the whole of the event, in order to detect the slightest variations. At the same time, details of the event should be recorded, especially any data relevant to the measurements that are being made (time, filter used, etc.). It is obviously essential to be prepared for any eventuality and to have drawn up a plan of action in advance.

4.8 Lunar eclipses

4.8.1 Eclipse conditions

An eclipse of the Moon takes place when the Moon passes through the Earth's shadow, the two bodies being then aligned with the Sun. Obviously, this configuration occurs only at Full Moon and, because of the inclination of the lunar orbit to the ecliptic, when the Moon is close to one of its nodes.

The geometrical conditions for an eclipse are determined by the dimensions of the two cones formed by the internal and external tangents to the Sun and the Earth (Fig. 4.32). The intersection of the cones by a plane P at the distance of the Moon gives two circles, one of which forms the umbra and the other the penumbra. The respective apparent (geocentric) radii of these are given by:

$$\sigma = \pi + \pi_\odot - R_\odot$$

$$\sigma' = \pi + \pi_\odot + R_\odot.$$

With $\pi_\odot = 8.794$ arc-seconds, and $\pi = 3422.62$ arc-seconds, the parallaxes of the Sun and Moon respectively, and the apparent angular radius of the Sun, $R_\odot = 960$ arc-seconds, we have:

$$\sigma = 41'11''$$

$$\sigma' = 1°13'11''.$$

The average apparent diameter of the Moon is $31'$, so it may obviously be completely covered by the Earth's umbra or penumbra. According to whether the Moon is completely or partially immersed in the umbral or penumbral cones, we have either total or partial umbral or penumbral eclipses (Fig. 4.33).

The magnitude of the eclipse (g) is the fraction of the lunar disk that is eclipsed in the middle of the event. It is given by:

$$g = (\sigma - d + R_M)/2R_M,$$

where R_M is the apparent radius of the Moon (937 arc-seconds) and d is the minimum distance of the centre of the Moon from the centre of the shadow. The dates and times, and the exact times of contacts between the Moon and the penumbra and umbra are given in ephemerides for each eclipse, as well as in various astronomical publications.

The Earth's atmosphere plays a considerable part in eclipses of the Moon, and greatly affects their appearance. If we consider the parallel rays of light R_S arising at one edge of the Sun (Fig. 4.34), these will be more strongly refracted the smaller

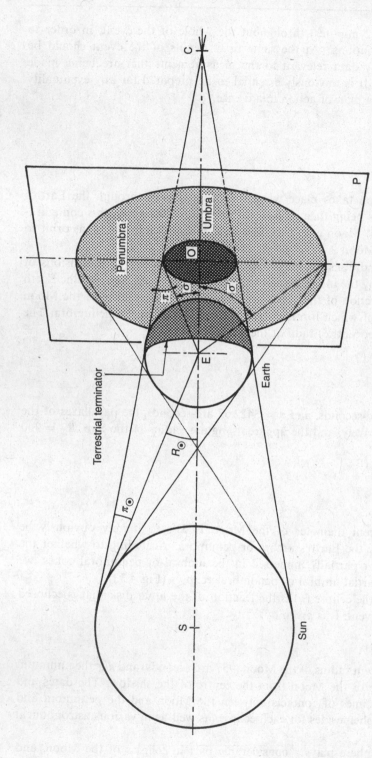

Fig. 4.32. *The geometry of a lunar eclipse. σ and σ' are, respectively, the radii of the umbra and penumbra on the imaginary plane P at the distance of the Moon.*

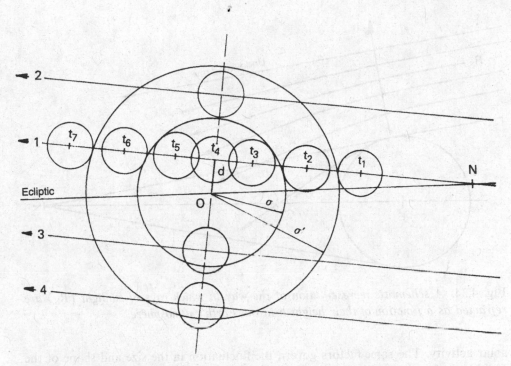

Fig. 4.33. *Different types of lunar eclipse. 1: total umbral eclipse; 2: total penumbral eclipse; 3: partial umbral eclipse; 4: partial penumbral eclipse. Times t_1 to t_7 are the times of contact between the edge of the Moon and the umbra and penumbra. d is the minimum distance between the centres of the Moon and the umbra.*

their height h, causing them to illuminate the inner part of the geometrical shadow-cone. For rays that pass close to the ground, the deviation is such that they will pass through point C' instead of C. As a result, the only region of total darkness occurs within a distance EC', which is equal to about 40 Earth radii. Because the Moon is always at a greater distance (at about 60 Earth radii), it is therefore never completely eclipsed.

Because absorption is stronger in the blue than in the red, the latter predominates in the light that is refracted into the shadow-cone. At totality the Moon takes on a characteristic reddish colour, which is more or less intense according to terrestrial atmospheric conditions at the time of the eclipse. The deviation of the Sun's rays may be considered to be negligible when h is of the order of 75 km.

In calculating the times of the Moon's contacts with the umbra, the Earth's radius is therefore increased by this amount, increasing σ by about 2%. The calculated times are always different from those observed, because the size and shape of the shadow-cone also vary slightly as a result of atmospheric conditions at the Earth's terminator at the time of the eclipse.

The primary factors responsible for the variations in the brightness and colour of the lunar disk are: cloud cover, the strength of the ozone layer, the greater or lesser amount of volcanic or meteoritic dust present around the terrestrial terminator, and

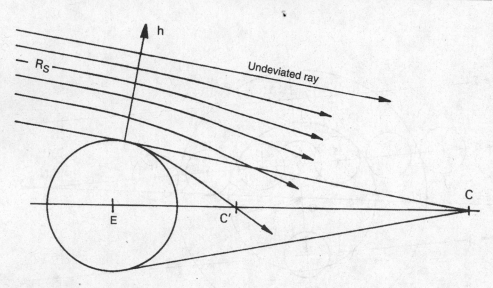

Fig. 4.34. *A schematic representation of the way in which rays of sunlight (R_S) are refracted as a function of their height h in the Earth's atmosphere.*

solar activity. The same factors govern the fluctuation in the size and shape of the Earth's shadow-cone that are observed from eclipse to eclipse.

4.8.2 Observation of lunar eclipses

In antiquity and for many centuries later, eclipses of the Moon were the principal means of determining the geometry and dimensions of the Earth–Moon–Sun system. They were also used to refine theories of the motion of the Moon. About 400 BC, Aristotle deduced that the Earth was round from the curvature of the Earth's shadow on the Moon. About 200 BC, Hipparchos determined the dimensions of the Earth–Moon system. The methods were repeatedly refined in succeeding centuries.

Eclipses also helped geographers, because they could be used to obtain more accurate longitudes on the Earth. In 1634, for example, the length of the Mediterranean sea shrank by 1000 km from the value previously accepted. Nowadays, of course, thanks to orbiting satellites, we have access to far more accurate methods than the observation of eclipses. Observing lunar eclipses remains of interest, because of the information that they can give about the state of the Earth's atmosphere.

4.8.3 The appearance of the lunar disk during eclipses

4.8.3.1 Phases of an eclipse
There are various phases in a lunar eclipse, and these are listed in Table 4.3 where the times are those shown in Fig. 4.33. All are present in a total eclipse. During the various phases different features may be observed.

Phases I and V are often neglected during total eclipses. The same applies to

Table 4.3. *Phases of a lunar eclipse*

Phase	Description
I	The Moon enters the penumbra, between times t_1 and t_2.
II	The Moon enters the umbra, between first external contact t_2 and interior contact t_3.
III	Total eclipse within the umbra, from t_3 to t_5, with the middle of the eclipse at t_4.
IV	The Moon leaves the umbra between second interior contact t_5 and exterior contact t_6.
V	The Moon leaves the penumbra between t_6 and t_7.

phase II at total or partial penumbral eclipses, which are often ignored completely. It is true that the visual appearance of the Moon remains essentially normal during these phases, but occasionally the density of the penumbra is greater than predicted. It is therefore advisable to follow these phases of an eclipse, particularly when the Moon is nearing the umbra.

During phases II and IV, the advancing and retreating umbra should be followed across the disk of the Moon, and accurate notes made of various features of the umbra, particularly:

- Edge: whether sharp or gradual;
- Shape: the edge of the umbra is normally circular, but flattening has sometimes been noted, as well as various irregularities. These are generally not real – they are either effects of the lunar topography or optical illusions – but such phenomena should be carefully followed and confirmed by timing the contacts of the umbra with surface features.
- Density and coloration: it is generally not easy to determine these at the beginning of the eclipse, because the brilliance of the uneclipsed portion of the Moon causes severe interference. The penumbral crescent is about 100–1000 times as bright as the part of the Moon within the umbra.

During phases II and IV the times at which surface features (usually craters) enter and leave the umbra are recorded, in order to determine the diameter of the umbra. Observations to be made during totality are described in the following sections.

4.8.3.2 Brightness of an eclipse

There are several means of determining the brightness of an eclipse. It would be possible, for example, to use the scale common among the Dayaks, one of the tribes of Borneo. According to them, one of three different monsters is trying to swallow the Moon, depending on whether the eclipse appears red (when it is the greediest monster: Rahu-Tambaga, the copper monster), very dark (Rahu-Bahuang), or greyish (Rahu-Ambon, the monster of fogs and mists).

More seriously, the relative brightness of an eclipse is determined visually by

Table 4.4. *Danjon lunar eclipse scale*

Number	Appearance of the Moon
0	Very dark eclipse. The Moon is difficult to see, especially in the middle of totality
1	Dark eclipse, with greyish to brown colour. Details are difficult to see on the disk
2	Dark red or rusty colour, with a dark zone in the centre of the umbra and lighter edges
3	Brick red. The umbra often has a brighter, yellowish border
4	Bright orange or coppery eclipse, with bright, bluish borders

using a simple scale established by Danjon (see Table 4.4). This scale is based on both the brightness and colour of an eclipse. The estimate should be made at the time of mid-eclipse, by observing the Moon's disk either with the naked eye, or with very low magnification ($< 20\times$) – a 7×50 binocular being ideal.

In partial umbral eclipses, estimation of the brightness is possible, but more difficult the larger the amount of the Moon remaining illuminated (in the penumbra). The darkness of the eclipse is then often exaggerated because of the effect of contrast. Difficulties in estimating the brightness can also arise when the Moon's disk shows a wide range of brightness and colour.

Another method of determining the brightness of an eclipse is to estimate the apparent stellar magnitude of the Moon, again at mid-eclipse. Various methods are used, one, for example, being based on the reduced image of the Moon as seen in a spherical reflecting surface. The simplest is to observe the Moon with reversed binoculars. The image of the Moon then appears almost point-like, which allows its magnitude to be compared more easily with nearby stars or planets. The presence in the sky of a bright planet like Jupiter helps considerably.

Eclipses are generally darker when the Earth's atmosphere is denser, i.e., when it is less transparent to refracted sunlight. Such eclipses coincide with major volcanic eruptions. Although on a planetary scale the amount of dust that is ejected is relatively small (amounting to $5-20\,\text{km}^3$ on average), thanks to upper-atmosphere winds, that amount is sufficient to cause an appreciable, and fairly even, increase in the opacity of the atmosphere, and a corresponding decrease in the brightness of the eclipsed Moon.

There is a distinct correlation between the brightness of eclipses (on the Danjon scale) and the 11-year cycle of solar activity. Beginning shortly after solar minimum, the brightness of eclipses increases progressively (from 0 to 4), falling abruptly to zero shortly after the subsequent minimum, before recommencing its cycle.

Shortly before minimum, corpuscular radiation from active centres on the Sun (from spots and faculae) is emitted in the plane of the ecliptic, because the active areas are then close to the solar equator. This increased radiation may cause scattering in the Earth's atmosphere, or even act directly on the lunar surface (through luminescent effects) and thus account for the bright eclipses that occur at that phase of the solar cycle.

4.8.3.3 Details in the umbra

The density of the umbra and its coloration are far from being homogeneous across the disk of the eclipsed Moon. Some of these variations are caused by the way in which the rays from the Sun are refracted into the umbral cone: there is, for example, a focussing effect close to the centre of the umbra. Others are caused by different atmospheric conditions that prevail as the Earth's rotation brings different areas of the atmosphere to the terminator. This can produce an asymmetry in the appearance of the umbra during the first and second halves of totality, or else the existence of abnormally dark (or bright) areas on the lunar disk. Detailed observations of the umbra should therefore be made during the total phase, noting down at regular intervals any variations in darkness and coloration across the lunar disk. The observations should be recorded using codes for intensity and colour, on sketches or drawings that have been prepared in advance with schematic outlines of the major formations.

Comparative studies may be made with red and blue filters. During totality, the brightness in the red at the centre of the umbra may be 100–1000 times that in the blue. These observations should be made using binoculars or a low magnification ($< 30\times$) to include the whole of the Moon in the field, and also to prevent a reduction in contrast.

It is also interesting to note the visibility and appearance of certain lunar formations within the umbra, including both the brightest (Aristarchus, Kepler, Tycho's rays, etc.) and the darkest (the maria or dark-floored craters). The latter will be easier to observe with larger instruments.

4.8.3.4 Photometric measurements of the umbra

Photometric study of the penumbra and umbra at eclipses is difficult. The depth of the shadow at a point on the Moon at any particular instant is given by the difference between the value measured during the eclipse and that measured outside eclipse. It will be immediately obvious that such measurements are very difficult, because it is impossible to use the standard procedure where differential measurements are made under identical (primarily atmospheric) observational conditions. This means that a specific allowance for the values of atmospheric extinction at the time of each measurement must be taken into account.

Because the ratio of the brightnesses of the Moon inside and outside eclipse is four to five orders of magnitude (between 1:10 000 and 1:100 000), the response of the detector – which obviously must be the same for both measurements – must be linear over a very large range.

An additional difficulty in measuring the depth of the shadow during partial phases of the eclipse is caused by scattered light from the part of the Moon that is outside the umbra. (The scattering occurs both in the Earth's atmosphere and inside the telescope.)

During the 1920s, Danjon perfected a visual photometer known as the cat's-eye photometer, which may be used to measure the difference in brightness of two points on the eclipsed Moon. Two prisms situated in front of the objective of a refractor create a second image of the Moon in the same field as the primary image, so that two points P and P′ on opposite sides of the lunar disk appear next to one another (Fig. 4.35). The diaphragm D (the cat's-eye diaphragm) allows the brightness of the point P′ to be adjusted until it is equal to that of P. The diaphragm setting then gives a measure of the difference between the two points.

When successive measurements are taken throughout the duration of an eclipse, P and P′ should be chosen such that the diameter PP′ points towards the centre of the umbra. The photometric curve of the eclipse is therefore a function of the distance to the centre of the umbra.

Measurements of the umbral eclipse are generally taken in the blue (460 nm) and in the red (620 nm). (As the penumbra is generally neutral in shade, any such measurements are made without a filter.) Because the distance separating P and P′ is very small, extinction may be neglected. Similarly, because the two images are projected against the same sky background, the effect of scattered light on the measurement is minimal.

This method has the disadvantage that only points on the limb of the Moon may be measured. This does not allow isophotes of the umbra to be established. It is also assumed that the transparency of the sky remains the same during the measurements.

4.8.4 Size of the umbra

4.8.4.1 Times of contact

The times of contact of the lunar limb with the umbra are calculated on the basis of a 2% increase in the radius of the umbra σ to allow for the effect of the atmosphere (Sect. 4.8.1). As the latter varies from eclipse to eclipse, there is always a difference between the observed and calculated times of contact.

Calculation of the radius of the umbra σ_u from the times of immersion and emersion of both lunar limbs (times t_2, t_3, t_5 and t_6 in Fig. 4.33), and of a series of craters, allows the average radius of the umbra to be determined, as well as its shape (provided enough measurements have been obtained).

The cause of variations in the radius of the umbra must be sought in varying absorption in the upper layers of the atmosphere (at around 90 km), which are, it seems, linked to the presence of greater or lesser amounts of meteoritic dust. (A correlation has been found between the percentage increase in σ and major meteor showers.) The times of contact, in UT – the time being checked from time signals before and after the event – are measured for the outer edge of craters (or the central peak, if there is one). These should be chosen to be fairly evenly spread across the surface of Moon, and for their distinct appearance or because they are easily identifiable. There is no generally accepted list, but Copernicus, Kepler, Tycho, Billy, Grimaldi, Plato, Atlas, and Langrenus may be mentioned in particular.

In determining times of contact, one difficulty that becomes apparent at the time of observation, is that of deciding on the exact edge of the umbra. Because this is

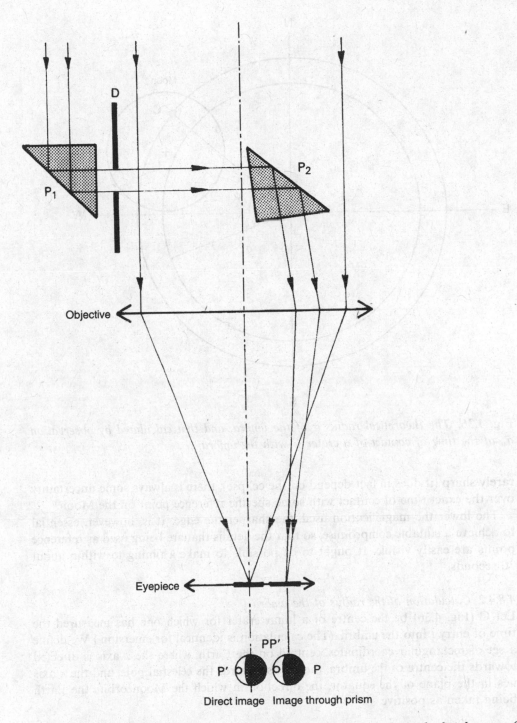

Fig. 4.35. *A diagram of Danjon's cat's-eye photometer, which allows the brightness of two points P and P' at opposite sides of the Moon to be compared.*

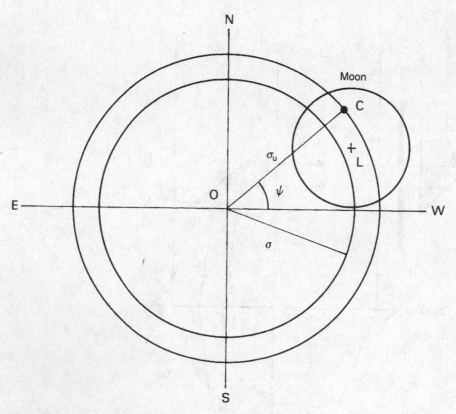

Fig. 4.36. *The theoretical radius σ of the umbra, and that calculated by observation σ$_u$ at the time of contact of a crater C with the umbra.*

rarely sharp (it does in fact depend on the eclipse), there is always some uncertainty over the exact time of contact with some specific reference point on the Moon.

The lower the magnification used, the sharper the edge. It is, however, essential to achieve a suitable compromise, so that the details that are being used as reference points are easily visible. It ought to be possible to make a timing to within about 30 seconds.

4.8.4.2 *Calculation of the radius of the umbra*
Let C (Fig. 4.36) be the centre of a lunar crater for which one has measured the time of entry t into the umbra. (The calculation is identical for emersion.) We define a set of rectangular coordinates, centred on the Earth, where the z axis is directed towards the centre of the umbra, the y axis towards the celestial pole, and the x axis lies in the plane of the equator, the direction in which the Moon orbits the Earth being taken as positive.

Using this system, the coordinates of the centre M of the Moon are:

$$x_M = \cos\delta_M \sin(\alpha_M - \alpha_a)/\sin\pi$$
$$y_M = [\sin(\delta_M - \delta_a)/\sin\pi] + 0.008\,726(\alpha_M - \alpha_a)x_M \sin\delta_a$$
$$z_M = [\cos(\delta_M - \delta_a)/\sin\pi] - 0.008\,726(\alpha_M - \alpha_a)x_M \cos\delta_a$$

where α_M, δ_M and α_a, δ_a are, respectively, the equatorial coordinates of the Moon and the antisolar point (point O in the Figure), and π is the parallax of the Moon.

The centre of the crater, C, with selenographic coordinates (λ, β) has the rectangular lunar coordinates:

$$x_u = R\cos\beta\sin\lambda$$
$$y_u = R\sin\beta$$
$$z_u = R\cos\beta\cos\lambda.$$

Transformation to the rectangular coordinates defined above is by means of the equations:

$$x = x_M + a_x x_u + b_x y_u + c_x z_u$$
$$y = y_M + a_y x_u + b_y t_u + c_y z_u,$$

where the cosine terms for the x, y and z axes in the first system, expressed in terms of the second system are:

$$a_x = -\cos\lambda_s \cos P + \sin\lambda_s \sin P \sin\beta_s$$
$$b_x = \sin P \cos\beta_s$$
$$c_x = \sin\lambda_s \cos P - \cos\lambda_s \sin P \sin\lambda_s$$
$$a_y = \cos\lambda_s \sin P - \sin\lambda_s \cos P \sin\beta_s$$
$$b_y = \cos P \cos\beta_s$$
$$c_y = -\sin\lambda_s \sin P - \cos\lambda_s \cos P \sin\beta_s.$$

In these expressions (λ_s, β_s) are the selenographic coordinates of the Sun and P is the position angle of the Moon's axis of rotation, which is given in ephemerides.

The radius of the umbra σ_u is then given by:

$$\sigma_u = \sqrt{x^2 + y^2}.$$

In discussing the results, the angle ψ, which is the position angle of the crater relative to the equator is also given:

$$\log\psi = y/|x|.$$

A more accurate calculation of σ_u would have to take into account libration and also the flattening of the Earth. (The Earth is flattened, so is its shadow!)

It would be a pity if your observations of an eclipse were to remain unknown. Unfortunately there is no organisation that specializes in the study of lunar eclipses.

The journal *Sky & Telescope*, published by the Sky Publishing Corporation, regularly carries notes about observing programmes, together with a summary of results.

Each observer should forward their observations with the following information:

- Times of contact (UT) of the limb and the umbra;
- Times of contact between craters and the umbra (with the coordinates of the points of contact);
- Estimates of the brightness and magnitude, and also full details of the colouration of the umbra, diagrams, photographs, and any other types of measurement;
- Coordinates of the observing site, instruments, films, etc. used.

Calculation of the diameter of the umbra for each contact is normally undertaken by those coordinating the observing programme (but there is nothing to stop you doing the calculation yourself).

5 Planetary surfaces

J. Dijon, J. Dragesco & R. Néel

5.1 Observation of the planets

5.1.1 The value of planetary observations

At the end of the 19th century and the beginning of the 20th, there was no clear distinction between amateurs and professionals, particularly in the field of the physical observation of the planets, because this was carried out visually with telescopes of 90–320 mm in diameter. In his two comprehensive volumes about Mars, Flammarion cited amateurs and professionals indiscriminately: Comas-Sola, Pickering, Denning, Williams, Lowell, Trouvelot, etc. The main thing was to know how to observe and to draw, the diameter of the instrument seemed secondary. For a long time afterwards, especially while planetary photography was in its infancy, the most renowned planetary observers included both amateurs and professionals: Schiaparelli, Cerulli, the Fournier brothers, Barnard, Antoniadi, Maggini, etc. It was only with the widespread use of photography and the introduction of very large telescopes that amateurs and professionals went their separate ways. On the one hand there were people like Phillips, Peek, Oriano, Saheki, etc., and on the other: Slipher, Camichel, Lyot, Ebisawa, Dollfus, Smith, Focas, Kuiper, etc.

Since the introduction of high-resolution planetary photography (at Pic du Midi, Flagstaff, and Catalina Observatories, and NASA's International Planetary Patrol Program), the role of amateurs has become increasingly reduced, particularly because the fiasco of the Martian canals was a hard blow to planetary artists!

Space-probes were soon to photograph Mars and Jupiter, etc., completely changing our knowledge of the planets with their close-up images. Since then, many amateur planetary observers, discouraged and seemingly out-classed, have gradually given up their work, and their number continues to decrease. Should amateurs abandon observation of planetary surfaces? Certainly not!

Contrary to what most amateurs think, a new era is opening up for those with expertise. Space-probes do not remain long in orbit around the planets that they observe. Monitoring the constantly changing phenomena that occur in planetary atmospheres is required, yet NASA's programme for systematically photographing Mars has been partially abandoned. Because professionals have better things to do, they have stopped systematic photography of Jupiter. So it falls to amateurs to monitor what is happening on the planets. This has become all the more interesting, because numerous amateurs possess powerful, high-quality instruments, and, thanks to progress in photographic techniques and materials, high-resolution photographs are now within their grasp.

More or less daily coverage of the atmospheric phenomena on Mars and Jupiter (and Saturn) may be found in the reports issued by the various groups of planetary observers. The various features change with time, and it is important for those changes to be followed: the cycle governing the occurrence of the dust-storms on Mars is more complex than originally thought, for example; the white ovals on Jupiter continue to fade – will they disappear one of these days? It will be amateurs that provide the answer!

It is therefore undeniable that the active monitoring of planets is interesting, useful and within the scope of experienced, well-equipped amateurs. But, on the other hand, it is pointless to build up false hopes. Most of the planets are not very interesting for either visual or photographic study. This is certainly the case with Mercury, whose unchanging surface is fairly well-known from the photographs returned by the Mariner 10 mission. It is also largely true for Venus, although it was here that the well-known amateur, Charles Boyer, discovered the 4-day retrograde rotation of the upper atmosphere. It is certainly the case with the most distant planets, Uranus, Neptune and Pluto.

The most accessible planet for amateurs, and the one that varies most, is Jupiter, which is at opposition every 13 months, and whose atmosphere is subject to very complex changes. It is important to note that the discoveries of the Voyager missions have invalidated none of the results obtained by observers on Earth (mostly amateurs).

The most fascinating planet is perhaps Mars, where, unfortunately, the well-known variations in the albedo markings are only caused by changes in dust cover, produced by complex wind patterns. At favourable oppositions, a good observer will easily see the appearance and disappearance of white and blue clouds, as well as the development of major dust storms.

The most spectacular planet is undoubtedly Saturn. Unfortunately, the planet is not as easy to observe as Jupiter, and its changes are less extreme and less significant.

To stress the very different significance that the various planets have for amateurs, this chapter departs from discussing the planets in the conventional arrangement of increasing order from the Sun. Here, the order in which they are presented and the amount of space devoted to each planet is a measure of their degree of interest. So, after a general discussion of factors affecting observation and methods that are common to all planets, Jupiter and Mars take up the major part of our discussion. Venus and Saturn, which may be the subject of interesting observations, but to a lesser degree, are discussed more briefly. Finally, the other planets are mentioned for the sake of completeness.

5.1.2 *General properties of the planets*

5.1.2.1 *Apparent diameters, and periods of visibility*

Inexperienced observers are always disappointed by the small apparent diameters of the planets. When compared with the Moon, for example, any planet appears minute. In fact, the only planets accessible with amateur-sized instruments are Jupiter, Saturn, Mars and Venus. The apparent diameters vary from 30.5–49.8 arc-seconds for Jupiter; 14.7–20.5 arc-seconds for Saturn; 3.5–25.1 arc-seconds for Mars; and 9.9–64.5 arc-seconds for Venus.

The planets are not always available for observation. Or if they are, their distance may mean that observation is of little interest. The inner (or inferior) planets are closer to the Sun than the Earth. Venus may only be effectively observed at its greatest elongation, i.e., when its distance from the Sun is a maximum. The planet then appears partially illuminated, like the Half Moon. (There are obviously two different elongations: one in the evening and the other in the morning.) The superior (or outer) planets are farther from the Sun than the Earth. Observation is best when they are at opposition, when the Sun and the planet are diametrically opposite in the sky for an observer on the Earth. At quadratures (western or eastern), the planets are 90° away from the Sun. At conjunctions, the Sun and the planet have the same longitude, and the planet is difficult to observe, being lost in the Sun's rays. At oppositions, the outer planets are also closest to the Earth and are visible throughout the night. For Jupiter and Saturn, all oppositions are comparable, but for Mars, which has a very eccentric orbit, oppositions that occur at perihelion are far more favourable than those when the planet is at aphelion.

5.1.3 Factors affecting observation

Despite the spectacular progress made in high-resolution astronomical photography, visual observation of the planets has retained its value whenever it is a question of studying the fine detail that may be detected with the best telescopes. For amateurs, visual observation of the planets has two additional advantages: the sheer pleasure derived from seeing a fine image of a planet, and also the technically undemanding nature of the observations. An excellent planetary drawing may be made with even an altazimuth mounting, with no equatorial drive.

5.1.3.1 The human eye
Let us begin by discussing that exceptionally versatile detector, the human eye.

The perception of contrast   Perception of contrast plays an important part in planetary observing. The eye is able to detect both differences in brightness and in colour in a planetary image. If two areas are of different brightness, the contrast is given by the relative difference in their brightness: $\gamma = (B - b)/B$, where B is the greater luminosity. If the areas have the same brightness, then $\gamma = 0$. On the other hand, if one is infinitely brighter than the other $\gamma = 1$. The eye cannot detect a contrast that is less than a certain limit, so we may call the lowest detectable contrast γ_0, which is also a function of the overall brightness. Although under good lighting conditions a contrast of 1/200 is detectable, at the retina's lower limit of sensitivity, γ_0 tends towards unity.

Visual acuity   Visual acuity or the eye's resolution, is of extreme importance for planetary observers. The resolution of very good eyesight approaches 1 arc-minute, under normal lighting. If the lighting or the contrast are poor then the resolution drops to 2 arc-minutes or less. But the problem is very different when it is not a question of separating two adjacent features (which is the definition of resolution), but more simply that of perceiving isolated details, which may be of various shapes and with greater or lesser contrast. According to W. H. Pickering (quoted by J. B. Sidgwick, 1961), and also experiments carried out by J. Dragesco, a good planetary observer can discern high-contrast, black circular spots on a light background when their diameters are approximately 30 arc-seconds (31 arc-seconds for a contrast of unity and 43 arc-seconds for a contrast of 0.5). A light spot on a black background may be seen even if its diameter is a low as 25 arc-seconds (33 arc-seconds for a contrast of 0.6). Yet a telegraph wire (a dark line on a light background) remains perceptible when the angle it subtends is only 1 arc-second (or even less).

These experiments prove that provided there is no atmospheric turbulence, it should be possible to see a division in Saturn's rings that is no more than 0.13 arc-seconds across with a 150-mm refractor, which has a nominal resolution of 0.8 arc-seconds.

Fabry and Arnult, using variable-contrast test cards, showed in 1937 that the eye's performance improves when the pupil is reduced (even at low light-levels). They maintain that the eye functions properly only when the pupillary aperture is between 0.5 and 0.7 mm. It is therefore counter-productive to use instrumental exit-pupils that are larger than this value. Contrary to the opinions stated by Danjon and Couder (among others), the diameter of the exit pupil should not really exceed 0.7 mm (which is equivalent to a magnification of 230× with a 150-mm telescope). Using high magnifications on large-aperture telescopes does not exploit the eye's efficiency to the full (unless very low-contrast objects are being observed).

In conclusion therefore, we would stress that it is important for inexperienced observers to test and to develop their eyesight, beginning by analysing their visual acuity, and then practicing the observation of planetary detail – if possible, under 'laboratory conditions'. A good early exercise, which is both instructive and useful, consists of observing and drawing the Moon with the naked eye (preferably during twilight). An experienced amateur is able to pick out a remarkable amount of detail, and any drawing that is made gives a lot of information about the observer's personal equation.

This may be taken even further: set up, experimentally, the sort of conditions that prevail when using a telescope. It would suffice to draw several complex dark markings on a circular disk, and then to observe the drawing with a small pocket telescope at a sufficiently long distance. (One could even stand a container of very hot water in front of the objective, thus mimicking bad atmospheric turbulence.) A drawing made under these conditions enables one to gain a good idea of the actual accuracy of an individual observation, because it is possible to compare the drawing with the original.

One point should be made, however, which is that beginners should not be discouraged. Eyesight does improve, and not only does it become better at perceiving fine detail, but it also becomes more skilled in interpreting that detail. Finally, we should mention that the eyesight of young people is apparently particularly suitable for telescopic observation, their eyes having good accommodation and more easily able to follow the constant slight changes in focus that are caused by atmospheric turbulence.

5.1.3.2 Seeing

Regardless of the optical quality of the instrument used, or the observer's visual acuity, it is seeing that plays the main part (often detrimental) in determining the amount of fine detail that may be seen on the planets. The atmosphere deflects the rays of light because it has local inhomogeneities, and therefore differences in refractive index, which are caused by the wind, eddies and temperature differences, the last being by far the most important. The wavefront reaching the objective is no longer essentially plane, and the normal at each point on this undulating surface gives a measure of atmospheric turbulence or scintillation. The atmospheric undulations have wavelengths λ that vary between 100 and 250 mm and form at heights of between 1 and 30 km. The image of a star therefore shifts by between 0.05 and 2 arc-seconds (sometimes more). If the eye is placed at the focus of a telescope that is aimed at a bright star, what have been described as 'flying shadows' are visible crossing the illuminated disk of the objective. To give an idea of the importance of the temperature that prevails in the turbulent layers, a 150-mm thick layer of air that is only 1°C warmer than its surroundings changes the optical path by $\lambda/4$!

Scintillation increases from zenith to horizon, of course, because the length of the light-path through the turbulent layers increases (Fig. 5.1). If turbulence is weak, the theoretical diffraction image is not greatly distorted, but shifts slightly in position. In general, however, seeing completely alters the diffraction disk. According to Danjon and Couder, for a low value of turbulence (0.25 arc-seconds), the useful magnification on a 2.5-m telescope drops from 1250× to only 230× – a magnification, moreover, that would give far too bright an image and an exit pupil of more than 10 mm, under which conditions the eye would not function to best advantage.

Danjon and Couder suggested a method of determining the stability of images, which consisted of observing a star at the zenith under high magnification (Fig. 5.2). Using this method, seeing may be estimated as 0.25 arc-second when the diffraction rings are seen to be continuous in a 140-mm telescope. (Under the same conditions, the rings appear broken in a 280-mm telescope; the image is greatly distorted in a 560-mm instrument, and dreadful in one of 1 m.)

Over most of France, for example, average seeing is equal to, or better than, 0.3–0.4 arc-second. In Haute-Provence, and also at the Pic du Midi (in particular), seeing conditions of around 0.2 arc-second are relatively frequent, and 0.1 arc-second is encountered on a number of nights in the year. In the daytime, average seeing is generally worse than 1 arc-second, and may reach 2 or even 3 arc-seconds in the early afternoon.

According to Meinel the best sites have average seeing of 0.1–0.3 arc-second, while seeing at low-altitude sites varies between 0.5 and 1 arc-second, but rarely reaches 2 arc-seconds at night. In any event, seeing is the planetary observer's worst enemy. It makes it difficult to employ high-resolution photography over most of France and at most low-altitude sites elsewhere.

5.1.3.3 Choice of an observing site

Finding sites where seeing is excellent becomes absolutely necessary when instruments exceed about 300 mm in diameter. Professional astronomers are even more

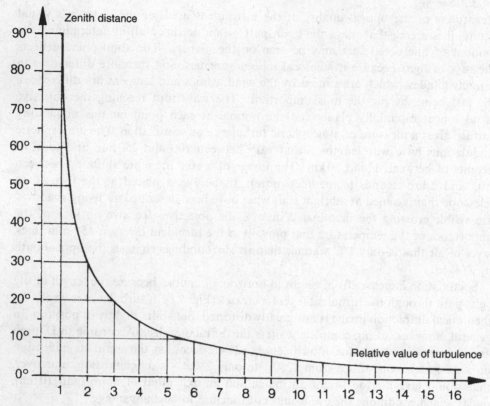

Fig. 5.1. *The effect of zenith distance on seeing, after A. Danjon and A. Couder. In this example the height of the layer has been assumed to be 3.5 km.*

demanding because they also require the greatest possible number of usable nights, high transparency, and the complete absence of light pollution. The modern tendency is to concentrate large telescopes at just a few observational centres that are particularly suitable: Hawaii, Chile, the Canary Islands, Spain, plus a few specific sites in the U.S.A.

The best images appear to be obtained from sites on dry plateaux at moderate altitudes (such as in Haute-Provence) or high, isolated mountains (as at Pic du Midi). Unfortunately, few amateurs are able to choose their observing site freely. Specific constraints linked either to one's home or place of work frequently mean that one has no real choice. Then only chance determines the seeing quality.

5.1.3.4 Instrumental effects

External, atmospheric turbulence is the most important, but not the only factor affecting planetary observation. Air currents within the telescope may also be very deleterious: a rise of 2–3°C in the temperature of an open telescope tube causes a distinct increase in the size of the diffraction disk. If the temperature within the tube is 7–8°C higher than that of the surrounding air, the telescopic image is unusable.

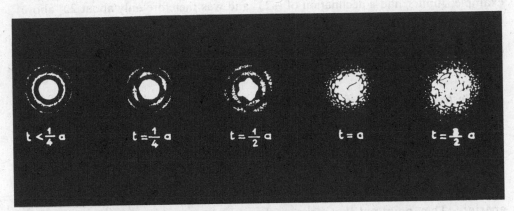

Fig. 5.2. *Determining seeing by the observation of a star's diffraction disk (after A. Danjon and A. Couder), where a is the diameter of the diffraction disk in any instrument of diameter D. (It may be shown that $a = 140/D$, where a is expressed in arc-seconds and D in millimetres.) Seeing may be expressed as a function of a by determining the changes that occur in the diffraction disk:*

$t < a/4$: *perfect image, with no perceptible distortion or movement;*

$t = a/4$: *the rings are complete, but exhibit moving condensations;*

$t = a/2$: *average turbulence; broken diffraction rings, central disk with wavy edges;*

$t = a$: *continuous motion, rings faint or absent;*

$t = 3a/2$: *an image that is rather like the image of a planet.*

It is very important, therefore, to avoid the telescope becoming heated during the day. A dome should be insulated so that the temperature does not rise to any great extent during the day. White paint, or even better, aluminium paint is, in any case, recommended for the whole observatory. The shutters should be opened one or two hours before nightfall so that the telescope may obtain perfect thermal equilibrium.

Refractors and catadioptric telescopes, having closed tubes, are much less subject to air currents than Newtonians and Cassegrains. A skeleton tube helps with the cooling of large-aperture telescopes. One possible source of air currents is also that of heat radiated by the observer or, above all, by careless visitors, who will stand underneath the objective or the end of the tube ...

5.1.3.5 Latitude of the site and declination

Because the planets are mostly confined to the ecliptic, it is obvious that observatories at high northern or southern latitudes are not particularly suitable for planetary observation. Seen from such sites, the planets generally appear low on the horizon, so seeing is obviously affected. On the other hand, observers in the tropics can see planets directly overhead.

Oppositions of planets that fall in the northern summer are not very favourable for observatories located at mid-northern latitudes (40–50°N), because the planets are then on the part of the ecliptic that is below the celestial equator. In 1984, for example, Jupiter had a declination of −23° and was therefore only about 20° above the horizon for an observer in southern France or the northern United States. On the other hand, the same observer would see the planet at altitudes of more than 60° at winter oppositions.

5.1.4 *Instruments suitable for planetary observation*

5.1.4.1 *Optimum aperture for planetary observation*

Given that seeing is a dominant factor in determining the resolution of astronomical instruments, one might ask whether it is really important to use very powerful telescopes for observation of the planets. Opinions are very divided on this point. Danjon and Couder maintain that it is always preferable to use the largest available aperture. They point out that stopping-down a telescope does reduce the effects of poor seeing but does not improve resolution, whereas it does, of course, reduce the amount of light collected. With an objective 840 mm in diameter (the large refractor at Meudon, for example), if the seeing is 0.04 arc-second one could fruitfully use a magnification of 1260×. But if seeing deteriorated to 0.25 arc-second, the magnification ought to be no more than 500× (and only 315× times if the seeing were only 0.5 arc-second). Most planetary observers have arrived at conclusions that are slightly different, because Danjon and Couder neglected two important factors:

(i) studies by Arnulf have shown that it is not advisable to use an exit pupil with a diameter greater than 1 mm, because of aberrations within the eye; and

(ii) using low magnifications with large diameters produces excessive brilliance in the resulting planetary images, causing the eye to be dazzled, which is very detrimental to detecting low-contrast features. (This may necessitate using a series of neutral density filters.)

In fact, as A. Dollfus has noted, in the majority of cases the best planetary drawings published show details whose average diameter varies around 0.4 arc-second, whatever the instrument used. Most observers therefore seem to agree that for planetary observation the optimum diameter is between 400 and 600 mm. Our fundamental knowledge of the physics of the planets and the Moon is primarily thanks to instruments with diameters between 300 and 840 mm. The largest reflectors do not seem to have added as much as one might have expected. (The problem is slightly different where photography is concerned.) The first maps of Jupiter's satellites were made with the 380- and 600-mm refractors at the Pic du Midi and not with the 2.5-m and 5-m reflectors.

As we have seen earlier, Pickering found that a 250-mm refractor is capable of showing a circular spot on a planet that is 0.2 arc-second in diameter, far smaller than the theoretical resolution (0.5 arc-second), because seeing an isolated object and separating two adjacent objects are completely different matters. It is well-known that the Cassini division in Saturn's rings, which subtends only 0.5 arc-second, may be seen with a good 60-mm refractor, which has a theoretical resolution of 2 arc-seconds.

One hundred years ago, Flammarion considered that a 108-mm refractor was the ideal instrument for an amateur, enabling high-quality planetary observations to be made. Yet the Planetary Section of the Société Astronomique de France has never received good planetary observations made with small 80–106-mm refractors. An excellent 102-mm fluorite refractor, which gives marvellous planetary images, does not show really fine detail, except occasionally the famous white ovals on Jupiter. On the other hand, very valuable observations have been obtained with 150-mm refractors (by L. Aerts using a Lichtenkneker instrument, and by Alexescu using a Zeiss). Many will remember the marvellous lunar and planetary photographs taken by G. Nemec at Munich, with a 200-mm folded refractor. Such a size enables useful observations to be made. We may also mention the refractor made for planetary observers by Chisten in the U.S.A.: the StarFire has a true apochromatic, three-lens objective, 178 mm (7 inches) in diameter and with a focal length of 1.6 m (F/D = 9). Although no longer being manufactured, it is undoubtedly an excellent piece of equipment despite being rather expensive. Few amateurs are able to afford refractors with diameters greater than 200 mm.

The ideal telescope for amateur planetary observers is therefore a Newtonian or Cassegrain reflector of 250 or 310 mm in diameter. An aperture of 310 mm is required for high-resolution photography. Ambitious amateurs sometimes install 410-mm reflectors and are disappointed at not seeing more details than those who make do with a slightly smaller aperture. At any ordinary site, even a 310-mm telescope is often somewhat too large for the available seeing. This is not to say that in particularly favourable sites a 520-mm reflector cannot give good results. We only have to look at G. Viscardy's work with the Moon to see what may be achieved. However, we feel that diameters larger than 520 mm are simply a snare and a delusion for amateurs. Unless, that is, they can occupy a site like the Pic du Midi!

5.1.4.2 Refractors, reflectors, or catadioptrics?

An old, but tenacious idea is that a refractor is the ideal instrument for planetary observation. This is based on the real advantages given by the lack of central obstruction and of a spider, with the resulting diffraction-limited image; the closed tube; and the relative ease (?) with which spherical surfaces may be figured. Unfortunately, refractors also have serious disadvantages such as very high price; significant, residual chromatic aberration (for large apertures and short focal ratios); excessive length; etc.

The situation has changed somewhat since the introduction of apochromatic, short-focal-length objectives (F/D = 8–12), whether they are doublets, using fluorite, or triplets, with more conventional glasses. But few amateurs can afford the only types that are of most interest: the 150-mm Takahashi or Goto fluorite instruments, or Chisten's 178-mm, StarFire refractor.

Reflectors therefore remain the most significant instruments for planetary observation. For visual observation, the Newtonian is difficult to beat, because it provides the observer with a fairly comfortable observing position, combined with a moderate price. It also enables one to obtain a high-quality parabolic mirror and to design a telescope with a very small central obstruction (15–20 %). Such an instrument is easy for an amateur to mount and may be used for excellent observations, and even for some high-resolution photography, provided the primary mirror is of good quality. (Don Parker, in Florida, remains the finest photographer of Mars and Jupiter, even though he only has a simple 310-mm Newtonian. This is, however, designed for high-resolution work.) Generally, a Cassegrain-type of mounting seems preferable for photography, because one immediately has a fairly long focal-length, and is thus able to obtain the final focal ratio required without any problems, using just the two mirrors and a Barlow lens. A Cassegrain for planetary work should be between 310 and 410 mm in diameter, and have a primary focal ratio of 5, and a secondary mirror giving a final ratio of 20–25, with a central obstruction amounting to 15–20 %. Its price is not excessive, especially if the observer is able to make a cradle-type mounting with a high-quality sector drive.

The catadioptric-type of telescopes (Maksutovs, Schmidt–Cassegrains) have become widely sold as amateur instruments. Their success is primarily thanks to their compactness and ease of use. But they are all-purpose instruments and are not high-resolution telescopes. Their optics are complex and cannot be produced to such a high standard as a simple Newtonian system. Their central obstruction is very large (as much as 40 %) and noticeably alters the diffraction image, reducing contrasts and seeming to be even more prone to poor seeing (because of the stronger first diffraction ring). In addition, these instruments are, above all, portable, and therefore light, so they are usually unsuitable for high-resolution photography (because of mounting vibration, the arguable accuracy of their drives, etc.).

5.1.5 Drawing and photography

5.1.5.1 A comparison of the two methods

Drawings have always been the basis of planetary studies. This is also the cheapest and easiest method. Unfortunately, you cannot become a planetary observer just by wishful thinking, and representing a complex, unstable, telescopic image on paper is never really easy. The great majority of planetary drawings produced around the world are poor, and a large number are frankly unusable. In addition, since Lowell and Douglass covered Mars and Venus with networks of imaginary, narrow canals, drawing has become discredited among professionals. This seems somewhat excessive, because there have always been fine artists who have been able to capture planetary detail accurately (Antoniadi, Ebisawa, and Dollfus among the professionals and Saheki, Osawa, McKim, Miyazaki and others among amateurs). Amateurs with a critical sense can soon determine whether their drawings are of value or not. Frequently, beginners have insufficient means of identifying their errors; this is why many drawings are more or less unusable.

Planetary photography is very often deceptive, being by no means easy, but it does allow objective records to be obtained. Since 1980, thanks to the appearance of Kodak TP 2415 film on the market, amateurs with good telescopes between 250 and 520 mm in diameter have been able to obtain excellent photographs of Mars and Jupiter.

5.1.5.2 *Making drawings*

Beginners, and all those who do not have sites that allow them to obtain good photographs, are forced to try drawing the planets. It should be noted that it is difficult to become a skilled planetary artist unless one has a certain facility for drawing. When one is sketching a planetary surface one is basically mapping it, so it is essential to be able to reproduce the details of the telescopic image as accurately as possible. Quite apart from having the natural gift, this also requires having had sufficient experience; both eye and hand can learn. It is important to beware of both the tendency to over-simplify and also the desire to produce 'works of art'. Even before looking at a planet through the eyepiece, it is possible to gain a good idea of one's own value as a potential planetary observer by carrying out various tests: drawing the Moon as seen by the naked eye, or 'artificial planets', etc.

It is best to start with a planetary body whose surface is well-known and relatively unchanging, such as the Moon or Mars. In this way it is easier to detect the errors that one is bound to make.

It is not easy to make drawings at the eyepiece. It is essential to try to be as comfortable as possible (preferably on a seat that is variable in height, which is reasonably easy to design and make). The paper should be fixed to either to a small drawing board, or else to a plywood base (using a spring clip screwed to the board). Some means of illuminating the drawing board needs to be provided, and this could be either the headband type, or a lamp fixed to the board with another spring clamp. The brightness of the light should be neither too dim nor so great that is causes dazzle, and it should be possible to switch it off at any time.

Drawings should be made on planetary blanks (which may be obtained from the various organisations coordinating planetary studies). A circular disk is used for Mars (Fig. 5.3), and a larger, flattened ellipse for Jupiter. If you do not feel confident of making a finished drawing at the eyepiece, it is better not to use these printed blanks, but to prepare your own on ordinary paper. Generally, soft pencils (B, BB and even 3B), erasers and 'stumps' are used for drawing. Once the equator and the central meridian have been indicated, it is necessary to work fast, allowing 15 minutes for Jupiter and 25 minutes at the very most for Mars. A typical report form is shown in Fig. 5.4. The principal features are first noted, details being added later. The exact time (in UT) of noting down the main features should be recorded (e.g., 1988 March 3, 02:10 to 02:25, features recorded at 02:15). The intensity of each spot may also be recorded, using the scale proposed by de Vaucouleurs (0 = brightest part of the image, such as the polar caps on Mars, 10 = point as dark as the sky background) [or the reversed scale where 0 = black and 10 = brilliant white as used by ALPO – Trans.]. Some coordinators of planetary observation programmes give considerable weight to such intensity estimates. In fact, it is not particularly easy to estimate them, and the figures given by observers often show considerable discrepancies.

Some observers feel that it is essential to make a final drawing at the eyepiece, but the majority prefer to make a 'clean' drawing from the sketches made at the telescope. As the eye becomes familiar with the image, more and more detail is seen, which means that the observer has to make frequent amendments, or else make annotations or detailed sketches in the margin of the main drawing. A clean copy

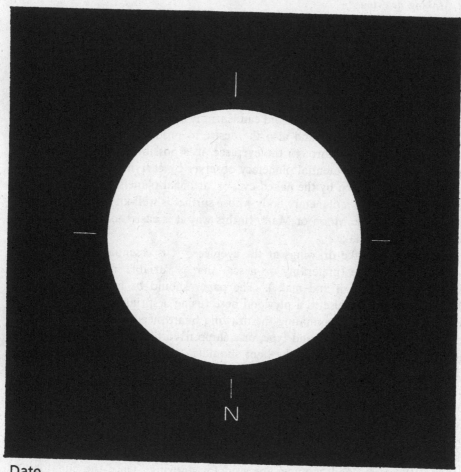

Date		$\omega =$	Start (UT)
		$\phi =$	Middle (UT)
$\eta =$		$V =$	End (UT)
Instrument		Refractor/Reflector	Aperture(mm)
			Magnification
Filter			Seeing quality
Turbulence		Transparency	Wind
Location			Observer
Remarks			

Fig. 5.3. *A Mars blank (SAF); scale 1:1*

BRITISH ASTRONOMICAL ASSOCIATION

JUPITER SECTION REPORT FORM

Name_____ Location_____

Date_____ Start_____UT Finish_____UT

Telescope_____ Magnif'n_____ Seeing_____

GENERAL NOTES

TRANSITS

Feature	Time (UT)	λ_1	λ_2

LEFT:

Time (UT)_____

CM: ω_1_____

CM: ω_2_____

Seeing_____

RIGHT:

Time (UT)_____

CM: ω_1_____

CM: ω_2_____

Seeing_____

Fig. 5.4. *A report form for recording observations of Jupiter as used by the British Astronomical Association but typical of those used world-wide.*

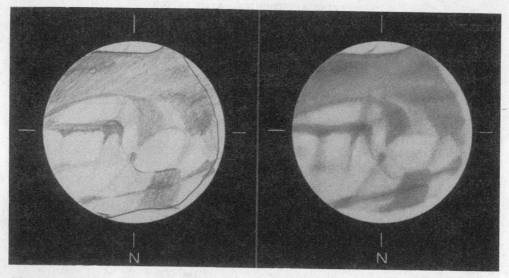

Fig. 5.5. *Planetary drawings:* left, *preliminary sketch, prepared at the telescope;* right, *the final drawing (made with* HB *to* 3H *pencils, stumps and erasers).*

may be made by tracing the original over a bright light, including the appropriate corrections, and tidying up the drawing so that, if necessary, it may be reproduced. For a final drawing harder pencils (HB) are used and stumps and erasers freely employed (Fig. 5.5).

A drawing is valueless unless it is accompanied by a full set of indispensable data such as: instrumental type and diameter; magnifications used; date and time (in UT, accurate to one minute in the case of Jupiter); and seeing conditions. With regard to the last item, some authors recommend the use of the ten-point Tombaugh–Smith scale (based on that of Pickering), where 1 = dreadful images, 5 = average, 10 = perfect. There is also Texereau's scale (I = dreadful, and V = perfect); Dollfus recommends the Antoniadi scale: from 1 (perfect) to 5 (bad). He also suggests – and this seems particularly interesting to us – the use of a scheme that takes the overall observing conditions into account (thus including the wind, transparency, the observer's comfort, etc.). Here one quotes an overall observational class on a scale of 1 to 5, where 1 indicates the best possible conditions. Some observers also note down the transparency of the sky, on an equally empirical scale. In general it suffices to note down transparency on an arbitrary scale of 1 to 5 (1 = planet scarcely visible through haze; 5 = absolute clarity). Apart from all this, the vital astronomical information should also be recorded: diameter of the planet; phase; phase angle; latitude of the centre; longitude of central meridian (ω); heliocentric longitude, etc. It is also useful for the observer to give written comments about any specific aspect of the observation.

It is essential to use coloured filters to improve the contrast and visibility of certain details. Gelatine Wratten filters are generally used, cut into small squares that are mounted with Canada balsam between two microscope slides. From the colours available in the Wratten range, those that are purest and have the sharpest cut-off are used: red (W25 or W29), yellow (W12), green (W53), blue (W80c or W47), and blue-violet (W49).

5.1.5.3 Planetary photography

In certain respects, this is the most difficult form of astrophotography. The difficulty arises from a completely random, and generally uncontrollable factor, the quality of the telescopic image. The eye, thanks to its essentially instantaneous ability to seize fleeting detail, can profitably use even a very turbulent image. Details of the image become clear little by little. Unfortunately, photographic film remains too slow to capture such images, because exposure times are generally between about 0.5 and 8 seconds, which means that the image must be stable for fairly long periods. High-resolution planetary photography is impossible without stable images (with seeing less than 0.4 arc-second). But the vast majority of sites are only capable of giving poor or bad images. The places where excellent planetary photography has proved possible may be counted on the fingers of both hands: Pic du Midi; La Turbie (near Monaco); Flagstaff, New Mexico; Catalina Observatory; Lick; Johannesburg; Tiede; etc. It is therefore quite pointless being the proud possessor of an excellent, large-aperture telescope, and having a faultless photographic technique, if the site is dreadful. For example, Jean Dragesco was highly successful during his time at Cotonou in Benin, where he was able to make hundreds of fine planetary photographs with a C14 telescope, whereas with the same telescope and with the same methods he was quite unable to obtain a single useful photograph at Butaré in Rwanda. Unfortunately, the same applies to the vast majority of astronomical sites, whether they are in the lowlands or up in the mountains. It is not even necessary to waste film unnecessarily, because visual observation will readily show whether scientifically useful, planetary photographs (showing details below 2 arc-seconds across) are likely to be obtained. It is sufficient to observe a planet (Mars or Jupiter, for preference) with a cross-hair eyepiece that gives an overall magnification of at least 300 times. The behaviour of the image is closely followed: if the planet shifts, relative to the graticule, by more than 3–4 arc-seconds over 2 seconds, obtaining high-resolution photographs is impossible. An experienced observer soon learns to judge the quality of a planetary image, and will know in advance if it is, or is not, possible to obtain a good photograph.

Other factors are also involved. It is essential to have a sufficiently powerful telescope that is of excellent optical quality. This might be a 310-mm Newtonian or Cassegrain, with perfect mirrors and small central obstruction. The famous Munich amateur, G. Nemec, obtained marvellous planetary and lunar photographs with a 200-mm, f/20 object glass, mounted as a folded refractor with two additional plane mirrors. In 1986, an excellent photograph of Mars was obtained in Japan, using a fluorite refractor of only 150 mm diameter. Refractors, even those of moderate aperture, are sometimes able to give valuable results. A high-quality reflector, 310 mm in diameter, is capable of recording photographically dark spots 0.4 arc-second across, and so it is able to obtain high-resolution planetary photographs. But it is essential for the optical quality to be beyond reproach. Don Parker, in Florida, was obtaining good planetary photographs, but he then had his 310-mm mirror re-figured. Immediately, the quality of his photographs improved. It is also important to have a stable mounting and as good a drive as possible.

Whether the telescope is a Newtonian or a Cassegrain, the image of a planet at the prime focus is always too small. There is a lot of disagreement about this point. Some people (such as Don Parker) like very long focal ratios, f/150–f/200, giving large planetary diameters (5–10 mm), which require relatively long exposures (between 3 and 8 seconds). Others (such as Georges Viscardy and Jean Dragesco) prefer to use low focal ratios, f/40–f/80, giving quite tiny planetary images (3–6 mm), and short exposure times: 0.5–2 seconds maximum. (This all applies to Mars and Jupiter, of course.) Given the exceptional resolution of Kodak TP 2415 film and its extremely fine grain, a small image is no handicap. (The image of Mars that Dragesco obtained on 1986 July 15, using the 1-m telescope at the Pic du Midi, with a focal ratio, of f/52, measures just 5 mm across; yet is remains one of the finest pictures of Mars ever taken.) Reasons for liking a short focal ratio are that accurate focussing (without which the image is unsharp) is easier, and that the short exposures are better able to tolerate a slight amount of turbulence or inaccuracies in the drive. (This is easy to test. Observe the planet though a cross-hair eyepiece giving a magnification of at least 300×. The planet should stay absolutely stationary for at least 2–3 seconds at a time.) Obtaining the required focal ratio (f/30–f/50 for Jupiter, f/50–f/80 for Mars), is done by a magnifying system, the type used being dependent on the initial focal ratio. If one uses a primary focal ratio of 6, Plössl or orthoscopic eyepieces that have short focal lengths of 6–8 mm should be used. If the primary focal ratio is longer, say f/25, then 2× or 3× Barlow lenses will suffice.

It is essential to be able to obtain a perfectly sharp focus. For this a special camera body has to be used (unless one has an instrument that is designed specifically for planetary photography, like Don Parker's). Focussing through a normal SLR body, with a pentaprism and a ground-glass screen is very uncertain. It should be possible to remove the pentaprism and the ground-glass screen, and replace them with a direct-focussing arrangement (as with the Nikon DW-4, for example), or a clear screen with cross-hairs (Nikon's F3M). The Olympus SLR body offers a compromise: the ground-glass screen may be replaced with a clear, reticulated screen and the 'Varimagni' finder used for focussing. But the ideal solution is that provided by Nikon, Canon, Pentax and Practica (direct focussing with a clear, reticulated screen) using bodies that are, unfortunately, frequently very heavy. It should not be forgotten, however, that it is always possible to use a Foucault test to focus on a bright star before turning the telescope to the planet to be photographed.

Two accessories are very useful for planetary photography: a motor drive (thus avoiding the necessity of touching the camera to wind on the film), and a data back, which can automatically register the date and time (in Universal Time, of course) on the negative.

With amateurs' light-weight telescopes, it is impractical to use the camera shutter for the exposure. The vibration caused when the mirror flips up out of the light-path may affect even a 310-mm telescope weighing more than 100 kg. It is therefore necessary to use a 'shutter' consisting of a sheet of light plywood, held over the mouth of the tube whilst the camera shutter is opened (set to B – time exposure). After waiting a moment for the vibrations to die away, the exposure (1/2 to 3 seconds) is then given using the plywood shutter (but without touching the telescope tube). It only remains to close the camera shutter and to note down the time in UT, before making another exposure. If the telescope is long, this exercise may often require a bit of acrobatics, and be quite tiring. Of course, heavy instruments (like Georges Viscardy's 520-mm or the 1.06-m at Pic du Midi) withstand vibrations from SLR bodies, which makes life much easier. (Don Parker has resolved the problem in another way by using an anti-vibration-mounted central shutter, on a special camera back.)

Planetary photographs are generally taken in black-and-white, which gives better resolution and contrast. The work of amateur astrophotographers has been made much easier since the introduction of the 'miracle' film, Kodak TP 2415, which has a very high resolution, high contrast, and imperceptible grain. Although its sensitivity is only 25 ISO when it is used for 'pictorial' photography (with the special developer Technidol), its sensitivity approaches 150 ISO, when developed in Kodak D19b for 6 minutes at 20°C (which is excellent for Jupiter); and even 200 ISO or more when developed in Kodak HC-110 (dilution 1 + 15), for 6 minutes at 20°C (giving lower contrast, which is excellent for Mars). Under the given conditions (f/40–f/60 for Jupiter, f/60–f/80 for Mars; TP 2415 film, developed to exhaustion), the exposure times range between 1/2 second and 3 seconds (depending on the atmospheric transparency and the telescope's transmission ratio). Some trials are required; it is, in any case, always a good idea to bracket the exposure time found by trial.

Most amateurs photograph Mars and Jupiter in integrated light, feeling that they do not have enough light to use colour filters. It is, however, repeatedly necessary to use 2 filters: a Wratten 29 (red) which increases contrast on Mars, and improves detection of fine detail on Jupiter; and a Wratten 49 (blue-violet), which is excellent for studying atmospheric detail on Mars, and which also accentuates reddish detail on Jupiter (Figs. 5.6 and 5.7). Unfortunately, the latter filter absorbs very strongly when used with TP 2415 film (giving 8-second exposures at f/52 for Mars!).

Colour photography is very tempting, because it shows the planets as they appear to the eye. Unfortunately, colour films have much lower contrast and resolution than TP 2415. The images are therefore much less detailed than in black-and-white. With 250–350 mm telescopes it is necessary to use faster (and therefore more grainy) films, such as Ektachrome and Fujichrome 200 or 400 ISO. If you develop these colour films yourself, you can over-develop them slightly to increase the contrast. Enlargement onto colour reversal paper is difficult and should be undertaken either by someone specializing in astronomical photography or by the person who has taken the transparencies. It is not advisable to use colour negative film as the resolution is even worse.

5.1.6 *Use of visual and photographic observations*

It is not sufficient just to accumulate planetary drawings and photographs. You must also know what to do with them. It is quite a good idea for amateurs to reduce their observations themselves (to a certain extent, at least). This may take the form of determining the positions of the main features on Jupiter (the Great Red Spot, the white ovals, disturbances, etc.), measuring the shrinking of the polar caps on

Fig. 5.6. *Photographs of Jupiter taken by J. Dragesco on 1986 July 20, using the 1.06-m, f/52 telescope at the Pic du Midi.*

Top: *image without filters, 03:13 UT; $\omega_1 = 307°$; $\omega_2 = 332.4°$.*

Centre: *image with a red filter (W29) at 03:14 UT; $\omega_1 = 307.6°$; $\omega_2 = 333°$. Note the dark patches on the southern border of the northern equatorial belt (NEB). These features are bluish.*

Bottom: *image with a blue filter (W49) at 03:15 UT; $\omega_1 = 308.2°$; $\omega_2 = 333.6°$. The dark patches have disappeared, but three belts now appear particularly dark, and are therefore reddish in colour. On the sunrise limb, the Great Red Spot may be seen, while on the sunset limb, white oval DE is visible in the south temperate belt (STB).*

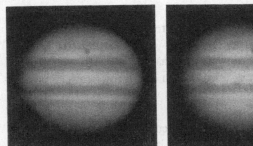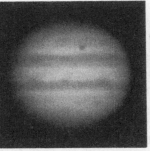

Fig. 5.7. *Photographs of Jupiter by T. Ishibashi, 1986 July 31, taken with a 21-cm reflector. Compare with Fig. 5.6.*
Left: *image without filters; 16:49 UT; $\omega_1 - 22.5°$; $\omega_2 - 319.7°$.*
Centre: *image obtained with a red filter; 16:37 UT; $\omega_1 = 15.2°$; $\omega_2 = 312.4°$.*
Right: *image with a blue filter: 16:42 UT; $\omega_1 = 18.2°$; $\omega_2 = 315.4°$.*
Note the different appearance of the two dark features at the northern edge of the North Temperate Belt. The white oval DE is also visible, together with Ganymede over the South Tropical Zone.

Mars, and so on. In trying to understand what has been recorded, observers will find their subject become even more engrossing. But it should be obvious that the results obtained will be of minor interest if one has only one set of observations to analyse. The value of any analysis increases with the number and quality of the records that are available. This is why it is essential for an observer to be affiliated to one or other of the international groups that handle analysis and publication of planetary observations.

There are various national groups, and certainly all amateurs should get in touch with at least one such organisation. However, this type of research ought to be carried out on an international scale. We therefore strongly recommend that observers should join such an international group. For about a century, the Mars, Jupiter and Saturn Sections of the BAA have carried out very important work, and published extremely high-quality reports. (Contact R. McKim for Mars; J. Rogers for Jupiter; and A. Heath for Saturn: c/o British Astronomical Association, Burlington House, Piccadilly, London W1V 9AG, U.K.)

Another organisation of international repute, ALPO, publishes *The Strolling Astronomer*, which is entirely devoted to amateur planetary work. The Mars Section is directed by Don Parker (12911 Lerida St., Coral Gables, Florida 33156, U.S.A.); for Jupiter contact P.W. Budine (P.O. Box 126, Plymouth, NY 13832, U.S.A.)

There are also other groups, the most important of which is that based in Japan, but language barriers mean that they are of limited interest. A good observer who forwards acceptable results to at least one of the internationally recognised organisations can feel that significant work has been achieved.

5.2 Jupiter

Even for a beginner or someone who is just curious, there is no problem in locating Jupiter in the sky – provided it is above the horizon, of course. It is brighter than any other body, which makes it obvious to even the least experienced eye. It may even be seen from the centre of towns and cities. When one knows roughly where it is and that it is visible in the evening, it may be found with the naked eye shortly after sunset.

It may also be found in daylight, using a pair of binoculars mounted on a tripod. For this, however, you need to know the time at which it crosses the meridian and its declination. This information may be found in any astronomical ephemeris. It is then only necessary to point the binoculars due South, at the time indicated, and at an altitude that is given by $h = 90° - \phi + \delta$, where ϕ is the latitude of the site, and δ is the planet's declination (positive or negative). The planet will appear as a tiny disk, with a quite distinct sharp edge, but does not show any detail.

At night, even the smallest telescope – provided it is braced against (say) a window-frame, or mounted on a tripod – will show a distinctly flattened disk with a dark belt on each side of the equator. The four satellites discovered by Galileo Galilei in 1610 are also easy to see with the smallest optical aid. It also seems that some people, with particularly keen eyesight are even able to detect Ganymede, the brightest of the four satellites, when it is at greatest elongation (Fig. 5.8).

5.2.1 Jupiter's apparent motion

The apparent motion of a planet is the movement as seen from Earth. With Jupiter, this movement is very easy to detect, especially around opposition, when it is most rapid. It is sufficient to take careful note of the planet's position relative to bright stars in its vicinity. In a week there will be no difficulty in detecting the planet's movement against the background stars. This is how early Man, who, of course, did not have the benefit of modern knowledge, recognized the existence of five 'wandering stars': Mercury, Venus, Mars, Jupiter, and Saturn. Unfortunately, few people nowadays realise that these five planets are visible to the naked eye as bright points of light. There is no real need for any detailed ephemeris to find them with a telescope. The general information often given in various journals (for example, that in 1987 October, Jupiter was in Pisces), is usually quite sufficient to locate the planet concerned.

Returning to the matter of Jupiter's apparent motion, this takes place very close to the ecliptic, because the inclination of the giant planet's orbit relative to that of the Earth is only 1.3°. Jupiter may therefore be observed in Taurus, between the Pleiades and the Hyades, and also close to some of the striking objects that lie close to the ecliptic: α Leonis (Regulus); α Virginis (Spica); α Librae (Zuben Elgenubi); M20 (the Trifid Nebula); the fine gaseous nebula M8 (the Lagoon Nebula); the wonderful globular cluster M22; and M44 (Praesepe), the well-known open cluster in Cancer. Anyone who is very patient would see Jupiter return to the same point on the sky, approximately 11 years and 314 days later, after having passed successively through all 13 constellations that are crossed by the ecliptic. This interval corresponds to the

Fig. 5.8. *Photograph of Jupiter by J. Dragesco, 1986 July 10, using the 1.06-m reflector at the Pic du Midi at f/52. This extraordinary photograph, which was obtained with a red (W29) filter, shows detail that is approximately 0.5 to 0.2 arc-second across. On the South Temperate Belt, white oval BC is at the sunset limb, and DE to the right of the central meridian. Ganymede is approaching the planet from the East.*

amount of time required for the planet to make one complete orbit of the Sun. Its motion is therefore very slow when compared with those of Mercury, Venus, Earth and Mars. Seen from the Sun, Jupiter moves about 30.3° every year. Oppositions occur about every 398 days. During the months that the planet may be observed around opposition, it stays within no more than one or two adjacent constellations. Its apparent motion, which is generally towards the east, becomes retrograde (i.e., towards the west) during the four months around the date of opposition. Then, about a month before eastern (evening) quadrature, it starts to move east again until a date about one month before western quadrature. Its motion is precisely what would be expected of a superior planet, and consists of a series of retrograde loops, which are about 10° wide in right ascension, that take place around the time of opposition. These loops occur at approximately every 36° in right ascension along the ecliptic. As Jupiter describes these loops, it comes into conjunction with

Fig. 5.9. *Conjunction of Jupiter and the Moon, 1980 January 7 at 04:00 UT. Photographed by J. Dragesco, using a 200-mm reflector at Cotonou in Benin.*

objects inside or just north and south of these loops, on three different occasions. These three conjunctions may present opportunities for obtaining some interesting photographs, when Jupiter is 'close' to some notable object, as was the case with M22 in 1984. Other photographs may also be attempted when, each month, the Moon appears to pass close to the giant planet (Fig. 5.9) or when other planets are in the vicinity.

5.2.2 The best time to observe Jupiter

In this section, we shall discuss observation of Jupiter from mid-northern latitudes such as those prevailing in metropolitan France, southern Canada, and the northern United States.

Let us first assume that the planet is at the highest point of the ecliptic. It is then in Gemini and its declination is approximately +23°. Jupiter rises in the northeast and sets more than 16 hours later in the northwest. It culminates at about 70° and as a result turbulence should be low, giving good observational conditions.

Unfortunately, any observing campaign has to be carried out around the winter solstice, and runs the grave risk of being interrupted by the cold and bad weather. The last winter opposition was in 1989 December, after which the planet began to move down towards the southern hemisphere, crossing the celestial equator in Virgo (as it had done in 1981). This happened in the spring of 1993, and the second half of the apparition occurred under relatively favourable weather conditions. Because Jupiter's declination was then about zero, the planet rose in the east and set in the west 12 hours later. It culminated at only 45°, so although its image was still fairly good, it began to deteriorate more and more as it got lower in declination. The lowest point will come in Sagittarius, at declination −23°, in 1994 June. The situation in 1996 will be similar. The planet will rise in the southeast and set 8 hours later in the southwest, and will culminate at about 20° only. The image will generally be very poor, and Jupiter will have to be observed around the time of meridian passage, so there will be just one or two hours available for observation in any one night. On the other hand, the weather should be fine. After this unfavourable opposition, the planet will slowly move back into the northern hemisphere, which it will enter about three years later in Pisces, as in the spring of 1987. The conditions at rising and setting, meridian passage, and length of visibility will be similar to those encountered 6 years earlier. Opposition will occur in September and October, so that both meteorological and seeing conditions throughout the observational period will be fairly favourable. Three years later, we shall be back to a winter opposition, with what should, in principle, be good images – but unfavourable weather.

5.2.3 Observation of Jupiter

Jupiter orbits the Sun at an average distance of 5.2 AU: this means that it is about 6.2 AU from us at conjunctions, and about 4.2 AU at oppositions. In the second case, the apparent diameter is a maximum, and varies between 45 and 50 arc-seconds, according to whether the planet is at aphelion, as in 1981 July, or near perihelion as in 1987 July. For a planetary image this is exceptionally large, and the distinctly flattened disk is always seen fully illuminated. The phase effect is detectable, especially around quadrature, from the fact that the eastern limb is brighter than the western limb, at western quadrature (in the morning). At eastern quadrature in the evening, the opposite effect is observed. At opposition, no phase effect is visible.

As seen from Earth, Jupiter's illuminated disk covers a greater area of the sky than those of all the other planets put together. Nevertheless, the disk still appears very small, even with a magnification of 300×. Such a magnification does, however, give an image with an apparent area that is 50 times as large as that of the Full Moon seen with the naked eye.

The detail visible on Jupiter depends on the instrument used. As we have seen, binoculars show a tiny, bright, featureless disk. A small spy-glass with a diameter of 30 mm, and magnifications of 15, 30 or 45×, will show the two equatorial belts, the North Equatorial Belt (NEB) and the South Equatorial Belt (SEB), above and below the equator respectively [for a non-inverted image – Trans.]. With a 60-mm telescope and a magnification of around 100×, a third belt may easily be seen between the

SEB and the southern limb. This is the South Temperate Belt (STB). The bright band between the SEB and the STB is the South Tropical Zone (STrZ). It is here that the Great Red Spot (GRS), first seen by Robert Hooke in 1664, is located. With the 60-mm telescope, the GRS will appear as a small spot, slightly darker than the surrounding area. It lies in an oval 'hollow' that cuts into the southern edge of the SEB. Some large plumes may be glimpsed in the zone between the two equatorial belts. The polar regions are darker and less varied than the rest of the planet. To see the most detail, however – subject to the effects of the Earth's atmosphere – it is necessary to use a telescope of at least 200 mm in diameter, which will give a magnification of 200–300 times. Results will usually be even better with a 300-mm telescope. Larger instruments, such as 400 and 500-mm reflectors, do not show any more detail most of the time, because of atmospheric turbulence. This is why large telescopes are not noted for giving particularly fine planetary images. Their role is simply to collect as much light as possible from faint objects. As far as planetary images are concerned, amateur-sized telescopes are almost as effective as the very largest telescopes. This always surprises people – some simply do not believe it – if they have no experience of the deleterious effects of poor seeing on telescopic images.

The most striking thing about Jupiter is the alternation of bright zones and dark belts. These features, which are more or less permanently visible in Jupiter's atmosphere, always remain at approximately the same latitudes and their nomenclature is well-defined (Figs. 5.10 and 5.11). Their presence may be explained by fairly satisfactory mathematical models. The cloud features that develop at any particular latitude have a rotation period that is characteristic of the belt or zone in which they form. Between latitudes of roughly 12°N and 12°S, cloud features are essentially stationary relative to a system of coordinates, known as System I, that rotates with a period of $9^h 50^m 30.0^s$. Features lying outside this region are essentially stationary relative to a second system of coordinates, known as System II, which rotates in $9^h 55^m 40.6^s$ (Figs. 5.12 and 5.13). There is effectively no transitional period between these two systems, so that the regions between them are subject to significant shear effects, which give rise to exceptionally violent, localized winds. There is, for example, a displacement of 7.6° per day between two neighbouring cloud features that belong to the different coordinate systems (Fig. 5.13). This shift, caused by the differential rotation, is easily seen from drawings or photographs of the planet made at an interval of two or three days.

5.2.4 Description of the various regions of the planet

The description that follows applies to the image of the planet as seen with a 200-mm or larger telescope, used under average seeing conditions. It assumes that the image is inverted, i.e., with south at the top and north at the bottom. East and west are defined using the same conventions as on Earth. In other words, an observer standing at the North Pole would 'see' the planet rotating from west to east. West is therefore to the right of the telescopic image, and east to the left. Any latitude

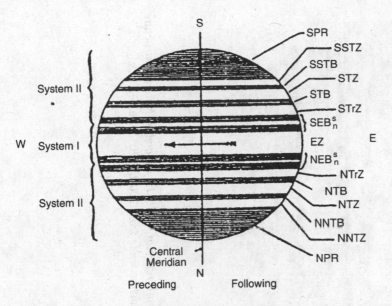

Fig. 5.10. *Drawing showing the zones and belts on Jupiter, as well as both System I and System II coordinates. Readers may wish to refer to this diagram when reading the description of the various features on the planet.*

mentioned later is the zenocentric latitude, which is defined as the angle between the radius that passes through the point concerned and the plane of the equator, measured in the plane of the meridian.

5.2.4.1 *The South Polar Region (SPR)*

This region appears as a dark, featureless hood, which extends from the southern limb to about latitude 50°S, or even 40°S. Within it, distinct features are exceptionally rare. The region just below it is a narrow bright zone, not always visible, known as the South South Temperate Zone (SSTZ). Under very exceptional circumstances it may exhibit certain circular, white, low-contrast spots, which are impossible to follow throughout an apparition. Farther down, we find a dark belt known as the South South Temperate Belt (SSTB).

5.2.4.2 *The South South Temperate Belt (SSTB)*

This belt lies at about latitude 40°S. It often goes unnoticed by many observers at times when it is close to the northern limit of the South Polar Region. It may appear as a narrow, greyish band, or as a series of dark segments, which may be followed to determine the rotation period of the South South Temperate current, which is approximately $9^h\,55^m\,5^s$. The visibility of the SSTB seems to follow a cycle in which it varies very considerably in intensity.

5.2.4.3 *The South Temperate Zone (STZ)*

This is a bright zone between the SSTB and another belt farther north, known as the South Temperate Belt (STB). Bright, circular, white spots often develop in the

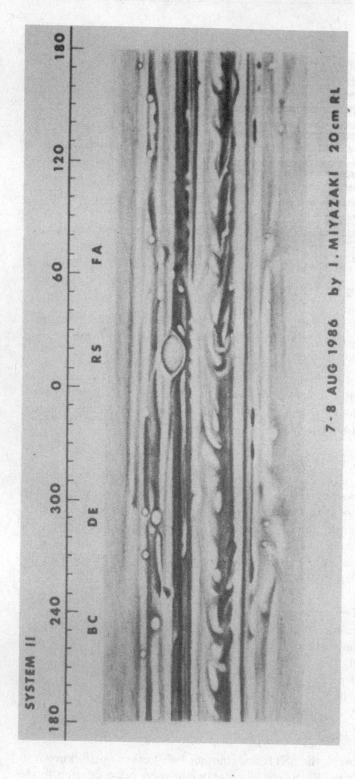

Fig. 5.11. Strip-chart of Jupiter by I. Miyazaki, obtained with a 200-mm altazimuth reflector, 1986 July 7 and 8. Note the structure within the belts and zones and the extremely fine detail. Compare this region, which extends from 220 to 320° with Fig. 5.8.

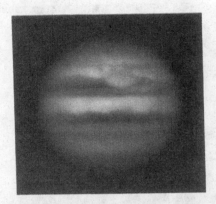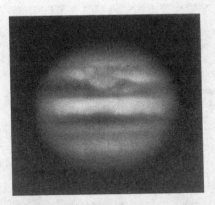

Fig. 5.12. *These two photographs, taken on 1985 August 4, 20 minutes apart, show Jupiter's rapid rotation. Left: 23:08 UT and right: 23:28 UT. The small white feature to the left of the GRS is white oval BC. The North Temperate Belt is particularly dark. Photographs by F. Nakagami, using a 290-mm, f/7.7 reflector.*

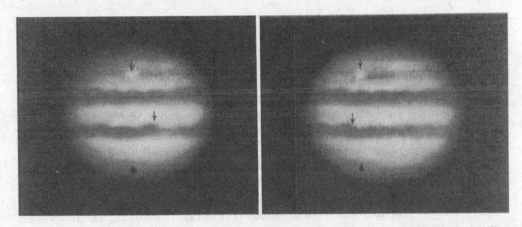

Fig. 5.13. *Photographs of Jupiter by J. Dragesco, using a Schmidt–Cassegrain, 355-mm telescope at Cotonou in Benin. Left: 1984 June 8, 02:16 UT ($\omega_1 = 125.7°$; $\omega_2 = 281.8°$). Right: 1984 June 11, 00:04 UT ($\omega_1 = 159.4°$; $\omega_2 = 293.3°$). The arrows indicate the difference in rotational velocity between the equatorial and temperate regions. The differential rotation amounts to 22.2°.*

South Temperate Zone, and move along the southern border of the STB (Figs. 5.8, 5.10 and 5.11). These white spots frequently appear like a string of beads around the planet. They also form part of the South South Temperate current. It needs good seeing for them to be seen, but it is possible to follow them fairly easily on photographs made on TP 2415 film. Their formation appears to be favoured by the presence of three white ovals in the South Temperate Belt (Fig. 5.14).

213

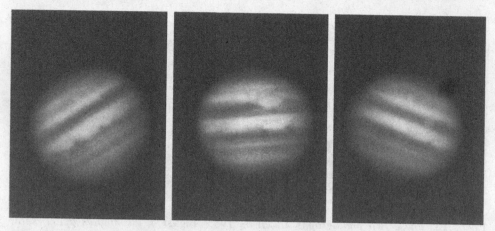

Fig. 5.14. *The three white ovals and the GRS, photographed by G. Viscardy in 1985, using a 520-mm reflector. Left: white oval FA, 1985 July 18, 00:39 UT. Centre: white oval BC and the GRS, 1985 August 14, 22:06 UT. Right: white oval DE, 1985 July 15, 00:20 UT. Note also the considerable strength of the North Temperate Belt.*

5.2.4.4 *The South Temperate Belt (STB)*

This is one of the three belts that may be seen easily, even with just a 60-mm telescope. It lies at 30°S, half-way between the equator and the southern limb. Its width varies considerably over the years and may attain as much as 10°, as in 1980. If conditions are good, it is possible to see that this belt is far from homogeneous. It often has dark patches separated by brighter areas, and it may appear double at certain longitudes. Above all, it contains the white oval spots, known as FA, BC and DE. These cyclonic features were discovered in 1930 on the southern edge of the belt. Since then, they have continued to shrink in size. The visibility of FA has become rather problematic since about 1983. In 1986 it could only just be seen in the vicinity of the Great Red Spot. The visibility of white oval BC is distinctly better and it occupies about 8° in longitude (Fig. 5.12). Towards the end of 1974 an immense bright area developed at its eastern edge. This is a cyclonic region, which appears as a fade of the STB between the white ovals FA and BC. This feature, which grew to 55° of longitude by 1975 July, continued to increase until it covered 140° by 1983. Since then it has tended to decay, allowing the full extent of the STB to reappear. White oval DE is very easy to see, because it lies fully within the STB, in a region where this is particularly dark. The average period of the white ovals varies slowly over the course of the years and seems to be approximately cyclic. It may be taken as indicating the period of the STB. In 1985 this period was $9^h 55^m 22^s$ as against $9^h 55^m 17^s$ in 1977–8. Frequently, other more-or-less permanent features, similar to the white ovals, are seen in the STB, as in 1977–8 and 1986.

5.2.4.5 *The South Tropical Zone (STrZ)*

This is the bright band between the STB and the SEB. This is where the Great Red Spot is found, and where the South Tropical Disturbances occur. The width of this

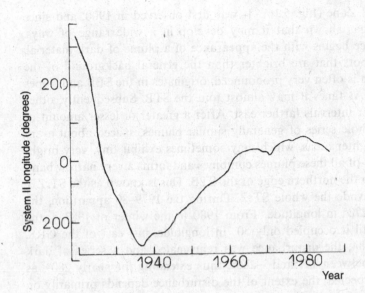

Fig. 5.15. *The drift in longitude (System II) of the GRS between 1920 and 1986.*

zone varies according to the strength of the South Equatorial Belt lying immediately to its north.

The Great Red Spot is undoubtedly the most famous feature on the giant planet. It is a gigantic cyclone, lying at about 20°S. In 1965 and 1966, Rees and Smith showed, photographically, that cloud features circulating on its outer edge made a complete, anticlockwise circuit in about 12 days. This rotation of the spot was confirmed by the Voyager probes in 1979 March and July. At present, the GRS extends about 20° in longitude, but the 'hollow' that it creates in the SEB may have a major axis of slightly more than 25°. The visibility of the GRS varies with time. From 1972 to 1975 it was very dark, even in tint, and completely distinct in the centre of the STrZ. Since 1975, it has become pale, but a little darker in its southern part. In 1984, it was almost invisible; its presence was indicated by the dark cloud-masses circulating around its edge, and it appeared like an enormous eye outlined in black. The visibility of the GRS appears to be directly related to the activity of the southern component of the SEB, as well as with the presence of the South Tropical Disturbance in its vicinity. The motion of the GRS is very slow, because its average period is very close to that of System II, i.e., 9^h 55m 40.6^s. Nevertheless, an independent motion does exist, and this is erratic, being sometimes westward (increasing longitude), and sometimes eastward (decreasing longitude). Between 1937 and 1980 the motion was westward, with a short interval of eastward motion between 1967 and 1971 (Fig. 5.15). The longitude of the GRS thus increased from −223° to +59° in 43 years. Since the beginning of 1980 May, the GRS has been slowly moving eastwards like the other cloud features that are normally present in this region of the planet. In 1986, it reached longitude 20°, where it was in 1926, 1964, 1969 and 1971. When will the motion again reverse? No one knows, but it is up to us to detect it.

The South Tropical Disturbance is an upper-atmosphere phenomenon that affects

Jupiter's South Tropical Zone (Fig. 5.16). It was first observed in 1900, and since that time observation has shown that it may develop in a wide range of ways. Typically, the disturbance begins with the appearance of a plume of dark material, often framed by two spots that are brighter than the general background of the STrZ. This plume, which is often very pronounced, originates in the SEB and slopes towards the southeast. At times it may almost join the STB. Subsequently, other plumes appear at regular intervals farther east. After a greater or lesser amount of time, there may be a whole series of generally similar plumes, spaced about every 15° and separated by lighter areas, which may sometimes exhibit tiny, very bright spots. The southern part of all these plumes combines and forms a very narrow band running along parallel to the northern edge of the STB. This is known as the STrZB. The disturbance may invade the whole STrZ. During the 1979–80 apparition, the phenomenon occupied 270° in longitude. From 1980 to the winter of 1983–4, the disturbance decayed until it occupied only 50° in longitude, just east of the GRS. At the beginning of 1984, the disturbance was rejuvenated and a series of dark plumes became visible between 17° and −260°, thus extending for nearly 280°, as in 1980. As might be expected, the extent of the disturbance depends primarily on the difference in velocity of the leading and trailing plumes. When the plumes reach the western edge of the GRS, they are affected by the circulation around it, and reach the eastern edge in about six days. The Red Spot then appears very pale and ringed by dark material. It seems that the plumes, which normally drift very slowly eastwards, are suddenly accelerated when they catch up with the Great Red Spot, advancing more than 20° in just six days. A very curious fact is that the Great Red Spot, which is an enormous feature, seems to be very sensitive to any disturbance. It has even been said that it behaves like a feather in the wind, which perhaps indicates just how sensitive it is to what would otherwise appear to be insignificant forces.

5.2.4.6 *The South Equatorial Belt (SEB)*

This belt may be observed in even a small, 30-mm telescope. It is about 12° wide. The SEB has two components, southern and northern (SEBs & SEBn), which may be separated by a narrow bright zone, the SEBz. The northern component, which runs parallel to the equator, is a permanent feature. The clouds forming it are carried along by the South Equatorial current, and their positions are measured in System I, with a rotation period of $9^h 50^m 30.0^s$. The southern component of the SEB is known to be subject to phases in which it fades, sometimes becoming almost invisible, or reduced to a very narrow, dark belt as in 1972, 1973 and 1974. The Great Red Spot then appears as a completely isolated feature in the middle of the very wide STrZ. The SEB then undergoes a revival, which may be very spectacular, as in 1975. Typically, these revivals take place as follows: a round, white spot appears on the southern edge of the northern component of the SEB. In front of this spot (i.e., on the eastern side), a dark plume appears, sloping towards the southwest. Its tip sheds small dark knots that eventually encircle the planet after about one hundred days. This causes a revival of the southern component of the SEB (Fig. 5.17). The white spot, which is the source of the revival, slowly drifts eastwards, whilst other dark plumes form at its leading edge. The base of these plumes is the SEBn, which becomes darker and darker (Fig. 5.17). After the great revival in 1975, the SEB has

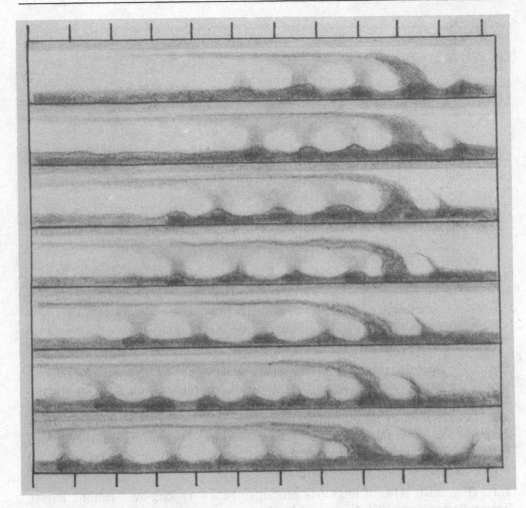

Fig. 5.16. *An example of the changes that may occur in the South Tropical Disturbance. Here, the leading plume, left, moves more rapidly than the trailing plume, right. The Disturbance extended towards the east as in 1979–80. The markers indicate 15°, and the interval between drawings is 15 days.*

always appeared in its entirety, with two components, because prominent centres of activity lying west of the Great Red Spot continue to feed material into it, as observed in 1980–1 and in 1985 (Fig. 5.18).

5.2.4.7 The Equatorial Zone (EZ)

This is the light zone, about 13° wide, that occupies the equatorial region of Jupiter. It is divided into two unequal parts by a narrow belt, which is sometimes difficult to see, called the Equatorial Belt (EB). The southern portion (EZs) is about 4.5° wide, as against 6.7° for the northern part (EZn). The EZs appears as a narrow, bright band, which occasionally shows dark projections that arise in the northern edge of the SEB and slope towards the northeast. Bright spots are also often seen in this

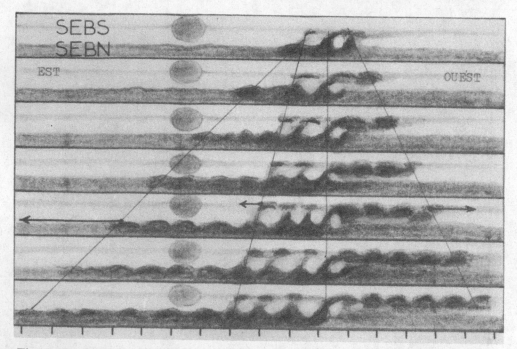

Fig. 5.17. *Diagram showing a classic rejuvenation of the SEB. The SEBs reforms, largely because a current carries dark material towards the west from an eruptive plume. The latter is almost stationary between the SEBs and the SEBn. Dark material also spreads towards the east along the SEBn, thus tending to reinforce it. These diagrams show that the SEB has a cyclonic structure. The markers indicate 20° and individual drawings were made every 5 days.*

zone (Fig. 5.10). One of them has proved to be a very long-lived feature, having been followed since 1975. It is carried in increasing System-I longitude by the South Equatorial Current, and passes the GRS about every 52 days. At present it is linked by a bright streak, known as the 'rift' to the white spots that lie in the SEB, west of the Red Spot (Fig. 5.10).

The northern Equatorial Zone is a highly complex area, containing many cloud features. Large equatorial plumes span it, but there are also circular white spots, which are sometimes extremely bright, and which are often found at the base of the plumes. All these features are essentially stationary in System-I coordinates, and form part of the North Equatorial current.

5.2.4.8 The North Equatorial Belt (NEB)

This is the prominent belt visible just below the equator. It may be seen in very small instruments – even with a small spy-glass. Like its southern counterpart, the NEB has two components. The southern one, the NEBs, is permanent. It is the source of the giant equatorial plumes, which appear to arise from dark, bluish spots that lie on its southern edge. The other component, the NEBn, may vary greatly in intensity because it is subject to significant fades that develop in its northern

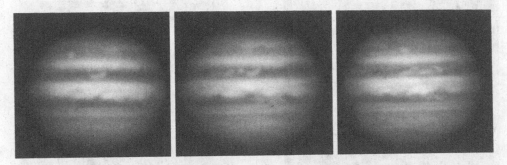

Fig. 5.18. *Eruption of the SEB in 1985, photographed by D. Parker, using a 310-mm reflector. Left: 1985 July 16, 06:37 UT. Centre: 1985 July 21, 06:08 UT. Right: 1985 July 30, 07:27 UT. Note the presence of white oval DE in the STB.*

portion. The NEBs and NEBn components quite frequently appear separated by a very narrow bright zone, the NEBz. This usually appears as an elongation of the long rifts that arise in the EZn or in the North Tropical Zone. When conditions are good, the extremely varied structure of the two equatorial belts is astonishing (Fig. 5.18).

5.2.4.9 The North Tropical Zone (NTrZ)

This zone lies immediately below the NEB. Its width varies over the years as a result of the activity of the neighbouring North Temperate Belt. Its tint also varies: when the NTrZ is narrow, it appears greyish, because then the North Temperate Belt is prominent and has a tendency to invade neighbouring areas, as happened in 1985 and 1986 (Fig. 5.18). When the NTrZ is wide it appears bright, because then the North Temperate Belt is almost invisible. White spots are often visible in the North Tropical Zone, as well as faint plumes arising from the NEB.

5.2.4.10 The North Temperate Belt (NTB)

We have just seen how the activity of this belt, which lies at about latitude 22°N, may modify the appearance of the whole region. In fact the NTB shows great variations in visibility, as it may pass from being almost completely absent to appearing as prominent as its southern counterpart, the STB. Revivals of the North Temperate Belt therefore occur, as was the case in 1975 and in the winter of 1984–5. During such episodes, dark material originating in one or more specific sources rapidly spreads out in decreasing longitude, encircling the globe in about forty days. Revivals of the NTB are thus very rapid events. In the 1985 and 1986 apparitions, for example, the NTB put on a fine display, appearing distinctly ochre in colour, whereas in previous years it had been very weak, and visible only in the form of isolated segments of dark material that were carried along by the North Tropical current.

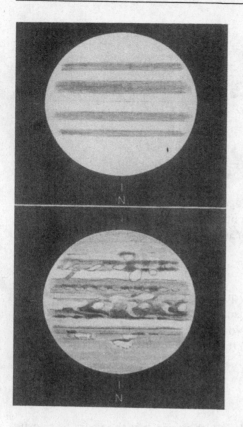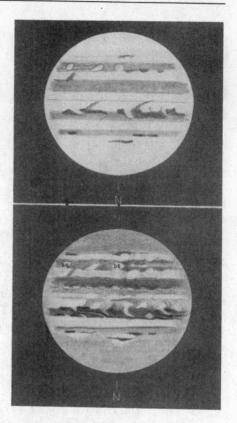

Fig. 5.19. *The four stages in obtaining a good drawing of Jupiter. It is possible to see most of the features that we have discussed in the final drawing.*

5.2.4.11 *The North Temperate Zone (NTZ)*

This narrow bright region, lying north of the NTB, does not, in general, show any very noticeable details, although isolated white clouds are sometimes visible. During the 1986 apparition, it developed four dark 'barges', which may be interpreted as holes in the white clouds that cover the zone (Fig. 5.19). Similar features were seen in 1979 (at the time of the Voyager probes) in the NTrZ, along the northern edge of the North Equatorial Belt.

5.2.4.12 *The North North Temperate Belt (NNTB)*

The description of this belt is almost a repetition of what has been said for the NTB. The visibility of the North North Temperate Belt undergoes cycles of activity during which it goes from being inconspicuous to appearing as a very prominent belt as in 1986. At that particular time, the whole of Jupiter's northern hemisphere, which is normally very quiet, was especially active and therefore extremely interesting to observe.

Table 5.1. *Latitudes of Jupiter's belts*

	STB S. edge	STB N. edge	SEB S. edge	SEB N. edge	NEB S. edge	NEB N. edge	NTB centre
ϕ	$-32°$	$-27°$	$-19°$	$-7°$	$+6°$	$+15°$	$+22°$

5.2.4.13 *The North Polar Region (NPR)*

This area appears as a form of hood of dark, evenly tinted material, very similar to that covering the southern polar regions. Very few details are ever seen in the NPR.

5.2.5 *Drawing Jupiter*

Not all the features just described are visible at any one time. Jupiter's image does, however, show such a vast amount of detail that it has to be recorded photographically, or by drawings, so that it may be studied properly. Figure 5.19 illustrates, specifically for Jupiter, what has been said in the general section about drawing.

(i) The four belts STB, SEB, NEB and NTB are drawn lightly on the blank in advance, to avoid gross errors in positioning. The distance d (in mm) between the edges of these belts and the centre of the disk may be calculated from the approximate equation:

$$d = 1/2 \text{ (equatorial diameter)} \times \sin \phi,$$

where ϕ is the zenocentric latitude, the values of which are given in Table (5.1).

Some slight retouches will quite possibly have to be made, because the position of the belts (as observed) varies slightly with time as a result of the 3.1° inclination of the planet's axis of rotation to the plane of its orbit. We therefore sometimes see the belts 'from above', sometimes 'face on' and sometimes 'from below'. These changes may be observed over the period of 12 years that Jupiter takes to orbit the Sun.

(ii) In the second stage of making a drawing, the major formations – such as the Great Red Spot, a white oval, or any equatorial plumes – that are easily visible through the eyepiece, are added. This should take no more than a few minutes, taking careful note of the positions of the features being recorded. The various exact times should be noted down as one goes along. This is one way of effectively 'stopping' the rotation of the planet.

(iii) The finer details visible through the telescope are then added. The disk should be examined, area by area, beginning with the regions closest to the evening terminator, because these will be the first to disappear from view. This part of the drawing may last 15–20 minutes. The fine detail is located by reference to the major features drawn in the second stage. Gradually the drawing becomes more detailed. The opportunity should also be taken

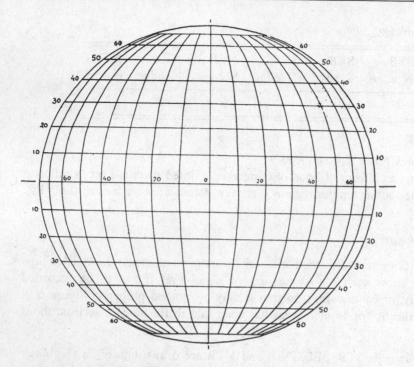

Fig. 5.20. *Latitude and longitude grid used to measure drawings.*

of timing as many central meridian transits as possible. This allows the longitudes of the features concerned to be calculated to an accuracy of a few degrees. The very narrow belts are also included; care being taken that their different intensities are correctly rendered.

(iv) The last stage takes place away from the telescope. The longitudes of the central meridian (in both Systems I and II) are calculated for the times at which the major features were recorded [in stage (ii)]. To add the finishing touches, check whether the features west of the central meridian, whose transit times have been recorded, are correctly placed. To do this, their longitudes are calculated. Then, by using a grid of latitude and longitude (Fig. 5.20), they may be accurately positioned on the blank, knowing the difference in longitude relative to the central meridian. The drawing is finished by making considerable use of a stump, especially at the limbs of the planet where no details are really visible. Finally, the detailed information required on the blank should be completed.

Any drawing obtained in this way will be of excellent quality, comparable with the best photographs. If twenty-odd observers, who belong to an observing association, each make about 30 observations of this type during an apparition, it will enable an accurate assessment to be made of the general behaviour of the planet's upper atmosphere (Fig. 5.21).

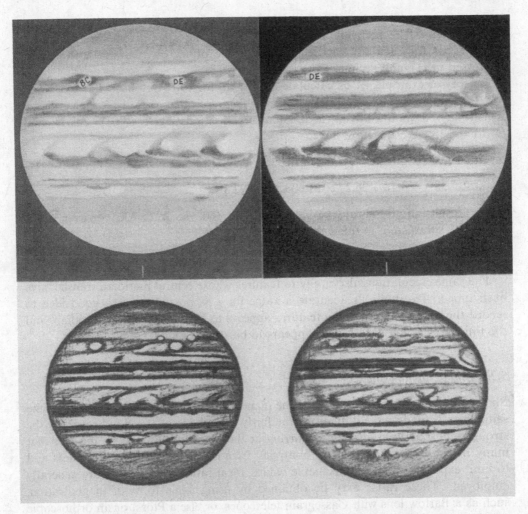

Fig. 5.21. *Drawings of Jupiter in 1986 compared.* Top: *by R. Néel on 1986 July 22, between 03:15 and 03:20 UT:* $\omega_1 = 266°$, $\omega_2 = 275°$; *and 1986 September 6, between 22:14 and 22:19 UT:* $\omega_1 = 311°$, $\omega_2 = 323°$. Bottom: *by I. Mayazaki on 1986 July 21, 17:24 UT:* $\omega_1 = 264°$, $\omega_2 = 277°$; *and 1986 August 9, 13:52 UT:* $\omega_1 = 318°$, $\omega_2 = 318.4°$ *Compare the lower, left-hand drawing with the photograph shown in Fig. 5.8.*

5.2.6 Calculating central-meridian longitudes

For the equatorial regions, where System I applies, the longitude of the central meridian, ω_1, at a time t (UT) is given by:

$$\omega_1 = \omega_1(0) + 36.58 \times t.$$

For regions where System II applies, we have, similarly:

$$\omega_2 = \omega_2(0) + 36.27 \times t.$$

223

The values $\omega_1(0)$ and $\omega_2(0)$ are the longitudes of the central meridian in System I and System II, respectively, at 00:00 UT. These values are given in astronomical ephemerides for each day of the year. The time of the observation, t, should be expressed in UT and decimal form.

By way of example, let us calculate ω_1 and ω_2 for 1987 October 17 at 21:38 UT. We have: $t = (21 + 38/60)\mathrm{hr} = 21.633\ldots\mathrm{hr}$. For that day the ephemerides – which may be obtained from the amateur organisations – give $\omega_1(0) = 205.78°$ and $\omega_2(0) = 8.18°$. We therefore have:

$$\omega_1 \text{ at } 21^\mathrm{h}38^\mathrm{m} = 205.78 + 36.58 \times 21.633\ldots + 997.1°$$

$$\omega_2 \text{ at } 21^\mathrm{h}38^\mathrm{m} = 8.18 + 36.27 \times 21.633\ldots + 792.8°.$$

The results should always lie between 0 and 360°, so 360°, 720°, or even 1080° should be subtracted. In this case the results are:

$$\omega_1 = 277.1° \text{ and } \omega_2 = 72.8°.$$

The same calculations also apply to features whose central meridian transits have been timed. To obtain as accurate a value for t as possible, it is a good idea to record the time t_1 at which the features appears to be on the central meridian, and the time t_2 at which it no longer appears to be so; t is then $(t_1 + t_2)/2$.

5.2.7 *Photographing Jupiter*

Photography of Jupiter is helped by the planet's large apparent diameter but, at the same time, is made more difficult by the fairly low albedo of the cloud cover, by the strong atmospheric absorption occurring at the limb and, finally, by the fact that many of the features are of low contrast. Because of the planet's brightness and to limit exposure times to reasonable values, focal ratios of f/50–f/70 are generally employed. These ratios may be obtained by using some form of magnification, such as a Barlow lens with Cassegrain telescopes, or else a Plössl or an orthoscopic eyepiece with Newtonian telescopes. For example, with an instrument with a relative aperture of 6, a final focal ratio of f/60 is obtained by magnifying the primary image 10 times (60/6).

With a focal ratio of f/60, and using TP 2415 film, exposure times range from 1/2 to 2 seconds, depending on atmospheric transparency. The exposure is, of course, given by a manual shutter over the telescope's aperture, or by a manually operated 'focal-plane' shutter, if it is possible to arrange for a gap between the eyepiece and the camera back. Such a 'focal-plane' shutter is required with large Newtonian or Cassegrain telescopes, where controlling the exposure by obscuring the aperture is impractical.

With amateur instruments, the use of filters is difficult because they require increased exposures. However, a red filter (a Wratten 25 gelatine) will enable more details to be recorded in the planet's dark regions. The exposure times will be increased by a factor of 2 or 3 with TP 2415 film. Generally there is little point in under-exposing the film, because of the amount of limb-darkening that is present.

TP 2415 film should, for preference, be developed for 6 minutes at 20°C in D19b, which increases the faint contrast found in Jupiter's atmosphere. Enlargement of black-and-white pictures of Jupiter is not at all easy, both because of the limb-darkening (especially with a red filter), and of the phase effect (when the planet is not at opposition). So a certain amount of correction has to be carried out in printing. For this a piece of card is used, with a slightly elliptical hole about 10–20 mm in diameter. Two exposures are made of the planet, the first with no dodging, and the second with the mask in place so that it just leaves the centre uncovered, ensuring that the limbs are under-exposed. Naturally, the mask has to be kept in motion all the time. This procedure is essential, but requires some practice to be satisfactory. It is also necessary to time the exposures made without and with the mask very accurately. For example, if 30 seconds gives good results for the centre of the disk, then about 16–17 seconds should be given without the mask, and then the remaining 13–14 seconds. The ratio of the two exposures will vary with each negative, and is mainly a function of its density. Negatives that are slightly over-exposed are easier to correct. Even when developed in D19b, pictures of Jupiter are low in contrast and should be printed on paper of grade 3 to 5, depending on how they are to be used. The diameter of the image should not exceed 40 mm.

Jupiter may also be photographed in colour, although transparencies have less scientific value. It is not easy, because of the low contrast of high-speed colour emulsions. The techniques used for making the exposure are the same as those for black-and-white films. Fast films such as Fujichrome 400 are used, and development should be slightly increased to improve the contrast and sensitivity.

Any photograph of Jupiter should be accompanied by precise information about the date and time of the exposure – if possible to the nearest second.

5.2.8 Measuring positions from photographs

When observations of Jupiter are submitted to a specific observing group, measurements are usually made by the person (or persons) who prepare the reports on the planet's activity. Figure 5.22 shows a basic method that uses an engineer's calliper gauge. The photographs are mounted on a piece of card, with a central square aperture that allows the image of the planet to be seen. The negative is lit from behind by some suitable means. (An adjustable desk-lamp with a 20 W opal bulb is quite sufficient.) The negative should be horizontal. With a fine-tipped pen, the points A, B and M are marked, where AB is the parallel of latitude passing through the point M, the longitude of which is to be established. AB and AM are then measured with a calliper gauge graduated in 1/50 mm. The longitude of the point M at the time of the exposure is then:

$$\omega(\text{M}, t) = \omega(0, t) + \arcsin(2\text{AM}/\text{AB} - 1) \pm 90/\pi(1 - \cos i) \pm 0.085i,$$

where $\omega(0, t)$ is the longitude of the central meridian at time t (UT), and i is the phase angle given in the ephemerides. The term $90/\pi(1 - \cos i)$ is a correction introduced here because the ephemerides give the longitude of the true central meridian. We are, however, making our measurements on a disk that shows a phase effect, so the meridian is shifted away from the true central meridian by a specific amount. The value of this is given by the correction term. This displacement is zero at

225

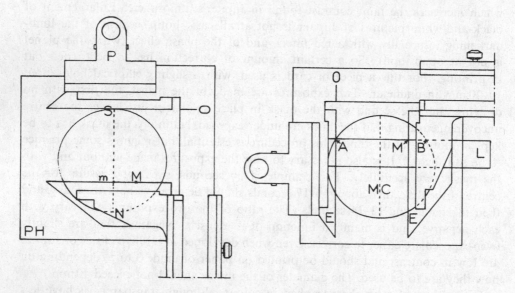

Fig. 5.22. *The basic method used to measure photographs. PH is the photograph, lit from behind, L is a strip of transparent plastic fixed to the left-hand jaw of the caliper gauge; it is used to align the belts on the planet. E indicates two transparent, right-angle subsidiary jaws fixed to the jaws of the caliper gauge, and projecting by 5 mm; they allow the exact positions of A, M and S to be seen clearly. All readings must then be reduced by 10.0 mm.*

opposition, but reaches about 0.5° at quadrature. It is positive before opposition and negative after it. The term 0.085i, known as photographic 'phase exaggeration' was introduced by Reese and Solberg. In measurements made on plates exposed by these two researchers in the 1960s it reached ±0.6° at quadrature. For TP 2415 film, its value is unknown; it is probably less than the value given here.

In general, the longitude ω of a cloud formation varies more or less linearly with time. It may be written in the form:

$$\omega(M, t) = \omega(M, t_0) + a(t - t_0).$$

If the longitudes $\omega(M, t_1)$, $\omega(M, t_2)$, $\omega(M, t_3) \ldots \omega(M, t_n)$ of the feature M at times t_1, t_2, $t_3 \ldots t_n$ are known from the measurement of photographs or central meridian transits, then it is possible to use the method of least squares to determine the constants $\omega(M, t_0)$ and a, which are, respectively, the longitude of M at opposition, which took place at time t_0; and the mean daily motion, expressed in degrees per day in either System I or System II. Knowledge of a allows the rotation period of the feature in question to be determined. It is positive if the longitude of the feature is increasing, and negative when it is decreasing.

To determine the zenocentric latitude ϕ of point M, SN and Sm are measured (where m is the projected height of M on the central meridian). From this:

$$\phi(M) = \phi(0) + \arctan\left[7/15(2Sm - SN)/SmSN - SM^2)^{\frac{1}{2}}\right],$$

where $\phi(0)$ is the latitude of the centre of the disk given in the ephemerides. All these repetitive calculations may be carried out with a programmable calculator that provides for calculation of least-squares. One can even imagine making the measurements with an electronic calliper gauge (yes, they do exist!), and entering the data directly into a microcomputer which will carry out all the necessary calculation and output the results to a printer or plotter.

Knowledge of the latitudes and the rotation periods allows the various features to be classified according to the various atmospheric currents that control their movement. Any such flow affects objects within a very narrow range of latitude that have very similar rotation periods. Twenty-four have been described, but perhaps a dozen may be determined over the course of an apparition. Taken overall, the mean period of all the features affected by a single current varies very little over the course of time.

5.3 Mars

5.3.1 The planet and its orbit

5.3.1.1 The frequency of oppositions

The average interval between two oppositions of Mars is 780 days (the synodic period). The considerable eccentricity of the planet's orbit (0.0934) means that the actual interval between successive oppositions may differ greatly from this theoretical value. The interval varies between 764 days for two oppositions close to perihelion and 810 days for two close to aphelion. Also because of the large eccentricity, the date of closest approach to the Earth does not necessarily coincide with the date of opposition. In 1984, for example, opposition was on May 11, whereas closest approach occurred 8 days later.

5.3.1.2 Apparent diameter

The distance between Earth and Mars varies greatly from opposition to opposition (Fig. 5.23). The planet comes to within 56 million km at perihelic oppositions, whereas at aphelic oppositions the distance is around 100 million km. Perihelic oppositions occur in August and September (Fig. 5.23). Under the most favourable circumstances the planet has an apparent diameter of 25.1 arc-seconds, and it may reach magnitude −3, as at the opposition on 1924 August 22. Mars is then in the constellations of Sagittarius or Capricornus, and therefore low on the horizon for European observers. This obviously means that seeing is very unfavourable. At an altitude of only 20°, turbulence is three times as bad as at the zenith. This is a serious handicap for any observer who does not have a truly excellent site.

Aphelic oppositions occur in February and March, in the constellations of Cancer or Leo. The planet appears much smaller, reaching perhaps 13.8 arc-seconds, as in 1980. Its magnitude is never more than about −1. The most favourable oppositions for observers at mid-northern latitudes are those following perihelic oppositions. The planet is sufficiently close to the Earth for it to have an apparent diameter of some 18–20 arc-seconds, and its altitude above the horizon (about 45°) is considerably

227

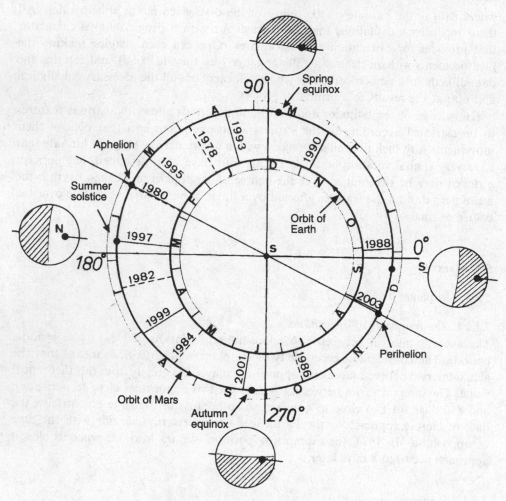

Fig. 5.23. *The orbits of Mars and the Earth. The angle shown is the heliocentric longitude η. The positions of Mars at the solstices and equinoxes are shown, together with the relative positions of Mars and the Earth at oppositions from 1978 to 2003. The terrestrial and Martian months are indicated around the orbits.*

better as far as seeing is concerned. Perihelic oppositions recur every 15 years, on average.

5.3.1.3 Axis of rotation

The axis of rotation is considerably inclined (23° 59′) relative to the normal to the plane of the orbit. This inclination (very similar to the Earth's 23° 27′), has two immediate consequences:

- The Martian climate has seasons that are as pronounced as those found on Earth;

- The particular area of Mars turned towards the Earth varies very considerably, depending on the date of opposition. At perihelic oppositions, we primarily see the South Pole, whereas at aphelic oppositions it is the North Pole that is visible (Fig. 5.23).

The large eccentricity, which has already been mentioned, causes the seasons to be of very unequal duration. In the northern hemisphere, spring and summer last 371 Martian days, whereas the autumn and winter only last 298 days. On the other hand, the southern summer is much hotter, because then the planet is close to perihelion.

5.3.2 Telescopic appearance

5.3.2.1 Phases and rotation

Because Mars is a superior planet, it can never show crescent phases like Venus. However, it does appear gibbous, and its phase angle (the Earth–planet–Sun angle) may reach 45°. This phase effect is easily observable, and affects the western side of the disk before opposition, and the eastern afterwards. The east–west direction is defined by the direction of the diurnal motion (west = preceding). To calculate the phase, reference should be made to published ephemerides.

Mars rotates in $24^h 37^m 22^s$ (the sidereal period), and therefore slightly slower than the Earth. During an observing session Mars turns by about 15° per hour, and zones of increasing longitude are observed as the night passes. The fairly slow rotation gives enough time to make any observation (drawing or photograph), and is not a factor that has to be taken into consideration as with Jupiter. On the other hand, if the planet is observed for several days in succession it becomes obvious that the visible area hardly changes. This is explained, of course, by the fact that the rotation periods of the Earth and Mars are so similar. Over 24 hours both planets make about one complete rotation, so that points on their surfaces are more or less back to the same places that they occupied the day before. Day by day, the central meridian on Mars shifts about ten degrees in decreasing longitude (towards the east on the surface of the planet). As a result, about a month of observation is needed to cover the whole planet (Fig. 5.24).

5.3.2.2 Maps of Mars

Mars is a terrestrial planet much like the Earth, and we are able to observe its surface directly through its atmosphere. The surface appears yellowish-orange in any telescope, which explains the distinct colour seen when the planet is observed with the naked eye. Apart from this overall colour, telescopes reveal a number of darker markings of a bluish-grey colour. In general, these markings have well-defined, permanent shapes. They have allowed the planet's rotational period to be determined accurately.

As with the Moon, early observers believed that these darker areas were seas. As a result the term 'mare' was applied to them as well, even though there is, in fact, not a single drop of liquid water anywhere on the surface of the planet. Because these markings are permanent, despite slight variations, mapping the surface is feasible.

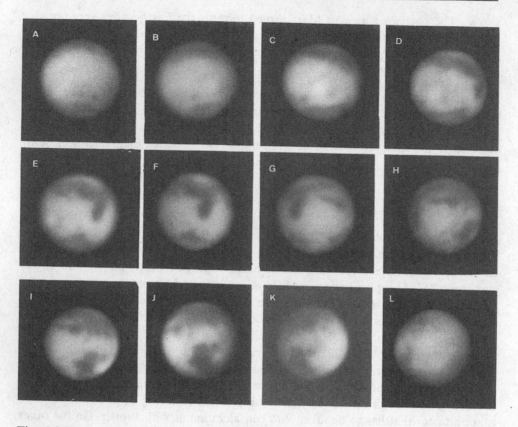

Fig. 5.24. *A complete rotation of Mars, photographed by D. Parker in 1984, with a 320-mm reflector, using* TP 2415 *film. Exposures 2.5 s, f/198. Left to right, top to bottom:*

A: *May 24, 05:24 UT,* $\omega = 147°$

B: *May 15, 03:16 UT,* $\omega = 194°$

C: *May 14, 04:17 UT,* $\omega = 219°$

D: *May 13, 05:00 UT,* $\omega = 238°$

E: *May 09, 04:19 UT,* $\omega = 263°$

F: *May 06, 03:30 UT,* $\omega = 277°$

G: *May 05, 05:54 UT,* $\omega = 321°$

H: *May 01, 05:00 UT,* $\omega = 343°$

I: *May 29, 05:34 UT,* $\omega = \ \ 9°$

J: *May 03, 03:37 UT,* $\omega = \ \ 32°$

K: *May 26, 06:57 UT,* $\omega = \ \ 56°$

L: *May 30, 06:20 UT,* $\omega = 107°$

It is not always easy to obtain accurate albedo maps of Mars. (Topographical or geological maps, such as the series published by the U.S. Geological Survey, are of little interest to planetary observers. Although albedo features are shown the maps are basically photomosaics and lack the reference grid.) The most famous map is that by Antoniadi, which consists of three very detailed maps (Antoniadi, 1930). Although very accurate, Antoniadi's map – which is practically unobtainable – has aged somewhat; the names of certain formations have changed over 50 years of observation.

The best map, which is very similar to Antoniadi's, is the one by Ebisawa (reproduced in McKim, 1986). A map adopted by the IAU, and list of features, appeared in *Nortons Star Atlas*, 17th edition, 1978. An 'unofficial' chart by J. Murray appears in *Norton's 2000.0*, ed. Ridpath, I., 1989. The old map by G. de Vaucouleurs appeared in several works, including de Vaucouleurs (1950). Other maps that are of interest, but not suitable for detailed reference, may be found in O'Leary *et al.*, (1990) and Briggs *et al.*, (1982).

A Mercator projection map of albedo markings made from photographs taken in July 1986 by J. Dragesco and D. Parker is shown in Fig. 5.25, together with the classical nomenclature for various Martian features on a 1971 base map (Fig. 5.26). Note the significant changes that have taken place during the interval between the two charts. We hope that this chart, although only rudimentary, will help future observers of the planet.

5.3.2.3 Suitable instrumentation

Because of its tiny apparent diameter, Mars remains a difficult telescopic object. A small 100-mm refractor allows one to recognize the principal formations and to follow the melting of the polar caps. To make interesting observations, however – such as following the planet's atmospheric features over a period that is not confined to just the time immediately before and after opposition – a high-quality refractor of 150-mm diameter, or a 200-mm reflector, is the minimum requirement. With such equipment, the phenomena described in later sections are visible. Nevertheless, the ideal instrument remains a reflector of 300 mm in diameter, because in a less-than-ideal site it is difficult to make best use of a more powerful instrument. On the other hand, a 200-mm is somewhat limited, especially if photography is intended.

5.3.3 Drawing Mars

There are two sizes of blanks for visual observation of Mars: the one used by ALPO and the BAA, which is only 50 mm (2 inches) in diameter, and the one used by the SAF, which is 70 mm across.

It is always advisable to prepare before observing, so that observers are not confused when they reach the eyepiece and thus waste time. With Mars, observers should have an accurate idea of what is likely to be seen. To this end, the equator should be drawn on the blank – this depends on the latitude of the centre of the disk, which is given in the ephemerides – together with the poles and an indication of the phase. Finally, one might check a map to see which region of Mars will be visible. The aim is not to copy the map, but to be prepared for noticing any anomalies: changes in the albedo markings, disappearance of a particular feature beneath clouds, dust-storms, etc.

In preparing a final drawing, a start should be made by covering the white disk with light grey shading, using a stump that has been rubbed on an area blackened with a 3B pencil. The idea is to show the average albedo of the Martian deserts, so that low-albedo features appear slightly darker, and a few strokes with an eraser suffice to render the brightest regions, such as the polar caps, clouds, mists, frost deposits, and dust storms. (The very brightest regions may be outlined by a lightly dotted border, to make them stand out better.)

It is important to record on the final drawing the date and time of observation,

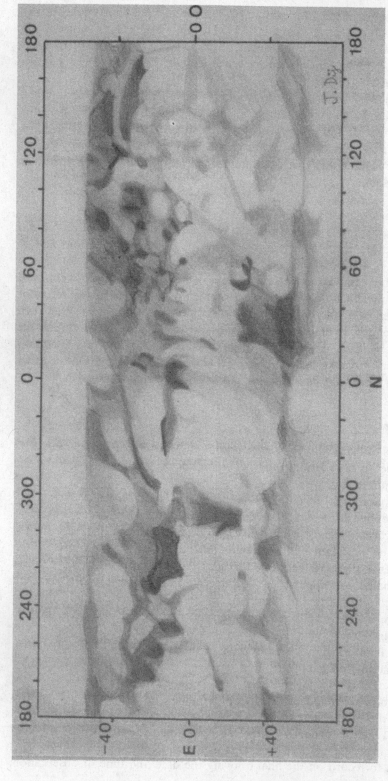

Fig. 5.25. *Map of albedo markings on Mars observed during the 1986 July opposition.*

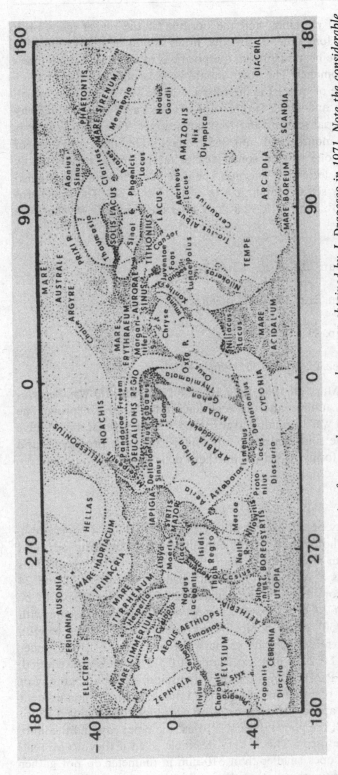

Fig. 5.26. *Nomenclature of the various martian features drawn on a base map obtained by J. Dragesco in 1971. Note the considerable changes between these two charts.*

the longitude of the central meridian, and the position of the planet in its orbit (by quoting its heliocentric longitude, η). Finally, notes should be made indicating which Martian features have been identified, the positions of any clouds, any filters used, and any apparent anomalies. The notes should be as complete as possible.

One concept that may be of interest is the Martian date (MD). This idea, which was proposed by C. Capen and D. C. Parker of ALPO, is based on the similarity between the familiar terrestrial seasons and Martian seasons. The Martian year is divided artificially into 12 months and into 365.25 days. This date is a form of mnemonic device enabling one to gain some idea of the Martian climate at any given date, by analogy with terrestrial seasons. The date of the Martian (northern) vernal equinox is arbitrarily set at March 20.8 ($\eta = 85°$, Fig. 5.23). The (northern) summer solstice is $85° + 90°$, June 21.6, 92.8 days after the equinox. We therefore have:

northern spring:
  $85° \leq \eta \leq 175°$: MD = Mar. 20.8 $+[(\eta - 85)/90] \times 92.8$
northern summer:
  $175° \leq \eta \leq 265°$: MD = Jun. 21.6 $+[(\eta - 175)/90] \times 93.4$
northern autumn:
  $265° \leq \eta \leq 355°$: MD = Sep. 23 $+(\eta - 265)$
northern winter:
  $355° \leq \eta \leq\ 85°$: MD = Dec. 22 $+[(\eta - 355)/90] \times 89.05$.

For example, observing Mars on 1986 June 10, $\eta \approx 270°$, the MD was September 28. It was the northern autumn and the beginning of the southern spring. The southern hemisphere, which was beginning to warm up after the winter, might have shown numerous clouds, the southern polar cap would have been prominent, and mists would have begun to form around the northern polar cap.

5.3.4 *Photographing Mars*

However fine a drawing may be, it never has the same objectivity as a photograph. Unfortunately, planetary photography is difficult under any circumstances, and photography of Mars is even more difficult than that of Jupiter or the Moon. This is because Mars always has a very small apparent diameter (between 13 and 24 arc-seconds) and requires considerable focal lengths (20–50 m), which have to be obtained either by the use of $2\times$ or $3\times$ Barlow lenses (with Cassegrain systems with large focal ratios of f/20–f/30), or by employing powerful (8–10-mm focal length), high-quality eyepieces (with relatively wide primary apertures of f/6–f/10). Under such conditions, and using TP 2415 film, exposures of 0.5–2 seconds are required. (We should point out that TP 2415 film is the only film suitable for photography of Mars, because it is the only one that has a high resolution, an almost non-existent grain, high contrast, and an appropriate speed.) With such high focal ratios it is not easy to focus on the image of the planet. It is best to use the polar cap and the limb.

Should coloured filters be used? This is very desirable, even if they do prolong the exposure. In fact, telescopes smaller than 310-mm in diameter do not gather enough light for them to use anything other than a fairly pale orange filter. Above

310 mm diameter it is also possible to use red Wratten 25 or 29 gelatine filters (which improve contrast and make the Martian atmosphere more transparent), as well as blue Wratten 80C filters, with which one can try to emphasize the clouds (Fig. 5.27 *top*). (These filters will double or triple the exposure times.) The ideal filter for recording white Martian clouds is the Wratten 49, which has, unfortunately, a very high absorption (Fig. 5.27 *bottom*). Given the exposure increase required (×60!), it may be used only with the largest telescopes and relatively fast apertures (approximately f/25).

Development of the film poses some problems. It is essential to use a developer that optimizes the film's sensitivity, while giving a suitable amount of contrast (which, with Mars, is not very high). HC-110 gives excellent results at a dilution of $1 + 15$, developing for 6 minutes at 20°C (normal agitation every 30 seconds). With image diameters of 25–35 mm, black-and-white enlargements should be made on glossy paper, grades 3 to 5. Colour photography is more difficult, because colour films have lower resolution and contrast than TP 2415. With small telescopes, a fairly high focal ratio has to be used (f/80 to f/100), and a moderately fast film (Fujichrome 400 or Ektachrome 200 or 400). Give exposures of 2–3 seconds (but trials are essential). It is best to develop the film oneself, giving a slight over-development to increase effective speed and contrast. (It is also possible to ask a laboratory to develop a 200 ISO film as if it had been exposed as a 400 ISO film.)

5.3.5 *The changing appearance of Mars*

As a guide to the observation of Mars, which is always difficult, the following sections describe the various phenomena that occur on the planet. Observation of the planet consists of following these changes, and the illustrations and results given show what may be achieved by amateurs. We shall describe the evolution of the polar caps, of the albedo markings on the Martian surface, and changes in the atmosphere (clouds and dust-storms).

5.3.5.1 *The Martian polar caps*

The early days of telescopic observation revealed (to Cassini in 1666) that there were white areas at the planet's poles. These features, which are some of the most prominent and easily observable features on Mars, show cyclic variations with the seasons (Fig. 5.28). In general, the cycle is the same for the two hemispheres.

The beginning of spring is marked by the appearance of the polar cap (which is hidden throughout the winter by mists – known as the polar hood), and by its subsequent melting.

Recession of the northern polar cap (NPC) occurs between $\eta = 145°$ and $\eta = 175°$. Melting is rapid, and the cap shrinks from its greatest extent, at roughly latitude +65°, to about +85°. It continues to melt slowly throughout the summer, and recedes to above +85° around $\eta = 250°$. Classically, the NPC breaks into separate portions as it melts (Fig. 5.29). The main breaks or 'rimae' are known as Rima Borealis, which separates the Olympia region from the rest of the cap, and Rima Tenuis, which cuts the NPC into two halves, passing almost directly across the pole. This feature was discovered by Schiaparelli in 1888 and was observed at aphelic

Fig. 5.27. *Two exceptional photographs obtained by J. Dragesco using the 1.06-m reflector at the Pic du Midi, 1986 July 15, on TP 2415 film. Top: 23:07 UT: $\omega = 67.6°$, red W29 filter; f/52; 1/4 s. Bottom: 23:09 UT: $\omega = 68°$; blue W49 filter; f/52; 8 s. These extraordinary photographs show Mars as it might be seen through a large telescope. Note the strength of the southern polar cap, which was in the course of melting ($\eta = 291.2°$, the middle of the southern spring), as well as the innumerable details in the region of Solis Lacus, Tithonis Lacus, etc. Apart from the southern polar cap, the blue photograph shows mist over the northern polar region (northern autumn), as well as clouds on the morning terminator, which are invisible in red.*

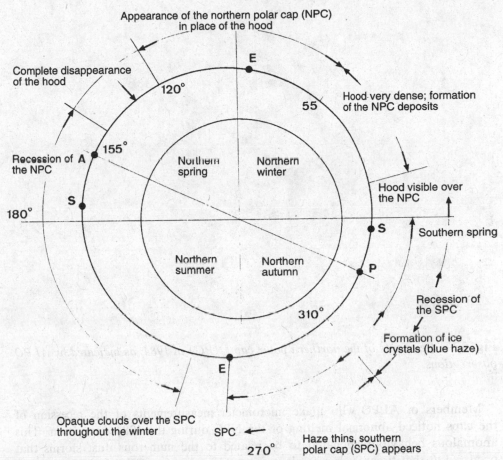

Fig. 5.28. *How changes in the polar caps are related to the position of Mars in its orbit.*

oppositions until 1918, after which it seemed to elude observers until 1980, when it was reported by the Mars sections of the BAA, SAF, etc.

The southern polar cap (SPC) begins to recede between $\eta = 310°$ and $\eta = 360°$, shrinking from latitude $-60°$ to $-85°$. Melting is faster than in the northern hemisphere. The single remnant is eccentric, and does not cover the pole itself. As they melt, the caps are surrounded by dark borders, which follow them as they recede and disappear at the end of spring.

Towards the end of summer, mists form over the polar regions. At first these are broken, but they rapidly spread, eventually covering the whole of the polar regions and part of the temperate zone. Beneath this persistent veil, the polar cap reforms during the autumn and winter.

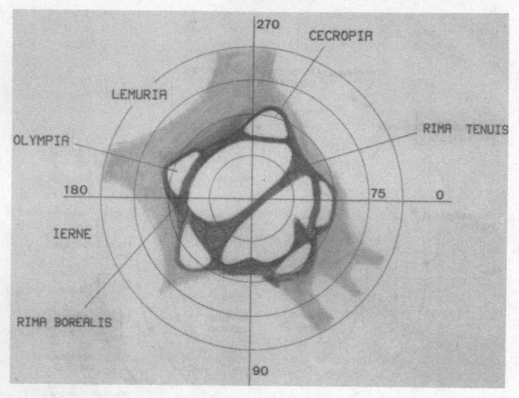

Fig. 5.29. *Remnants of the northern polar cap (NPC) in 1984, as indicated by ALPO observations.*

Members of ALPO who make micrometer measurements of the recession of the caps noticed abnormal melting of the NPC during the 1984 opposition. This anomalous behaviour appeared to be related to the numerous dust storms that occurred at that time. Such a result is outstanding, because following the recession of a polar cap on a disk that has a diameter of less than 10 arc-seconds requires a first-class site, instrument and observer!

5.3.5.2 Changes in the appearance of the surface
The dark areas of the surface undergo changes: these variations are either seasonal and thus periodic, or unpredictable. The latter are known as secular variations. For these reasons, observation of the surface always has a topical element. Areas of the Martian surface generally recognized as being subject to these variations are as

follows (cf. the map shown in Figs. 5.25 and 5.26):

Solis Lacus, Thaumasia	(90°, −30°)
Mare Serpentis, Hellespontus	(320°, −35°)
Pandorae Fretum	(340°, −25°)
Syrtis Major, Nepenthes	(290°, +10°)
Moeris Lacus	(265°, + 8°)
Lunae Palus	(60°, +20°)
Mare Acidalium	(30°, +45°)

Secular variations, which may persist for years, consist of unpredictable changes – sometimes very rapid – in the apparent shape of certain dark markings. For example, Solis Lacus changes shape, size, and intensity. On the other hand, its surroundings (Tithonius Lacus and Mare Erythraeum) generally show no changes in their appearance.

Other regions have historically shown significant changes, including the appearance of new markings, notably the areas of Sinus Gomer (230°, −5°), which appeared in 1924; Nodus Laocoöntis (245°, +20°), which appeared in 1935; and Daedalia Claritas (105°, −30°), where a new marking appeared in 1973.

Seasonal variations are changes that occur more or less regularly every Martian year (Fig. 5.30). These changes may affect the apparent shape of certain dark markings, such as Syrtis Major, which appears wide just after aphelion and narrow after perihelion.

Such changes also take the form of variation in the degree of darkening seen in some of the markings (Fig. 5.31). When it is winter in one hemisphere, its dark markings are at their palest. At the beginning of spring, the receding polar cap is surrounded by a dark border. This border marks the beginning of the 'wave of darkening', which reaches the equator around the middle of spring, and which ends at about 20° latitude in the other hemisphere towards the end of spring. For example, Solis Lacus is very pronounced between $\eta = 280°$ and $\eta = 110°$ (Fig. 5.30). This phenomenon has been observed at recent oppositions: in 1986 ($\eta = 294°$), Solis Lacus was very prominent (Figs. 5.25 and 5.33), whereas it was not particularly visible in 1984 ($\eta = 230°$).

For many years this wave of darkening was attributed to the diffusion of moisture produced by melting of the polar caps. There also appears to be a correlation between the darkening and the seasonal increase in the amount of solar radiation received by the surface. But not all Martian features react in the same way. Whatever the cause, this darkening is certainly real, and observable with the telescopes available to amateurs.

During the last decade or so, we have seen major changes in the area of Deltonon Sinus, Mare Serpentis, and Sinus Sabaeus (at about 310°, −20°), as shown in Fig. 5.32. During the 1977–8 opposition ($\eta = 121°$), the region appeared more or less normal: Pandorae Fretum and Hellespontus were faint; Syrtis Major was fairly narrow. On the other hand the area of Moeris Lacus and Nepenthes-Thoth was abnormally faint, and Deltonon Sinus was abnormally variable (being very faint at the beginning of the opposition).

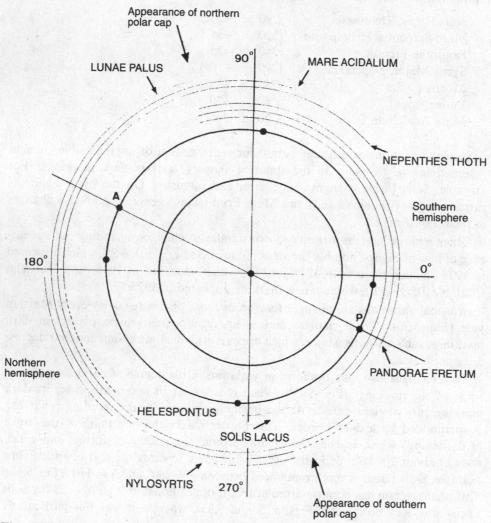

Fig. 5.30. *Cycles of darkening for certain Martian features. The arcs indicate the periods when the regions concerned are darkest. When it is winter in a particular hemisphere the features are pale.*

In 1979–80 ($\eta = 154°$), the complete disappearance of Deltonon Sinus was observed, as well as the weakening of the eastern part of Sinus Sabaeus. Nepenthes-Thoth and Meoris Lacus remained faint. This tendency was confirmed at the 1981–2 opposition ($\eta = 193°$), when Syrtis Major was abnormally wide and complex, and Deltonon Sinus and the eastern part of Sinus Sabaeus were almost invisible. These changes continued in 1984 and 1986. Syrtis Major was still very wide, Nepenthes-Thoth and Moeris Lacus were invisible, and the area of Deltonon Sinus and Sinus Sabaeus were unrecognizable. Are we witnessing a very marked, seasonal phenomenon, or long-term changes? Only the future will tell.

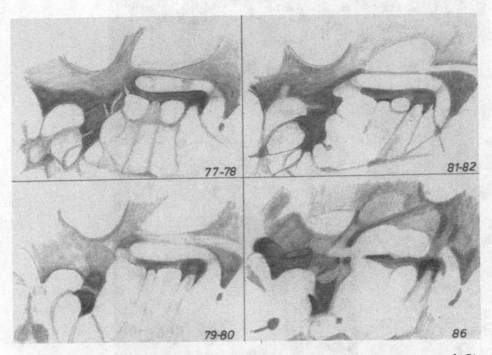

Fig. 5.31. *Changes in the region around Syrtis Major, Deltoton Sinus and Sinus Sabaeus, between 1978 and 1986.*

5.3.5.3 Atmospheric phenomena

Like the Earth, Mars has an atmosphere. Its composition is very different to Earth's, and mainly consists of carbon dioxide. The Martian atmosphere is subject to two types of events, associated with white or yellow clouds. The appearance and development of yellow clouds, which may cover the whole of the planet, are typically Martian phenomena, which have no equivalent on Earth.

Appearance and evolution of white clouds Under normal circumstances, the atmosphere of the planet is transparent in the visible region, but selectively scatters shorter wavelengths like a light haze. When the ice-crystal veils are sufficiently dense, they appear as whitish clouds, which may hide surface features. These clouds are sometimes very bright, and may appear as projections on the Martian terminator (Fig. 5.33). These clouds form preferentially over bright regions. When they cover just the bright deserts, the latter appear much brighter than the dark markings in blue light, as if they were being observed in the red. This gives the impression that the planet's atmosphere is completely free of any veiling, whereas it is nothing of the sort. This is the phenomenon known as the 'blue clearing'.

When, on the other hand, the atmosphere is actually completely clear, the disk appears more or less featureless in blue light, because the contrast between surface features is very low in that wavelength range.

Mists occur frequently while the polar caps are melting in spring and reforming

241

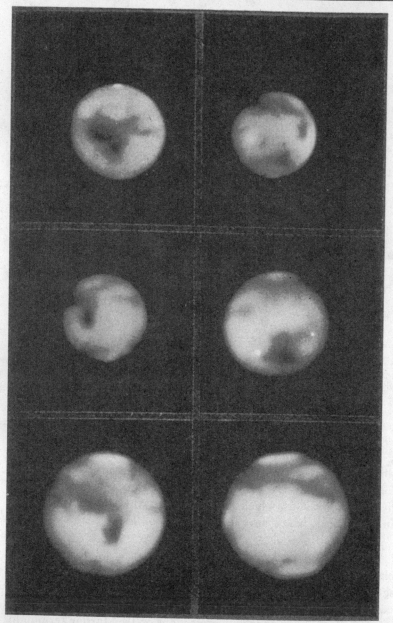

Fig. 5.32. *Some exceptional photographs of Mars obtained by amateurs:*
G. Viscardy, St Martin-de-Peille, France, 520-mm reflector, f/54:
1973 October 10, 23:15 UT: ω = 297° (southern spring),
1982 March 10, 00:32 UT: ω = 247°; TP 2415, 2 s (northern spring).
J. Dragesco, Cotonou, Benin, C14, f/60, TP 2415:
1984 April 29, 01:28 UT: ω = 309° (northern summer),
1984 April 23, 21:45 UT: ω = 35° (northern summer).
D.C. Parker, Coral Gables, Florida, 320-mm reflector, f/198, TP 2415:
1986 July 04, 06:24 UT: ω = 280°,
1986 July 12, 04:52 UT: ω = 186° (southern spring).

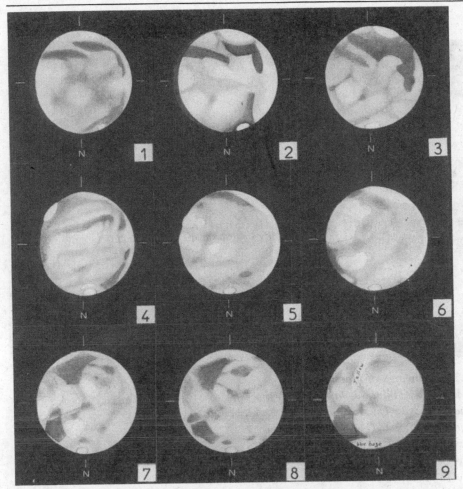

Fig. 5.33. *Drawings obtained during the 1984 opposition:*

1: R. Néel, 310-mm reflector, 1984 March 31, ω = 220° (low-lying site, poor seeing).

2: J. Dijon, 500-mm reflector, 1984 March 31, ω = 245° (low-lying site, poor seeing).

3: J. Dragesco, C14 (350 mm), 1984 April 01, ω = 240° (African site, good seeing).
[These drawings of the same region, all obtained at about the same date, show the importance of site and seeing conditions for visual work.]

4: J. Dragesco, 1984 March 22, 04:15 UT, ω = 334°: note that the north polar cap is in two portions, and also the cloud over Hellas, which appears as a bulge on the limb.

5: J. Dragesco, 1984 April 12, ω = 142°: note the orographic clouds over the Tharsis volcanic region, as well as the morning mists at the planet's western limb.

6: J. Dragesco, 1984 April 18, ω = 84°: a number of clouds are present.

7: 1984 April 20, ω = 64° (middle of the northern summer).

8: 1984 May 19, ω = 77°.

9: 1984 June 24, ω = 74° (end of the northern summer).

⌐Drawings of the same region over a period of three months. In (7), there are persistent mists over Tempe, Candor, and the south polar cap (where it is winter). In (8), these mists are disappearing, but the first clouds are appearing around the northern polar cap. In (9), localized mists have disappeared. There is a dust storm over Aurorae Sinus, and the northern polar cap is hidden beneath the hood.⌐

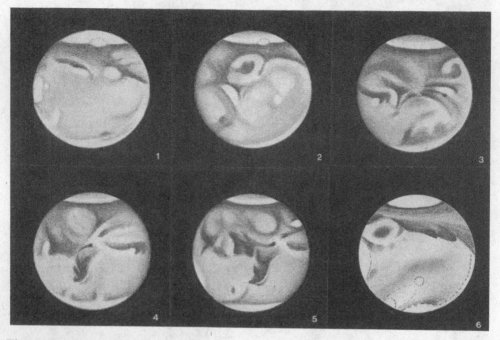

Fig. 5.34. *Drawings of Mars, 1986*
Isao Miyazaki (Okinawa, Japan), 200-mm reflector:
1: June 21, 15:26 UT, $\omega = 169°$.
2: June 28, 14:22 UT, $\omega = 90°$.
3: July 05, 14:29 UT, $\omega = 30°$.
4: July 14, 13:58 UT, $\omega = 302°$.
5: July 18, 14:40 UT, $\omega = 277°$.
Richard McKim, 1.06-m reflector, Pic du Midi:
6: July 10, 23:45 UT, $\omega = 121°$.
⌜*These drawing may be compared with those obtained at the 1984 opposition (Fig. 5.33). Note: the prominence of the northern polar cap, and the dark border; in (1) and (2) the division in the cap; the presence of clouds over the northern polar cap and the morning terminator; Solis Lacus, and the faint appearance of Syrtis Major and Hellas. These drawings, obtained with a 200-mm reflector, are excellent. They may be compared with the photographs of the 1986 opposition published here.*⌟

in autumn (Fig. 5.34). Thin hazes are then observed at the morning and evening terminators, betraying the existence of water vapour in the atmosphere. This water vapour condenses into cloud during the Martian night and is evaporated by solar radiation during the day. When summer arrives, these hazes are less frequent and occur only around certain relief features (such as the Tharsis volcanoes, and Olympus Mons). These orographic clouds occur as characteristic W-shaped or 'domino-like' features (Fig. 5.33).

Yellow clouds and dust-storms Another meteorological phenomenon is the occurrence of yellow veils. These clouds may remain localized, or may spread around the whole of the planet, when it is obvious that we are observing a major dust-storm. These dust-storms may be very extensive. We only have to recall the problem that arose when Mariner 9 arrived at Mars during the great dust-storm of 1971. The probe was quite unable to begin mapping the planet: the surface was completely invisible!

Yellow veils may be observed at any time of the year, but only turn into great storms during the southern spring and summer. This is when Mars is closest to the Sun, at perihelion, and the southern tropics are subject to intense heating. Major dust-storms generally begin in a small number of specific regions in the southern hemisphere, namely:

- The Hellas region (1909, 1911, 1924, 1939, 1956, and 1971);
- The Noachis, Hellespontus and Iapygia region (1969);
- The Solis Lacus region (1973).

Initially, the feature appears very bright and white, before it turns yellow. The cloud, which consists of water vapour, creates a strong ascending current of air, which injects large quantities of dust into the planet's atmosphere.

The appearance of this type of event follows a typical pattern, corresponding to the three stages in the development of a dust-storm:

- The initially white, and then yellow, feature;
- The disappearance of albedo features beneath the cloud of dust; this disappearance may involve the whole planet, as in 1971 or 1973;
- Bluish-white clouds that appear around the edge of the dust-storm when it reaches its greatest extent.

The movement of these large dust-storms in the southern hemisphere generally takes place in an east–west direction. Various observers have reported that before a dust-storm develops the contrast between areas of the surface is very pronounced, and the atmosphere appears very calm and clear: truly the calm before the storm!

The major, global events are not fully predictable. For example, predictions were made that there would be a major event in 1988. As it happened, although there were two large, regional events, no global dust-storm occurred as in 1956, 1971, or 1973.

Early in 1984, when the planet was still far from opposition and was beginning the northern summer, major atmospheric activity was observed. This activity took the form of successive dust-storms, the first of which began at the end of November 1983 (Fig. 5.35). On November 13, white clouds were observed over Libya, and Isidis Regio, as well as over the region of Eridania and Amazonis. On November 24 a yellow cloud covered Mare Sirenum and began to move east. This cloud avoided the regions of Aurorae Sinus, and Margaritifer Sinus, and then spread out, moving south over Argyre I, and invading Chryse in the north on December 4. This 'pincer movement' extended over Noachis in the south, sending a finger down the length of Syrtis Major on December 11. In the south, the region of Eden was invaded. This

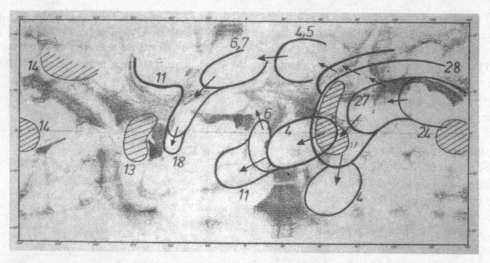

Fig. 5.35. *Dust storm between 1983 November 13 and December 18.*

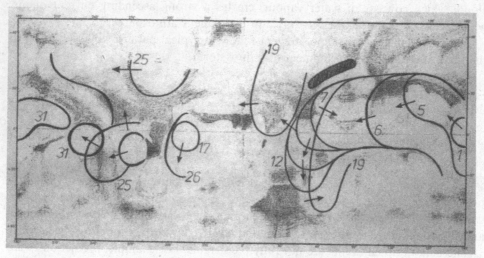

Fig. 5.36. *Dust storm between 1984 January 17 and February 19.*

dust-storm practically encircled the planet in a month, developing on both sides of the equator, but completely spared certain areas.

Another major dust-storm was observed in 1984 Jan.–Feb. (Fig. 5.36). It began, again in Isidis Regio and Aeria, around January 18. It spread eastwards, invading Mare Thyrrhenum and Zephyria at the end of the month. It then developed rapidly between February 5 and 12, covering Mare Sirenum, Solis Lacus, and Aurorae Sinus. Between February 12 and 19, it split in two, causing Aurorae Sinus to reappear between Margaritifer Sinus and Mare Erythraeum (which were invisible), and the areas of Tempe and Candor, which were also veiled by dust.

Following such atmospheric activity requires a large number of visual or photographic observations. We hope that this brief description, together with the results presented (all drawn from amateur records), will have convinced the amateur that the observation of Mars is fascinating and always unpredictable.

5.4 The other planets

5.4.1 Saturn

To the naked eye, Saturn appears as a first-magnitude object, with a very characteristic yellowish tint. This relatively low magnitude is a result of the planet's considerable distance from the Sun (1.5 thousand million km). Its spectacular telescopic appearance, thanks to its system of rings, which are visible in even the smallest 60-mm telescope, means that it is the beginner's favourite planet. In fact, observation of Saturn is a fairly thankless task, because very little ever happens!

5.4.1.1 Saturn's appearance

At its closest, the planet subtends 20 arc-seconds, and the rings then have an apparent diameter similar to that of Jupiter, about 43 arc-seconds. A telescope reveals the strongly flattened disk, with cloud bands running parallel to the equator. These light zones and dark belts are low in contrast and although some amateurs have reported, and even drawn, numerous dark spots and even equatorial plumes like those on Jupiter, observation with large telescopes and the photographs returned by the Voyager probes makes it quite obvious that such detail is illusory – always a problem with visual observation. Unlike Jupiter, atmospheric activity on Saturn is weak, or at least is beyond the low resolution of telescopes used from the ground.

Two atmospheric phenomena are, however, recognized:

- One is seasonal, and associated with the rings;
- The other is sporadic, and may justify constant surveillance of the planet by amateurs.

In the first of these, a light equatorial zone is observed, and this systematically occurs above the rings. This zone is not fixed: it crosses the equator every fifteen years, appearing on the same side as the illuminated face of the rings.

In the second case, Jovian-type activity is observed in the North Tropical Zone (NTrZ). Sporadically, persistent white spots appear (as in 1876, 1903, 1933, 1946, 1960 and 1990), and then migrate into the NTrZ within a few weeks. Because of their rarity, observation of these spots is of great interest. Their meridian transits allow their rotation period around Saturn to be determined. This is one of the few ways that we have of measuring the rotation periods of Saturn's various cloud belts. This is why Saturn should be subject to constant surveillance. Such an observation has every chance of being unique, given the small number of observers monitoring the planet. It would be as well to give some thought as to whom to contact if such an event is ever observed.

5.4.1.2 *The appearance of the rings*

Despite recent discoveries by space-probes and from the ground of other systems of rings around Jupiter, Uranus, and Neptune, Saturn and its system of rings still remains the most spectacular object in the Solar System.

The considerable inclination of the planet relative to its orbit (26° 44′) results in the rings showing a very varied appearance over the course of a Saturnian year, which lasts 29 terrestrial years. The relationship between the degree of opening of the rings and the planet's orbital position was determined by Huygens in 1654. Because the plane of the rings coincides with that of the equator, so at the equinoxes the rings are lit from the side and because of their extreme thinness (about 1 km), practically disappear. At the solstices, on the other hand, the rings are open to their fullest extent. The last disappearance of the rings took place in 1980; accurate drawings of the strange, and varied appearance at times just before and after disappearance are not devoid of a certain interest. A similar disappearance will occur again in 1995. The rings were open to their greatest extent in 1987, and at the time of writing the North Pole is visible.

Classically, as seen from Earth, three rings are visible. A fourth, Ring D, which is very close to the planet and of low surface brightness, is not visible with amateur equipment. Moving inwards towards the planet, we have:

- Ring A: Its diameter is 283 000 km, and it is about 15 500 km in width. It is the second brightest ring, and is divided into two parts by the Encke Division. This is often visible in small instruments, but is very difficult to photograph. This does not apply to the famous Cassini Division, discovered in 1675, which separates Ring A from Ring B. This division, which subtends an angle of about 0.7 arc-second, is visible in a 80-mm refractor, because of its high contrast and linear nature. It forms a test for the quality of photographs of Saturn (Fig. 5.37).

   The ring sometimes seems irregular to visual observers. Such variations have been confirmed by several observers and followed, enabling the rotation period of that part of the ring to be deduced.

- Ring B: This is the brightest and widest (26 000 km) ring. Its brightness is uneven: its inner portion is darker than the outer. With good seeing and a large telescope it is possible to observe several divisions in the inner part of the ring. The largest of these may be detected with a telescope about 300 mm in diameter, but the visibility of these divisions is very variable. A gap, sometimes known as the Lyot Division, separates Ring B from Ring C.

- Ring C: Ring C, or the Crêpe Ring, is largely transparent (stars and Saturn may be seen through it). Its brightness increases from the inner edge towards the outside, and its width is 16 400 km. A fairly large telescope is needed to be able to appreciate it properly, but it is visible in a 200-mm.

5.4.1.3 *Photography of Saturn*

Saturn is difficult to photograph because of its low albedo and its small apparent diameter. A 310-mm telescope should allow the Cassini Division in the rings, the presence of Ring C, and the main bands on the planet to be recorded fairly

Fig. 5.37. Top to bottom:

1: *Photograph of Saturn in 1973 by G. Viscardy, using a 520-mm reflector. Composite photograph on ORWO film.*

2: *Photograph of Saturn by G. Viscardy, 520-mm reflector, TP 2415 film; exposure: 3 s.*

3: *Drawing by R. Néel, using a 200-mm reflector, 1976 March 31; magnification: 240×.*

Note that detail visible with a 200-mm telescope requires a 520-mm to be photographed.

easily. This type of record is not of very great scientific interest: it is more of a technical challenge. Given the planet's low brightness and the need to use fairly long focal ratios (f/80 to f/100), TP 2415 film is not recommended, because it is not sufficiently fast and is rather too contrasty. We prefer one of the other modern films, particularly the new Kodak Tmax 400, which combines high speed (800–1000 ISO, after development for 6 minutes in D19b), with a fine grain-size and adequate resolution. At f/100, exposure times ought to be between 3 and 6 seconds. Curiously, Saturn often responds well to being photographed in colour. Fujichrome or Ektachrome 400 are appropriate with a slight over-development.

Finally, we would urge amateurs to devote a little more of their time to the observation of Saturn. It is a marvellous telescopic object, which is very suitable for gaining experience both visually and photographically. In addition, constant monitoring of the planet may lead to unique and important observations.

5.4.2 Venus

Venus is a very accessible planet, both because of its large apparent size (amounting to as much as 1 arc-minute) and its high albedo. Unfortunately, the cloud layers covering the planet do not show any detail in the visible region. It is now known that some of the vague markings that certain observers thought they had seen were only contrast effects. Photography of the phases of Venus is relatively easy, particularly during the crescent phases in the evening and early morning, when it is almost full daylight (Fig. 5.38). Atmospheric turbulence is always bad, because of the planet's low altitude, so it is essential to take a large number of exposures and choose the best. Given the high albedo of Venus, and the impossibility of making exposures of less than 0.5 second (when manually covering the end of the tube), fairly large focal ratios should be chosen, such as f/80–f/100. The shortest possible exposure should be given on TP 2415 film, developed in HC-110 (at dilution 1 + 30) for a maximum of 4 minutes.

Much more interesting is photographing the clouds in ultraviolet light. Cloud formations in the form of recumbent Y or ψ shapes become visible (Boyer, 1965). It was these that allowed Charles Boyer to detect the 4-day retrograde rotation (Fig. 5.39). A reflector is required, with a diameter of 250–310 mm, and a focal ratio of about f/80, for the half-Moon phases (at maximum evening and morning elongations). Experiments may be made with a Wratten 49 gelatine filter, but results are largely dependent on the transparency of the Earth's atmosphere to ultraviolet light. To improve chances of success, true ultraviolet filters should be used (which are opaque to white light), such as Schott's UG2 and UV5. To focus, a transparent glass filter that is the same thickness as the UV filter should be used. This is replaced by the UV filter just before taking the photographs. Because these filters absorb strongly, exposures have to be fairly long (1–4 seconds), and the TP 2415 film should be developed in D19b developer (6 minutes at 20°C).

Some observing groups feel that it is important to make observations of the date of dichotomy – the exact date of Venus' first or last quarter – because a discrepancy has been found between the theoretical date and that observed. The scientific value of such observations has not yet been established, however.

Fig. 5.38. *Crescent Venus, photographed by J. Dragesco, using a 350-mm, Schmidt-Cassegrain telescope.*

5.4.3 Other planets, of little scientific interest

5.4.3.1 Mercury

As we said in the introduction to this chapter, the surface of Mercury is reasonably well-known from space-probe information, and cannot be observed from Earth. Mercury is always difficult to observe: so it is quite natural to want to make use of a favourable elongation to observe it visually, or to photograph the tiny crescent or semicircular disk.

Transits of Mercury across the disk of the Sun are more interesting. (The last took place on 1986 November 13; the next will occur on 1993 November 6.) It is suggested that these should be photographed using the techniques described for photographing sunspots, as in Fig. 5.40.

5.4.3.2 Uranus

All that one can hope to observe of Uranus, under the best conditions, is the tiny bluish disk of the planet. One can also try photographing three or four of the satellites (at the telescope's prime focus).

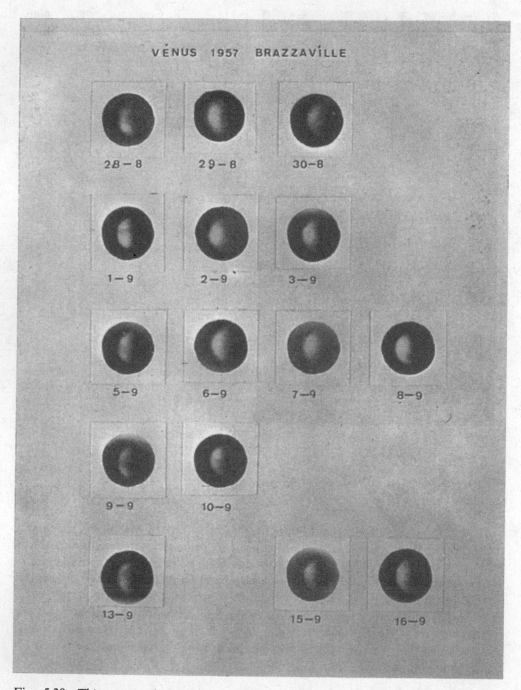

Fig. 5.39. *This series of photographs of Venus was obtained in 1957 by C. Boyer, and enabled him to discover the 4-day rotation period of the atmosphere. Note a dark feature that is visible every four days (left-hand column). Telescope: 260-mm reflector, Ilford Pan F film; Wratten 34 ultraviolet filter; exposure 1 s.*

Fig. 5.40. *Transit of Mercury, 1986 November 17, 05:04 UT. Photograph by J. Dragesco from Butaré in Rwanda.*

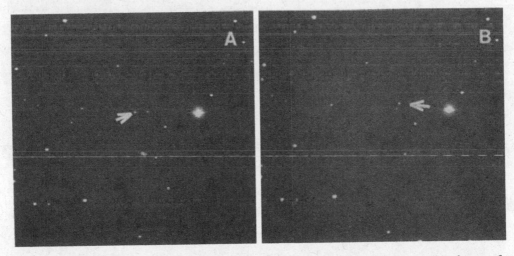

Fig. 5.41. *Apparent motion of Pluto. Photographs by R. Néel, taken at prime focus of a 310-mm, f/6 reflector. Left: 1980 April 12, 22:00–22:22 UT. Right: 1980 April 13, 22:04–22:33 UT.*

5.4.3.3 Neptune

Neptune is even less accessible. Visual observation is of little interest, so all that remains is to try photographing the planet and its two satellites (although Nereid is of magnitude 18.7, and thus only detectable with 310-mm telescopes, with exposures of more than 30 minutes).

5.4.3.4 Pluto

This planet is beyond the amateur's grasp. An experienced astrophotographer might like to try photographing the planet's motion against the background of stars (Fig. 5.41).

6 Planetary satellites

B. Morando

6.1 Satellite systems

6.1.1 General

Planetary satellites are faint bodies that generally offer few opportunities for amateurs. Only the Galilean satellites of Jupiter, which will be discussed in detail, can be the subjects of interesting and useful observations. We shall, however, briefly mention all the satellites in the Solar System.

Table 6.1 shows, for each planet, the number of satellites that are observable by amateurs (column 1); the number that are only observable by professional instruments (column 2); and those that have been discovered by space-probes but which are not detectable from the Earth (column 3). This last column may see additions in the form of further discoveries in the future.

Mercury and Venus have no satellites. The Earth's satellite, the Moon, is the subject of Chap. 4. The two satellites of Mars, Phobos and Deimos, have largest dimensions of 27 km and 15 km respectively. Their orbits are essentially circular and lie in Mars' equatorial plane, with, respectively, radii of 2.76 times and 6.9 times the equatorial radius of Mars, and periods of $7^h 39^m$ and $30^h 17^m$. As a result, for an observer on Mars, Phobos, whose sidereal rotation period is shorter than a Martian day ($24^h 37^m$), would rise in the West and set in the East. The faintness of the satellites (only 11.6 and 12.7 at perihelic oppositions of Mars), and the fact that they are so close to the planet mean that, in general, they are not observable with the types of instruments available to amateurs.

Uranus has five major satellites, which are, in order of distance from the planet, Miranda, Ariel, Umbriel, Titania and Oberon. The orbits are essentially circular and lie in the equatorial plane of Uranus, except that of Miranda, which is inclined at 3.4° to that plane and has an eccentricity of 0.027. Their distances from Uranus are 129, 191, 266, 436, and 584 million kilometres, respectively. The radius of Uranus itself 25 400 km. The magnitudes of the satellites lie between 14 and 16.5. Voyager 2 discovered ten further, smaller satellites.

The two satellites of Neptune known before the Voyager encounter, Triton and Nereid, have orbits that are inclined at about 25° to the planet's equator, although Triton's orbital motion is retrograde. Triton's orbit is circular, but Nereid's has a high eccentricity of 0.75. Triton has a diameter of 3800 km, and is thus larger than the Moon, but Nereid is just 200 km across. Their magnitudes are 13.7 and 18.7 respectively. [The five (or possibly six) tiny satellites discovered by Voyager are far too faint to be detectable with amateur equipment. – Trans.]

Finally, Pluto has one satellite, Charon (magnitude 16.8), which is less than one arc second from Pluto. Its period is 6.387 days. Between 1985 and 1991 the Earth was close to the plane of Charon's orbit around Pluto. Mutual occultations were

Table 6.1. *Satellites of the planets*

Planet	Observable by amateurs	Observable by professionals	Discovered by space-probes
Mercury	–	–	–
Venus	–	–	–
Earth	Moon	–	–
Mars	0	2	0
Jupiter	4	9	3
Saturn	7	2	9
Uranus	0	5	10
Neptune	2	2	6
Pluto	0	1	–

Table 6.2. *The Galilean satellites of Jupiter*

Parameter	Io	Europa	Ganymede	Callisto
Radius (km)	1816	1563	2638	2410
Albedo	0.61	0.64	0.42	0.20
Magnitude at opposition	4.8	5.2	4.5	5.5
Greatest elongation	2′17″	3′40″	5′48″	10′13″
Sidereal period (days)	1.769 14	3.551 18	7.154 55	16.688 99

therefore possible and a considerable number were observed from which a lot of information about the two bodies could be derived.

6.1.2 *The satellites of Jupiter*

Jupiter has, at present, 16 known satellites. They may be classified into four groups. The first and most interesting of these is that comprising the Galilean satellites. These were discovered by Galileo on 1610 January 7 and are four in number. They have approximately circular orbits, lying in the equatorial plane of Jupiter. In order of distance from Jupiter they are: Io, Europa, Ganymede, and Callisto. They are also known by the Roman numeral designations: I, II, III, and IV, respectively. They are bright objects, easy to observe, but the diameter of the largest (Ganymede) does not exceed 1.7 arc-seconds. Specific details are given in Table 6.2.

The second group consists of satellites XIII, VI, X, and VII, which have orbits inclined at about 25° to Jupiter's orbit, direct motion, and orbital radii of around 11 million kilometres. Their diameters lie between 10 and 180 km, and their magnitudes between 16.8 and 20.

The third group of four satellites (XII, XI, VIII, and IX) orbits at about 22

Table 6.3. *The major satellites of Saturn*

Parameter	Tethys	Dione	Rhea	Titan
Radius (km)	530	560	765	2575
Albedo	0.88	0.65	0.67	0.21
Magnitude at opposition	10.2	10.4	9.7	8.3
Greatest elongation	0'48"	1'01"	1'25"	3'17"
Sidereal period (days)	1.887 80	2.736 91	4.517 50	15.945 42

million kilometres from Jupiter. The orbits are retrograde and inclined at about 20°. The satellites have diameters of between 20 and 40 km, and magnitudes between 17 and 18.9. Finally, the fourth group consists of four satellites very close to Jupiter. Of these, only one, Amalthea (satellite V) was discovered from Earth. It has a circular orbit, 181 000 km in radius in Jupiter's equatorial plane. Its largest diameter is 270 km, and it is no brighter than magnitude 14.1. Satellites XIV, XV, and XVI were discovered by Voyager 1. They have circular orbits with radii of less than 200 000 km, lying in Jupiter's equatorial plane. They are small, being between 20 and 80 km, and their magnitudes are between 16 and 18.9.

6.1.3 The satellites of Saturn

Saturn is well-known for its rings, which are visible in even a 60-mm telescope. It has 17 [18] known satellites, only four of which are brighter than magnitude 10.4 when Saturn is at opposition. Table 6.3 gives their main characteristics.

To these may be added Mimas, Enceladus, and Iapetus, which are at the limit of observation for amateurs. The magnitudes when Saturn is at opposition are 12.9, 11.7 and 11 respectively. Hyperion has a magnitude of 14.2, and Phoebe, whose orbit is retrograde, a magnitude of 16.5.

Finally, eight satellites were discovered by Voyager 2 in 1980, including the famous 'shepherd' satellites that stabilize the rings. They are quite invisible from Earth. [A ninth satellite, lying within the Encke division, was discovered in 1990. – Trans.]

6.2 Positions of the Galilean satellites

Because the periods of the orbits of the four Galilean satellites are very short, the bodies cannot be immediately identified, even if they were observed the previous night. A very simple and practical method of identifying them, however, is to plot a graph of each satellite's position with respect to Jupiter's disk. With time running vertically on the plot, joining the positions of the satellites gives four intertwined sine-curves corresponding to the satellites' positions at any time (Fig. 6.1). The distances of the satellites from the planet are in proportion to their true distances and the scale is indicated by the two vertical lines that represent the diameter of Jupiter. The orientation is such that East is to the right, and South at top: the relative positions are therefore those seen from the northern hemisphere at meridian passage in an inverting telescope. It is easy to read from the graph whether the satellites are passing in front of, or behind, the planet.

These diagrams appear in numerous publications, in particular in the *Astronomical Almanac*, published each year by the U.S. Naval Observatory and the U.K. Nautical Almanac Office. The most accurate diagrams are those given in the *Astronomical Phenomena for the year ...* from the same sources. These publications also give details of various events involving the Galilean satellites that will be discussed a little later.

6.3 Ephemerides for the Galilean satellites

Ephemerides may be calculated for these satellites. Such positions are given in rectangular coordinates relative to Jupiter in the plane tangent to the celestial sphere at the centre of Jupiter's disk; the x-axis is parallel to the equator, and the y-axis increases towards the north celestial pole. We therefore have:

$$x = \Delta\alpha \cos\delta \qquad y = \Delta\delta,$$

where $\Delta\alpha$ is the difference in right ascension between the satellite concerned and Jupiter, and $\Delta\delta$ the difference in declination, δ being the declination of Jupiter itself. There are various publications where these differential coordinates may be found: the *Astronomical Almanac* (which contains data for satellites of Mars, Jupiter, Saturn and Uranus), and particularly the *Connaissance des Temps*, published by the Bureau des Longitudes, which represents these coordinates by Tchebychev polynomials with a precision of about one tenth of a second of arc, and the *Ephemerides for the satellites of Jupiter, Saturn and Uranus for the current year*, a supplement to the *Connaissance des Temps*, which is also published by the Bureau des Longitudes. In this latter publication the differential coordinates are represented by equations with periodic terms that must be multiplied by the appropriate value of time. For more information, reference should be made to the explanatory material in these publications. Charts of satellite positions and times of mutual phenomena are given in the yearly BAA *Handbook* and the Royal Astronomical Society of Canada (RASC) *Observer's Handbook*.

Highly accurate ephemerides for the satellites are not necessary for finding these bodies, it usually suffices to identify them as described in the earlier paragraph. However, if one wants to make use of observations to improve the theory of their motion, the positions derived from observation must be compared with those predicted by the ephemerides, obtaining $O-C$ ('Observed minus Calculated') values. This is why the most accurate available ephemerides are published. Mathematical treatment of $O-C$ values to improve theory is not easy, but amateurs interested in celestial mechanics and who have a computer can take the plunge.

6.4 Galilean satellite phenomena

As it orbits Jupiter, a satellite may, to a terrestrial observer, undergo one of the following four types of event (Fig. 6.2):

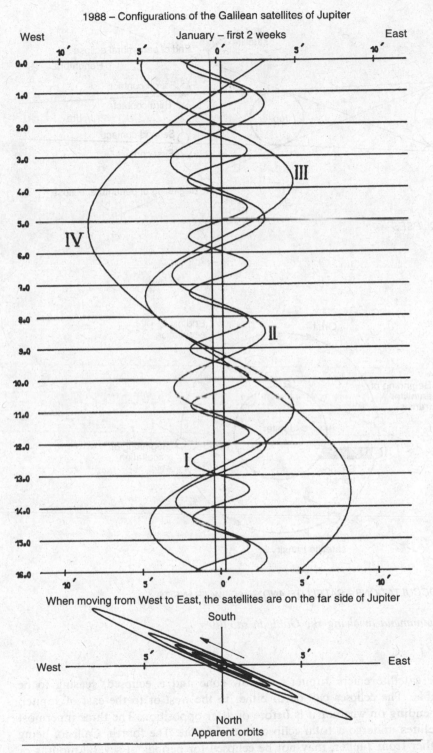

Fig. 6.1. *A diagram typical of those included with most ephemerides, showing the positions of the Galilean satellites.*

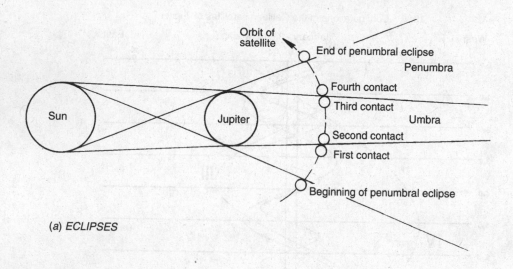

(a) ECLIPSES

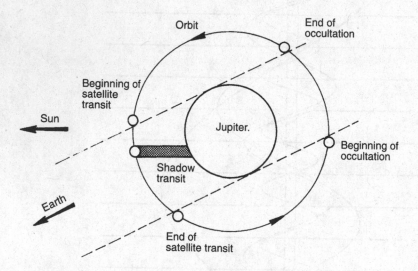

(b) OCCULTATIONS – SATELLITE AND SHADOW TRANSITS

Fig. 6.2. *Phenomena involving the Galilean satellites.*

- The satellite enters Jupiter's shadow cone and is eclipsed, ceasing to be visible. The eclipses can occur either to the west or to the east of Jupiter, depending on whether it is before or after opposition. The three innermost satellites undergo a total eclipse on each orbit. The fourth, Callisto, being farther from Jupiter, may not be eclipsed for periods of several months or even years.

- A satellite, at superior conjunction, disappears behind the disk of Jupiter. These events, known as occultations, occur on every orbit for the three inner satellites, but again, with the fourth there are periods when it is not occulted.

- The shadow of a satellite is projected onto Jupiter's disk, when certain points on Jupiter undergo an eclipse of the Sun. As the satellite moves, so does the shadow. These events are known as shadow transits. They occur under the same circumstances as the previous events.

- Finally, at inferior conjunction, a satellite may be seen against the disk of Jupiter. These events are known as transits. Again the conditions governing their occurrence are the same as before.

The maximum duration of one of these events is $2^h 18^m$ for Io, $2^h 53^m$ for Europa, $3^h 40^m$ for Ganymede, and $4^h 52^m$ for Callisto. These maximum durations occur when the Earth and the satellite are both in the same plane as Jupiter's equator. They are longer the farther the satellite is from Jupiter. Its period is greater (in accordance with Kepler's third law), so its angular velocity is smaller, and it takes longer to cross the shadow cone or to pass in front of, or behind, Jupiter. The same applies to shadow transits.

Eclipses have been, and still are, the only events whose observation may be of scientific use. They do, in fact, often occur when the satellite involved is some distance from Jupiter, which means that observation is fairly easy. They may be timed with a reasonable degree of accuracy that was, for many years, quite adequate. The other events are, however, more difficult to observe, even with good modern instruments. It is not feasible to time them for astrometric purposes. Observation of eclipses of the Galilean satellites was, until the end of the 19th century, the only method of determining the various physical characteristics of the orbits. Astronomers are not agreed, however, on the relevance of making such observations nowadays; methods of reduction need to be substantially revised. This is because Jupiter's shadow cone is not sharply defined, because the planet itself does not have a distinct edge, but is surrounded by an atmosphere, the density of which varies with depth in a way that is not only unknown, but is in any case difficult to model. Theoretical investigations are being made at present, and if they succeed, may rekindle interest in the observation of eclipses. Nevertheless, past observations of eclipses are the only data available for calculating the long-period terms in the orbits. This is why lists of old observations, such as those collected by Delambre, are very valuable to astronomers.

The events are listed in the published ephemerides that give the configurations of the satellites, and which have been mentioned earlier. The most accurate predictions are those given in the *Connaissance des Temps*, which gives the times of the beginning and end of each event, distinguishing between exterior and interior contacts with the shadow cone or with the cone of visibility. For eclipses, the times of the satellite's contact with the penumbral cone are also given. For further details, reference should be made to the explanations given in the works mentioned.

To make useful observations of eclipses, the times of contact should be noted as accurately as possible. If one is observing visually, an accuracy of a few seconds is all that one can expect. The events are all the more difficult to observe because they are not instantaneous, as the satellites are not point sources. The duration of immersion and emersion varies between 3 and 9 minutes depending on the satellite. Once again, the most distant satellite has the longer time. It would be preferable to obtain a light-curve by photoelectric methods. This technique will be described later in connection with the observation of mutual events.

6.5 Mutual phenomena of the Galilean satellites

6.5.1 *The interest in these events*

In the last decade, observation of the mutual phenomena of Jupiter's Galilean satellites has been shown to be the most accurate method of improving our knowledge of their orbits. Such an improvement has been essential for space research, in particular for the Galileo project, which is to study the Jovian system. The accuracy with which the position of the Galilean satellites can be determined from the study of eclipses by Jupiter is about 0.2 arc-seconds, which amounts to an error of around 750 km in their positions in space. Photographic observations with instruments of 2.5 m focal length give errors of the same order, but with focal lengths of 10 m, an accuracy of 0.06 arc-second can be attained, which amounts to 230 km. Greater accuracy cannot be expected, because it is limited by film-grain and atmospheric turbulence. Observation of mutual phenomena, however, can further reduce the error by a factor of 2, leading to an error in position of about one hundred kilometres in objects that are, at opposition, some six hundred million kilometres from Earth. Mutual phenomena are sharply defined, because the satellites are spherical, and without atmospheres. It is sufficient to record the time of mid-event accurately, which is possible with photoelectric methods. The accuracy in timing mutual phenomena is in fact limited by uncertainties arising from variations in the sky background, and by unknown variations in the albedo of the satellites' surfaces.

This is why campaigns for the observation of these phenomena – which occur only every six years – have been organised on the last two occasions. In France, the Bureau des Longitudes has promoted two campaigns in 1979 (PHEMU-79) and 1985 (PHEMU-85). Amateurs played an important part in these programmes, which is why we intend to discuss them in detail.

6.5.2 *A description of the events*

The orbits of the Galilean satellites have very small inclinations to the plane of Jupiter's equator. During Jupiter's sidereal period of twelve years, the Earth and the Sun spend six years above this plane and six years below it. As a result, the orbits of the Galilean satellites appear as very elongated, concentric ellipses. Every six years, however, the Earth and Sun cross the plane. Two satellites may then be aligned with the Earth, one satellite occulting the other, or one may even enter the shadow cone cast by the other, and thus be eclipsed (Fig. 6.3). The phenomena are therefore mutual occultations and mutual eclipses. For an eclipse to occur the jovicentric declination of the Sun should be less than a certain value, which depends

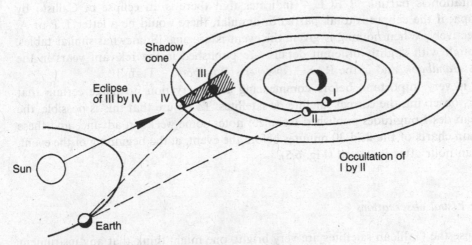

Fig. 6.3. *Mutual phenomena of the Galilean satellites.*

on the size of the satellite. Similarly for occultations, the jovicentric declination of the Earth should be less than a certain value, which depends on the radii and the relative positions of the satellites involved. From this it may be seen that prediction of the mutual events requires an accurate knowledge of the values for the radii of these satellites, and of their orbital elements. This explains why it has only been possible to predict these phenomena in the last fifteen years, following work carried out in France by the Bureau des Longitudes and in the U.S.A. by the Jet Propulsion Laboratory. The values for the radii that are used are those provided by the Pioneer and Voyager space-probes.

In a mutual occultation, one sees two satellites gradually approach one another until only a single spot of light can be seen, after which the two gradually separate. The moment when the two satellites appear to merge depends on the resolution of the instrument being used, but a time arrives when one sees just one point of light, the brightness of which decreases, reaches a minimum, and then increases again. During an eclipse, one sees a single satellite, whose magnitude decreases, reaches a minimum, and then returns to its initial value. The duration of these events may vary from a few minutes to about an hour. The amplitude may vary considerably: it is zero for grazing events, but may be very pronounced for some eclipses; with a total eclipse the satellite concerned disappears completely.

6.5.3 Prediction of mutual phenomena

One or two years before each series of mutual phenomena, predictions are published in specialized journals, such as *Astronomy and Astrophysics*. Detailed predictions are published in the *Astronomical Almanac* and also elsewhere. Figure 6.4 reproduces one of the tables that were published by the Bureau des Longitudes in Paris. The column headed 'phenomena' is read as follows: '2 OCC 3 P' indicates that the second satellite (Europa) occults the third (Ganymede), the P indicating that the occultation is partial. '2 ECL 4' indicates that there is an eclipse of Callisto by Europa; if the eclipse is total, partial or annular, there would be a letter T, P or A, respectively. When nothing is given, the event is grazing. [Somewhat similar tables, although with slightly different details, are published in the relevant years in the BAA *Handbook* and in the RASC *Observer's Handbook*. – Trans.]

It is very important, before commencing an observation, to make certain that one is observing the correct satellite or satellites. To ensure that this is possible, the Bureau des Longitudes published technical notes some weeks in advance and these contain charts of the field 30 minutes before the event, at the beginning of the event, and an hour after it started (Fig. 6.5).

6.5.4 *Visual observations*

Because the Galilean satellites are very bright, one might think that any instrument could be used for the observation of mutual events. In reality, an aperture of 200 mm is required to obtain useful observations, but this often proves to be quite adequate, as was shown by some of the results obtained during the PHEMU-85 campaign. A driven, equatorial mounting is required and, because the Jovian system moves with respect to the background stars and tracking is never perfect, a system for guiding and correction is essential.

A fundamentally important aspect of this sort of observation that is often neglected by amateurs – which means that their observations are not usable – is how the time is recorded. It is absolutely essential to record the different phases of the phenomena in Coordinated Universal Time (UTC). This is the time given by the speaking clock at about 01:00 in winter and 02:00 in summer. [Radio time signals are an even better source of accurate timing, especially if methods similar to those for occultations (Chap. 9) are adopted. – Trans.]

If it is not possible to obtain a time signal at the observing site, one may be used to set a good-quality quartz watch an hour, at most, before the event. It is advisable, and quite feasible, to determine the time to an accuracy of less than one second, if possible to within 0.5 second. In addition, it is advisable to begin observing some time in advance (about a quarter of an hour) because the predictions are not precise. Indeed, the aim of observing mutual phenomena is to use the differences between the observed and calculated times to improve the ephemerides. It should, incidentally, be noted that the predictions, with which the observations are to be compared, are calculated using Ephemeris Time (ET), or rather in a very similar system, Terrestrial Time (TT), which in practice coincides with International Atomic Time (TAI) adjusted by +32.184 s. In mid-1992, the difference TT − UTC was 59 s. [For a discussion of time systems, see Appendix 1, Vol. 2. – Trans.]

It is important to prepare carefully for an observation, because even though it may appear simple beforehand, it soon becomes obvious that there are many things to be done, and the event will not wait for anyone. It is, in effect, essential to carry out a rehearsal of the event so that one will not be taken by surprise, and also to be prepared for possible unexpected eventualities at the last minute, which might otherwise jeopardize the observation. During the observation it is a good idea to

Fig. 6.4. *Typical table of mutual phenomena of the Galilean satellites.*

1985 November

Date (ET) of maximum Yr M D	Phenomena	Start of event H M S	Start of shadow/occultation H M S	Start of totality H M S	Maximum H M S	Magnitude	Duration S	Distance from Jupiter (R_J) S (RJ)	Longitude of Jupiter at midday HR MN	Longitude of Sun at midnight HR MN	Latitude of Jupiter at the zenith DEG (')	Latitude of Sun at the nadir DEG (')	Notes
1985 11 2	2 1 OCC 2	20 39 1.	20 40 34.		18 31 10.	0.C	0	5.5	+ 0 35	− 5 10	−18 55	+14 54	(2)
1985 11 2	2 4 ECL 2 P				20 42 2.	10.7%	182	6.6	+ 2 42	− 3 4	−18 55	+14 56	
1985 11 3	3 4 OCC 2 P	5 52 2.	5 52 2.		5 59 10.	0.904	862	9.3	+11 56	+ 6 8	−18 54	+15 3	(3)
1985 11 3	2 2 OCC 2 T	2 42 1.	2 42 1.	2 52 20.	2 53 35.	1.C35	1385	0.9	+11 49	+ 4 0	−18 53	+15 19	
1985 11 4	2 1 ECL 1 A	3 45 14.	3 45 14.		3 46 32.	10.8%	156	2.8	+ 9 51	+ 4 58	−18 53	+15 20	
1985 11 4	1 2 ECL 1 A	18 17 14.	21 1 43.		18 20 42.	0.X		2.8	+ 9 27	+ 5 26	−18 52	+15 31	(2)
1985 11 4	3 1 ECL 3 A	20 59 55.			21 3 13.	3.5%	177	4.2	+ 3 10	+ 2 43	−18 52	+15 33	
1985 11 5	4 1 ECL 1	3 11 30.			3 24 39.	0.X		4.6	+ 9 22	+ 8 24	−18 51	+15 38	(2)
1985 11 5	4 4 ECL 1	8 8 24.			8 8 24.	6.6%	247	5.8	+ 9 39	+ 8 24	−18 51	+15 42	(2)
1985 11 6	4 2 ECL 2 A	12 49 6.	12 53 23.		12 55 25.	0.C		6.4	+ 9 58	+10 54	−18 49	+15 45	(2)
1985 11 6	1 4 OCC 1				7 42 56.	4.9%	140	6.6	+ 9 53	+10 5	−18 49	+15 59	
1985 11 6	6 2 ECL 2 P	9 49 13.	9 50 54.		9 52 6.	10.9%	163	3.3	+ 0 48	+ 5 22	−18 47	+16 1	(2)
1985 11 7	3 1 ECL 1 P	16 51 1.	16 52 60.		16 53 27.	5.9%	223	5.8	+ 4 42	+ 1 22	−18 46	+16 24	
1985 11 9	2 3 ECL 1 P	20 42 28.	22 24 11.		22 26 4.	0.C		7.1	+ 7 27	+ 0 58	−18 44	+16 28	(2)
1985 11 9	2 2 OCC 2				20 42 53.	0.X	74	5.4	+ 3 22	+ 0 48	−18 43	+16 47	(2)
1985 11 11	1 2 ECL 2 P	22 59 27.	23 1 27.		20 55 2.	0.7%	167	6.5	+ 5 27	+ 0 44	−18 43	+17 3	(2)
1985 11 13	1 4 ECL 1 A	5 57 58.	5 59 3.		23 2 6.	11.1%		3.8	−11 29	+ 6 13	−18 40	+17 24	(2)
1985 11 14	2 2 OCC 2	12 9 11.			20 13 45.	0.C		7.3	+ 2 54	+ 3 24	−18 39	+17 34	(2)
1985 11 14	2 2 OCC 2	2 36 0.			12 6 49.	0.X		5.3	+ 7 5	+10 29	−18 36	+18 0	(2)
1985 11 15	1 2 ECL 1 A				12 11 32.	0.X		6.4	+ 9 18	+11 35	−18 34	+18 11	(2)
1985 11 15	1 4 OCC 2	19 6 3.			19 47 29.	1.3%	169	13.7	+ 9 18	+ 2 51	−18 34	+18 21	(2)
1985 11 16	3 1 OCC 1 A	21 57 18.			21 58 47.	0.C86	177	3.5	+ 4 42	+ 1 47	−18 33	+18 23	(2)
1985 11 17	3 3 ECL 1 P	14 21 20.	14 24 24.		14 50 23.	2.8%	236	5.9	+ 8 30	+ 2 0	−18 32	+18 26	(2)
1985 11 18	3 3 ECL 1 P	14 24 33.	14 24 33.		14 34 56.	96.7%	1244	2.0	+ 2 51	+ 0 23	−18 31	+18 34	(2)
1985 11 18	1 2 OCC 2	4 5 27.	23 46 1.		23 52 26.	91.2%	759	5.8	+ 6 31	+ 4 20	−18 30	+18 40	(2)
1985 11 20	2 1 ECL 2 A				23 18 56.	0.C	0	6.8	+10 54	+ 0 17	−18 28	+18 42	(2)
1985 11 20	4 3 ECL 3 P	1 18 58.	8 13 7.		1 21 6.	11.5%	170	6.3	+ 6 19	+ 1 34	−18 28	+18 54	(2)
1985 11 21	2 1 OCC 1	8 12 2.			8 14 34.	0.X		3.6	+ 8 11	+ 8 26	−18 26	+18 55	(2)
1985 11 22	2 4 ECL 1	18 33 18.	18 35 30.		18 35 30.	0.238	786	1.6	+ 8 51	+ 9 37	−18 25	+19 14	(2)
1985 11 22	4 3 OCC 2 P	9 23 21.	9 23 21.		9 29 46.	0.C25	72	5.1	+ 7 33	+11 15	−18 24	+19 20	(2)
1985 11 23	2 2 OCC 2	12 30 12.	12 30 48.		12 30 48.	0.X		6.2	+ 2 25	+ 9 17	−18 20	+19 43	
1985 11 23	1 2 ECL 1 P	14 28 23.	15 3 7.		14 30 11.	0.308	1024	3.5	+ 2 27	+ 8 38	−18 20	+19 44	(2)
1985 11 24	2 1 OCC 1 A	21 19 9.	21 20 15.		15 11 49.	11.7%	169	3.7	+ 1 48	+ 2 26	−18 17	+19 46	(2)
1985 11 24	3 3 OCC 1	5 45 32.	21 21 37.		21 21 37.	6.3%	387	5.5	+ 4 27	+ 1 25	−18 16	+20 5	
1985 11 24	3 1 ECL 1 P	12 32 38.	5 51 11.		5 55 31.	80.2%	535	5.6	+ 8 21	+ 1 59	−18 15	+20 11	(3)
1985 11 25	3 2 ECL 1 P	3 28 10.	12 36 45.		3 44 53.	49.1%	1431	5.3	−11 44	+ 3 41	−18 14	+20 19	
1985 11 27	3 3 ECL 1 P	7 28 14.	1 41 58.		7 32 13.	0.X	460	1.1	+10 41	+ 7 41	−18 13	+20 21	
1985 11 27	2 2 OCC 2 P	1 41 58.			1 42 60.	0.C78	122	6.5	+ 8 58	+ 1 55	−18 11	+20 30	(2)
1985 11 28	1 2 ECL 2 A	3 37 55.	10 27 26.		3 39 18.	12.0%	166	6.1	+10 54	+ 3 51	−18 11	+20 31	(2)
1985 11 29	2 1 OCC 2 P	10 26 20.	14 53 38.		10 28 47.	0.140	157	4.9	+ 6 13	+10 39	−18 8	+20 46	(2)
1985 11 29	2 1 ECL 1 A	14 53 38.	23 34 42.		14 54 57.	0.X	162	3.8	+ 1 38	+ 8 54	−18 2	+21 12	
1985 11 30	3 3 OCC 1 P	16 47 8.	4 49 30.		16 48 1.	12.2%	659	6.0	+ 0 15	+ 0 14	−17 58	+21 25	(2)
1985 11 30	4 3 ECL 1 A	23 34 34.	16 13 4.		23 54 57.	0.543	821	5.9	−11 37	+ 7 35	−17 58	+21 28	
1985 11 29	3 3 OCC 1 P	4 49 30.			16 19 54.	0.233	267	0.3	+ 0 12	+ 6 56	−17 56	+21 32	
1985 11 30	1 2 ECL 3 P	6 45 22.	6 47 54.		6 50 6.	14.3%		4.8	+ 9 37	+ 6 56	−17 55	+21 38	(3)
1985 11 30	2 3 ECL 2 P	10 49 55.	10 52 46.		10 54 26.	9.5%	198	6.2	− 5 32	+11 1	−17 54	+21 40	

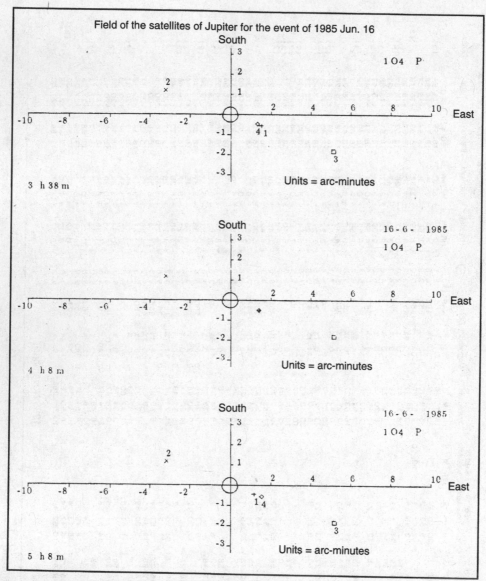

Fig. 6.5. *Typical field charts of satellite phenomena.*

use a tape recorder for any comments by the observer or observers, with audible time-signals.

After the event, a full report should be drawn up, giving: observer's name, observing site, type of event observed, date, details of the equipment used (aperture, f/ratio, etc.), estimated reliability of the observation, seeing, etc. Then the results should be given and the manner in which these are presented will depend on the type of observation that was made. We shall return to this point shortly.

Experienced observers of variable stars may try to observe these events, or at

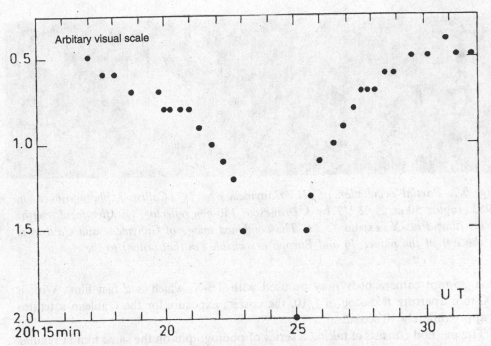

Fig. 6.6. *Light-curve obtained during an eclipse of I by III (Io by Ganymede) in 1979. The luminosity scale is calibrated in arbitrary units. The abscissa shows time.*

least eclipses, visually. The brightness of the eclipsed satellite is compared with that of another satellite. The Argelander method is not often usable, because there is normally a lack of suitable comparisons in the field of view. In addition, whereas a variable-star estimate may be allowed to take 30-odd seconds, here one must make up one's mind in just a few seconds, because the magnitude varies very rapidly.

In observing an occultation visually, it is a good idea to defocus the telescope slightly, as this allows the two satellites to be seen, when very close to one another, as a single area of light, the brightness of which can then be more easily estimated. Figure 6.6 shows the light-curve of Io as it was occulted by Ganymede in 1979. The observation was made with the 830-mm refractor at the Meudon Observatory. Even with small instruments it is possible to time the minimum to within one or two seconds.

6.5.5 *Photographic observations*

It is possible, but difficult, to make photographic observations of mutual phenomena that are fairly long (for example, with durations of 10 minutes or more) and that have moderate amplitudes. These would, in any case, be less accurate than photoelectric observations made with the same instrument (and which will be described in the next section). But not everyone has a photoelectric photometer, and photographic observations can give useful results (Fig. 6.7).

Fig. 6.7. *Partial occultation of III (Ganymede) by IV (Callisto), photographed on 1985 August 30 at 21:13 UT by J. Dragesco; 150-mm reflector; effective focal length: 3 m; film: Plus-X; exposure: 3 s. The combined image of Ganymede and Callisto is to the left of the planet; Io and Europa are visible (in that order) to the right.*

A 35-mm camera body may be used with Tri-X, which is a fast film. With a 200-mm aperture telescope of f/10, the correct exposure for the Galilean satellites is approximately 10 seconds.

The method consists of taking a series of photographs on the same film at regular intervals. These are chosen with regard to the total duration of the event, making, for example, a ten-second exposure every 40 seconds. If the event lasts about ten minutes, a record of the whole event may be obtained on a single cassette of 36 exposures. For long events, one will have to change cassettes, which may be done when (say) one-third and two-thirds of the event have elapsed.

Reduction of the photographs is done by measuring the density of each image on the film by using a microdensitometer. Assuming that one has obtained images falling on the linear portion of the film's characteristic curve, it is possible to determine the time of minimum brightness. If one wants to check this, or if one wants to draw a light-curve showing the fall in magnitude as a function of time, the film must be calibrated. To do this it is necessary to take some photographs (say five) on the beginning of the film, and before the event occurs, of a satellite of known magnitude. These frames should be exposed for various times, beginning with twice the exposure chosen to record the event itself, then the same time, one-half, one-quarter, etc.

If no densitometer is available, the diameter of each image on the film may be measured; the smaller this diameter, the less light the film has recorded. This method may be carried out by projecting the negatives onto a rigid screen (a white-painted wall, for example). However, this type of procedure should not be expected to be very accurate.

6.5.6 *The declination-drift photographic method*

6.5.6.1 *The basic principle*
The basic idea is to photograph the satellites involved in the event on one single frame, which covers the whole duration of the event, by letting the image trail smoothly across the film during the exposure. The photographic image of the satellite is no longer a point, but a straight line; each point making up this line represents the image of the satellite at a specific time. If the magnitude of the satellite varies with time, the degree of darkening of the film varies correspondingly along the trail. By measuring the density along the trail, the light-curve of the satellite versus time may be derived. It will immediately be apparent that two precautions are necessary:

- The photographic image must carry time markers, which will provide the time scale for the light-curve;
- The drift must be completely regular. Any variations in the speed at which the image moves will produce unwanted changes in the density of the negative. This second aspect involves not only the technique that is used to give the required drift, but also the equatorial mounting's drive, which should be obtained from a very stable motor. Naturally, the operator must avoid making any corrections – even in right ascension – during the exposure.

6.5.6.2 *Motion in declination*
The first difficulty that we encounter is the presence of Jupiter alongside the satellite. The planet also produces a trail on the film, but this is so over-exposed that diffusion of the light within the emulsion causes the trail to have a width that is considerably in excess of Jupiter's diameter. So there is the risk of finding that the satellite's trail is drowned in that of the planet. It is therefore important to separate the two trails as much as possible; for this it is necessary for the direction of trailing to be more or less at right angles to the plane of the satellite orbits. At the time of the two PHEMU campaigns, this plane, which was seen side-on, made an angle of 26° with the East–West direction. The ideal direction of trailing would therefore have been at 26° to the declination axis. Such trailing might be obtained by a combination of motion in declination and an alteration in the speed of the telescope drive. The motion in declination is the most significant component, however, and it may be used alone, without causing any perceptible change in the quality of the image. This is why we call this method the 'declination-drift' method, and we will mainly discuss its application.

Trailing in declination is easy to achieve with a telescope fitted with a declination drive. A few telescopes have such a variable-speed drive, primarily designed to compensate for the Moon's proper motion. Owners of such telescopes will have little problem in obtaining the required trailing. Other telescopes have declination-drives designed for making corrections during long-exposure photographs. These could doubtless be use by modifying their electrical supply so that their speed remains constant, but adjustable.

However, the most frequent case is that of telescopes that have no declination

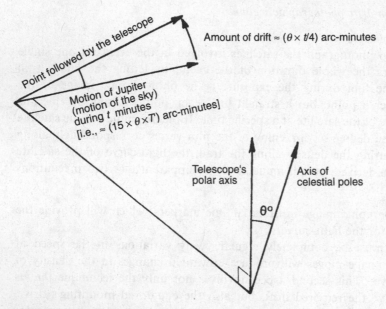

Amount of drift ≈ $(\theta \times t/4)$ **arc-minutes**

Point followed by the telescope

Motion of Jupiter
(motion of the sky)
during t **minutes**
[i.e., ≈ $(15 \times \theta \times T)$ **arc-minutes]**

Telescope's
polar axis

Axis of
celestial poles

$\theta°$

Fig. 6.8. *Relationship between the angle of the polar axis and the amount of drift.*

drive. The best solution then is to adjust the inclination of the mounting's polar axis. As is well-known, if the axis of a mounting is not perfectly aligned with the celestial pole, a drift in declination occurs, indeed this property is used in setting up a mounting using Bigourdan's method.

This solution, which is the only one suitable for an instrument without a declination drive, does have two disadvantages. On the one hand, anyone having a permanently mounted telescope who has carefully aligned the instrument may well hesitate before touching the adjustment of the axis, for fear of having to spend some time in correctly aligning the instrument again. On the other hand, the amount of trailing that results depends on Jupiter's hour angle; a different orientation should be used for each event. The trailing is a maximum when Jupiter lies in a direction perpendicular to the plane in which the axis is shifted. (The trailing varies as the cosine of the angle between the direction of Jupiter and that perpendicular; it is zero when Jupiter and the shift of the polar axis lie in the same plane.) Because the planet is always fairly close to the celestial equator, we may assume that the maximum amount of trailing is equivalent to 1 arc-minute for every 4 minutes of time, per degree of inclination of the polar axis (Fig. 6.8).

6.5.6.3 Taking the photograph
The most suitable film is Kodak TP 2415, which will give the narrowest trail of Jupiter. (It will be recalled that this width depends primarily on the way in which an over-exposed image is scattered within the emulsion). It has sufficient sensitivity to capture the image of the satellite, as well as a very fine grain, which is needed to obtain the accuracy necessary for the later reduction. The film should not be hypersensitized; the image of the satellite does, in fact, drift rapidly, and remains

on a particular portion of the film for just a few seconds. The TP 2415 should be developed to give maximum contrast (in D19b, Dektol or low-dilution HC 110), so that minor changes in brightness of the satellite are more easily visible.

The density of the satellite's trail on the negative should have a value of between 0.6 and 2.5 and lie on the straight portion of the characteristic curve; reduction will then be simpler. In practice, it is even preferable for the density not to exceed 1.5. The greater the over-exposure of the satellite's trail, the wider it becomes and the more the image at any given point is affected by the images of neighbouring points. In addition, it is advisable to work with the lowest possible luminance, so as to avoid any unnecessary enlargement of the image of Jupiter. (Luminance is the quantity of light received per unit area of film surface; it is the product of the illumination at the emulsion, times the length of exposure. Here, it is thus the product of the amount of light provided by the satellite (and captured by the telescope's objective), times the period that the image of the satellite takes to move across a unit element of the film surface.) Given the contrast required – the film should be developed to a gamma of approximately 3 – this implies a factor of just 2 between the maximum and minimum luminance given by the image of the satellite on the photographic film. For a given exposure time, this luminance is inversely proportional to the length of the trail, and thus inversely proportional to both the focal length and the speed of the trailing (expressed as an angle per unit time).

For example, with a focal length of 3.3 m, trailing at 2′/min produces a trail 10 mm long on the film if the photograph lasts 5 minutes. If, for a given telescope, the trail thus obtained has a density of 1.5 (the maximum value), the minimum value that we are prepared to accept (i.e., 0.6, which corresponds to a luminance twice as weak) will be obtained with a focal length that is twice as great (6.6 m) with identical drift, or with a drift that is twice as large (4′/min), with the same focal length.

We have seen that Jupiter should have the lowest possible luminance. This is inversely proportional to the trailing speed (as for the satellite), but is inversely proportional to the *square* of the focal length, because Jupiter is an extended object, whereas the satellite is a point source. In addition, the distance on the film between the trail of Jupiter and that of the satellite is proportional to the focal length. From these points we see that it is advantageous to use the longest possible focal length, which in most cases will mean the use of some system of magnification (a Barlow lens or an eyepiece).

Once the focal length is chosen, the speed of trailing is determined so as to obtain the correct density for the satellite's trail. The best way of determining the appropriate trail is to make several trials on the Galilean satellites before the mutual events take place. If the trailing is obtained by altering the orientation of the polar axis, the correct value for this may similarly be best found by trial and error, the calculation described above having been used to establish the appropriate amount.

In practice, the focal length is limited by considerations of the field size (it is of interest for the satellites not involved in the event to be photographed at the same time), and by the fact that the longer the focal length, the lower the amount of trailing required. This is more difficult to obtain accurately, either because the declination motor is running too slowly, when its speed is less constant, or because the angle at which the polar axis is offset is small, and difficult to measure.

Finally the displacement of the image of the satellite across the film should be sufficiently large to provide a suitable time-resolution for the reduction. On Kodak TP 2415 film, a displacement of 2 mm per minute of time should allow an accuracy of about one second of time to be achieved, which is satisfactory. Such trailing is possible with a small amateur instrument of 100-mm diameter or even less. For a given luminance value at the film, the displacement of the image is proportional to the available illumination, i.e., to the square of the diameter of the objective. A larger telescope (200–300 mm) would therefore allow easy reduction of the photograph.

On the other hand, if the displacement is large and the event lasts a long time, it may be necessary to take several photos to cover the whole of the event, because the satellite will rapidly move across the field. Considerable care should be taken not to use the edges of the field where vignetting may cause unexpected variations in the apparent magnitude of the satellite. We have seen that it is important to cover the whole of the event, not just the middle of it, even if we want to determine just the time of minimum.

6.5.6.4 *Making the reductions*

Photographing the event has produced an image on the film that consists of a trail, each point of which represents the satellite at a given time. Reduction consists of measuring the density of the trail point by point using a microdensitometer.

To be completely rigorous, the density values obtained should then be converted into luminance values by making use of the film's characteristic curve. But if we only wish to determine the time of minimum magnitude, which corresponds to the point with the least density, then the whole reduction can be carried out from the density curve. This avoids the work of converting the values and, above all, of having to calibrate the film systematically.

External effects may falsify the results: the passage of a light veil of cirrus cloud, or irregularities in the telescope drive may give rise to unexpected variations in the apparent magnitude of the satellite. This is why it is always useful to have the image of another Galilean satellite, not involved in the event, on the same frame. This obviously has a constant magnitude. If its image is analyzed in the same way as that of the satellite concerned and variations are detected, these are of instrumental or atmospheric origin. One may assume that they have affected all the objects in the field equally, so it is possible to derive corrections that can be applied, point by point, to the curve obtained for the event itself.

The microdensitometer is described in Chap. 15, Sect. 15.6.2.2. Few amateurs have access to such expensive equipment, however. It would appear that amateurs will have to entrust their negatives for the necessary length of time to a well-equipped, professional centre (such as the Bureau des Longitudes), that will carry out the reduction.

Amateurs could, however, with a certain loss of accuracy, develop reduction methods appropriate to the means at their disposal. For example, the photograph of the event could be printed several times onto high-contrast paper (grade 5), giving increasing exposures. The satellite's trail will appear narrower, the greater the degree of over-exposure in the print. Beyond a certain value, the part of the trail with the least density (on the negative) will disappear. By measuring the point where the trail is broken, the approximate time of minimum may be derived.

6.5.6.5 *Time-reference markers*
As we have just described, examination of the photograph enables the point 'of minimum light to be determined. It then becomes necessary to know when the satellite was at that point. Time markers should therefore be incorporated in the photograph as it is being exposed. The simplest method is to interrupt the exposure by covering the aperture of the tube (without touching it and causing any vibration, of course). One may, for example, interrupt the exposure for two seconds every minute, not forgetting to double the length of the first break, so that the beginning of the trail can be distinguished from the end. On the film the image of the satellite then appears as a broken line. Because the time of each break is known, it is possible to interpolate the time at any particular point of the trail when making the reduction.

6.5.6.6 *Initial trials*
The method that has just been described has not yet been tried in practice; we hope that it will be employed by readers during the next set of mutual events. As far as we know, only preliminary tests designed to prove the general principles of the method have yet been carried out (by Patrick Martinez in 1985). The instrument used was a 90-mm refractor with a focal length of 1300 mm, on a portable, driven, equatorial mounting. The trailing in declination was obtained by misalignment of the polar axis. The rates of trailing thus obtained were between 1 and 3 mm per minute, for effective focal lengths of 3–5 m. These tests produced the following results:

- Despite the small size of the instrument, it proved possible to record the images of satellites relatively close to the planet (down to a distance of 2 Jovian radii from the centre). In addition, the rate of trailing was sufficiently large (3 mm/min) for us to expect to be able to determine positions that correspond to a difference of approximately one second in time;
- Misalignment of the polar axis is an effective way of obtaining a suitable drift in declination; numerous trials are required, however, to determine how to obtain a trail of the required density on the film;
- Covering the objective produces timing marks that are easy to use when the time comes to reduce the observation;
- The quality of the drive and the stability of the mounting are of paramount importance; during the trials the main problems encountered were vibration of the instrument by the wind, and irregularities in the driving rate of the motor (Fig. 6.9).

The method described should allow good results to be obtained by amateurs who do not have a photometer. Although it does not present any particular difficulties, it is obviously somewhat tricky to apply, and requires a number of evenings spent in preliminary trials.

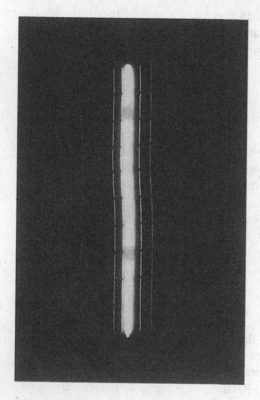

Fig. 6.9. *An early attempt at photographing the satellites of Jupiter by driving in declination, using Kodak TP 2415 film with a 90-mm refractor. The trails left by the four satellites can be clearly seen. The closest satellite, only 4 Jovian radii from the centre of the planet, is at the edge of the planet's halo. The exposure was interrupted for 5 seconds every minute; the reference markers are easily visible. Note the waviness of the trails, caused by inaccuracies in the drive. Photo by P. Martinez.*

6.5.7 Photoelectric photometry

This method is the one that gives the most accurate results (Figs. 6.10 and 6.11), but it requires relatively complicated equipment, which amateurs are now, however, beginning to acquire.

A photoelectric photometer is a detector that may be placed at the focus of a telescope instead of an eyepiece or a camera. A diaphragm is used to limit the size of the field that is being examined; all the photons arising from that area fall onto a sensitive surface from which they dislodge electrons. These electrons then, by an appropriate amplification system, give an output current that is strictly proportional to the incident illumination. This is one of the advantages of photoelectric photometry over the photographic plate, whose response curve is only approximately linear between certain density limits. In addition, the information gained by a photoelectric photometer is immediately available: no development, fixing, or drying is required.

The current that forms the output signal may be employed in either of two ways. It may be converted into an analogue signal, for example by using it to move a pen over a chart that is driven at a constant speed. The light-curve of the event will thus be plotted directly. The signal may also be digitized: in other words, it is sampled at regular intervals and converted into a numerical value that is proportional to the current. The numbers thus obtained are stored on diskette and may be used to recreate the light-curve on a microcomputer screen. The values may also be used

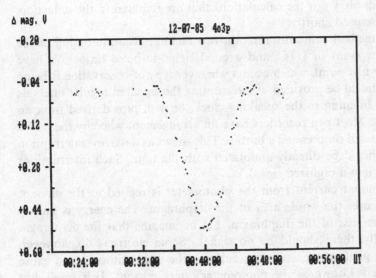

Fig. 6.10. *Light-curve showing the partial occultation of III (Ganymede) by IV (Callisto), on 1985 July 12, obtained by the Grup d'Estudis Astronomics of Barcelona, using a 200-mm Schmidt–Cassegrain telescope fitted with a photo-diode photometer.*

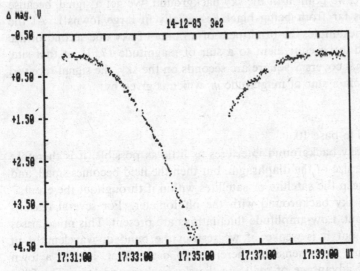

Fig. 6.11. *Light-curve showing the eclipse of II (Europa) by III (Ganymede), on 1985 December 14, obtained by the Grup d'Estudis Astronomics of Barcelona, using a 410-mm reflector fitted with a photomultiplier photometer.*

in a program that will carry out the calculations that are required in the reduction, and which will be discussed shortly.

Success with this method primarily rests on two factors: accurate timing of the various phases of the event in UTC, and a good signal-to-noise ratio. We have already discussed the first point, which occurs whatever type of observation is being made. Some means should be provided of interrupting the signal at regular intervals of time, either when listening to the speaking clock, or with pips derived from an accurate quartz clock. Most pen recorders have an arrangement whereby the pen is momentarily disconnected on pressing a button. This shows as a narrow interruption in the trace, which should be directly annotated with the time. Such interruptions may also be inserted into a digitized signal.

The level of the output current from the photometer is related to the amount of radiation incident over the whole area of the diaphragm. The energy is said to be integrated over the area of the diaphragm. Let us imagine that the diaphragm is closed; theoretically, there should be no signal. Some electrons do, however, still escape from the photometer's sensitive surface (the photocathode) and cause a weak current. This is known as the photometer's dark current. It is small, but it is measurable, and steps may be taken in the construction of the photometer to reduce it as much as possible. It must also be taken into account in reducing observations.

Let us now imagine that we open the diaphragm, but instead of bringing a satellite to the aperture, we point it at the sky background. We get a signal, because the sky background is far from being black (especially in large towns!). At the Paris Observatory, for example, the brightness of an area of sky one second of arc square has been found to be equivalent to a star of magnitude 17. If, at that site, we use a diaphragm that covers S square arc-seconds on the sky, the signal received corresponds to that from a star of magnitude m, which is given by:

$$m - 17 = -2.5 \log(S)$$

where the logarithm is to base 10.

To ensure that the sky background interferes as little as possible, it is therefore necessary to reduce the size of the diaphragm, but then the field becomes small, and it may be difficult to keep the satellite or satellites within it throughout the event.

Let us measure the sky background with the photometer. For several reasons the signal is not constant. Low-amplitude fluctuations are present. This noise arises from various sources, but it is weaker if we restrict the band of wavelengths in which we are observing. It is sensible, therefore, to use a filter, and in a town this may also have the advantage of reducing the sky-background light. The 500–530 nm bandwidth appears to be suitable. Unfortunately, filters with such a narrow bandwidth are expensive and absorb a lot of light, so they are only justifiable for use with large-aperture telescopes.

Slow, but large-amplitude alterations in the sky-background brightness may also occur. If, during an event, the satellite or satellites are close to, and either approaching or receding from Jupiter, stray light from the planet may affect the sky background. A similar effect arises if the event occurs in the evening or morning twilight when the sky is either darkening or becoming brighter. To make allowance for these effects, the sky background should be measured before the event, and then at regular intervals whilst it takes place if it is fairly long (more than a quarter of an hour), and then after it has finished. During the event, it is necessary to unclamp the telescope and rapidly measure the sky background at four points around the main field. Moving the telescope can be avoided by having a diaphragm that has one aperture for the main field, which may be covered – using a flexible cable release – while secondary apertures are opened to observe the sky background.

If the main aim in observing these mutual events is just that of determining the time of minimum light, then there is no real need to express the values that form the light-curve in magnitudes; any set of values that are proportional to the signal received would serve. If, however, one wants to determine the actual magnitudes, a star of known magnitude, close to Jupiter, should be observed before the event. The spectral type should be G8 or a very closely related class. Such spectral classes are similar to those of the Galilean satellites when observed in the 400–500 nm band. During the PHEMU-85 campaign, for example, α Aquilae, α^1 Capricorni, θ Aquarii and a few other stars were observed.

6.5.8 Reduction of the observations

The reduction of a raw observation involves applying a certain number of corrections so that it can be compared with other, independent observations, or with predictions that had been made. Reduction of a photographic plate to derive the positions of stars included on it, for example, involves converting the measured x and y rectangular coordinates to values of right ascension and declination at a specific epoch (1950.0, the epoch of date, etc.).

The corrections that have to be made obviously depend on the type of event observed, and the nature of the observations (visual, photographic, etc.). We shall just summarize the problems that arise in photoelectric observations of mutual satellite events.

After the event we have a table that consist of two columns: one containing the time, the other values, which, as we have seen, are proportional to the amount of radiation that was received. We could plot these values with time as the abscissa, and energy as the ordinate; obtaining a rough light-curve, which would be quite different from the true light-curve, and which should certainly not be used to determine the time of minimum. What we have to do is to make an allowance for the variations in observing conditions that might have occurred during the event. The brightness of the sky background may have changed. If it was measured from time to time, its value may be found by interpolation for the times at which the measurements of the satellite or satellites were made, and these figures may then be subtracted from the measured values.

In addition, the altitude of Jupiter above the horizon (or its complement, the zenith distance) varied over the duration of the event. As a result the thickness of the atmosphere through which the light had to travel also varied, affecting the amount of radiation received. To incorporate the appropriate correction, the zenith

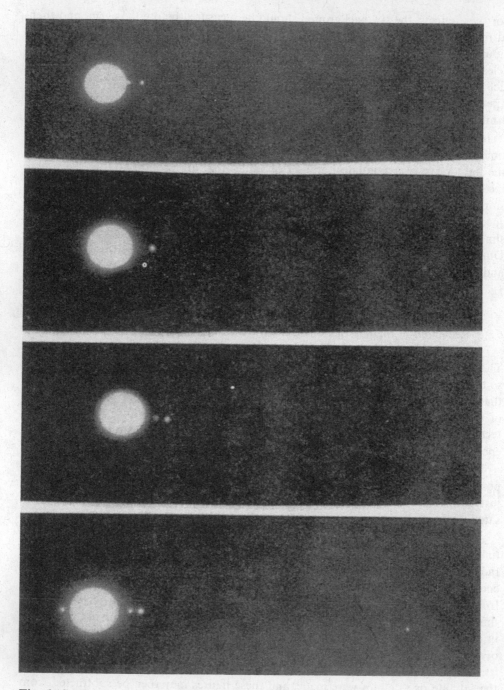

Fig. 6.12. *Eclipse of Europa (closest to Jupiter on the right) by Ganymede (just to its right), on 1985 September 4, photographed by Jacques Silvain. From top to bottom: 21:15, 21:51, 22:00, and 22:19 UT; 308-mm reflector; 3M 1000 ASA film; exposures: 1 s (first three photographs), and 2 s (lower photograph).*

distance Z should be calculated for the time of each measurement. This may be done with the following equation:

$$\cos Z = \sin l \sin \delta + \cos l \cos \delta \cos H,$$

where l is the latitude of the observing site, δ is Jupiter's declination, and H is its hour angle at the time of observation. The hour angle is given by:

$$H = T - \alpha,$$

where T is local sidereal time, and α is right ascension. Next, what is known as the 'air mass' X, is calculated, using Bemporad's equation:

$$X = \sec Z - 0.001\,816\,7(\sec Z - 1) - 0.002\,875(\sec Z - 1)^2$$
$$-0.008\,083(\sec Z - 1)^3.$$

If S is brightness of the sky background, corrected as described earlier, the magnitude of the satellite or the blend of the two satellites, outside the atmosphere, at a specific time, and omitting a correcting constant, is given by:

$$m = -2.5 \log S - kX$$

where k is an extinction coefficient, which should be estimated by the observer. It is generally taken to be equal to 0.2.

We are now in a position to draw a light-curve, and for this we need to know how to draw a regular curve from a finite number of points. There are various methods of doing this, but it is not, properly speaking, a problem of reduction.

The reduction of photographic observations of mutual events may require calibration of the film, as explained in Sect. 6.5.5. Reduction always involves some tricky problems. Some observers introduce corrections, but give no explanation of what they have done. Anyone working with such data then finds that they have aberrant results, and have to reject the observations in question. It is always best to send the raw observations, provided that full details are included, rather than make inappropriate reductions.

6.5.9 *The PHEMU-79 and PHEMU-85 campaigns*

After a few trials in 1973, the Bureau des Longitudes in Paris decided to organize an observing campaign to cover the mutual events in 1979. Unfortunately, the Sun crossed the plane of Jupiter's equator in July, and the Earth crossed the same plane at the beginning of August. Jupiter's conjunction with the Sun occurred on August 13, so most of the events, which occurred around conjunction, were unobservable. Nevertheless, nine events were observed between January 14 and April 18 at seven French observatories and at the European Southern Observatory (ESO) in Chile. Excellent light-curves were obtained, which showed the superiority of the method for astrometry (*Astronomy and Astrophysics*, 1982). Amateurs, who observed some of the events for pure curiosity, played no part in this campaign.

In 1985, on the other hand, the conditions were far more favourable. Opposition of Jupiter occurred on August 4, the Earth being very close to the planet's equatorial plane between June 15 and August 15, and again at the end of December. The Sun crossed the equatorial plane at the beginning of October. As a result, 361 events were theoretically observable from Earth, 74 from Europe and 31 from Chile and Brazil. Obviously meteorological conditions and other factors entered into it, but 85 events were observed, for which 200 light-curves were recorded. Participation from amateur observers, envisaged from the very outset, was particularly strong, and their observations, which were visual, photographic and photoelectric, were, in general, of a very high quality.

A colloquium was held at Bagnères-de-Bigorre at the end of April 1986, which brought amateur and professional participants together for an initial exchange of views on the results. The report of this colloquium has been published in the Supplement to the *Annales de Physique* (1987). But final analysis of the PHEMU-85 campaign results had not been published as this work went to press.

The 1991 events were less favourable than those in 1985, because they occurred close to Jupiter's conjunction with the Sun. However, amateurs did secure some useful results.

7   The minor planets

J. Lecacheux

7.1 General description of the minor planets

7.1.1 Early discoveries and principal orbits

The first discoveries of minor planets (otherwise known as 'asteroids') were accidental and at haphazard intervals, resulting from observations of other astronomical objects. The first, 1 Ceres, was found on 1801 January 1 by Piazzi at Palermo in Sicily. By 1845, just before the first systematic searches began, only five minor planets had been recorded. It was quite obvious that they were very small planetary bodies orbiting between Mars and Jupiter, at distances of between 2 and 3.5 AU (Astronomical Units) from the Sun. Ten years later, forty objects had been reported and discoveries were becoming more and more frequent, averaging about five per year by 1865. It became obvious that there was a swarm of widely dispersed minor planets occupying the region where the Titius–Bode 'law' predicted a 'missing' planet. Were these small fragments of an ancient planet that had been destroyed by some catastrophe?

With the introduction of photographic techniques, even though they were still in their infancy, the rate of discovery increased to about fifteen minor planets per year by about 1895, twenty-five by 1910, and forty around 1930 ...

The introduction of the photographic plate and corresponding improvements in optical instruments allowed fainter and fainter minor planets (i.e., smaller and smaller ones), to be discovered. We now have approximate dimensions (to an accuracy of about 10 %) for most of the larger minor planets, and are therefore able to examine the way in which these discoveries occurred in the 19th century and during the first part of the 20th. Minor planet 1 Ceres measures 1020 km across and is by far the largest of the minor planets, so it is fitting that it was the first to be discovered. Next after Ceres in size are 2 Pallas and 4 Vesta, discovered in 1802 and 1807 respectively: they both measure about 540 km across. The objects discovered visually between 1845 and 1875 averaged 120 km in diameter. The largest found during that period was 10 Hygeia, which is 410 km across, whereas the smallest were less than 40 km in diameter. However, the average size of minor planets discovered photographically around 1920 was no more than about 25 km, and most of those discovered nowadays are only a few kilometres across. However, we should not let this idea of a gradual change in the average size deceive us. A few large-sized minor planets managed to evade discovery for a long time. For example, minor planets larger than 300 km continued to be found, only ceasing with the discovery of 704 Interamnia in 1910 – or, if you prefer, of 2060 Chiron in 1977. The complete catalogue of minor planets larger than 150 km in size (Chiron excepted) was not complete until 1930. As we shall see later, a complete catalogue of all bodies larger than 100 km across (numbering about 220) is, unbelievably, only just about complete

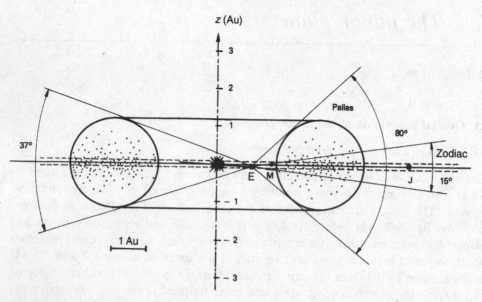

Fig. 7.1. *Schematic section at a scale of* 5×10^{-14} *of the toroidal cloud of minor planets known as the 'Main Belt'. The broken lines indicate the thin disk within which the principal planets (Venus to Neptune) have their orbits. The abscissa (the reference plane) is taken to be the plane of Jupiter's orbit, the ordinate axis is the system's rotational axis. The letters E, M and J indicate Earth, Mars and Jupiter respectively. In this figure, the vertical concentration of the minor planets with respect to the reference plane correctly represents reality. The high-inclination object 2 Pallas is shown at its extreme height of* $z = +1.83 \, AU$ *above the reference plane.*

now. Table 7.1 lists the minor planets larger than 200 km across in decreasing order of size. [Recent observations have, of course, now shown that Chiron is a comet, rather than a minor planet. – Trans.]

The first type of study of minor planets was an investigation of their orbits, especially as celestial mechanics was then a discipline that was undergoing rapid development. It was not long before statistical investigations could be made of the orbital elements that had been calculated. To give a concrete example, let us take the 322 minor planets discovered before 1892 – i.e., with refractors and visual observations. All except three belong to what we now call the 'Main Belt'. Collectively they form a representative sub-set of orbits. Figure 7.1 shows the Main Belt at a scale of 5×10^{-14}.

Although the explanation was a matter of conjecture, it was soon noted that the orbits of the minor planets are, in general, considerably more eccentric than those of the planets. Omitting Mercury, which is close to the Sun, and Pluto, which was unknown in the 19th century, the orbits of the major planets have an average eccentricity of 0.04, which indicates that they are ellipses that differ very little from circles. The 322 objects in our sample, however, have an average eccentricity of

Table 7.1. *Minor planets with diameters greater than 200 km*

Rank by size	No.	Name	Year of discovery	Mean dia. (km)	Type*	
1	1	Ceres	1801	1023	C	(C)
2	2	Pallas	1802	539	B	(U)
3	4	Vesta	1807	536	V	(U)
4	10	Hygiea	1849	409	C	(C)
**	2060	Chiron	1977	300–400?	B	
5	511	Davida	1903	309 (350)	C	(C)
6	704	Interamnia	1910	305	F	(U)
7	87	Sylvia	1866	281	P	(CMEU)
8	65	Cybele	1861	280	P	(C)
9	15	Eunomia	1851	262	S	(S)
10	3	Juno	1804	257	S	(S)
11	52	Europa	1858	250	CF	(C)
12	16	Psyche	1852	249	M	(M)
13	31	Euphrosyne	1854	247	C	(C)
14	95	Arethusa	1867	232	C	(C)
15	41	Daphne	1856	228	C	(C)
16	45	Eugenia	1857	227	FC	(U)
17	451	Patientia	1899	225	CU	(C)
18	624	Hektor	1907	221	D	(U)
19	324	Bamberga	1892	221	CP	(C)
20	19	Fortuna	1852	219	G	(C)
21	88	Thisbe	1866	210	CF	(C)
22	165	Loreley	1876	208	CD	(C)
23	121	Hermione	1872	204	C	(C)
24	6	Hebe	1847	201	S	(S)

* The classification in parentheses is that given in the TRIAD file (Bowell *et al.*, 1979), as well as an improved classification proposed in 1981. Because of the uncertainty in the diameters, too great a significance should not be given to the exact order shown. In addition, most objects are tri-axial ellipsoids with unequal axes. The diameter quoted is an average of the largest and smallest diameters. There are therefore several minor planets whose major axes exceed 200 km, but which are excluded from this table.

** 2060 Chiron is included here for completeness, because it is not an ordinary minor planet: its orbit lies between those of Saturn and Uranus. [It has now been found to have a coma, and must thus be classified as a comet. – Trans.]

0.15. Only 25 % of minor planets have eccentricities as small as those of the planets, whereas 6 % have eccentricities greater than 0.25.

It was also established that minor planets tend to have inclinations greater than those of the planets. It is customary to describe inclinations relative to the plane of the ecliptic, but in reality it would be more sensible to take a neighbouring plane: that of Jupiter's orbit. Because of its enormous mass and its relative proximity, Jupiter is the planet that exerts most influence on the minor planets. The orbits of the planets (Mercury and Pluto again excepted) have an average inclination of 1.3° to that of Jupiter. As they orbit the Sun, the other planets oscillate slightly on either side of Jupiter's orbital plane. On average, however, they depart from it by just 0.1 AU, which is a very small amount, considering that the overall system spans 60 AU. In contrast, our 322 orbits have an average inclination of 8°. To be more precise, 11 % of the orbits have inclinations exceeding 15°, and a few even exceed 25°. Inclinations that are essentially zero (less than 1°) are curiously non-existent, which probably indicates a specific instability in orbits that are co-planar with that of Jupiter. The average distance of minor planets from Jupiter's orbital plane is 0.27 AU. The number of objects per unit volume naturally decreases away from this plane. Some 30 % of the bodies may reach distances of more than 0.5 AU from the plane, and 5 % more than 1 AU. One exceptional object among the 322, 2 Pallas, even slightly exceeds 1.8 AU from the plane once in every orbit. Figure 7.1 is a realistic impression of the vertical distribution of the minor planets at any moment in time.

It is interesting to draw up a histogram of the semi-major axes *a* of the orbits of bodies in the Main Belt (Fig. 7.2). It will be seen that values close to 2.4 AU, 2.75 AU, and 3.15 AU are particularly frequent. On the other hand, values close to 2.55 AU and 2.85 AU are decidedly under-abundant. These unpopulated regions are called 'Kirkwood gaps'. The standard explanation is that Jupiter's gravity eliminates orbits that are a simple fraction of its own orbit through resonant perturbations. For example, objects having orbits with a semi-major axis of 2.50 AU make exactly three orbits of the Sun while Jupiter makes one: it is a 3:1 resonance. Those with a semi-major axis of 2.82 AU make five orbits while Jupiter makes two. For *a* = 3.28 AU the ratio is 2:1, etc. At these critical distances, certain small, regular perturbations accumulate over the course of time, so that any bodies occurring there would soon be forced into more stable, neighbouring orbits. Such an explanation is, however, rather simplistic. It does not, for example, give any explanation for the distribution that is actually found, in particular for the relative heights of the observed peaks and troughs. Then again, certain resonances correspond to secondary peaks in the histogram, and not to gaps. This is the case with the 3:2 resonance, for example. Beyond the Main Belt, resonances paradoxically play a stabilizing role. Despite a lot of research, and the help of computers, we have still not succeeded in arriving at a general theory that will account for the behaviour of orbits that occur close to specific resonances. Only a few individual cases have been satisfactorily explained.

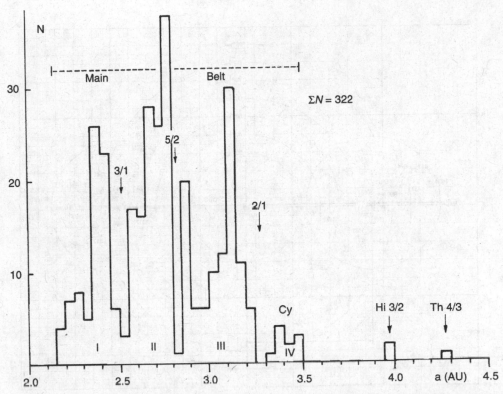

Fig. 7.2. *Histogram of the semi-major axes a for 322 objects discovered before 1892. The most-heavily populated regions I, II and III, which form the Main Belt, are separated by the Kirkwood gaps. There is a secondary maximum at IV on the outer edge at around 3.4 AU: this consists of Cybele-type objects (the main body is 65 Cybele). Isolated objects occur outside the Main Belt: these are 153 Hilda and 190 Ismene (Hilda-type orbits in a 3:2 stable resonance), and 279 Thule, in a stable 4:3 resonance.*

7.1.2 Unusual types of orbit

By 1891, two or three distant minor planets, such as 153 Hilda and 279 Thule, had been discovered whose orbits, because of the stabilizing resonances that we have just mentioned, were quite distinct from those in the Main Belt. In 1898, 433 Eros was discovered and this was found to be a small, short-period minor planet that came well within the orbit of Mars. It was capable of making a closer approach to the Earth–Moon system than any other body known at that time, certain comets excepted. It was the first of a group of very small minor planets to be discovered; this group is now known as the 'Aten–Apollo–Amor' family, and will be discussed in detail, because of its significance for us on Earth.

In the same year (1898), 434 Hungaria was discovered. This was the first of a family of small bodies, lying at the inner edge of the Main Belt, all of whose orbits are similar. At each orbit they pass just outside the orbit of Mars.

285

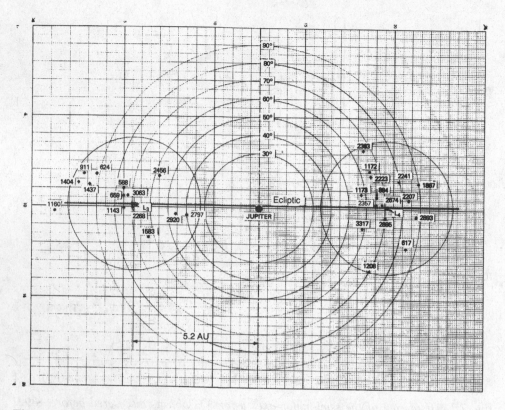

Fig. 7.3. *Apparent positions of the Trojan minor planets against the sky on 1987 October 18, which was the date of Jupiter's opposition. A limiting magnitude of V = +17.0 has been taken here. The brightest object (624 Hektor) is magnitude +15.4. The horizontal line is the ecliptic. The circles centred on Jupiter are 10° apart. The minor planets are indicated by their numbers.*

In 1906–7 the first of the Trojan minor planets appeared in the catalogues: 588 Achilles, 617 Patroclus, 624 Hektor, etc. These are in a stable 1:1 resonance with Jupiter. They have orbits that are very similar to that of Jupiter with exactly the same orbital period, and they make wide oscillations, with periods of 11.8 years, about what are known as Jupiter's L_3 and L_4 Lagrangian points. In the 17th century, celestial mechanics had predicted the existence of these two theoretical points, which lie in Jupiter's orbit, one 60° in front of the planet, and the other 60° behind it. Several Trojans are fair-sized minor planets: near L_3, the largest, 624 Hektor, is 220 km across, and 1437 Diomedes, 1143 Odysseus and 911 Agamemnon are all larger than 150 km. Near L_4, 617 Patroclus is 160 km across. Figures 7.3 and 7.4 show the two swarms of Trojan minor planets as seen from the Earth at the date of Jupiter's opposition in 1987.

In 1920, 944 Hidalgo was discovered. This is a very strange, tiny object, the first (and one of the few known) with a period longer than that of Jupiter. Its aphelion

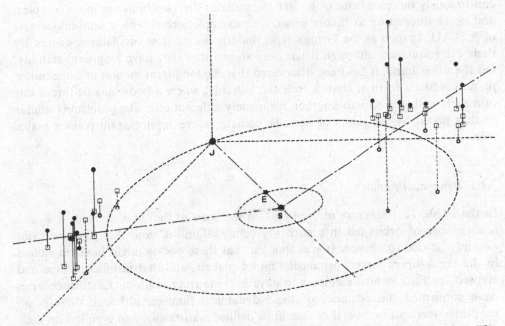

Fig. 7.4. *The same configuration as Fig. 7.3 seen in an arbitrary perspective. The letters S, E and J indicate the Sun, Earth and Jupiter respectively. The chain-dotted lines indicate the Lagrangian points. To indicate the distances from the plane of the ecliptic, each minor planet has been joined to its projected position (square) by a full or dotted line.*

N.B. – The Trojans are not satellites of Jupiter: they appear to orbit points L₃ and L₄ over 12-year periods. Some may depart considerably from Jupiter's orbital plane. The record is held by 2146 Stentor (not shown here), a Trojan 25 km across, which reaches a distance $z = +3.5\,AU$ at each revolution.

lies at the same distance as Saturn's orbit. In fact, with an eccentricity of 0.65, its orbit strongly resembles that of a short-period comet. We shall return to the question of minor planets with comet-like orbits. We may, however, mention here that more and more of them are now being discovered.

Much of the spectacular success of celestial mechanics in the 19th century (by Le Verrier, Adams, Newcomb, etc.) was partly because of a fortuitous simplification of the problem. This was the fact that the orbits of the planets have low eccentricities and small relative inclinations, and this greatly helped in solving the equations. With minor planets, the orbital eccentricities and inclinations are considerable and major resonant perturbations are involved, so theoretical analysis becomes much more complicated. It was only when numerical simulations could be made using computers that definite conclusions could be drawn about the evolution of minor-planet orbits over tens of thousands of revolutions. It has since been established that most of the minor planets have orbits where the elements a, e, and i (semi-major axis, eccentricity and inclination) oscillate slightly around stable mean values, and where the argument of perihelion ω, and the longitude of the ascending node Ω slowly and continuously increase from 0° to 360°. Sometimes the variations are more complex, and more interesting to theoreticians. For example, orbits with a semi-major axis of 3.24 AU, known as the Griqua type, undergo large, slow oscillations, caused by their 2:1 resonance, although it has been shown that they have long-term stability. On the other hand, it has been discovered that the long-term motion of some minor planets is like a form of chaotic 'celestial billiards', where a body may be forced out of a meta-stable orbit into another, completely different one. The problem is similar to that posed by the orbits of periodic comets, where Jupiter again plays a major role.

7.1.3 Dynamical families

In the 1920s, K. Hirayama noticed that on the basis of the three elements *a*, *e*, and *i*, a number of orbits fell into narrowly defined 'families', and that this could not possibly be due to chance. It was thought that these bodies might have originated in the break-up of larger, primordial minor planets, and that insufficient time had elapsed for their orbital elements to have become truly random. Later discoveries have confirmed the existence of these dynamical families, although there is no general agreement on how they should be defined statistically, and even less on their physical origin. Some of the principal families will be mentioned here. In general, a family consists of a principal minor planet, after which the group is named, and a cluster of much smaller, associated minor planets. Within the family, the angles ω and Ω are randomly distributed around the Sun, but the other orbital elements are quite obviously related.

The family associated with 8 Flora is perhaps the best known. It is also the largest, with more than 420 members, although it is poorly defined statistically. The family associated with 24 Themis (200 km in diameter) is far more distinct. It contains 150 members, three of which are over 100 km in size (90 Antiope, 268 Adorea, and 171 Ophelia). Equally well-defined, and with roughly the same number of members is the family containing 221 Eos (105 km). We mentioned the Hungaria group of objects earlier, and this has certain signs of being a family. We cannot enter into a prolonged discussion here of the one hundred-odd families that have been proposed by various authors. We will simply say that the existence of Hirayama families lends credence to the idea that certain minor planets may have been disrupted by collisions in the past. This brings us to consideration of the physical properties of minor planets.

7.1.4 The physical nature of minor planets

The 1970s saw the introduction of methods of studying the minor planets themselves, and not just their orbits. In a few years, our knowledge of them has advanced rapidly. The combination of infrared photometry at wavelengths of 10 and 20 microns, and photometry in visible light, for example, enables us to determine fairly accurately the diameters and albedoes of hundreds of minor planets, so much so that by 1973 it was possible to devise a classification scheme, a 'taxonomy' that very rapidly became the standard. The catalogue that describes the main characteristics of the minor planets (in particular their classification), and which is often cited, is called the TRIAD catalogue. Versions appear in both *Asteroids* (Bowell *et al.*, 1979), and *Asteroids 2* (Binzel, R. P., *et al.*, 1989).

Overall, and by analogy with the meteorites now in museum collections, the major divisions of minor planets are:

- C-type minor planets: extremely dark (albedo 0.04 ± 0.02); neutral in colour ($B - V$ between 0.65 and 0.75); optically similar to the meteorites known as carbonaceous chondrites, which consist of very primitive, hydrated, carbonaceous silicates and are reliably thought to date back to the origin of the Solar System;

- S-type minor planets: less dark (albedo 0.15 ± 0.07); slightly reddish in colour ($B - V$ between 0.80 and 0.85); somewhat resembling the stony meteorites, which consist of silicates with high or moderate iron contents (the lithosiderites, and H-type chondrites);

- M-type minor planets: the same albedo as S-type bodies, but neutral in colour and with a characteristic rise in their reflectivity in the red; they strongly resemble the nickel-iron meteorites;

- a few minor types such as type E (light-grey objects with an albedo of 0.4), and type R (light-coloured, quite reddish objects);

- finally, objects that cannot be classified in any of the categories just described, which are designated U (for 'unknown'), and which are fairly numerous (amounting to about 20 %).

Naturally, such a classification is of limited relevance. Photometric methods only tell us about the surface layer of the body, and not about the material in the interior, which, by its very nature, remains unknown. It is probably rather rash to assume that the surface is always representative of the interior. Some minor planets may also have areas on their surface with distinctly different optical properties, which would confuse the analysis. This certainly applies to 4 Vesta, which appears, in any case, to be an unusual object with a light-coloured basaltic surface. Finally, and this is a more fundamental criticism, it is by no means certain that the analogies with meteorites are completely justified. For example, it is by no means scientifically proven that 16 Psyche is a lump of metal 250 kilometres across. To be certain, we would have to visit it!

In recent years, photometric and spectroscopic observations in the near infrared have introduced an additional level of complexity to the classification, particularly by enabling distinctions to be drawn amongst the darkest objects. For example, a new D-type now applies to surfaces that are exceptionally dark (albedo 0.03 ± 0.01) and reddish ($B - V$ approximately 0.75). These are particularly frequent in the outer part of the Main Belt (on the Hilda objects, and the Trojans), on certain satellites of Saturn and Uranus, and even on cometary nuclei. This may be a sign of surface deposits of organic material.

The relative proportions of the various classes change with distance from the Sun. In the outer third of the Main Belt, for example, objects of type C predominate. On the other hand, at the inner edge objects of type S are roughly as abundant as all the other types taken together. A major question being debated by specialists is whether this is really a fundamental difference in composition of minor planets with their distance from the Sun, or whether it is simply a question that the way in which surfaces evolve varies with distance from the Sun (or distance from Jupiter). It is not a simple question. For example, people have investigated whether bodies belonging to a particular dynamical family are all of the same type. The answer is no: homogeneous families are the exception. In most families, even in those that seem most likely to have arisen from the fragmentation of a single primordial body, a mixture of S, C and U types has been found. It is almost as if the parent body itself was originally of heterogeneous composition.

A hundred years ago it was thought that minor planets were simply rocky bodies isolated in space. In recent years, however, another theory has replaced this idea. This is of a pile of debris that is only bound together by its own gravity. At the beginning of this section we saw that minor-planet orbits have very significant eccentricities and inclinations. The relative velocities of objects in the Main Belt therefore attain at least 5 km/s. Collisions, when they occur, are therefore very violent and destructive. It is thought that the majority of minor planets must have undergone several cycles of destruction and reconstruction in the first few hundred million years of their history, when the Solar System still contained numerous planetesimals. Objects larger than 350 km were able to form into more-or-less spherical bodies through their own internal gravity, perhaps differentiating into concentric layers of different composition like those found in the planets. Smaller objects, about 200 km across, may well have remained as heterogeneous aggregates with no internal differentiation or cohesion. The natural shape for any such aggregate that rotates is an ellipsoid with three unequal axes, known as a Jacobi ellipsoid. Photometric measurements, which we discuss in Sect. 7.5, have shown that minor planets between 50 and 250 km do consist of triaxial ellipsoids. The smallest minor planets that are just a few kilometres across may simply be irregular debris from collisions.

One of the aims of future space missions to study minor planets – such as NASA's CRAF (Comet Rendezvous Asteroid Flyby) and the 'Vesta' mission for 1994 that is being discussed by French and Russian scientists – would be to check this type of theory. We should, however, note that the surfaces of the minor planets are undoubtedly covered by a very thick layer of pulverised material (a regolith), probably much thicker than that on the Moon, which may even be several hundred metres in depth. Individual features of the interior risk being completely masked by such a thick layer, in which material from all parts of the minor planet would be thoroughly mixed.

7.1.5 Meteorites and Apollo-type minor planets

One question that still remains very difficult to answer is that of the real origin of meteorites. We know that the fine dust and most of the small particles that hit our atmosphere derive, in the main, from comets, but we also know that meteorites of an appreciable size must come, by some means or other, from the region populated by the minor planets. Analysis of the isotopes produced by cosmic radiation have shown that stony meteorites have spent no more than a few tens of millions of

Fig. 7.5. *The minor planet 433 Eros, photographed by M. Verdenet on 1975 January 11 at 23:15 UT, when it was at magnitude 8.2 in the constellation of Lynx, and was only 0.15 AU from the Earth. Telephoto lens 135-mm focal length, f/2.8; Tri-X film; 5-minute exposure.*

years between their last fragmentation and arriving on Earth. The crystallographic structure of some meteorites shows evidence of their violent birth. On the other hand, carbonaceous chondrites consist of very old material that has hardly changed for 4500 million years.

When we discussed the discovery of 433 Eros (Fig. 7.5), we said that we now know of several, small minor planets that have orbits that closely approach the Earth. These bodies, which may brush past the terrestrial planets are divided, fairly arbitrarily, into various categories:

- Objects with orbits whose perihelia lie just outside the orbit of Mars: we have mentioned the Hungaria-type objects, but there are many others;
- Objects that have perihelion distances less than that of Mars (1.38 AU) and which are often called 'Mars-crossers' (MC). 1036 Ganymede appears to be the largest of these. [1991 DA is an extreme example that falls within this group. It has perihelion inside the orbit of Mars, aphelion at 22.2 AU (beyond the orbit of Uranus) and an inclination of 62°. – Trans.]

The second category may be sub-divided into at least four further groups:

- Those that do not approach closer to the Sun than 1.30 AU.
- Those whose orbits range from just outside the orbit of the Earth to about 1.30 AU. The division between these and the objects in the previous group is purely arbitrary. The prototype object is 1221 Amor, a tiny body discovered in 1932, but 433 Eros, which was discovered thirty-four years earlier and is the largest of the group, could equally well have been chosen.
- Those that come within the orbit of the Earth, and whose periods are longer than a year. The prototype is 1862 Apollo, which was also discovered in 1932.
- Objects similar to the last group, but which have periods of less than a year. The representative object, 2062 Aten, was discovered in 1976.

Because of planetary perturbations, the orbits may evolve quite rapidly. Some objects may change categories: an Amor object may, for example, become an Apollo-type object, or vice versa. This is why the whole group is best described as the 'Aten–Apollo–Amor', or A–A–A objects. Frequently, when there is no way of distinguishing between the Apollo objects (taken in the strict sense), and Amor or Aten objects, they are simply called 'Apollo objects'. They are often known in English-speaking countries as 'Earth-crossers' or 'Earth-grazers' (EGA).

The terminology, regrettably, remains very confusing. Some authors take the aphelion distance of Mars (1.67 AU) as a dividing line. To others, a 'Mars-crosser' should, by definition, intersect the plane of the orbit of Mars inside the planet's orbit. This may mean that (when the line of apsides is more-or-less at right angles to the line of nodes) an Amor object is not, strictly speaking, a 'Mars-crosser'. The opposite may apply: a 'Mars-crosser' may, strictly speaking, not be an Amor object (for example, 1982 DA). Then again, some 'Mars-crossers' may belong to the Main Belt (an example is 887 Alinda, which is in a 3:1 resonance), and others may not (e.g., 1951 Lick, with $a = 1.39$ AU). Moreover, to some people, Alinda is an Amor object (perihelion at 1.10 AU), but not a Lick-type (perihelion at 1.305 AU). Similarly, an Apollo object is not perforce an 'Earth-crosser' in the strict sense. 'Earth-grazer' is not always synonymous with 'Earth-crosser', and so on. Such subtleties appear unwarranted. The designation Aten–Apollo–Amor has the advantage of being the most generalized.

To date, some one hundred A–A–A objects have been catalogued, but there must be considerable numbers of others. Their significance lies in the fact that they may collide with the Earth. During recent years it has been realised that the evolution of life on our planet has probably been influenced by collisions with A–A–A-type minor planets some 5–10 km in diameter. There should have been about twenty collisions of this type since early Precambrian times, and a large one apparently occurred at the end of the Mesozoic. Some collisions may, it is true, have been caused by cometary nuclei. Many large meteorites, and even larger numbers of small ones, are probably tiny A–A–A objects, or fragments detached from them.

This does not completely explain the origin of meteorites, however, because the origin of the A–A–A objects themselves remains somewhat of an enigma. The fact that S-type, C-type and M-type objects have been discovered among them shows that the problem is complicated. It has been found that the majority of A–A–A objects have orbits with aphelia in the Main Belt, but that 25 % of cases do not obey this rule. The Aten and some Apollo–Amor objects have aphelia well within the orbit of Mars, yet in recent years an increasing number of objects have been discovered with very elongated orbits whose aphelia lie close to the orbit of Jupiter. Again, the A–A–A-type orbits are dynamically unstable over periods of a few million years, and the objects themselves should be destroyed by collisions with Earth, Venus or Mars. The orbits that we observe are therefore very recent by comparison with those prevailing in the rest of the Solar System, so there is undoubtedly some mechanism that creates new objects of this type. One possible hypothesis is that some A–A–A objects are the nuclei of extinct, short-period comets. For example, the aphelion of Comet Encke lies in the region of the Hilda-type minor planets. It is inactive at that distance, and could be confused with a genuine Apollo-type object. Another example is Comet Machholz (1986e), which seems relatively inactive, and has a period of 5.3 years and high orbital inclination ($i = 60°$). It passes well inside the orbit of Mercury, but its aphelion lies at distances similar to those of Jupiter and the Trojans. The astonishing Apollo-type object 3200 Phaeton, discovered in 1983, should also be mentioned. This appears to be the source of the Geminid meteors and may have been a comet fairly recently. At present, Phaeton, which has a perihelion distance of 0.14 AU, holds the record as being the minor planet with the smallest perihelion distance. Finally, there is 2201 Oljato, an Apollo-type, where slight, current, cometary activity has been suspected indirectly from the results of an experiment carried by a space-probe.

7.2 Locating and discovering minor planets

The number of amateur astronomers who regularly observe minor planets remains very small. Of course, unlike planets or comets, no minor planet – except occasionally 1 Ceres – ever shows any appreciable diameter in a telescope. All that is seen is a tiny, star-like point of light, which is distinguishable only by its fairly rapid motion against the background stars. This is particularly noticeable when it passes a star cluster or a nebula, and such appulses are announced from time to time in journals such as *Sky & Telescope*. Figure 7.6 shows a nice illustration of such an event. Another fairly painless way of locating a minor planet is to take advantage of any predicted stellar occultations. Chapter 9 gives more specific details of this type of event.

We will begin by giving some practical advice about locating a specific minor planet in the sky from catalogue information. Later we will describe a strategy for discovering unknown minor planets.

7.2.1 Locating minor planets

First, it should be noted that at any given time 75 % of minor planets lie within the Zodiac, which is defined as being a band with an apparent width of 15 degrees, centred on the ecliptic. A further 20 % lie between 7.5 and 15 degrees away from the ecliptic. The final 5 % are therefore more than 15 degrees away. We have seen, in Fig. 7.1, that the Earth is eccentrically located in the central hole of the toroidally shaped Main Belt. This subtends an apparent angle of 80 degrees in the direction opposite to the Sun, i.e., where minor planets are at opposition and where they may therefore be found as much as 40 degrees away from the ecliptic.

It is useful to have some idea of the span of time required for a minor planet that has just been observed to be visible again under similar, favourable conditions. We know that oppositions of Jupiter recur at regular intervals of 13 months, and that those of Mars return at somewhat variable intervals of about 26 months. The average period between oppositions is known as the synodic period. Table 7.2 gives the intervals between successive oppositions of minor planets in the Main Belt at different distances from the Sun. The period for a given body generally varies because the orbit is usually markedly eccentric, but overall the periods lie between

Fig. 7.6. *The minor planets 11 Parthenope (trail at top, magnitude B = 11.4) and 877 Walkure (trail at bottom, magnitude B = 15.3), close to the galaxy NGC 4527 on the night of 1984 March 6/7. Photograph taken by Hervé Le Tallec between 23:30 and 00:40 UT on 103a-E film, at the prime focus of an f/10, 200-mm Schmidt–Cassegrain. The black streak was caused by accidental damage to the negative.*

13 months and 21 months, the most typical value being 16 months. That being said, we should still note that in practice not all oppositions are equally favourable. For a site in the northern hemisphere, one might, for example, want the ecliptic latitude of the minor planet to be positive so that it would be as high as possible in the sky, and therefore visible for a reasonable amount of time each night. A perihelic opposition might also be desirable, as in the case of Mars, when the object would be at maximum brightness. It might be an advantage for the object to cross the Milky Way, maximizing the chance of a stellar occultation occurring. Similarly, the object's axis of rotation must not point towards the Earth if we want to determine its light-curve (*see* Sect. 7.5 on photometry).

These various additional constraints mean that conditions favourable for a specific observational programme or for a specific minor planet may not, in fact, recur for another two or three synodic periods. Because this may be 2×13 months, or 3×16 months, we may actually have to wait 3 or 4 years. Sometimes an unsuccessful observing campaign on an interesting minor planet may be repeated only some 2 years later. This is one feature of minor planets that distinguishes them from many astronomical objects and even from most of the planets. It should always be borne in mind. Table 7.2 also gives brief details of the retrograde motion (only in

Table 7.2. *Oppositions of minor planets*

Object	a (AU)	Interval between two oppositions (months)	Retrograde motion near opposition		
			Duration (days)	Amplitude in longitude (°)	Maximum angular velocity in longitude (″/hr)
MARS	1.523 7	26 (±1)	62–80	11–18	41–57
MINOR PLANETS					
Hungaria region (e = 0.1)	1.90 ± 0.05	19.5 (±1.5)	72–88	12–17	40–47
Main Belt (e ≤ 0.25):					
Inner edge	2.15	18 (±2.5)	56–103	5–7	24–41
Region I	2.1 ± 0.1	7.5 (±2.5)	65–106	7–16	26–37
Region II	2.7 ± 0.1	16 (±2)	72–111	8–14	27–33
Region III	3.15 ± 0.05	15.5 (±1.5)	80–117	9–13	27–28
Outer edge	3.45 ± 0.1	14.5 (±1.5)	83–120	9–13	25–27
JUPITER	5.202 6	13	118–123	10	20

longitude) near the date of opposition. Objects with high orbital inclinations, such as 2 Pallas, 31 Euphrosyne, etc., generally follow sweeping S- or Z-shaped curves or closed loops in constellations outside the Zodiac.

We have not included in this discussion the tiny Apollo objects, which rush rapidly past the Earth, and which may approach from any direction. With them, common multiples of their sidereal period and that of the Earth govern close encounters, and therefore visibility. There may be a considerable range of cases, from those that have frequent returns to those that have widely spaced apparitions. One might have to wait 15 years and more for a favourable return of a particular object.

To locate a listed minor planet, tables of positions and a good atlas have to be used. Ephemerides of some of the brightest objects are given in various widely available yearbooks, such as the BAA *Handbook* and the RASC *Observer's Handbook*. Many journals (such as *Sky & Telescope*) regularly give the positions of minor planets that are favourably placed in the sky. A specialized publication is the *Minor Planet Bulletin*, which gives detailed predictions that are suitable for advanced amateurs.

In all these publications, positions are generally given every ten days, so either a Bessel or polynomial interpolation is generally required to obtain an accurate position. It is perhaps worth pointing out that the simplest method of making a non-linear interpolation is not to switch on a computer, but to draw a freehand curve through the positions plotted on millimetre graph paper. Such a path drawn by eye is no more arbitrary than any particular mathematical formula that one might decide to use.

The dedicated observer or professional who requires the positions of numerous minor planets will find the information in a unique and invaluable publication: the annual volume entitled *Ephemerides of Minor Planets* (EMP) published by the Institute of Theoretical Astronomy (ITA) in Leningrad. The EMP gives accurate orbital elements and 2000.0 positions [The change from 1950.0 positions was made at the beginning of 1992. – Trans.] to 0.5 arc-minute for all minor planets that have well-defined orbits (4265 in the 1991 edition), for every ten days during the 70 days around opposition. The 80 brightest objects are even given to 0.05 arc-minute for six months around opposition, with their distances from the Earth and the Sun, and their magnitudes. The main A–A–A objects observable are the subject of a special listing at the end of the volume. The 4265 objects listed in the 1991 volume of the EMP did not include every single minor planet then known, but only those whose orbits had been well-defined by the end of 1989 November. [Approximately another 400 had been designated by the end of 1990. – Trans.] They had been given unique identity numbers, which will serve to identify them for all time. Thousands of other objects are known that either have provisional orbits or do not have any calculated orbit at all. We shall come back to these in Sect. 7.3. [*See also* Sect. 8.3.4.1 regarding the form of time (TT) used to specify ephemerides, and Appendix 1, Vol. 2. – Trans.]

Anyone specializing in positional measurements will make considerable use of the *Minor Planet Circulars / Minor Planets and Comets* (MPC), published by the Minor Planet Center at Cambridge, Mass., directed by Brian Marsden. These bimonthly lists give, in order of object, all the measured positions that have been communicated by workers from all over the world; new identifications of objects observed several times and not previously recognized; new names; new orbital elements; and improved ephemerides. It is a sort of newsletter for specialists. Astronomers interested in Apollo objects will also find the MPC of considerable interest. They will also eagerly await the *International Astronomical Union Circulars* (IAUC), which are the airmail version of the astronomical telegrams. When an Apollo object or a comet is discovered, short-term ephemerides and preliminary orbital elements are swiftly made available through the IAUC.

It should be noted in passing that, prior to the 1992 ephemerides, 1950.0 coordinates were universally used for Solar-System objects, in particular in all NASA's internal and external documents, as well as in the MPC and the IAUC. The change to Epoch 2000.0 was delayed because of the lack of suitable astrometric catalogues for use by professional astronomers.

Some amateurs and groups of amateurs receive the EMP through the Minor Planet Center. Failing that, it may be consulted in an observatory or other astronomical library, and the required pages photocopied. Both the MPC and EMP are easy to obtain, because it is only a question of subscribing directly to the service.

A different method of obtaining the ephemerides, which is both powerful and convenient, is to program a portable microcomputer to calculate the positions. All the necessary spherical astronomy and constants have been gathered together in the book by Jean Meeus, *Astronomical Formulae for Calculators*. All one has to do is to enter the orbital elements (taken perhaps from the EMP), the date and time. The machine will then solve Kepler's equation by iteration, and after a few seconds, will give the right ascension and declination, even if the eccentricity is high. In practice, it needs at least 8 kb of memory to be able to operate without too many problems, but with 16 kb and an internal clock, it is also possible to calculate auxiliary quantities such as the velocity vector of the minor planet, the local hour angle in real time, etc. It should be noted that dialects of BASIC that are limited to single-precision calculations (internal accuracy to seven figures) are not quite adequate to obtain the last decimal place given in the ephemerides published in the EMP, although the accuracy is probably sufficient for finding purposes. It would be tedious to repeatedly enter orbital elements for the same object, so it is worth storing a file of data on a suitable magnetic tape or floppy disk. That way, entering (say) 'PHAETON, PIC, 1987 Oct.22, 04 : 15' would suffice to obtain the response '$08^h32^m54.2^s$, $+24°57'35''$, V = $+17.3$, culmination (South) at 06:36 UT at 72°, nautical twilight at 05:18 UT'. We should emphasize that for the greatest accuracy it is essential to use relatively recent orbital elements. Planetary perturbations cause deformations and slow precession of elliptical orbits. Elements for the current year are practically perfect, but those that are two or three years old lead to unacceptable errors. With the A–A–A objects that pass very close to the Earth, the slightest deviation is very apparent, and errors are sometimes degrees or tens of degrees on the sky if the very latest orbital elements are not employed.

Once the coordinates have been obtained, it is next necessary to find the object on the sky. For some of the brightest minor planets, one of the charts from *Atlas Eclipticalis*, *Atlas Borealis* or *Atlas Australis* by A. Becvár, which are complete down to V = $+9$, often suffices. Some may prefer a smaller-scale equivalent, *The AAVSO Variable Star Atlas*. *Atlas Coeli* by Becvár (now out-of-print) and *Atlas 2000.0* by Tirion are unsuitable, because their limiting magnitudes are too high. [*Uranometria 2000.0* by Tirion, Rappaport and Lovi has a limiting magnitude of 9.5. – Trans.] Table 7.3 gives the visual magnitude V at perihelic oppositions of all minor planets that may, at times, become brighter than magnitude $+9.25$. We may note in passing that at least four minor planets may be seen with the naked eye at perihelic oppositions: these are 4 Vesta, 2 Pallas, 7 Iris and 1 Ceres. Vesta comes top of the list because its rocky surface is light in colour, whereas the other large minor planets have dark surfaces. Table 7.3 differs profoundly from Table 7.1, and it will be seen to include primarily objects of type S. This is why bright minor planets have a reputation for appearing slightly orange in colour when seen through small telescopes. However, it is not wise to place too much reliance on this fact when trying to identify them. Objects of type S have roughly the same colour as a star of class K0 (like Arcturus), which is not particularly striking. No minor planet, not even the exceptionally red ones such as the R-type object, 349 Dembowska (which may attain magnitude 9.5 at maximum), has a surface as red as that of Mars.

Below magnitude $+10$ (which therefore means in most cases), there are two possible methods of identification: the first involves trying to determine which of a selected group of stars has moved, either by taking careful note of their positions,

Table 7.3. *Minor planets brighter than visual magnitude +9.25*

No.	Name	Maximum V magnitude at perihelic oppositions[a]	Type*		Closest approach to Earth (AU)	Maximum apparent diameter (")
4	Vesta	5.5	V	(U)	1.13	0.65
2	Pallas	6.7	B	(U)	1.22	0.61
7	Iris	6.9	S	(S)	0.85	0.30
1	Ceres	6.9	G	(C)	1.59	0.89
3	Juno	7.5	S	(S)	1.03	0.34
18	Melpomene	7.6	S	(S)	0.81	0.25
6	Hebe	7.7	S	(S)	0.97	0.28
15	Eunomia	8.0	S	(S)	1.20	0.30
8	Flora	8.0	S	(S)	0.87	0.25
324	Bamberga	8.2	CP	(C)	0.78	0.39
9	Metis	8.5	S	(S)	0.82	0.18
192	Nausikaa	8.5	S	(S)	0.81	0.15
20	Massalia	8.5	S	(S)	1.08	0.16
27	Euterpe	8.6	S	(S)	0.96	0.14
12	Victoria	8.7	S	(S)	0.82	0.23
532	Herculina	8.8	S	(S)	1.35	0.18
23	Thalia	8.9	S	(S)	1.06	0.13
29	Amphitrite	8.9	S	(S)	1.39	0.19
89	Julia	8.9	S	(S)	1.11	0.18
14	Irene	8.9	S	(S)	1.18	0.18
44	Nysa	9.0	E	(E)	1.08	0.09
11	Parthenope	9.1	S	(S)	1.19	0.15
19	Fortuna	9.1	G	(C)	1.06	0.28
43	Ariadne	9.1	S	(S)	0.82	0.13
42	Isis	9.1	S	(S)	0.89	0.13
10	Hygiea	9.1	C	(C)	1.75	0.32
216	Kleopatra	9.2	M	(CMEU)	1.11	0.11
5	Astraea	9.2	S	(S)	1.10	0.16
471	Papagena	9.2	S	(S)	1.26	0.14
16	Psyche	9.2	M	(M)	1.54	0.22

[a] Some A–A–A minor planets may exceed the limiting magnitude given here during their occasional, fast-moving, close approaches to the Earth.
* TRIAD data as Table 7.1.

or by quickly developing a short-exposure photograph of the field. The second method means using a photographic atlas of the sky. We will say no more of the first method, which may sometimes succeed immediately, and sometimes waste a lot of time. We will concentrate on using an atlas. Amateur astronomers frequently use H. Vehrenberg's *Atlas Stellarum*, which has a scale of 30 mm per degree and which goes down to magnitude 14. Other atlases sometimes used are Vehrenberg's *Falkauer* atlas, which is older and cruder, or that by C. Papadopoulos, which has the advantage of showing the relative brightness of stars as they appear visually, and not as they are recorded by a panchromatic emulsion. We must also mention the existence of an old, unobtainable atlas, the *Lick Atlas*, made some time ago with orthochromatic emulsions, which reaches magnitude B = 15, and which is easier to use than *Atlas Stellarum*. Professional astronomers most frequently use the *Palomar Sky Survey* (PSS or POSS), which has a scale of 53.6 mm per degree and a limiting magnitude of +21, but which is unfortunately completely outside the price range of any ordinary amateur, or even of any moderate-sized group. The PSS also suffers from the fact that neighbouring plates have a very small overlap.

Most of the time both amateurs and professionals make photocopies of the relevant plates in these atlases before observing. Because the stars are exceptionally small on the original plates, their visibility often suffers, especially under the faint lighting normally used at the eyepiece. Photocopiers that are able to enlarge originals are very useful in this respect. As a specific example, perhaps we may describe how we easily located faint minor planets such as 2060 Chiron (magnitude 16.5), and rapidly moving objects, like the Earth-grazing nucleus of the weakly active comet Sugano–Saigusa–Fujikawa (1983e, magnitude 15, motion 1 arc-second per second of time) with the 2-m telescope at the Pic du Midi. Before observing, we made a photograph of the field from the PSS atlas with a 50-mm, 1:3 macro lens onto Kodak Panatomic film (32 ISO). This positive (the atlas is a negative) was then printed at a scale of 180 mm/degree onto sheets of 240 × 300 mm, grade 4 paper. This enabled even the faintest stars to be reproduced. The coordinates of two bright stars in the field were then taken from the *Smithsonian Astrophysical Observatory Catalogue* (SAOC). In general, several stars were available, which meant that we had a choice. In addition, a magnetic-card, programmable calculator held the 1950.0 coordinates of the object at five equidistant intervals of time, taken from the published ephemerides. On entering the date and time, the 220-step program made separate parabolic interpolations in right ascension and declination, solved the spherical triangle formed by the two stars and the object, and gave an output in the form of 'object at so many millimetres from star 1, and so many millimetres from star 2'. These values were plotted with a pair of dividers. This method could also be used with *Atlas Stellarum*. It should be possible to obtain a positional accuracy of 0.5 arc-minute, which is far better than that obtainable with the 15 arc-minute-square transparent graticules supplied with the atlas. But we must admit that such photographic preparation is fairly onerous. In the near future we may be able to avoid this. We plan to use a small video camera to image the atlas and feed a microcomputer that simultaneously calculates the ephemeris. Routine finding of faint stars will then become a pleasurable process. The SAO catalogue (of 259 000 stars down to magnitude 9.8) does not yet exist in a form readable by microcomputer, but may be expected to become available at some time in the future. Limited to declinations between $-30°$ and $+90°$, this catalogue would occupy between 7 and 10 Mb on a hard disk, which is already within the range of a well-equipped individual.

7.2.2 *The number of observable minor planets*

To evaluate the potential for discovering new minor planets, we must have a clear idea of the statistical distribution of their stellar magnitudes at any given time. The absolute magnitude of a minor planet is defined as the B (photographic) magnitude that it would have at a distance of one astronomical unit from the Sun, as seen from the centre of the Sun. (Commission 20 of the IAU has recently recommended that absolute visual magnitudes, denoted H, should be used.) This definition does not depend on the position of the Earth, which is why the magnitude is denoted 'absolute'. The EMP orbital catalogue gives absolute magnitudes for all listed objects, so it is easy to use them in a numerical simulation. We have, for example, calculated the apparent V magnitudes of quite a large number of main-belt objects, for an arbitrary date – for the record this was 1987 Jly 24 – ensuring that a wide range of elongations from the Sun were represented. The difference $\Delta m = V_{apparent} - B_{absolute}$ was then plotted as a function of the angular distance of the Sun to give Fig. 7.7. It will be seen that Δm is around $+2(\pm 1)$ for objects at opposition, and that it increases to a maximum of about $+4.5(\pm 1)$ at about 60° from the Sun, and then falls to about $+4$ at conjunction. The increase from $+2$ to $+4.5$ reflects two simultaneous effects: the progressive increase in the distance from Earth after opposition, and the decrease in brightness because of the minor planet's increasing phase angle. Remember that the phase angle is, by definition, the Sun–Object–Earth angle in the plane containing the three bodies, and that it reaches a maximum at quadrature. Jupiter's phase angle never exceeds 12°, but with many main-belt minor planets it may reach 20–30°. In the antisolar region, the fall in brightness with the increase in phase is rapid because of what is known as the 'opposition effect'. (To within about 0.3 magnitudes, minor planets behave rather like cat's-eye reflectors.) Away from opposition, the decline in brightness is about $0.035(\pm 0.015)$ magnitude for every degree increase in phase angle. These phase effects are, in practice, very sensitive, and account for the general shape of Fig. 7.7.

We now need to determine the statistical distribution of absolute magnitudes for the 3300 objects in the EMP 87 catalogue. A histogram of magnitudes for a representative sample is shown in Fig. 7.8.

Turning to quantitative results, let us first consider an area of the sky $10° \times 10°$ centred on the antisolar point. On average, it contains four catalogued objects brighter than visual magnitude $+12.0$, eight objects brighter than $+13.0$, or 15 objects brighter than magnitude $+14.0$. (Note that we cannot say 'magnitude $+14.0$, etc.' for reasons that we will soon discuss.) Now assume that we slide that square 45° East or West along the ecliptic, so that it is centred 135° away from the Sun. On average, we will now find only one object brighter than magnitude $+12.0$ within the square, three brighter than magnitude $+13.0$, and eight brighter than magnitude $+14.0$. By moving just 45° in longitude away from the antisolar direction, the

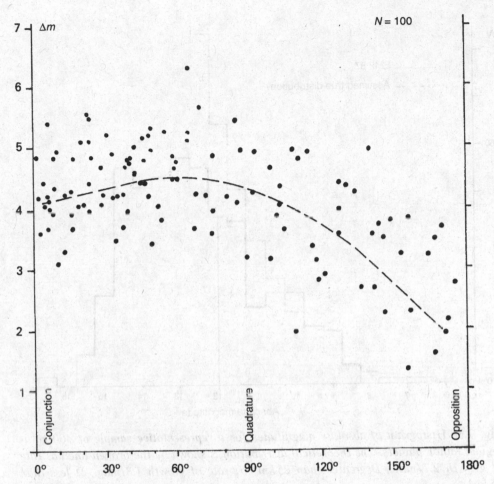

Fig. 7.7. *The difference between the apparent V magnitude and the absolute B magnitude as a function of the angular elongation of the Sun. Calculated for 1987 July 24, using a sample of 100 objects.*

number of objects above a given, fixed, limiting magnitude, has decreased by a factor of 2 or 3. At 90° from the Sun (at quadrature), the result would be even worse, with the difference reaching a factor of 5 at least. So this is the first, very important conclusion to be drawn:

- To see the maximum number of faint minor planets with a given instrument, observations should be concentrated on the zodiacal region around the antisolar point.

Let us now turn from listed minor planets to unknown ones, paying close attention to Fig. 7.8. Up to now we have implicitly assumed that all existing minor planets with absolute magnitudes brighter than +12.0 – i.e., larger than about 30 km – are catalogued in the EMP. To within 10 % this assumption is valid, at least to a first

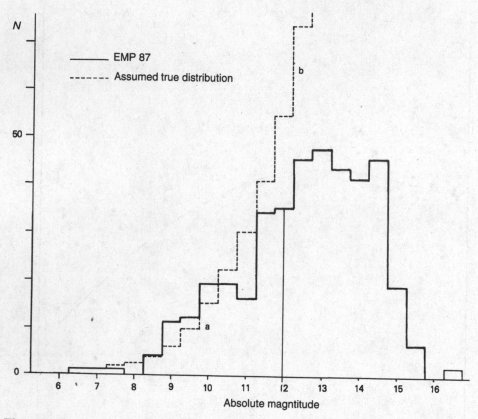

Fig. 7.8. *Histogram of absolute magnitudes for a representative sample of 400 cata-*
logued minor planets. The theoretical distribution is shown by the broken line (a: rate
of growth 2.7/mag., D greater than 25 km; b: rate of growth 1.8/mag., D less than
25 km).

approximation. But Fig. 7.8 shows that after an almost exponential increase on
the left-hand side, the frequency curve's rate of increase falls when the absolute
magnitude reaches +12. Yet our current knowledge and certain photographic
searches that have been made suggest that the number of objects actually continues
to increase exponentially by a factor of about 1.8 per unit absolute magnitude.
This means that only 70 % of objects that exist with magnitudes between +12.0
and +13.0 are included in EMP 87, and only 33 % of objects with magnitudes
between +13.0 and +14.0. The right-hand side of the histogram in Fig. 7.8 declines
rapidly, and the deficit, which is already noticeable, becomes even greater. Beyond
magnitude +16 (which implies a diameter of about 5 km), at least 99.5 % of minor
planets are still missing from the EMP. The actual, total population of objects larger
than 5 km in size doubtless exceeds 10 000.

Finally, let us return to our 10° × 10° square centred on the antisolar point. We
have seen that it contains about 15 minor planets with apparent V magnitudes
brighter than +14.0, and that these are all catalogued, with perhaps some rare

exceptions. But there are 14 minor planets with apparent magnitudes between +14 and +15, of which four are missing from the EMP and one or two of which may be still unknown. There are 26 objects between magnitude +15 and +16, three-quarters of which are not included in the EMP, and half of which are unknown. There are 48 objects with magnitudes between +16 and +17, most of which are unknown. In *Pulsar* (**653**), Alain Maury published an example of a photograph taken with a large Schmidt telescope just 5° away from the antisolar point. The limiting magnitude of the plate must be approximately V = +20. Maury found three unknown minor planets per square degree, which agrees perfectly with the estimate made above.

V magnitudes between +15 and +17 are well within the range of amateur astronomers who specialize in photography, so our second major conclusion is:

- It is still relatively easy to discover new minor planets photographically.

We hope that any amateur astronomer who is looking for an unusual programme will have been convinced by these figures that there is a very real possibility of discovering new minor planets. This possibility has been rarely discussed in books and popular works until now, which is why we have emphasized the point. In fact, it is undoubtedly easier for an amateur to discover several minor planets than it is to discover a single comet, nova, or extragalactic supernova. To succeed, however, the observer must adopt an effective and organized method of working. This is what we intend to discuss next.

7.3 Searching for faint minor planets

No one nowadays could announce 'I have discovered a new minor planet on such-and-such a date: I propose to call it such-and-such, and here is its preliminary orbit.' This romantic picture of the solitary discoverer was in fact the norm in the 19th century. It may still apply in the case of certain faint, unexpected comets, but certainly does not occur with minor planets, because they exist in such an extremely large number. Things happen differently nowadays. Photographs of the sky are examined to locate the maximum number of minor planets (we shall see later how this is done). The positions of the minor planets are measured (this will also be discussed) and the list of results is communicated to the Minor Planet Center in Cambridge, Massachusetts. This group, which operates under the auspices of IAU Commission 20 and is directed by Brian Marsden, maintains an up-to-date, central computer file of all minor-planet positions that have been measured since 1939, as well as many older ones dating as far back as 1891. It also holds a complete file of known orbits. This second file is far larger than the list in the EMP because it contains a large number of limited-accuracy preliminary orbits.

For each observation, the Center's computer searches to see if the object can be identified with one of the orbits held on file. As we have seen, at the antisolar point there is less than one known object per square degree, so the risk of accidental confusion is still very low. Two possibilities arise, depending on whether the computer does, or does not, recognize the object. If it does recognize it, the observation is automatically included in the computation of an improved orbit, which will be published in the MPC, and then later used, for example, in the EMP.

If the computer does not recognize it, the suspected object is given an identification code that depends on the date of the observation. This alphabetical code is allocated according to the half-month (days 1–15, and days 16–end), the letter I being omitted. For example 1983 TB was the second object (B=2) recorded in the twentieth (T=20) half-month in 1983.

The Center then awaits new observations of the object. When they arrive (at least three are needed, preferably a few more), a preliminary orbit is calculated, and the computer tries to allow for planetary perturbations and, by checking back in time, whether any earlier isolated observations, one or more oppositions previously, might be attributed to the same object. Often this proves to be the case: it retrieves isolated measurements, or even a preliminary orbit calculated at a previous opposition. The *Circulars* then publish identifications of the following sort: 1973 SZ_3 = 1978 EM_2 = 1986 XA_1. Frequently, however, no identification is found, so the observer learns, on receiving the MPC, that they have *perhaps* discovered a new object. Patience is still required, however. After three or more oppositions that give orbits agreeing within the limits of observational error – and we have seen that this may require three to five years – the orbit is taken to be confirmed. The minor planet is given an official number. For example, 1983 TB was designated 3200 in 1985. The potential discoverer becomes the official discoverer and is then permitted to choose a name for the object. In this way, 1983 TB = 3200 became 3200 Phaeton, a definitive, officially recognized name. Naturally, this does not preclude someone discovering one or more pre-discovery images on archival plates. On 1974 December 15, Phaeton was at magnitude +10.5 somewhere between Cassiopeia and Andromeda, so it is possible that someone might find it on a plate taken on that date. Naturally this would not detract from the work of the team that actually discovered it in 1983.

The basic work that an amateur discoverer of minor planets has to carry out therefore mainly consists of taking photographs and measuring them. But it is obviously possible for a keen programmer to set up a personal system for calculating orbits, incorporating perturbations and the identifications of objects that have not been resolved by Brian Marsden's group. There are a few individuals around the world who are having some success doing just this, and the Minor Planet Center is quite willing to sell copies of its master magnetic tape with the file of position measurements. This file is currently growing at the rate of 13 % per year, which shows an exponential explosion in interest in minor planets around the world. (The number of minor planets that receive official numbers is currently growing at 8 % per year. This proves that the detection of new objects is progressing faster than knowledge of orbits.) The 1987 edition contained almost 500 000 entries. But adopting such a course means undertaking a vast deal of work, which has its own consequences. It means becoming highly specialized, and will certainly leave little time for making thorough astronomical observations. So you are forced to make a choice. Because we are more concerned with observation rather than computation, we shall concentrate on discussing photographing the sky.

7.3.1 *Photography*

Section 7.2 has shown that it is important to combine a low limiting magnitude, which implies an adequate focal length, with the largest possible field-size. We have, in fact, to reckon on 0.5 unknown objects per square degree at the antisolar point down to magnitude +17, and 0.1 unknown objects per square degree, at most, away from that ideal point. So a compromise has to be made between field-size and focal length, and this is rather tricky to resolve.

The most successful professional workers use either large Schmidt telescopes or astrographs that have objectives of 400–500-mm diameter, focal lengths of about 2 m, and fields of about 15 square degrees.

A modern, amateur, f/5 Newtonian telescope does not cover more than about 0.3 square degrees at best, and is not particularly appropriate because nine exposures out of ten will show no minor planets. A specially designed reflector, with a focal ratio of f/3.5, would probably be disappointing because of the strong, off-axis coma. In any case, it would certainly not have a field as large as even 1 square degree. It would seem, however, that Japanese minor-planet observers are doing good work with telescopes with focal ratios of f/4 to f/6, with focal lengths of between 1–2 m. Because of the low efficiency they undoubtedly take an enormous number of photographs. Anyone hoping to see what might be done with a very fast telescope with a focal length of 2 m, might be tempted to request an allocation of time for minor-planet searching at the Pic du Midi. There, a 600-mm, f/3.5, Newtonian telescope is reserved for amateur use. The best time would be the end of the autumn. (Contact the Association des utilisateurs du T60, Observatoire du Pic du Midi et Toulouse, 14 avenue Edourd-Belin, 31400 Toulouse, France.)

Obviously the preferred types of equipment for amateurs are Schmidt or Maksutov telescopes, because both optical designs make an acceptable compromise between the conflicting demands just mentioned. Some amateurs have purchased one of the small Schmidt telescopes of either 200-mm or 350-mm aperture that are made by the Celestron company. A handful of brave individuals, who are both good optical workers and good engineers, have even built instruments themselves. The field of a Schmidt typically covers 25, or sometimes even 50, square degrees. The scale of amateur photographs is likely to be around 15 mm/degree (which may be contrasted with the 35–50 mm/degree obtained with professional instruments). Such an instrument is expensive and powerful, but rather difficult to use regularly, especially because of the problems arising from having to bow the film to fit the curved focal plane. Other amateurs employ simpler cameras, using a conventional lens 150–200 mm in diameter as objective. These are used on an equatorial mounting. So-called 'flat field' models have a focal ratio of around f/4 and a flat field with a scale of about 10 mm/degree. They give good results, with an image-quality that in theory is attainable only with Schmidt telescopes. Minor planets down to magnitude V = +15.5 at least are certainly within reach of a 200-mm aperture, f/4 camera with 30-minute exposures.

It is quite obvious that currently, hypersensitized Kodak TP 2415 film is most frequently used by all types of observers, because of its exceptional performance (Fig. 7.9). Readers who want to know more about methods of hypersensitization should consult books that deal with astronomical photography (*see also* Chap. 15, Sect. 15.4.4.3).

It is important to remember that because of minor planets' proper motions, the limiting magnitude attained is considerably worse than for stars. Table 7.2 indicates that the angular velocities of Main-Belt objects at opposition are around 30 arc-seconds/hour, in the direction of decreasing ecliptic longitudes. With the sort of focal length typically found in amateur instruments and for exposures of around 40 minutes, minor planets produce trails on the film that are about 100 microns long, three times as long as they are wide. For this reason the effective photographic integration time for minor planets that are moving is, *de facto*, limited to about 15 minutes. One way of partially offsetting for this limitation is to drive the telescope slowly to counteract their motion and, if possible, parallel to the ecliptic. Stars will then form trails parallel to the ecliptic and minor planets will appear more like point sources. This method is even more useful when one wants to photograph minor planets that are not retrograding, i.e., when they lie well east or west of the antisolar point. Their motion is then twice as fast as it is at opposition, and their brightness is also reduced by the phase effect. It should hardly be necessary to stress the necessity for careful guiding with a supplementary eyepiece and high magnification under these conditions. [The techniques and special equipment used for minor-planet work are identical to those employed in photographing comets, a full discussion of which may be found in the next chapter, Sect. 8.7. – Trans.]

A fundamental point that any serious observer always bears in mind is that *every exposure must be duplicated*. A known or unknown minor planet will often be noticed because the image on the film is slightly elongated in comparison with the circular star images. (The opposite will obviously apply if the telescope has been driven to compensate for the minor planet's motion, as just described.) It is essential to be able to eliminate other possible causes: an unresolved double star, a spiral galaxy seen edge-on, or just a simple flaw in the emulsion. As we shall see shortly, two exposures also allow discoveries to be made that would go unnoticed with just a single image. Another absolutely imperative rule is to note down scrupulously the *exact time in UT* of the beginning and end of the exposure, as well as those of any possible interruptions by cloud. It is a good idea to adopt the practice of always rounding times to the nearest one-tenth of a minute when writing them down in the observing log – in other words, times must be accurate to 3 seconds. The easy availability of quartz watches means that any failure to do so is unforgivable. On the other hand, adjustment of clocks to Summer Time has been the source of innumerable errors in transcription by both professional and amateur astronomers over the years. We shall see that timing an observation to 0.1 minute may become important if one is lucky enough to capture an Apollo object close to the Earth.

A blink- or stereo-comparator is an essential piece of equipment, which enables all the minor planets on the plates to be detected. The first concern of any amateur who wants to take up this sort of observation should be the construction of a simple form of blink-comparator (*see* Chap. 16, 'Plate comparators'). This operates on the principle of alternately viewing two photographs, taken with the same exposure, while using binocular vision to fuse the two images. The illumination of the left- and right-hand images is varied rapidly. Any detail present on one of the plates and

Fig. 7.9. 15 Eunomia passing close to the galaxy M33, on 1985 September 19. Photograph taken by C. Viladrich on hypersensitized Kodak TP 2415 film, using a flat-field camera, 180-mm aperture, focal length 760 mm; exposure 60 minutes. Eunomia was magnitude V = 8.5, 1.31 AU from Earth, and was moving at 29 arc-seconds per hour.

not on the other shows a strong blink. Anything that is present on both plates, but which moved between exposures, will jump backwards and forwards in front of, or behind, the apparent plane of the sky. The effect is spectacular, extremely noticeable and an indispensable aid to locating suspect objects. Variable stars are detected at the same time, and some of those may well be unknown and not recorded in the catalogues. Using the principles of stereoscopic vision, observers-cum-instrument-makers have produced highly ingenious and extremely varied equipment of differing degrees of sophistication. An amateur who is conversant with computing but not so good at precision engineering might try linking a video camera to a microcomputer, and then displaying the differences between two digitized images on a VDU. The limited size of current video detectors, however, means that this idea, although wonderful in principle, is not yet very satisfactory in practice. A small piece of sheet film $25 \, cm^2$ in area is capable of resolving 2500×2500 individual elements (pixels). Any computerized search would have to divide the overall image into small areas that were scanned separately. It is also very difficult to ensure that the computer-processing is able to discern faint effects, which are immediately detectable to anyone who views the film under a good microscope. Optical methods are still the best method of detecting the changes that interest us.

7.3.2 Possible search programmes

Intensive professional programmes exist at various large and small observatories around the world. The large Schmidt telescopes at Siding Spring in Australia (the Anglo–Australian Schmidt) and at Mount Palomar in California have been responsible for fantastic numbers of both measurements (thousands per year) and discoveries. The large, French, Schmidt telescope near Grasse, in the Alpes-Maritimes, is now photographing and discovering minor planets and did not take long to become a world-class instrument. In recent years most discoveries have been made at Lowell Observatory in Arizona, Mount Palomar in California, the Crimean Astrophysical Observatory, the European Southern Observatory at La Silla in Chile, and the Kiso station of the Tokyo Observatory. Less conventional instruments such as the infrared-sensing satellite IRAS have discovered some particularly interesting new minor planets. A charge-transfer (CCD) camera, called 'Spacewatch', that uses the Earth's rotation to sweep the sky is now operational in the USA. Minor-planet astronomy is burgeoning nearly everywhere at the moment.

7.3.2.1 Minor planets between Mars and Jupiter
In the face of such widely scattered, professional instrumentation, amateurs tend to think that they are outclassed. What chance is there, they think, of beating so many well-equipped specialists, especially if the most favourable region of the sky, as discussed above, is only $90° \times 15°$, or 60 times the field of a powerful Schmidt telescope? In an article in the journal *Mercury* in 1985, about the observation of minor planets by amateurs, for example, J. U. Gunther cited the discovery of 2090 Mizuho by the Japanese amateur T. Urata as a commendable, but isolated instance (Gunther, 1985). 2090 Mizuho was at magnitude $V = +15.3$ at the time of its discovery, and may rise to $V = +13.0$ at perihelic oppositions. A glance

through recent issues of the MPC does indeed show that few discoveries are by non-professionals. A survey of 400 objects that have recently received official names indicates that only about 2 % of discoveries are by amateurs. The Japanese, including T. Seki, Urata and several other compatriots, are almost the only ones upholding the amateur tradition. Some of Seki's latest discoveries have been 2961 Katsurahama, 3150 Tosa, 3182 Shimanto, 3262 Miune, and 3431 Nakano. Among the 400 names, only one discovery had been made by non-Japanese amateurs: that of 3344 Modena by observers from Bologna (the Observatorio San Vittore) in Italy. [More recently, although the Japanese have continued to dominate the scene, an increasing number of minor planets have been discovered from Italy, 14 by Colombini *et al.* at Bologna, and four by Baur at Chions. From Stakenbridge in the United Kingdom, Brian Manning has made several discoveries, including 1989 TE (earlier observed as 1982 TB), 1989 VX, 1989 WN$_1$, 1990 FJ (now designated 4506 Hendrie), 1991 BG (4751 Alicemanning), and 1991 AF$_1$ (eventually traced back to 1934). He has also recovered 1968 OF (1989 TF); and found numerous other objects that have received provisional designations. – Trans.]

But the figures just given show that most amateurs are quite wrong in believing that the outcome is a foregone conclusion. Professional specialists are in reality completely swamped by the size of the task, so that nothing would please them more than to receive a helping hand! Take the example of the Trojan minor planets. All the books and articles agree that their discovery is not within the reach of amateurs, and in general they cite only 624 Hektor, which may at times reach magnitude V = +13.7, as being the only 'readily accessible' object. Perhaps someone can explain why we have had to wait so long for the discovery of the Trojans 2241 (1979 WM), 3063 Makhaon (1983 PV), and 3317 Paris (1984 KF)! These are all minor planets of 90–130 km in diameter, and thus relatively large. Under the most favourable conditions they may reach V = +15.3, V = +15.6 and V = +14.8, respectively, and yet so few observations were made of them that they were not named until 1986! The truth is that for a long time – as their magnitudes show – they have been within reach of amateurs, who were unfortunately preoccupied with photographing the Messier objects for the umpteenth time. Professionals doubtless missed these Trojans because they did not look far enough away from Jupiter's orbit. The orbit of 3317 Paris, for example, has an inclination of 27°, which is by no means the record for a Trojan (cf. 2146 Stentor: 38°).

What is the best strategy for discovering new minor planets between Mars and Jupiter? At the risk of disappointing the reader, we will simply say that this remains an open question. Attempting to reply would, at the very least, involve a very thorough statistical analysis of the MPC records to determine the methods of the most active discoverers, and probably some systematic experiments on the sky as well, because the reply must to a certain extent be a function of the instrumentation employed. Until someone does carry out such an investigation, here is a suggestion: sweep perpendicular to, and 15° or more above the ecliptic. You will lose slightly more than 0.5 magnitude because of the phase effect. You will also see a twentieth of the number of already-known minor planets than if you were looking at the ecliptic (which will save you from being overloaded during checking of the plates). You will, perforce, be restricted to objects with high orbital inclinations, but at least you will detect minor planets that have often been missed by observers who concentrate on quantity. They will be objects that professionals would perhaps not discover until they were at the nodes of their orbits, that is to say some weeks – or even years – later. It might also be as well to shift slightly towards the East so as to anticipate opposition.

7.3.2.2 A–A–A objects or 'Earth grazers'

Most popular works more readily recommend that amateurs attempt the discovery of Aten–Apollo–Amor objects that brush past the Earth, whereas this sort of task is, in fact, *far more difficult than the discovery of ordinary minor planets*. It may be difficult, but how exciting! A very instructive table of catalogued, 20th-century, close-approaches that came within 0.05 AU of the Earth has been published (Combes and Meeus, 1986). Until recently the record was held by the object called Hermes, which has since been lost, and which was observed at only 0.005 AU from Earth on 1937 October 30. [On 1991 January 18, the Spacewatch telescope detected 1991 BA – at magnitude $H = 28.5$ it is the faintest object ever discovered – which approached to a geocentric distance of 0.0011 AU (170 000 km). It is estimated to be just 5–10 m across. – Trans.] We reproduce here (Table 7.4) the portion of the table given by Combes and Meeus for objects between 1960 and 1986, adding the apparent magnitude V and the maximum angular velocity against the sky at the time of closest approach.

[The number of known, close-approach minor planets has approximately doubled since 1986, when the first edition of this book was prepared. Among notable recent discoveries we may mention those of 4179 Toutatis = 1989 AC, which makes a record predicted close approach to the Earth of 3 million km in 1992 and then less than 2 million km in 2004; the record non-predicted Earth approach for a numbered minor planet of 0.7 million km in the case of 4581 Asclepius = 1989 FC just before discovery; and the apparent duplicity of 4769 Castalia = 1989 PB. – Trans.]

It will be seen that, on average, one object with a magnitude between +10 and +13, and therefore easily observable, comes within 0.05 AU of Earth every 5 years. Its maximum angular velocity against the sky is then around one degree per hour, i.e., one arc-second per second of time. We now have to extrapolate from this sample to the number of objects that are fainter or that pass slightly farther from us, and finally, to evaluate the number of objects that pass by without being detected.

Taking the current limit of detection by professionals as $V = +17$, we make the following assumptions to see if we can obtain a simple numerical model: The largest objects, which have an absolute magnitude of +15, and thus a diameter of about 5 km, are rare, but are detectable out to 0.25 AU, or even more when the phase angle is favourable; the smallest bodies have an absolute magnitude of +22, which means a diameter of about 200 m, and are numerous, but are hardly visible beyond 0.01 AU; the number/magnitude relationship is the same as that found in the Main Belt.

We find that the potential discovery frequency is better than one per month if the limiting magnitude is taken as $V = +17$. In practice, the few astronomers who specialize in searching the sky are far from reporting so many discoveries by at least a factor of 2. The magnitude distribution indicates that the majority of the

Table 7.4. *Close approaches by A–A–A minor planets to the Earth*

Object	Closest approach to Earth				Observations*
	Distance (AU)	Date	Mag. (V)	Velocity (°/hr)	
6344 P-L	0.038	1960 Oct. 12	16.5	0.42	Yes (d)
1566 Icarus	0.042	1968 Jun. 14	12.6	0.96	Yes (p)
3200 Phaeton	0.043	1974 Dec. 15	10.5	1.03	No (d: 1963)
2120 Tantalus	0.047	1975 Dec. 26	11.7	0.99	Yes (1975 YA)
2340 Hathor	0.0077	1976 Oct. 20	13.4	2.36	Yes (1976 UA)
2135 Aristaeus	0.032	1977 Apr. 01	14.0	0.78	Yes (1977 HA)
3361 Orpheus	0.031	1978 Apr. 13	–	0.35	No (d: 1982)
1979 XB	0.037	1979 Dec. 17	16.6	0.87	Yes (d)
1982 DB	0.028	1982 Jan. 23	16.3	0.27	Yes (d)
3361 Orpheus	0.031	1982 Apr. 13	15.2	0.36	Yes (1982 HR)
1982 XB	0.038	1982 Dec. 22	13.7	0.21	Yes (p)
1984 KD	0.031	1984 Jun. 19	12.0	0.49	Yes (d)
2101 Adonis	0.045	1984 Jly. 09	14.8	0.72	Yes (p)
1986 JK	0.029	1986 May 25	13.3	0.70	Yes (d)
1620 Geographos	0.033	1994 Aug. 25	10.3	0.51	

* An initial designation indicates that the minor planet was discovered during that apparition; where none is given, 'd' indicates discovery, and 'p' that it was a predicted passage.

A–A–A objects that do not reach magnitude V = +14 at any time during their apparition currently go undetected. This is why professional astronomers try to involve amateurs. If one assumes that the best-equipped amateurs have a detection limit that is two magnitudes brighter than the professionals, but that on the other hand there are enough of them to cover the whole sky properly, we may conclude that all the amateurs together could discover about three Earth-grazers per year. The fact that they do not find any of them – one exception is again a Japanese, M. Kizawa, who discovered 1986 DA – proves that the opportunities offered by such a search programme is not recognized in amateur circles.

There is no preferred direction in the sky when it comes to searching for A–A–A objects. Some pass to sunward of us, which does not help their detection because the phase effect reduces their apparent magnitude; others cross the Earth's orbit directly in front of, or behind, us, so their elongation from the Sun remains almost constant at around 90° throughout their passage; and finally others pass on the opposite side to the Sun and therefore have optimum conditions of illumination, which improves their chances of being detected. In general, the apparent paths of A–A–A objects are nearly half of a great circle on the sky. Initially one may appear almost stationary in a particular constellation, but it increases rapidly in brightness.

Then it suddenly seems to accelerate, and reaches its greatest angular velocity 90° away from the direction in which it originally approached. Eventually, its speed drops and it disappears in the distance on the opposite side of the sky to that where it was first seen. There is no preferred season for these events, and naturally there are as many passages at Full Moon as there are at New Moon.

There is really nothing to choose between certain cometary nuclei that brush past the Earth and true A–A–A minor planets. Two of the comets in 1983, IRAS–Araki–Alcock and Sugano–Saigusa–Fujikawa, appeared just as rapidly, and had the same size and mass as typical Apollo objects, so they could almost be included in Table 7.4.

The characteristics just described govern the practical advice that can be given to those who want to specialize in searching for Apollo minor planets, and more generally to all those who are already photographing minor planets and comets, or even variable stars, and who do not want to be silly enough to miss an Apollo object on their photographs. First, it is absolutely essential to examine every photograph by the end of the afternoon following the observations, at the very latest. If by chance an unexpected object has left a narrow, straight trail on a photograph, it is important to be ready to take new photographs the next night. This is where the precaution of *taking two photographs* pays dividends. They allow one to eliminate spurious trails left by flashing artificial satellites – a source of confusion that is becoming more and more frequent – and also meteors with short paths. Two photographs give an indication of the direction of movement and permit an acceleration or a deceleration in angular velocity to be detected. The day before its closest approach, for example, an A–A–A object that will pass by at 4 million kilometres accelerates at 2.5 % per hour. If that acceleration is not taken into account, the instrument might be pointed seven degrees away from the correct direction on the following night, when it would obviously risk failing to detect the object. If the closest approach is less than 4 million kilometres, this non-linear effect would, of course, be even greater. Finally, as we have seen, a pair of images enables the field to be examined in detail stereoscopically. During an exposure it is also advisable to interrupt the light path for 2 minutes between two accurate time pips, after about one-third of the exposure time. If unforeseen factors or clouds prevent the second exposure from being made, it will at least be possible to determine the direction of movement. In addition, the sharp, precisely timed break enables the position of the object with respect to the stars to be measured very accurately. When the weather conditions warrant the risk, it is advisable to wait 2 or 3 hours before taking the second of the two photographs. Measurement of angular velocity will then be much more accurate. A perfectionist (or very experienced observer) will probably take three exposures during a night, rather than just two.

To check one's own performance and become experienced in this sort of work, beginners should start by observing faint Apollo objects whose predictions are given in the ephemerides. In some years there are several. A list of close approaches coming within 0.20 AU was, for example, given in an issue of the *Journal* of the BAA (Townsend and Rogers, 1986).

When the object is very fast, the limiting magnitude that may be attained is not particularly low. A typical focal length for this work is between 300 and 900 mm.

Fig. 7.10. *The minor planet 1372 Haremari (magnitude 15.5 approx.), photographed by R. Chanal on 1986 December 2 between 22:10 and 22:40 UT; 410-mm reflector, f/4.83; hypersensitized Kodak TP 2415 film. The exposure was restricted to 30 minutes in order to obtain an accurate astrometric measurement. The motion of the minor planet has, however, caused a short trail.*

With angular velocities of between 0.2 and 2 degrees per hour, the rate at which the image moves across the emulsion lies between 1 and 30 millimetres per hour (Fig.7.10). This means that the individual exposure time of any point on the trail is often very short. Frequently the trails will be grossly under-exposed and at the limit of visibility on the emulsion. They are only detectable because a slightly greater number of grains of silver than normal seem to lie in a straight line, i.e., in a trail that has a definite beginning and end. Longer trails, corresponding to 10-minute exposures, will assist recognition by providing what would under normal circumstances be redundant information. On the other hand, exposures should not be too long, if several areas of the sky are to be surveyed. In most cases it is absolutely crucial for an uncertain trail to be checked with a second exposure.

This problem of the fast rate of motion of the image across the film discriminates against the discovery of highly inclined orbits. An A–A–A object that moves in a low-inclination orbit and which slowly passes the Earth always has an apparent motion that is much less than an object that rushes past at right-angles to the plane of the ecliptic.

The smallest piece of equipment that may be sensibly used for such a search

313

programme would seem to be a long, fast telephoto lens, for example a 300-mm f/2.8, on an equatorial mounting. The need for fast optics is paramount here, because we need to combine a wide field with short exposures. For the past ten years the most successful instruments, with nearly two-thirds of all reported discoveries, have been the two Schmidt telescopes at Mount Palomar, the larger of which is 1200-mm, f/2.5, and the smaller 460-mm, f/2. The names of Eleanor Helin, and of Carolyn and Eugene Shoemaker are particularly associated with this work. Between 1983 July and 1986 July, for example, using the 460-mm Schmidt, the Shoemakers discovered nine new A–A–A objects between magnitudes +14 and +17.5 and eight comets! Such a success rate implies extremely efficient organisation.

It goes without saying that if one is completely confident of any particular detection, the details should be relayed without delay by telegram, telex, or electronic mail to the Central Bureau for Astronomical Telegrams in Cambridge, Massachusetts. The address, telex, electronic mail, and telephone numbers are the same as those for the Minor Planet Center, and are to be found on every IAUC. Given the rapidity with which an A–A–A object appears and disappears, it is fair to assume that the urgency is ten times as great as it is for a comet. But do take care not to be prematurely discredited by imaginary discoveries! If one has a photograph showing a superb trail, but that photograph is unfortunately unique, it is much wiser not to say anything.

7.4 Astrometry, or position measurement

We have already described in the previous section how the discovery of new minor planets is generally the result of patient and methodical work checking negative photographs. We now propose to discuss this in greater detail. Once the observations have been made, the major part of the work for the observer consists of measuring the plates, reducing the measurements in a standard manner, and finally sending the lists of calculated coordinates to the Minor Planet Center. The latter checks them, publishes them, and uses the best to improve the orbital data.

A beginner is quite wrong in thinking that there is no point in remeasuring the coordinates of minor planets discovered a century ago, especially those of 4 Vesta or 2 Pallas, which traditionally appear in all tables of ephemerides alongside those of Mars or Venus. Is it not true that space technology or giant radars are now able to give the positions of the planets relative to one another to within a few kilometres? Unfortunately, these are only 'one-off' experiments, so finding the positions of the planets in this way on a regular basis is still just a pipe-dream. Observers of minor-planet occultations find each week that the reality is far less glamorous. They find that large, bright, minor planets between, say, magnitudes +8 and +12, are typically two or three minutes early or late in occulting the star. Because the angular velocities are often around 30 arc-seconds per hour, or 1 arc-second every 2 minutes, that means that the published differential coordinates between the star and the minor planet are often wrong by 1 or 2 arc-seconds. Should the minor planet or the star be blamed? It is true that people calculating occultations use star catalogues like the AGK3, which was prepared in 1960 and which is now sorely in need of revision. The proper motion of stars is typically 3 arc-seconds per century, and catalogued values are frequently very poor (sometimes in error by as much as 100% or even more!), so the predicted positions of stars often differ from the true positions by 0.5 arc-second in a purely arbitrary direction. Indeed, it is often noted that the track of the minor planet is shifted laterally relative to the star by some 0.3 or 0.5 arc-second from that predicted. But all the evidence also points to minor planets being early or late in their paths by amounts that are frequently much greater than the lateral shift. So it is not the star that is responsible for the greater part of the difference measured. (A survey of 39 bright minor planets made in 1983 with the transit instrument at Bordeaux Observatory found that a typical error in the position of the object was 1.7 arc-seconds.) Even so, people who calculate occultations have a 'black-list' of minor planets that have been found to be more than 4 arc-seconds, or even more than 8 arc-seconds away from their prediction positions, and whose orbital elements require urgent revision. Pending better data, the theoreticians resign themselves to omitting these objects from their published predictions. But despite this some badly behaved minor planets slip through the net and cause a bit of a stir among occultation observers. A few years ago, for example, we found that 332 Siri was 28 arc-seconds from its predicted position. Yet 332 Siri has been known since 1892, reaches magnitude +13.0 at perihelic oppositions and never moves beyond the ecliptic. It is obvious that no one had measured 332 Siri's position for years! Even worse: there is an object 471 Papagena, which is on the official list of objects to be observed by Hipparcos (which, as we know, was designed to measure the positions of stars extremely accurately). Such a minor planet should, in principle, have been the subject of any observing campaign by the best instruments. 471 Papagena is an important object because it measures 130 km across. It is included in Table 7.3. It was nevertheless 2 arc-seconds out (being early) at a recent occultation.

Are we then to conclude that those who are measuring minor-planet positions, or even the Minor Planet Center itself, are being somewhat careless? Is the latter perhaps accepting too many inaccurate measurements? Not at all! The truth is rather different: it is simply that the few people specializing in this work are completely overwhelmed. There are far too few of them, and far too many minor planets. The MPC frequently publishes reductions that have just been made of plates that are 10 years old. Simply no one, for example, had the time to check 332 Siri, although it had undoubtedly been recorded on a number of plates that were waiting to be measured. Guess where 332 Siri was at its previous opposition in 1985 May: it was in Libra, about 7 degrees from Saturn. Any astronomer, amateur or not, who thought of checking its position then would have had no trouble in detecting the obvious error in its ephemeris.

This brings us to a discussion of astrometric methods. Amateurs can set themselves two goals, which have different problems. They may be content with trying to detect position errors of (say) more than 5 arc-seconds. The case of 322 Siri is by no means unique. More than 100 minor planets may have similar errors, especially among the fainter objects. Amateurs can also make very accurate measurements and, after serving their apprenticeship, join the small group of specialists who regularly add to the file at the Minor Planet Center.

A suitable instrument for astrometry is, in principle, more or less the same as one that is used for discovering comets or Apollo objects. It should definitely not have too short a focal length. 1 second of arc at the focus of an instrument of 1-m focal length is 5 μm, which seems to be the minimum if one wants to retain a reasonable amount of accuracy. It also needs to have a fairly wide field so that it includes several reference stars. The FK4 fundamental catalogue, which is the most accurate (and is shortly to be replaced by the FK5), contains, on average, only one (bright) star within a circle 6 degrees in diameter, so it cannot be used directly. The most frequently used catalogue remains the AGK3, published by the Astronomische Gesellschaft, the eight volumes of which contain more than 180 000 stars with declinations higher than −2.5°, and the faintest of which have photographic magnitudes of +11. Farther south, we have to resign ourselves to using the SAOC, which is not consistent in its accuracy, or even individual catalogues like the excellent ZC, which is confined to the zodiacal constellations.

The AGK3 contains, on average, eight or nine stars per square degree, and the SAOC six stars per square degree, although obviously the numbers vary slightly with distance from the Milky Way. In practical terms, serious reduction aimed at securing really accurate positions requires the use of about ten reference stars. Our astrometric instrument should therefore have a field free from major aberrations that is at least 1.5 degrees in diameter, which again seems to eliminate most ordinary amateur telescopes. Professionals use refractors of 250–500-mm diameter, with focal lengths of 2–4 m, which are traditionally known as astrographs (the last major use of refractors), or even Schmidt telescopes, with focal lengths of 3 m or less, but which give smaller images and wider fields than refractors. On the other hand, one large, specialized, 1.5-m, f/10 telescope is used at the U.S. Naval Observatory in Arizona.

Let us now turn to measurement of plates and first to the accuracy that may be attained by a very simple 'shoestring' method. This method hardly seems worthy of being called 'scientific', and is never described in books, but it is, in fact, quite adequate for general use, especially when the work must be done quickly. Working on a print enlarged to a scale of 20 arc-seconds per mm, a good-quality graduated ruler (preferably an engineer's rule) will give the distances between stars to rather better than 10 arc-seconds accuracy, especially if care is taken to align the graduations with the stars by using a watchmaker's eyeglass. A linen-tester with a scale graduated in tenths of a millimetre, used directly on the negative over a light-box, will give comparable, and apparently even more accurate results, but which are more tiring visually. If the useful portion of the photograph is very small, it is also possible to use a binocular microscope with a simple filar eyepiece. This was how Saturn's satellite Dione B (magnitude +17.5) was measured at the Pic du Midi the day it was discovered: no more sophisticated equipment was available. Projecting the negative with an enlarger onto a macrophotograph of the Palomar Sky Survey (POSS) made to a scale of 5 arc-seconds per mm, gives, without any complicated calculations, an accuracy of 5–8 arc-seconds, which is comparable with the diameters of the faint stars in the atlas. The major limitation is then caused by the fact that in the edition of the POSS on photographic paper the east–west scale differs from the north–south scale by about 1 %. (It is anamorphic – distorted – because of the non-uniform texture of the paper.) Unfortunately, most amateurs require special permission to have access to the POSS.

All these stop-gap methods are very valuable when you have just discovered a nova, an Apollo minor planet, or a comet. You can then telegraph approximate coordinates to Brian Marsden, and these will at least allow him to alert subscribers to the astronomical telegrams. In less urgent cases, avoid reporting approximate coordinates, and reserve them for your own use. Brian Marsden complains of receiving far too many approximate determinations.

We can now turn to the methodical and accurate work required from anyone intending to specialize in plate measurement. Amateurs must be content with instruments that are a third or a fifth of the size of those used by professionals, and which are therefore a tenth or a hundredth of the cost. As a result, amateurs rarely have a focal-plane scale of less than 150 arc-seconds per mm, which is hardly enough for demanding astrometric measurements. The accuracy will not be a third or a fifth of that obtained by professionals, however. The latter are, in fact, limited by a whole series of factors, particularly the internal errors in catalogues (which we have already mentioned), but also by atmospheric effects caused by observatory sites and domes. A typical accuracy obtained by most professional installations is 0.3 arc-second for a fixed star of magnitude +9, or 0.5 arc-second for a moving body of the same brightness. Errors of more than 1 arc-second are still encountered with faint objects or with the diffuse comas of comets. Amateurs, using small photographs, seem to arrive at an accuracy of 1 arc-second at best. We have to admit that the cost-effectiveness ratio is definitely in favour of the latter!

It must be understood that in order to obtain accurate, reproducible results, amateurs must have good equipment and a certain financial commitment. They may have to use very expensive photographic plates, rather than film, to order them in batches, and to use them quite rapidly. It is important for photographs to be stable to better than 10 μm over an area some centimetres square, and this may only be attained with a rigid base. This accounts for the interest in flat-field telescopes when compared with Schmidts. The owner of a small Schmidt telescope also runs the risk of encountering some technical problems. Glass plates cannot be used because of the curvature of the focal plane. Circles must be cut from sheet film, and held in place by a vacuum in the classic manner. The modern, plastic film-base material, Estar, seems to have better dimensional stability that the old cellulose acetate material, but its properties for astrometry do not appear to have been fully investigated as yet.

A plate-measuring machine must also be obtained or constructed. Professional measuring machines are designed for 160×160 mm or 300×300 mm photographic plates and have precision guides some tens of centimetres long, precision screws with trapezoidal threads of the same length, and elaborate read-outs. Nowadays they are often motorized and controlled by a microcomputer, like the well-known P.D.S. machine from Perkin-Elmer, which is more of a microdensitometer, but which 'astrometrists' prefer to use because it has an accuracy of about 3 μm. Amateurs are not able to indulge in such luxuries and have to restrict themselves to movements of about 40 mm, consistent with the available field. Complete X-Y tables that are accurate to ± 3 μm may be purchased from a number of commercial sources – for a fairly large sum. Competent engineers could doubtless incorporate micrometer screws identical to the ones used in these tables, but bought separately, in a base of their own design. Electronics experts would fit stepping motors and encoders. For positioning stars to within a few microns, a microscope with a graticule and several magnifications is required, together with a suitable means of adjusting the illumination, which is all-important. It is essential to remember that one is often trying to measure details on the negative that are at the limits of visibility. The designers of professional instruments have not always thought of this, and the user is often irritated by this sort of 'minor detail' that has been forgotten. For those who do not want to attempt to rival large observatories in accuracy, and who have relatively modest financial means, it is quite possible to devise a system for obtaining good measurements by using an enlarger. More about this will be found in Chap. 17.

A final possibility, and the least costly of all, would be to obtain permission to use one of the old measuring machines that may be found in nearly every observatory, and which are hardly used by astronomers any more. It may occasionally be necessary to renovate one before it can be used. At the time of the international Carte du Ciel project astrometric activity was widespread. Astrographs 330 mm in diameter and with focal lengths of 3.4 m still exist, more or less in working order, at Toulouse, Paris, and Besançon in France. The ones at Bordeaux and Nice (the latter a different design) are still in use. All such 'old-fashioned' instruments could be refurbished and put to useful work if the will were there.

Improvised methods cannot be used in serious work on photographic reduction. Microcomputers are almost universally employed, but they have no need to be big. A well-tested program is required to calculate what are known as the plate constants (i.e., the precise orientation of the plate, its scale, etc.) by the standard least-squares method, from the X, Y coordinates of the ten catalogue stars. The positions of the stars [previously 1950.0, but now generally 2000.0 – Trans.] therefore have to be converted to coordinates of date and corrected for atmospheric refraction. We do not have the space here to go into all the details that would be covered by an astronomy course. Specialized textbooks may be found in observatory libraries. Practical advice has been published in the 'Astrometry' chapter of the *International Halley Watch manual* and in other works. In France, a publication devoted to amateur/professional liaison has also made available software written in Microsoft BASIC, which may be adapted for various machines, to observers who measured the position of Comet Halley. The program calculated 1950.0 astrometric coordinates for the object being examined, as well as the internal accuracy and all the residuals. Any incorrect reference stars were detected and discounted in the calculation, thus protecting the operator from gross accidental errors. [Note that coordinates for epoch 2000.0 would be used for positions from the beginning of 1992. – Trans.]

The rather dry material published in the MPC shows that a number of amateurs regularly make high-quality contributions to minor-planet astrometry. Apart from T. Seki, who has already been mentioned, there are some ten observers in the Shimizu area, 150 km from Tokyo, who collaborate closely with T. Urata. In the U.S.A., there is W. S. Penhallow, whose instruments are 200 km from New York. Unlike photometry or deep-sky photography, astrometry loses none of its accuracy when undertaken close to an urban centre. [In recent years, B. Manning (who has discovered several minor planets), D. Buczynski, and H. B. Ridley in the United Kingdom have also made a considerable number of high-quality astrometric measurements. – Trans.]

The most advanced level to which an amateur can aspire – one could say the most 'professional' – is that of helping the Minor Planet Center to improve orbits. This certainly requires at least a basic knowledge of celestial mechanics. If you have made a series of measurement of a particular object; if you have devised software for calculating orbits that is soundly based on celestial mechanics; and if you have transferred the data held on file at the Minor Planet Center to disk, there is nothing to stop you from suggesting improved elements that agree with both old and more recent observations. You will be taken all the more seriously if your values are truly accurate, i.e., if the Minor Planet Center confirms your results, and they are also upheld by subsequent observations. With a sophisticated personal computer it is also possible to plunge into numerical integrations that calculate the changes in orbital elements caused by planetary perturbations. This allows the identification of objects photographed a long time ago (sometimes as much as 50 years before), with newly detected objects. Such research cannot be undertaken fully automatically, so the investigator needs some of the qualities of a detective and also a certain amount of luck. Frequently the work amounts to a virtual second discovery. Such work requires close contacts with professional astronomers specializing in this field, because it would be a waste of time trying to develop effective and accurate methods all on one's own. In any case, we should perhaps emphasize again that if observers undertake this sort of activity, which is undeniably fascinating, they will find that, for lack of time, they tend to be diverted from regular, night-time observation. Professionals work in teams of technicians and researchers: some work during the night, others develop the plates during the morning, and still others make the reductions during the afternoon. An individual amateur may well feel isolated. Astrometry of minor planets or comets should perhaps be a group project, drawing together people with different skills.

7.5 Photometry of rotating minor planets

To end this chapter we will discuss a completely different way of studying minor planets, which is also suitable for amateurs, and which requires less specialized equipment than the studies just described. This is the investigation of minor planets' periodic variations in brightness.

These variations in brightness have major repercussions on the physical theory of minor planets because they are related to the shape of the bodies. We have already said that with the exception of three or four of the largest, which are essentially spheres, many minor planets appear to be tri-axial ellipsoids or, in other words, rather like a rugby (or American) football. However a rugby ball is thrown, it appears to turn in the air. In reality, the motion around its centre of gravity comprises a true rotation and a free oscillation (known as precession or nutation) of the axis of rotation. Study of the variations in brightness of minor planets has shown that, with rare exceptions (e.g. the small body 1220 Crocus), the free oscillation is non-existent. It has been eliminated by viscous damping over millions of years, so that all that remains, as with the planets, is a uniform rotation around the shortest of the three axes. (We know that the Earth also has a precessional motion with a period of 25 000 years, caused by the effects of the Moon and the Sun, but that is a forced oscillation created by a torque, which is a different physical effect.)

319

A minor planet therefore has an instantaneous rotation vector that is fixed in space: this points towards a stationary North Pole on the celestial sphere; its modulus is the angular velocity, which if we prefer we may express as the sidereal period. In a single rotation the area of the elliptical disk as seen from Earth varies, and goes through two maxima and two minima. The amount of sunlight reflected towards the observer varies in proportion to the apparent area, giving rise to the periodic variation in brightness. Naturally, shadows tend to complicate the matter when the phase angle is not zero. The practical problem that astronomers face is to determine the rotation vector for each minor planet, and the ratio between the lengths of the three axes.

Theoreticians studying the Solar System would like to know the statistical distribution of all the rotation vectors. They hope to be able to draw some conclusions from this about the history of the minor planets and in particular the significance of earlier mutual collisions. We shall see that there is a lot for observers to do before the theoreticians have enough data.

First, let us describe how the variations in brightness appear to the observers. We may assume that they have gained experience in observing variable stars, and now want to diversify.

We should not imagine that the three axes of most minor planets are almost equal and that, as a result, the variation in the surface area, and thus of the brightness, is difficult to detect. Quite the contrary applies, which is what makes things interesting. It is true that if by bad luck we observe a minor planet when its axis is pointing towards us, we shall see a disappointingly constant brightness, because the apparent area does not vary. But if we choose a date when the axis is approximately perpendicular to our line of sight, we will often find a periodic variation in the luminosity by several tenths of a magnitude. The amplitude of the light-curve of a minor planet therefore changes from one epoch to another. There are favourable oppositions and unfavourable ones. This is one important difference from most variable stars.

Table 7.5 gives a list of objects, down to a magnitude of V = +14.5 at aphelic opposition, with a maximum amplitude greater than 0.5 mag; i.e., objects that reach this amplitude at two points of their orbit where the rotational axis is favourably oriented.

Table 7.5 calls for several comments. First, the limiting magnitude of V = +14.5 at aphelic opposition was chosen to include objects that could be seen throughout their orbits by someone possessing a 300-mm telescope, and making use of *Atlas Stellarum*. We have therefore excluded some large-amplitude minor planets such as 288 Glauke and 622 Esther, which reach magnitude +12 at perihelic oppositions, but which fade below our limit at aphelic oppositions. The second point is that Table 7.5 is undoubtedly incomplete, because the information available to the author is by no means exhaustive. Results are widely scattered throughout the literature, and are sometimes to be found in journals that have a very restricted circulation. It should be said that a large number of minor planets, including some of the brightest ones, have still not been measured by the few teams of professionals that specialize in this sort of work. When this list was compiled in mid-1987, the periods of about 550 minor planets were known, which was very few when compared with

Table 7.5. *Minor planets with amplitudes greater than 0.5 mag. and magnitudes above +14.5 at aphelic oppositions*

Object	Maximum amplitude (mag.)	Period (hrs)	Mean mag. at perihelic & aphelic oppositions	Dia. (km)	Type	Notes
1. Main-Belt or Trojan minor planets:						
216 Kleopatra	1.4	5.39	9.2, 12.0	87	M	Table 7.3
624 Hektor	1.1	8.53	14.3, 14.5	221	D	Trojan, Table 7.1
63 Ausonia	0.95	9.30	9.7, 11.2	81	S	
753 Tiflis	0.8	9.84	11.5, 14.3	23	S	
201 Penelope	> 0.75	3.75	10.5, 12.6	62	M	
434 Hungaria	0.70	26.5	12.1, 13.8	10	E	Hungaria
182 Elsa	0.7	80	10.8, 12.9	37	S	
317 Roxane	0.67	8.17	11.4, 12.5	19	E	
13 Ariadne	0.66	5.75	9.1, 11.3	79	S	Table 7.3
792 Metcalfia	0.62	9.18	12.9, 14.4	–	–	
250 Bettina	> 0.60	5.11	11.3, 12.6	37	M	
82 Alkmene	0.55	13.00	10.4, 12.9	59	S	
39 Laetitia	0.54	5.14	9.3, 10.6	159	S	
15 Eunomia	0.53	6.08	7.9, 9.8	262	S	Tables 7.1 and 7.3
11 Hysa	0.52	6.42	9.0, 10.7	70	E	Table 7.3
1245 Calvinia	0.52	4.8	13.4, 14.3	35	S	
152 Atalia	0.50	5.28	12.5, 13.3	54	S	
694 Ekard	0.50	5.93	10.9, 14.4	99	CP	
186 Celuta	> 0.5	> 12	11.0, 12.8	49	S	
675 Ludmilla	0.5	7.75	10.7, 13.0	–	S	
116 Sirona	0.5	13.7	10.7, 12.2	77	S	
344 Desiderata	0.5	10.7	9.6, 13.2	134	C	
2. Aten–Apollo–Amor objects:						
1620 Geographos	2.0	5.23				Apollo
1865 Cerberus	1.6	6.8				Apollo
433 Eros	1.5	5.27				Amor: 40 × 15 × 15 km
1981 QA	1.2	148				Amor
3288 Seleucus	1.0	75				Amor
1985 PA	1.0	190 ?				Apollo
1864 Daedalus	0.85	8.6				Apollo
2368 Beltrovata	0.85	5.9				Amor
1685 Toro	0.8	10.20				Apollo
1978 CA	0.8	3.76				Apollo
1862 Apollo	0.6	3.06				Apollo
1580 Betulia	0.5	6.1				Amor
2608 Seneca	0.5	8.0				Amor

the 3330 objects catalogued in the EMP. The number of known light-curves is currently increasing by about 18 % per year, however, which again goes to show the increase in interest in these objects around the world. The list in Table 7.5 should be substantially increased in the near future. Nevertheless, most of the 550 objects investigated are far from having been properly followed all round their orbits, so it is quite possible that some of the amplitudes given in the table have been considerably under-estimated. For example, the amplitude '> 0.60' indicated for 250 Bettina means that this minor planet should doubtless be higher in the list. The amplitude of 0.5 mag, chosen here as a limit, is also a rather severe limitation. A variation of 0.5 magnitude is very easy to detect visually or by photography. If the limit had been set at 0.3 mag, Table 7.5 would have been four times as long. Undoubtedly the vast majority of minor planets have maximum amplitudes that are detectable by photoelectric photometry, reaching (say) at least 0.05 mag. In other words, not many of the minor planets are spherical bodies.

From Table 7.5 we see that there are few C-type and similar minor planets. This reveals a real tendency for carbonaceous minor planets to be less elongated than their type S, M or E counterparts. The exceptionally large amplitudes of 216 Kleopatra, of the 'chief' Trojan, 624 Hektor, and perhaps also 63 Ausonia should be noted. Some astronomers think that these particular minor planets are double, i.e., that they consist of two ellipsoidal bodies that always turn the same face towards one another. It rests with observers of stellar occultations to tell us whether this is true. Amateurs who study variable stars know that, for geometrical reasons, eclipsing, contact binary stars (the W-UMa stars, class EW) cannot have amplitudes exceeding 0.75 mag. On the other hand, we now know that Pluto is double, yet its amplitude is not much more than 0.3 mag. So the three objects mentioned are not easy to explain, and continue to excite our curiosity.

Table 7.5 also shows that many of the A–A–A or 'Earth-grazing' minor planets have very large amplitudes, considerably larger than ordinary minor planets. This table includes most of the A–A–A objects that have been measured with photometers. The A–A–A objects are certainly small, elongated objects that are very irregular in shape. 1620 Geographos, which holds the record for amplitude at present, seems to be a lump of rock that is cigar-shaped: its length of 10 km is six times its width. To those who are tempted by the idea of following the variations in brightness of some of these 'Earth-grazers' as they brush past the Earth, we would point out that the rapid change in the line of sight, and the alteration in the conditions of illumination and phase, cause both the shape and the amplitude of the curves, and even the apparent period, to change from day to day. Artificial-satellite observers are used to such changes, but on an even more rapid time-scale. Figure 7.11 shows a fine light-curve for 433 Eros that was recorded visually by the GEOS variable-star group. But some A–A–A objects have periods of rotation that are even shorter than that of Eros. 1984 KD, which is not on the list because its amplitude does not exceed 0.26 mag., rotates in 1.97 hours. In 1984, 3200 Phaeton went from minimum to maximum, an amplitude of nearly 0.4 mag., in about 20 minutes. Unfortunately, because of insufficient observations, it was not possible to determine its period.

We need to look in greater detail at the important question of the periods of minor planets and the role that amateurs may, for some years, still continue to

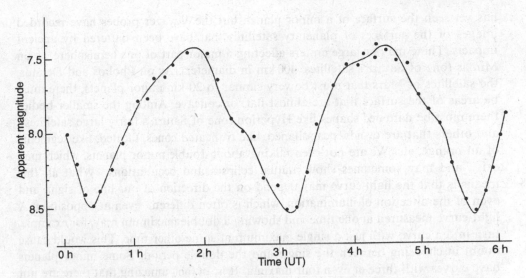

Fig. 7.11. *Visual light-curve of 433 Eros obtained by observers of the European group GEOS, during the minor planet's close approach (0.15 AU) to the Earth in 1985 January. The minor planet's mean magnitude was +8.0, and it was moving at 0.05 degree per hour against the sky.*

play in accurately determining them. A large minor planet that is only bound together by its internal cohesion is unable to rotate as fast as a tiny object such as 1984 KD without being disrupted by centrifugal forces. It has been calculated that a rapid rotator like 201 Penelope (with a period of $3^h 45^m$) is almost on the point of disintegrating. The situation is even more critical for 321 Florentina, a minor planet that is only a third of the size of Penelope, and which rotates in just 2 hours 52 minutes. Having said that, it is noticeable that periods between 5 and 12 hours are by far the most frequent among large- and medium-sized minor planets, indeed they form 60% of all the known cases. Neither are periods between 12 and 21 hours rare, and they form another 25% of the cases. Finally, there is an extended tail to the distribution with very long periods significantly longer than 1 day. One example may be mentioned, 182 Elsa, with a period of about 3.5 days, while a completely exceptional case is 288 Glauke, which has a period of one-and-a-half to two months!

Despite the hundreds of well-established periods, known to a high degree of accuracy, there still remain many periods among the larger minor planets given in Tables 7.2 and 7.3 that are poorly measured and of doubtful accuracy.

Observers of short-period variable stars such as the dwarf Cepheids or contact eclipsing binaries will not be surprised to learn that minor-planet periods that are longer than one night, and which bear some simple relationship to 24 hours, are very often poorly known or subject to considerable error. But other phenomena further complicate the issue. A minor planet is never a perfect tri-axial ellipsoid that would give, as we have seen, a light-curve consisting of two identical halves, with identical maxima and minima. In many cases there are significant departures (of more than 10 km) from the ellipsoid that best represents the surface. No one has yet seen the surface of a minor planet, but the Voyager probes have recorded images of the surfaces of planetary satellites that have been battered by ancient impacts. There may be large craters affecting a major part of one hemisphere, as on Mimas (one of Saturn's satellites, 400 km in diameter). As on Phobos and Deimos, the satellites of Mars that must be very similar to 20-km minor planets, there must be areas of the surface that are almost flat, or concave. Among the smaller bodies there must be flattened shapes like Hyperion (one of Saturn's fairly large satellites), and others that are ovoids, pear-shaped, like truncated cones, faceted, like segments of an orange, etc. We are not even talking about double minor planets, which may exist, and may sometimes show mutual eclipses and occultations. What all this means is that the light-curve may depend on the direction of our line of sight, and even of the direction of illumination, which is often different, even at opposition. A light-curve measured at one time and showing a double maximum may, for example, turn into a curve with just a single maximum at some other time. This would cause doubt in choosing between the single and the double period. Some minor planets have curves with three or even four maxima. It is, in fact, amazing that there are not more ambiguous cases, and that the simple ellipsoidal model often gives satisfactory results. It is worth noting that a team of Italian astronomers is about to publish an atlas of light-curves, which will undoubtedly be very valuable.

Now let us turn to practical observation. Several of the objects in Table 7.5 may be followed visually to gain experience. As may be seen, 15 Eunomia is probably the only minor planet whose light-curve may be followed comfortably with binoculars. 216 Kleopatra, with a period of 5.4 hours should be spectacular in a small telescope. With a 250-mm telescope, 624 Hektor should be visible on certain nights (V = 13.9) and invisible on others (V = 15.0). Note that three periods equal 25.6 hours, so one should be able to detect it at maximum again without too much difficulty.

But visual observation is not recommended for scientific work except when a bright, unknown, Apollo object appears, suddenly heralded by telegrams and copies of the IAUC. (Try keeping a magnitude-13 object that is rushing past at 3600 arc-seconds per hour in the centre of a 40 arc-second photometer diaphragm, and you will begin to appreciate the advantages of the human eye.) Although the period of 433 Eros was already known in 1975, the GEOS group would have been able to establish its length to within 1 % (Fig. 7.11).

Wide-field photography with a small Schmidt telescope could be of great service by detecting, a month or two before opposition, new minor planets that have never been measured by anyone, and whose earlier oppositions were either not perihelic or too far South for proper observation. Measurements may perhaps have been made on one or two occasions when the rotational axis was not favourably placed. When professionals decide to obtain light-curves of faint minor planets, they generally determine the dates of their observing sessions by taking a few objects given in the EMP that are very close to opposition. They then make their measurements without knowing in advance what they are likely to see. Note that, in general, photometric workers choose minor planets that are close to opposition, regardless of which objects they are, because this is the only way of minimizing the problematic phase effect, and also so that they have the longest possible observation-times every night.

It would not be difficult to detect the ten or so minor planets between magnitudes

13 and 15 that are listed in the EMP, when they are beginning their retrograde motion, by taking a few Schmidt-telescope photographs, spaced 45–90 minutes apart, of a field centred a few degrees to the East of the antisolar point. The magnitude differences between each of these minor planets and several nearby reference stars could then be measured. Because these would be differential measurements, it would not be necessary to worry about the problem of absolute calibration, which is always difficult, or about atmospheric absorption. By this method it should be possible to detect the two or three objects that show the largest variations in brightness, and to prompt photometrists to concentrate their efforts on these around opposition two or three lunations later. In some cases, it may even be possible to make a preliminary estimate of the amplitude of the variations and of the period. The Swedish astronomer C. I. Lagerkvist has described tests made in 1973 with the large (1-m corrector plate) Schmidt telescope at Kvistaberg near Stockholm, using 103a-G plates, and using an iris photometer to measure stars on the plates. Out of a total of 20 objects measured, Lagerkvist was the first to detect a faint minor planet with an amplitude of 0.7 mag. 1789 Dobrovolsky. Extrapolating these results to a small Schmidt telescope with a 300-mm corrector plate and modern TP 2415 film, we might expect minor planets to be measurable to within 0.1 mag. down to about magnitude $V = 15.0$. A small iris photometer is certainly a tool that amateurs can build for themselves.

A group of amateurs with a Cassegrain or Newtonian telescope of about 400-mm diameter, on a site away from towns and with a good photoelectric photometer, would be able to get started with measuring minor-planet light-curves. The *Minor Planet Bulletin* (MPB) gives a lot of information about this, as well as about minor planets that are interesting each particular season. Observers will find that the MPB is essential reading.

One can hope to attain magnitude $V = +12.5$ with a reasonable signal-to-noise ratio if the photometer is of the photon-counting type, and one has good clear skies. In a few years it will possibly be too late to do useful work measuring 12th-magnitude minor planets, but at the moment this is still possible, and work to improve the many poorly known periods is strongly recommended. We would not advise this sort of work for individuals, but rather for a group of two or three observers who can observe as a team, and who can, therefore, take turns at the eyepiece and at the recording equipment. This type of observation is quite exhausting. It means spending 6–8 hours making repeated measurements, with an approximately 5-minute cycle, of the minor planet, the sky background, two comparison stars in the same field as the minor planet, the sky background, etc. The telescope is constantly moving from one object to another, and it has to be centred repeatedly. Sometimes one is measuring not one minor planet but two or three that are close to one another in the sky. Sometimes all the measurements are repeated in two colours, B and V. Several times during the night the sequence has to be interrupted to measure a standard star, which allows the photometry to be linked to the Johnson scale and also checks the stability of the sky conditions. Less emphasis is placed on finding a straight-line relationship for atmospheric extinction (Bouguer's law) than in classic stellar photometry, because we are dealing with differential measurements between objects within a small field. However, increased attention should be given to it when, as sometimes happens, measurements are continued down to low altitudes in an attempt to observe the minor planet's maximum. The maximum must not be found to occur at an incorrect time because of unexpected atmospheric effects. One precaution that promotes accuracy is to choose, at the beginning of the session, two stars close to the minor planet that have a (B − V) index of between +0.6 and +0.9. This will ensure that extraneous effects caused by a difference of colour do not occur as the air mass increases (*see* Chap. 19, 'Photoelectric photometry').

It must be admitted that this type of investigation requires considerable motivation and persistence. Professional astronomers are often helped by having a photometer controlled by a microprocessor, which automatically handles the long sequences of repetitive measurements. But the automatic photometers that they use, and which have generally been designed by stellar astronomers, are not capable of following an object that is moving at a uniform rate with respect to fixed reference stars. This is why continual intervention by the astronomer is required. The ideal photometer would probably include fibre-optic [or fluid-optic – Trans.] guides transmitting the light to four photomultipliers. Three light-guides would be adjusted manually at the beginning of the session to cover two field stars and an area of the background sky. The fourth would be held by a moveable carrier, driven at a uniform rate by a motor and lead-screw to follow the minor-planet's apparent motion.

In the fairly near future, a few groups of amateurs are about to begin stellar photometry without diaphragms, using CCD video cameras. A simple, 8-bit microcomputer of the Apple II sort is quite capable of handling the files created by a mosaic of 100×100 sensitive elements (i.e., 10 000 pixels), whereas an IBM-compatible with a disk drive can handle images consisting of 400×400 points (160 000 pixels). The camera itself is smaller and lighter than any other photometer or camera body. The main difficulty (apart from computing problems) is that the refrigeration of the camera to below −40°C is essential. The advantage of this method over photoelectric photometry is that the software can subtract the exact amount of sky background, because it is being measured over thousands of individual points. With exposures of about 2 minutes on an amateur telescope of 400 mm in diameter, it should be possible to obtain accurate measurements of 15th-magnitude minor planets, which conventional photometry could only attain with a professional, 1-m telescope. So there is no need for those who are fascinated by the problem of minor-planet rotations to fear that before long there will be nothing for them to study.

What can amateurs do with the good light-curves that they obtain, like that shown in Fig. 7.12? Certainly they can be published in the Minor Planet Bulletin, because professionals specializing in this field read that journal and will not fail to extract any useful information from it. Observers should also consider joining the Asteroids and Remote Planets Section of the British Astronomical Association.

Can amateurs analyse their own data mathematically, like the professionals, to derive the orientation of the pole and the dimensions of the ellipsoid? In principle, yes, because the mathematical methods are far from being too esoteric. In a recent GEOS publication devoted to amateur photoelectric photometry, for example, G. Balmino has described one of the methods that could be used. There does exist a practical problem, however – quite apart from the difficulty of finding

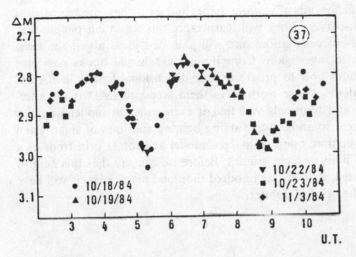

Fig. 7.12. *The composite light-curve of 37 Fides measured over five nights in 1984 with a 280-mm reflector by K. W. Ziegler and W. B. Florence (Gila Observatory, Arizona). Reproduced by permission of the Minor Planet Bulletin, after MPB* **12**, *(3), 1985. The photometer was an uncooled silicon photo-diode, and 37 Fides was then varying between V = +9.5 and V = +10.0. Comparable results could doubtless be obtained down to magnitude V = +12 with a 400-mm telescope and a photon-counting, cooled, photoelectric photometer.*

all the necessary references, which are likely to be widely scattered throughout the literature – and this is having all the measurements that are required to begin calculating a solution. According to the best estimates made from the literature, only forty-odd minor planets have been sufficiently well observed all round their orbits for the direction of their poles to have been accurately calculated. This is a very small number, showing that even with the current 18 % increase in observational activity each year, there is still, from the specialists' points of view, a serious lack of light-curves. That being said, all those specialists have unreduced light-curves that they obtained in earlier years, and which they hope to complete soon with one or two additional observations. (They also hope that none of their distinguished colleagues will beat them to it.) So if you, as an amateur, provide good, unreduced observations to a professional, you may enable them to publish a new pole of rotation, for which you will undoubtedly be given due credit. But if you want to determine a pole of rotation all on your own, you will certainly require far more data than the best-placed professional astronomer, and certainly he or she will not go out of their way to offer you their data. So perhaps what is needed is a public data bank for physical data on minor planets, somewhat like the information held by the Minor Planet Center. The overall efficiency would be greatly increased. Perhaps this should be suggested to the Centre de Données Stellaires [Stellar Data Centre] at Strasbourg Observatory.

Despite these organisational and personal aspects, undoubtedly amateurs that have microcomputers with graphic screens will find a lot of fascination in numerically simulating the rotation of ellipsoids of varying shapes and seeing how the geometrical effects influence the light-curves. They will learn more this way than people who stick to classical methods of calculation, and will gain new ideas about the most promising minor planets to investigate. Computing journals and books now give plenty of information about how to program three-dimensional figures in BASIC and PASCAL. Any amateurs whose work gives them access to CAD (Computer-Aided Design) software and terminals will find it even easier to model irregular minor planets and set them rotating with varying lighting and lines of sight. Even without a numerical simulation, one can make a model and rotate it in front of a photometer. There is no limit to these studies. Before we can say that this field of research has been exhausted, at least one hundred thousand minor planets will have been discovered within the orbit of Jupiter.

8 Comets

J. C. Merlin

8.1 Introduction

It may seem surprising that amateurs still have a role to play in the observation of comets. Pessimists might even assume that the close approach of the European space-probe Giotto to within 600 km of the nucleus of Halley's Comet in 1986 March sounded the death knell of cometary observation from the ground. But Earth-based observatories remain of major importance in most fields of cometary work. Launching a space-probe costs far more than the construction of a large telescope, and observational programmes carried out from ground level cover a much longer span of time. In addition, the large Schmidt telescopes now in use on all five continents produce a rich crop of discoveries, and photograph new comets every month.

The true situation is far more encouraging that one might imagine from a simple overall view. In parallel with the development of professional techniques, the equipment used by amateurs has also evolved, both as regards size and performance for a given aperture. We can assume that this evolution will continue over the next few decades in ways that are still unforeseen. It has indeed been said that amateurs observed Comet Halley in 1986 with instruments comparable with those of professionals in 1910. There are two corollaries to this inevitable trend. First, amateurs are tending to specialize in areas that have been relinquished by professionals, and are thus becoming fully-fledged specialists. Second, the organisations to which amateurs submit their raw observations are becoming more and more exacting in the quality of data that they will accept, which tends to encourage higher standards among the amateurs.

Specializing in a specific, limited field of observation is, for an amateur, the sign of a considerable change in attitude. Devoting one's observing time to a single category of object is always fruitful. Preparations for observing go like clockwork, time at the telescope is fully occupied, and the results obtained are of equal quality. The experience acquired over the years leads to such amateur observers being regarded as sound and warranting attention when unique observations, such as the discovery of a comet, for example, are made. Such credibility means that they are regarded as equals by professionals, such as radio astronomers, who might require specific, vital information about the position, magnitude and activity of new comets before turning their radio telescopes onto them. An error of just a few minutes of arc in the predicted position, or a fainter magnitude than expected, and hours of signal integration might be wasted.

Major national observatories do not employ people who specialize in estimating magnitudes or making observations at the eyepiece – even if large telescopes still have such accessories. Over the years, the preoccupations of professional astronomers

Fig. 8.1. *Comet Arend–Roland 1957 III on 1957 April 25. This photograph shows the antitail that appeared when the Earth intersected the comet's orbital plane. Lick Observatory photo.*

have turned more and more towards astrophysics, and several aspects of research are only covered by amateurs. Financial restrictions have also led to the closure of some sites or equipment that were devoted to Solar-System studies (systematic study, astrometry, monitoring, etc.). For some decades to come the field will remain open to amateurs.

8.2 Comet observation now and in the past

The study of comets took on its modern form only about thirty years ago. Two major events, which were more in the way of working hypotheses than discoveries, seem to have triggered developments:

- Jan Oort's theory about the existence of a reservoir of cometary nuclei at the edge of the Solar System (between 10 000 and 100 000 AU from the Sun);
- Fred Whipple's 'dirty snowball' model.

These two theories embody the main questions that concern observers: the origin, movement and evolution of comets. Amateur observers find this reflected in their preferred fields of work, whether those are searching for new comets, the study of their motion by astrometry, or classical observations (magnitude estimates, drawings, etc.).

Until the 17th and 18th centuries, astronomers were mainly preoccupied with the motion of comets. Their apparent paths, which took them across constellations far from the ecliptic, where planets never appeared, and even sometimes as far as the polar regions, remained an enigma for a long time. By observing the comet of 1577, Tycho Brahe established that comets were way beyond the Moon and were therefore not atmospheric phenomena, as was generally believed at the time. Kepler thought that comets followed straight paths across the Solar System. It was Isaac Newton, the discoverer of the law of universal gravitation, who, at the end of the 17th century, finally linked the motion of comets with the Sun. He determined that most of the observed comets followed very elongated, almost parabolic, orbits, with the Sun at one focus. But posterity has remembered the name of Halley above all others. Friend and contemporary of Newton, Edmond Halley applied the ideas developed by Newton to a score of comets that had been well-observed in previous centuries. He assumed that they followed parabolic orbits – as do modern workers when they begin to calculate the orbital elements (Fig. 8.2) of a newly-discovered comet. We know the rest. Halley discovered that several sets of orbital elements (inclination, argument of perihelion, longitude of the ascending node, etc.) belonging to comets observed at approximately 75-year intervals were essentially the same. He deduced that several passages of the same object were involved, and predicted that it would next return in 1758–9. Sixteen years after Halley's death, a German amateur astronomer, Johann Palitzsch, discovered the expected comet at Christmas 1758, guided by calculations made by Halley and his successors. The first periodic comet had been found. The international scientific community recognized Halley's achievement and decided to give his name to the comet. We now call it P/Halley (the 'P/' indicating that we are dealing with a periodic comet with a period of less

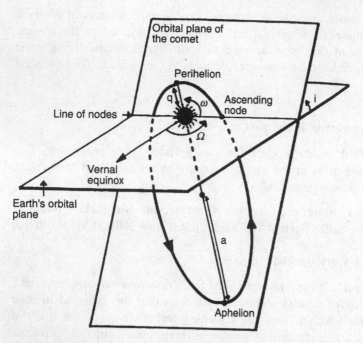

Fig. 8.2. *Definition of the orbital elements of a comet. The orbit of any body (planet, minor planet, or comet) orbiting the Sun is defined by the following parameters:*
q = Sun–comet distance at the time of perihelion
e = eccentricity; e = 1 for parabolic orbits. For an ellipse (as in the diagram), the distance q and the eccentricity e are linked by:q = a (1 − e) where a is the semi-major axis of the orbit. The aphelion distance Q is equal to a (1 + e).
i = inclination of the orbit with respect to the plane of the Earth's orbit (the ecliptic), measured from 0° to 180°. If i is greater than 90°, the orbit is retrograde.
Ω = longitude of the ascending node, measured from 0° to 360° from the vernal equinox (the origin of both equatorial and ecliptic coordinate systems).
ω = argument of perihelion, measured from 0° to 360° from the ascending node.
(From Halley, le roman des comètes, Anny-Chantal Levasseur-Regourd and Philippe de La Cotardière, 1985, Editions Denoël, by kind permission of the authors.)

than 200 years). By way of example, the first five periodic comets for which the returns were predicted and observed were: Halley, Encke, Beila, Faye and d'Arrest. Modern methods of calculation allow the orbits of new comets to be determined very quickly: whether they are short-period ellipses or not. For this we need to have a certain number of positions of astrometric quality (i.e., with a precision of about 1 arc-second) and, in principle, spread over a span of a few weeks. As Newton realized, close to the Sun, highly elongated elliptical orbits cannot, in practice, be distinguished from parabolic orbits. It is here that comets are brightest and thus most easily observed.

Examination of the orbital elements of comets show that they all originate within the Solar System. *No comet of interstellar origin has ever been seen.* The orbits that are found fall into three categories:

- elliptical: short-period orbits, i.e., those indicated by 'P/', when $P \leq 200$ years; and long-period, $P > 200$ years, when they can be distinguished from parabolic orbits. The periods may be as much as 10–15 million years. (The comet having the orbit with the least eccentricity – in other words, the closest to a circle – is P/Schwassmann-Wachmann 1, where $e = 0.05$ at present.);
- parabolic: comets for which it has not proved possible to determine the orbital elements with sufficient accuracy or certainty, generally for lack of astrometric positions. For calculation purposes their eccentricity is assumed to be unity. It is important to realise that it is not easy to determine the orbit of a comet accurately when the period is over 10 000 years, and when it is only observed for a year or two around perihelion – or even for just a few months;
- hyperbolic: comets that initially follow parabolic or quasi-parabolic paths ($e \leq 1$) and which are ejected from the Solar System after having been strongly perturbed by a close encounter with a planet, particularly with Jupiter or Saturn.

Perturbations caused by the planets, especially Jupiter, have a considerable effect on the number of periodic comets that are observed. For convenience, we shall call orbits that are not classified with 'P/', 'long-period', whether they are elliptical or parabolic. We shall call those designated 'P/', 'short-period' or simply 'periodic'.

It is interesting to recall the theory developed by Jan Oort around 1950. This Dutch astronomer noted, when studying the motions of some fifty-odd very-long-period comets, that the aphelia (the most distant points from the Sun) generally occurred in a shell, centred on the Sun, and lying between 10 000 and 100 000 AU, with the strongest concentration between 40 000 and 50 000 AU. He deduced from this that there must be an enormous reservoir of comets at the edge of the Solar System, and that this must be 'storing' a fantastic number (10^{10}– 10^{12}) of cometary nuclei. They would be remnants from the formation of the Solar System. For various reasons, for example because of the gravitational effect of a passing star some light-years away from the Solar System, inert nuclei would be ejected from the reservoir by perturbations. Some would fall inwards towards the Sun, towards the small region of space that contains the planets. The most distant of these is, of course, Pluto, and its distance from the Sun is only 30 AU. This is very small when compared with the distance separating us from the Oort Cloud.

Comets that approach the Sun for the first time are known as new comets (in the Oort sense). They are generally the finest comets that we are lucky enough to see. On average, three or four new comets are discovered each year. It is among these unforeseen comets that amateurs have a chance of making their names. Between 40 and 50 % of long-period comets are discovered visually by amateur astronomers.

Unlike most short-period comets, which are trapped close to the ecliptic (or at least near the main plane of the planetary orbits), long-period comets have orbits with every possible inclination between 0° and 180° to the ecliptic. In other words, both direct and retrograde orbits occur. In addition, the distribution of inclinations is practically isotropic, without any marked concentration. This shows that the Oort Cloud is probably homogeneous and spherical in shape.

Analysis of the oldest systematic records, in particular those left by Chinese astronomers, does not show any perceptible change in the number of comets observed each year. It is generally considered that ancient records are comprehensive in their coverage of comets down to about magnitude 2. The increase in the number of comets observed has mainly been caused by searches being carried out to lower magnitudes, largely thanks to the introduction of wide-field photographic telescopes (Schmidt and Maksutov types in particular) about 50 years ago.

8.3 The observation of known comets

Although the discovery of a new comet is generally the outcome of hundreds of hours of careful searching, announcement of such an event implies that the observer has obeyed certain criteria in other, earlier, less spectacular observations. Without, for the moment, stressing the need for more general observational experience, we consider that it is very important for observers to have had a certain amount of practice in classical cometary observation. What credibility can one give to announcements by observers who think they have discovered new comets, if they fail to give: the position of a suspected object with adequate accuracy; the direction of motion so that others can pick it up; or an estimate of its brightness so that one can judge what size of telescope is required to see it? The IAU Central Bureau of Telegrams regularly receives false alarms that often come from observers who are acting in good faith, but who have not followed certain elementary principles. On the other hand, it is by no means a rarity for unknown observers to announce a valid discovery, but to do so in either an incomplete manner or late. The object discovered is not officially recognised because the name of the discoverer bears no weight. Time goes by, and the object may become unobservable because of the Full Moon, or its small elongation from the Sun, or it may even disappear before other observers can recover it and confirm its existence.

Each year has its tally of new or expected comets (the latter being returns of periodic comets). Depending on circumstances, between two and six comets reach or exceed magnitude 10 each year, and are therefore accessible with even the smallest amateur instruments. There is certainly no lack of subjects on which to practice. Use of instruments of between 150 and 200 mm in size considerably increases the number of comets available and allows a much wider range of brightness and types to be covered. A three-year observational programme will show the variety of cometary phenomena that are within reach of one's equipment. We might indeed be lucky enough to be able to pick up a very bright comet, or one that comes very close to the Earth, without any instrumental aid whatsoever.

8.3.1 *Cometary nomenclature*

Comets are the only celestial objects that systematically receive the names of the astronomers, professional or amateur, who discover them. A comet's designation includes, at most, three names, which are those of the first three persons to discover it, in chronological order if the discoveries were independent. At the peak of the Japanese amateurs' activity in comet searching, five observers independently discovered the same comet within an hour of each other; only the first three were honoured in the object's official designation. Observers are sometimes late in identifying their discoveries or in notifying the details. The first discoverer is then the person who first announced the discovery. After the discovery has been confirmed and the Central Bureau for Telegrams has given the object an official designation, it is too late to argue. The IAU will not subsequently change an object's name.

Some comets are named after astronomers who lived at different times. This is the case for example with Comet Denning–Fujikawa. This was discovered in 1881 October by W. Denning, but was then lost for a century, despite returning every 9 years. It was accidentally recovered by the Japanese amateur S. Fujikawa in 1978 October. His name comes second. There are numerous other examples: P/Pons–Brooks, P/Pons–Winnecke, P/Swift–Gehrels, P/Tuttle–Giacobini–Kresák and, more recently, P/du Toit–Hartley and P/Peters–Hartley. As will be seen, all these comets are short-period ones ($P \leq 200$ years). There are also cases, like that of P/Halley, where the comet is named after the astronomer who calculated its orbital parameters and identified earlier passages of the same comet. These are P/Lexell, P/Encke and P/Crommelin. P/Lexell ($P = 5.6$ years) was discovered by Charles Messier in 1770 and was then lost after being strongly perturbed by Jupiter, according to the calculations of Anders Lexell in 1779. P/Encke was rediscovered on several occasions – because of its very short period (the shortest known: 3.3 years) – by Pierre Méchain in 1786, Caroline Herschel in 1795, Jean-Louis Pons, Huth and Bouvard in 1805, and then Pons again in 1818. Johann Encke showed that these were four returns of the same comet. P/Crommelin ($P = 28$ years) was discovered by Pons in 1818, by Coggia and Winnecke in 1873, and by the South-African amateur Forbes in 1928. The three apparitions were linked by Andrew Crommelin in 1928.

Over a period of more than 200 years, ever since systematic study has been carried out with instruments rather than with the naked eye, more than 600 comets have been observed and, in the case of known short-period comets, re-observed. For example, Comet P/Encke's 54th perihelion passage was observed in 1987. When a new comet is discovered and eventually confirmed, or when a periodic comet is recovered, the IAU Central Bureau gives it a provisional designation, which consists of the current year followed by a letter of the alphabet, from a to z, in chronological order of discovery or recovery. Comet Wilson, discovered at Mount Palomar by Christine Wilson on 1986 August 5, was the twelfth comet in 1986, and therefore received the provisional designation 1986l. A week later, on 1986 August 12, Comet P/Grigg–Skjellerup, which was expected at its fifteenth return, was recovered at magnitude 22 (obviously not by accident) by K. Birkle using the 3.5-m telescope at Calar Alto (the Spanish–German observatory in Spain). P/Grigg–Skjellerup was given the provisional designation 1986m. Neither comet passed perihelion in 1987.

[In some years, as in 1987, more than 26 comets are discovered. The letters are then repeated with a subscript '1', e.g. Comet Shoemaker = $1987g_1$. – Trans.]

After a delay of about 2 years, which is basically the time required for any discoveries that are going to be made to actually occur, the IAU gives a definitive designation to all the comets in order of their perihelion passages, at least for those comets that were sufficiently well-observed for their orbital elements to be known with sufficient accuracy. Some comets have not received final designations, because it was not possible to calculate orbital elements that agreed with the available astrometric positions, which were either too few or subject to too much error. This was the case with Comet Kowal, 1979h, for example, where we only have three astrometric positions between 1979 July 24 and July 27. However, it should be said that this was a very faint object (magnitude 19) and it could not be detected from any observatory other than Mount Palomar, where it was discovered.

A comet's final designation consists of the year of perihelion passage followed by a number (in Roman numerals) giving the order in which it passed perihelion. The comet discovered by E. Bowell on 1980 February 11 was given the provisional designation 1980b, the second comet in 1980. It only passed perihelion in 1982 March, and is therefore now known as 1982 I, the first comet to pass perihelion in 1982. Exceptions do, of course, occur, mainly as a result of late examination of plates. Some comets are even detected years after the plates were taken. It may also happen that a faint, strongly condensed, comet with an almost stellar appearance, may be confused with a minor planet. Its weak nebulosity is only identified after detailed analysis. There have been several examples of this sort in recent years, in particular P/Skiff–Kosai, which was detected in 1986 on plates taken at Palomar in 1977. For some years we have also seen delayed discoveries from photographs taken for artificial-satellite tracking. In all these cases, comets discovered 'late' are given the first free number available for their year of perihelion passage.

It should be noted that an individual astronomer often discovers several periodic comets. The discoverer's name is then followed by a number in order to distinguish one from another or from subsequent returns. We have, for example, P/Wild 1, P/Wild 2, P/Schwassmann-Wachmann 1, P/Schwassmann-Wachmann 2, etc. Non-periodic comets ($P > 200$ years) discovered by the same individual or group, are not known by anything other than their final designation, for example 1959 IV, 1959 VI, 1963 III and 1965 IX, which are all Comet Alcock. (There is hardly any risk of confusion at the next perihelion passages – in a few million years time!) As far as possible, it is strongly recommended that, whatever the comet, the final designation is used in preference to the provisional one.

8.3.2 *True and false nuclei*

The heart of a comet, where all the activity occurs, and from which the material that gives rise to the observable phenomena is ejected, is the nucleus, an extremely small body. The average size of a cometary nucleus is a few kilometres. Making certain assumptions about the nature of the surface, infrared observations allow estimates of the size to be made. The diameter of the smallest cometary nuclei is only a few hundred metres, for example that of Comet Sugano–Saigusa–Fujikawa, 1983 V, is thought to be about 200 m. At the other extreme, there are rare objects, like the nucleus of P/Schwassmann–Wachmann, the 'diameter' of which is doubtless larger than 50 km. It is possible that some distant objects, beyond the orbit of Saturn, and currently catalogued as minor planets, may actually be inactive comets: the minor planet 2060 Chiron, for example, the diameter of which is certainly more than 300 km. [The observation of a coma has, of course, now confirmed that Chiron is a comet. – Trans.]

The only proper measurement of the size of a cometary nucleus is that obtained by the Giotto space-probe, which transmitted images of the nucleus of Comet P/Halley in 1986 March. The nucleus of this outstanding comet is a very dark, oblong object, measuring 15 km in its greatest dimension and about 8 km along the two other, perpendicular axes.

Nuclei consist of a mixture of frozen gases and dust of varying porosity and homogeneity. It is possible that some of them contain a rocky centre that resembles carbonaceous chondrite meteorites, but this is by no means certain, especially now that we know that the average density of the nucleus of P/Halley is less than 0.4. The fact that some minor planets follow orbits that resemble those of comets, with high eccentricity or inclinations, or are associated with meteor showers, suggests that they may be old cometary nuclei that have lost their volatile materials after hundreds of perihelion passages. But this is far from being definite. Similarities between orbital elements are often fortuitous. When a more thorough analysis is made, it is found that the orbits of minor planets are far more stable than those of comets. There is still considerable controversy about this among specialists.

Cometary nuclei are unobservable from Earth. When a comet passes fairly close to the Earth, as happened with Comet IRAS–Araki–Alcock in 1983, we might expect that the angular size of the nucleus would be sufficient for it to be resolved using interferometric techniques with large telescopes. If we assume, for example, that the nucleus of Comet IRAS–Araki–Alcock is 1 km in diameter, its apparent size was of the order of 0.05 arc-second in 1983 May when it was at perigee, 4.5 million km from Earth. Unfortunately, proximity to Earth also involves proximity to the Sun, which means that a comet's nucleus is generally hidden by a very dense, bright cometary atmosphere, which consists of gas and dust released from the nucleus and reflects sunlight. The tiny, brilliant, star-like point that is often observed with high magnifications through optical telescopes, within the bright central region of a comet, is the false nucleus. A typical size is 1000 km, which is much greater than the actual size of the solid nucleus.

The only hope we have of observing the true nucleus is to observe it far from the Sun, when it is inactive and reflects sunlight like an ordinary minor planet. It is then a very faint object, generally fainter than magnitude 18, which we can expect to observe at 3 or 4 AU from the Sun. When Comet Halley was recovered on 1982 October 16 by Jewitt and Danielson, using the 5-m [200-inch] telescope at Mount Palomar, it was 11 AU from the Sun, beyond the orbit of Saturn. It was a tiny point of light at magnitude 24. The difficulties encountered in determining its period of rotation from variations in its brightness suggest that it was quite possible that some activity was present. [The outburst observed in 1991 February, when the comet was at 14.3 AU tends to confirm this suspicion. – Trans.]

In what follows, we prefer to use the term 'photometric centre' for the brilliant point of light that is observed visually or photographically within the head of a comet. It is interesting to realise that comparison of the in-situ measurements of Comet Halley's nucleus made by Giotto with astrometric measurements made from Earth revealed that the centre of gravity of the nucleus was about 1100 km from the photometric centre. This distance, which applied at 1 AU from the Sun, varies inversely as the square of the distance from the Sun. When a comet is remote, the nucleus coincides far more closely with the photometric centre. The latter in fact coincides with very bright jets of material. This is easy to see from the images returned by Giotto. Experience has shown that the rotational periods of comets have been more frequently determined by studying phenomena caused by activity, such as amateurs try to record in drawings, rather than by direct measurement of changes of brightness.

8.3.3 *The various features observable in comets*

Figure 8.3 shows the terms for various features that may be observed in comets. The false nucleus may be a few arc-seconds across, or even, in very active comets and under high magnification, appear complex, elongated, or irregular, and the origin of the brightest jets. These jets are generally directed towards the Sun, but curve round towards the rear, approximately in the direction of the tail, appearing like luminous fountains. The contrast between these features and the background sky is generally greater than that shown by the arms of even the finest spiral galaxies. The eye's particular advantages are exploited to the full in cometary observation. The eye is capable of appreciating a large range of contrasts, whereas a photographic plate is soon saturated. Naturally, photography does enable faint features to be captured with long exposures, but the brightest regions of the head appear as a large, uniformly illuminated, area that is nothing like reality. It is no longer possible to distinguish fine detail near the nucleus, i.e., those details that are most concerned with the cometary activity. It is generally considered that the area that is richest in information lies within a radius of no more than 1 arc-minute of the nucleus. Because of projection effects against the sky, jets and rays may appear to point in almost any direction. Very careful drawings of their orientation are useful for determining the spatial orientation of the rotational axis of the nucleus, the rotation period, and even for mapping the surface by identifying areas of permanent activity.

The vast atmosphere of gas and dust that surrounds the nucleus, the head (or coma), rarely appears perfectly circular. Even very old, periodic comets, which have passed close to the Sun many times, and have lost most of their dust, show asymmetric comae. The most famous example is perhaps Comet Encke, which fairly systematically shows a parabolic-shaped coma opening towards the Sun after having passed perihelion. It is rather the exception that proves the rule. More frequently a coma appears as a parabola opening towards the axis of the tail, on the opposite side to the Sun. The side nearest the Sun is fairly well-defined, and not far from the nucleus. These different types of coma correspond to very different types of activity.

The shape of a comet's head also varies with heliocentric distance. Heads develop

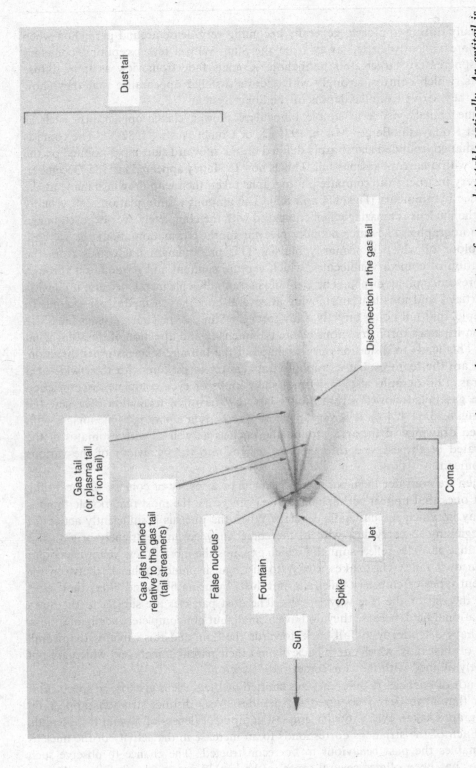

Fig. 8.3. *The morphology of comets. A schematic diagram showing the various cometary features detectable optically. An antitail is not observable at the same time as a fan-shaped array of dust tails. When the conditions for seeing an antitail are correct (Fig. 8.1), all the ordinary dust tails are superimposed on one another along our line of sight. Diagram by J. C. Merlin.*

Dust tail

Disconection in the gas tail

Gas tail
(or plasma tail,
or ion tail)

Gas jets inclined
relative to the gas tail
(tail streamers)

Coma

False nucleus

Fountain

Spike

Jet

Sun

gradually during approach, generally becoming very dense around perihelion when gases are violently ejected away from the Sun. With a few exceptions, cometary atmospheres are larger after perihelion. Comets fade from view as faint diffuse objects, which contrast strongly with the quasi-stellar appearance that they show when they arrive from the depths of the Solar System.

Some comets visible in simple binoculars show a classic appearance, such as Comet Kobayashi–Berger–Milon, 1975 IX or Comet Austin, 1982 VI. The coma is pear-shaped, rounded and sharply defined on the sunward side, more pointed on the other, with a narrow, gaseous tail. This is how P/Halley appeared in 1985 December. It is very instructive to compare photographs taken then with drawings made at the eyepiece by amateurs (Figs. 8.4 and 8.5). The amount of information seen visually near the nucleus is amazing when compared with the completely over-exposed image on photographs. The latter actually give far more information about the gas tail, the colour of which is primarily bluish. This predominant tint comes from the ionization of cometary molecules, which capture sunlight and re-radiate it through a fluorescence phenomenon. The gas tail is essentially ephemeral, because it consists of particles and ions that are moving at several hundreds of metres per second. In principle, that tail points directly away from the Sun. As the comet crosses the Solar System it passes through regions where the intensity and direction of the solar wind vary. This leads to distortions in the shape of the tail and changes in its direction. These are the features that astrophotographers try to capture with their wide-field cameras. Photographs also sometimes show kinks or even complete disconnection of the gas tail, followed several hours later by the rapid formation of a new tail (*see* Fig. 8.33, p. 424 *ff*). It is very interesting to correlate photographic images with detailed drawings of the area around the nucleus as well as with estimates of the integrated brightness, and thus gain an insight into the behaviour of the various phenomena.

When we consider the next, more spectacular, category of comets, the beautiful, bright ones that appear perhaps once every ten years, the most remarkable thing is the way in which the dust tails develop. When the nucleus is sufficiently active for the degassing to carry dust with it, the comet may become extremely bright. The freed dust particles reflect sunlight, which explains the more or less yellowish colour seen in dust tails. This makes it easy to distinguish photographically between the different types of cometary tails (gas and dust) by using films and filters that select either the red or the blue wavelengths. The dust particles are subject to the Sun's gravitational field, because they do have a small, but not completely negligible, mass. They have a tendency to 'fall back' towards the Sun, and this explains the overall shape of dust tails, which curve back behind their parent comets and which are not precisely aligned with the Sun–comet radius vector.

Clouds of particles of different sizes emitted by the nucleus give rise to spectacular events that have been recognized for decades. Very distinct striations appear on photographs taken even with the simplest equipment (lenses of 50-mm focal length, etc.). Each of the striations corresponds to an increase in the activity of the nucleus and enables the past behaviour to be reconstructed. The chance to observe such features has been offered several times in the last 30 years, as with Comet Arend–Roland, 1957 III (*see* Fig. 8.1), Comet West, 1976 VI (Fig. 8.6) and, more recently,

Fig. 8.4. P/Halley photographed on 1985 December 13. Exposure 6 minutes, beginning at 18:09 UT; hypersensitized TP 2415 film; 200/300-mm Schmidt operating at f/1.5. On photographs, the beginning of activity by P/Halley was primarily marked by development of the gas tail and by growth of the head. Photograph by Michael Jäger.

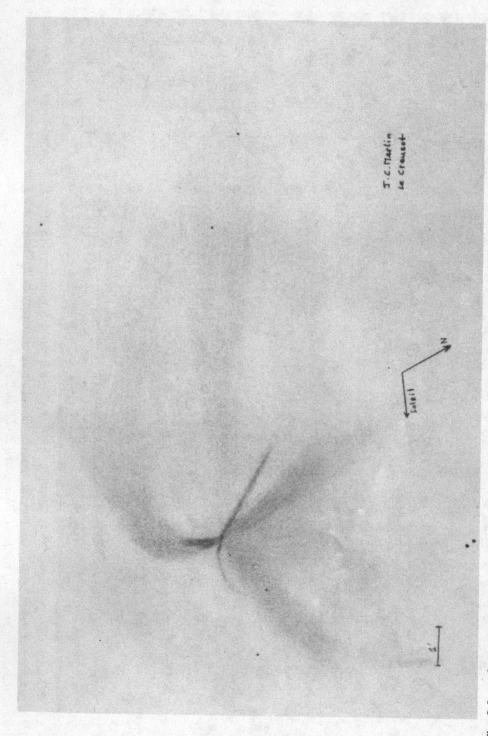

Fig. 8.5. *A drawing of Comet Halley made with a 400-mm reflector, focal length 2 m, with magnifications of 81× and 254×, 1985 December 12 at 18:25 UT. When approaching perihelion, the head of P/Halley showed striking fountain-like jets. Drawing by J. C. Merlin.*

Fig. 8.6. *Comet West, 1976 VI, on 1976 March 5. Exposure 5 minutes, beginning at 04:45 UT on Plus X film using a Maksutov telescope, 140/180-mm aperture, 280-mm focal length. After perihelion (1976 February 25), Comet West showed a series of spectacular dust tails and synchrones. Each was caused by dust emitted from the nucleus at very specific times. Photograph by G. Klaus.*

Halley's Comet shortly after its perihelion passage in 1986 February. The fact that such features appear in a periodic comet shows that Comet P/Halley is an exceptional object, although there hardly seems any need to stress that.

Observation of these normal tails, which are generally directed away from the Sun, may be grouped with the study of antitails and other features that point towards the Sun. A jet pointing towards the Sun is relatively frequently observed, even in medium-sized comets. This phenomenon often occurs in short-period comets where the rotation axis has become fixed relative to the Sun. It is quite common for the rotation axis to point towards the Sun at perihelion (as in P/Encke, P/Crommelin, and perhaps P/Giacobini–Zinner), which leads to a quasi-permanent area of activity

343

at the pole of the nucleus. Monitoring the orientation of a jet of this sort would give useful information about the position of the axis and secular changes in its orientation (because of precession, etc.). Antitails have a different origin. They tend to occur in comets that emit a lot of dust. The dust particles spread out in the plane of the comet's orbit. When the Earth intercepts the comet's orbital plane, we see sunlight reflected by all the dust along a line between the Sun and the nucleus. We therefore see a considerable increase in brightness of the tail in the direction opposite to the Sun, and also in front of the nucleus, towards the Sun. This is what gives rise to the name antitail. This feature is not a result of renewed activity. It is mainly a projection effect. The most famous antitail was that shown by Comet Arend–Roland (*see* Fig. 8.1). In such comets it is possible to observe the whole range of cometary phenomena.

It is hardly possible to describe a typical comet, whether bright or not. Just about century ago, a farmer in Blauwberg (South Africa) discovered a brilliant naked-eye comet only 14° from the Sun: the Great Southern Comet 1887 I. The tail stretched over more than 50° by the end of 1887 January, but the most astonishing thing was that this comet had no coma or any sign of a condensation. The head was merely a slight diffuse nebulosity, which did not bear comparison with the brightness of the tail.

This survey of the wide range of forms that comets may take should have whetted our appetites for learning how to record them as accurately as possible on paper or in photographs. To ensure that we obtain worthwhile results, we need first to learn how to prepare for our observing sessions.

8.3.4 Preparing for observation

To observe a comet we require its ephemeris and an atlas of the sky, on which we can plot its apparent motion or determine its position. So we require a certain amount of current information, and also some reference material.

8.3.4.1 Ephemerides

Ephemerides of comets are published by various organisations. The amateur who wants to obtain information direct from the source can contact the IAU Central Bureau of Telegrams. Practically as many amateurs as professionals subscribe to the *IAU Circulars*. These *Circulars* are sent by airmail all over the world and announce new comet discoveries, their orbital elements, and, for most comets, ephemerides every 5 or every 10 days throughout their period of visibility. This information is obtained and made available through other services in Europe, especially by *The Astronomer* magazine in the United Kingdom, which provides both specialized circulars and an electronic mail service. The IAU offers various more comprehensive and faster services according to what one wants. If you want to hear of astronomical discoveries within hours, it is possible to subscribe to the astronomical telegram service. This is effective but fairly expensive: $11.00 per telegram. The subscriber gives a deposit for a certain length of time and the account is debited by $11.00 each time a telegram is sent. The service provided by *The Astronomer* is probably more suitable for most persons in Europe, because circulars are issued after most telegrams, and the delay is only 2 or 3 days rather than the 5–10 days required by the airmail service. Even faster information is available, worldwide, through the electronic mail or telephone services. Contact *The Astronomer* for further details.

For more detailed information about Solar-System objects such as comets and minor planets, one can contact the Minor Planet Center, at the same address as the IAU Central Bureau for Astronomical Telegrams. The *Minor Planet Circulars* contain all the observations obtained all over the world, the latest orbital elements, and ephemerides that are more comprehensive than those given in the *IAU Circulars*, in particular the ephemerides of expected periodic comets, before they are actually recovered. This is a monthly service, which publishes about 1000 circulars per year for an annual subscription of about $180.00 in 1989. The *IAU Circulars* are sent by electronic mail, at no extra charge, to all subscribers who can be contacted easily on a recognized network.

There is also a world-wide computer network. For an additional fee (equal to that charged for the *IAU Circulars*), one can access certain files at the IAU Central Bureau and Minor Planet Center directly – provided one has suitable computer equipment, of course.

Expected periodic comets do not require the same degree of urgency. The orbital elements published in the MPC take planetary perturbations into account; they are published 6 months to a year before the assumed start of the period of 'visibility' or of activity. The period of visibility becomes more and more difficult to define, because some comets, which are to be subject to intensive study, are detected a long way from the Sun (P/Halley, P/Giacobini–Zinner, P/Grigg–Skjellerup, etc.), at magnitudes around 22–24. Most of the time, periodic comets are recovered at magnitudes of 18–20. Recovery is followed by adjustment of the orbital elements, which may be published in IAU *Circulars*, but this is not done on a regular basis.

The simplest and cheapest method of obtaining full ephemerides for periodic comets that are expected to reach at least magnitude 12, is to obtain the annual BAA *Handbook*. Other publications issued by various organisations also give a selection of the brightest comets expected each year.

It is worth stressing the importance of receiving information from various sources. Amateurs should not be worried about receiving the ephemeris for a single comet from more than one source. When it is an object like Comet IRAS–Araki–Alcock, which passed very close to the Earth in 1983, and which was observable for only about two weeks in Europe, one appreciates the value of being quickly informed of what is happening, and the valuable service offered by the various telephone and electronic mail services that are available. We would encourage anyone who is interested in making regular observations of comets to subscribe to the IAU *Circulars* at least. These regular bulletins also include observations of comets (magnitude estimates), which give an indication of the behaviour of current objects and about the activity of periodic comets.

Learning to read an ephemeris Figure 8.7 shows an example of an IAU *Circular*, giving the orbital elements and ephemeris of a comet. The reader should refer to Fig. 8.2 for the definition of the orbital parameters.

[Note that from 1991 December 24, there was a change in the form of time used

```
                                              Circular No. 4295

     Central Bureau for Astronomical Telegrams
            INTERNATIONAL ASTRONOMICAL UNION

        Postal Address: Central Bureau for Astronomical Telegrams
     Smithsonian Astrophysical Observatory, Cambridge, MA 02138, U.S.A.
       TWX 710-320-6842 ASTROGRAM CAM    Telephone 617-495-7244/7440/7444

                      COMET LEVY (1987a)
       David Levy, Lunar and Planetary Laboratory, reports his dis-
    covery of a comet. Observations are available as follows:

       1987 UT          α₁₉₅₀      δ₁₉₅₀     m₁    Observer

       Jan.  5.5      17ʰ17ᵐ5    +11°20'    10    Levy
             7.5      17 17.0    + 9 50     11      "
             8.56     17 17.0    + 9 02     11:   Morris

    D. Levy (Tucson, AZ). 0.40-m reflector. Object diffuse with
      slight condensation, no tail. Poor seeing and twilight inter-
      ference on Jan. 5.
    C. S. Morris (Little Rock, CA). 0.26-m reflector. Twilight.

                  PERIODIC COMET LOVAS 2 (1986p)
       J. V. Scotti, Lunar and Planetary Laboratory, has succeeded
    in finding diffuse images of this comet (cf. IAUC 4291) with the
    SPACEWATCH camera 0.91-m telescope on Kitt Peak:

       1987 UT           α₁₉₅₀           δ₁₉₅₀           m₁

       Jan.  3.14921    2ʰ21ᵐ20ˢ79     +15°34'00".6
             3.21661    2 21 25.53     +15 34 21.6       17
             3.23582    2 21 26.96     +15 34 25.2

       The following improved orbital elements are from 19 observa-
    tions 1986 Nov. 30-1987 Jan. 3:

          T = 1986 Sept. 1.503 ET
          ω =  70°637                     e =   0.58891
          Ω = 283.037   ] 1950.0          a =   3.52222 AU
          i =   1.519                     n° =  0.149101
          q =   1.44795 AU                P =   6.61 years

       1987 ET       α₁₉₅₀      δ₁₉₅₀       Δ        r       m₁

       Jan.  5     2ʰ23ᵐ71    +15°42'9    1.317    1.950    16.5
             15     2 37.20    +16 33.6
             25     2 51.68    +17 27.2    1.643    2.080    17.3
       Feb.  4     3 06.94    +18 21.5
             14     3 22.85    +19 14.6    1.996    2.212    17.9
             24     3 39.23    +20 04.9
       Mar.  6     3 56.00    +20 51.3    2.363    2.346    18.6

    1987 January 8                           Brian G. Marsden
```

Fig. 8.7. *An example of an IAU Circular. Circular 4295 dated 1987 January 8, reports the discovery of a new comet by David Levy (Comet 1987a). The discovery was confirmed a few days later by D. Levy himself and by C. S. Morris. This circular also gives astrometric positions for Comet P/Lovas 2 (1986p). New (elliptical) orbital elements have been calculated; using these new elements an ephemeris is given for the next two months.*

for calculating ephemerides. Prior to that date, ephemerides were given in Ephemeris Time (ET). Subsequently, ephemeris data has been given in Terrestrial Time (TT). This is a very close approximation to Terrestrial Dynamical Time (TDT) – which is used for observations from the Earth. In practice, TT may be considered as identical to ET, and the difference $TT - UT = 32 + N$ seconds, where N is the difference between TAI (atomic time) and UTC (Coordinated Universal Time). N is also an exact integer depending on the number of leap seconds occasionally inserted on June 30 or December 31. From 1992 July 1 this difference was 27 seconds, and the difference between TT and UT was thus 59 seconds. Observations continue to be given in UT, as previously. (*See also* Appendix 1, Vol. 2.) – Trans.]

Each line of the ephemeris gives the following information:

- date (year, month, day): the data are generally given for 00:00 TT;
- right ascension (α) and declination (δ) for epoch 1950.0 [Note that the epoch is 2000.0 for *Circulars* issued after 1991 December 24. – Trans.];
- Δ = Earth–comet distance in AU;
- r = Sun–comet distance in AU (the radius vector or heliocentric distance);
- m_1 = predicted total visual magnitude (see below).

For distant or more or less inactive comets, the magnitude quoted is sometimes that of the nucleus (m_2), which is treated like a minor planet that merely reflects sunlight, where the photometric parameter, $n = 2$ (see below). The difference between m_1 and m_2 is very large for bright comets (from 3–10 magnitudes). The remarks made earlier about the visibility of the nucleus suggest that there is little point in paying much attention to the m_2 magnitudes, because they rarely bear any relation to what it actually observed.

Some specialized ephemerides give other information:

- apparent coordinates (α and δ) that incorporate precession to the epoch of the ephemerides;
- elongation: the Sun Earth–comet angle, in degrees, which is very useful in determining the visibility of the comet; observable comets have elongations between about 15° (bright comets) and 180° (opposition);
- phase angle: the Sun–comet–Earth angle, in degrees;
- tail position angle: the direction of the tail against the sky background (Fig. 8.8), measured from north (0°) through east (90°).

Cometary magnitudes The brightness of a comet is generally considered to vary in accordance with a law such as:

$$B = B_0 \Delta^{-2} r^{-n},$$

where Δ is the Earth-comet distance in AU, r is the Sun-comet distance in AU, and B_0 is the absolute brightness when $\Delta = r = 1\,\text{AU}$.

The Δ^{-2} term reflects the fact that the brightness varies inversely as the square of the distance from the Earth. Similarly, the r^{-n} term indicates that the brightness varies inversely as a power n of the distance from the Sun (the heliocentric distance). The quantity n is sometimes known as the activity parameter. When a body, such as a minor planet, merely passively reflects sunlight, $n = 2$. Most comets have values of n of about 4–6. When a new comet is discovered, the predicted brightness is calculated on the assumption that $n = 4$.

The equation just given may be converted to magnitudes as follows:

$$m_1 = H_1 + 5 \log \Delta + 2.5 n \log r,$$

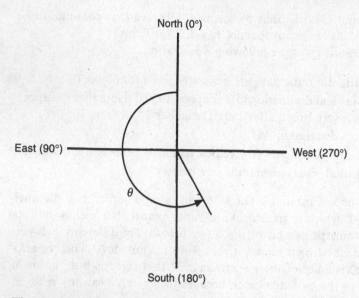

Fig. 8.8. *The definition of position angle* θ. *The position angle (PA) is measured from* 0° *to* 360° *from celestial north, going through east. The direction shown here corresponds to a PA of* 210°.

where m_1 is the apparent total magnitude (generally visual) when the comet is Δ AU from the Earth and r AU from the Sun. H_1 is the absolute total magnitude for $\Delta = r = 1$ AU.

Predictions of nuclear magnitudes m_2 are based on the assumption that $n = 2$.

The law expressed in the second of these equations is basically empirical, and cannot be used to derive any physical information about the comet. At the most, is it sometimes possible to relate the absolute total magnitude, H_1, to the rate of production of certain molecules, as observed at certain specific wavelengths by radio astronomy, for example.

8.3.4.2 Atlases and catalogues

Once having sorted out the ephemeris, we need to consider the question of the basic reference material that an observer of comets requires. Certain specific aspects of reference material will be covered in the appropriate later sections (dealing with magnitude, drawings, etc.). Initially, however, we require a sufficiently detailed atlas for us to prepare for observing.

At present, it would appear that of the available atlases the one that best suits the work of comet observers is the *AAVSO Atlas*, published by Sky Publishing Corporation. This atlas has been computer-plotted from the *SAO Catalog* (SAOC). It includes all the stars in that catalogue (about 260 000), which means that the position and magnitude (visual or photovisual) of all the stars in the *AAVSO Atlas* are known, and that they may be found in the SAOC. The limiting magnitude is around 9.5, with some fainter stars included, and various intentional omissions in the magnitude range 8–9. The SAOC is a compilation of various older catalogues, which partly explains why it is not homogeneous. An observer who 'discovers' that an 8.5-magnitude star is missing will have some doubts about the atlas, and will be prompted to consider obtaining a detailed, homogeneous, photographic atlas. This is bound to happen to an assiduous observer. Apart from stars, the *AAVSO Atlas* indicates the positions of 1500 nebulae, clusters and galaxies, as well as about 2000 variable stars followed by amateurs. It was originally conceived for variable-star work. The scale of the first edition (published in 1980) is 15 mm per degree but in the revised, second edition (1990) this has been reduced to 11 mm per degree, with a smaller overall chart size. In both editions each chart covers an area of about $20° \times 18°$, which means that the path of a comet over several days or weeks may be plotted. The overlap between one chart and the next amounts to several degrees. The charts have a grid for epoch 1950.0. Observers should be careful to correct the positions of particular comets to this epoch; any good ephemeris always specifies the reference epoch.

The various details given on the *AAVSO Atlas* are all of use to those studying comets:

- nebulae, clusters, etc., which may be confused with comets;
- variable stars, whose comparison sequences may be used to estimate the magnitude of comets;
- specific indication of the magnitudes of certain stars, which is very useful for very bright comets.

The charts from the *AAVSO Atlas* make excellent finder charts. Because each chart shows a large part of the sky, it is very easy to find one's way around with 50 to 60-mm binoculars. This atlas may be used for finding comets down to about magnitude 10–12, depending on the instrument used. In any case it guarantees that one can find suitable reference stars (in the SAOC) to estimate the brightness of comets above magnitude 9.5–10.

[Another atlas that has now become available is *Uranometria 2000.0*, by Tirion, Rappaport and Lovi, which includes stars down to magnitude 9.5. Its scale is 18.5 mm per degree. The main problem reported by observers is that it consists of a single, bound volume, rather than a collection of individual sheets, and is therefore difficult to use (and protect) at the telescope. The fact that facing charts overlap at their *outer* edges is also a disadvantage. – Trans.]

It is also worth mentioning the photographic atlases produced by the German amateur and publisher, Hans Vehrenberg. These two atlases have grids for epoch 1950.0. The most elaborate is *Atlas Stellarum*, which is a set of photographic prints, showing all stars down to magnitude 14–14.5, with a uniform scale of 30 mm per degree. A simpler version exists, the *Atlas Falkauer*, with a limiting magnitude of about 13, and a very interesting scale for those who have the first edition of the *AAVSO Atlas* – 15 mm per degree! Both *Stellarum* and *Falkauer* show black stars on a white background, which is easier to read than the traditional positive prints.

Observers with fair-sized instruments will eventually want to detect even fainter comets. It is obvious that suitable reference material is required to use a 250-mm or larger telescope effectively. Sooner or later a photographic atlas will be required. The paths of the fainter comets may then be plotted directly onto the very detailed charts, which contain a large number of suitable reference stars. There is one photographic atlas specifically designed to show photovisual magnitudes that are very close to visual magnitudes. This is the *Papadopoulos Atlas* (published by Pergamon Press), which shows stars down to approximately m_{pv} 13. We mention this atlas because of a problem that arises with Vehrenberg's atlases (*Stellarum* and *Falkauer*), and which may cause a great deal of bother to visual or photographic observers. These two atlases show photographic magnitudes (m_{pg}), being made on blue-sensitive plates. Stars of magnitude 11–12 that are not on the *Atlas Stellarum* (i.e., with m_{pg} below 14.5) are frequently observed visually. These are very red stars, with large colour indices; they are not necessarily novae. Be warned!

We do not mean to imply that we reject all the other available atlases *en bloc*. It would seem to be a good idea to consider any atlases that:

(i) show a large part of the sky on each chart. This enables paths of comets to be plotted over the longest possible period of time;

(ii) do not stop at individual constellation boundaries;

(iii) have a good overlap between one chart and the next. *Atlas Coeli* by Becvár (epoch 1950.0) was excellent. It is now out-of-print, and has been replaced by Wil Tirion's *Atlas 2000.0*, which shows stars down to magnitude 8. This was the first atlas to have 2000.0 coordinates. The fact that *Atlas Coeli* was prepared by a comet-hunter, however, suggests that this type of atlas is more suitable for comet searching than it is for conventional observation of known comets.

8.3.4.3 *Plotting paths*

Now let us turn to the first stage in using an atlas. The less energetic will probably be content to plot the position of a comet day by day, but will almost certainly have problems because the times of observation will rarely coincide with one of the ephemeris dates. An accurate (non-linear) interpolation is desirable, or even calculation of the position from the orbital elements, if one has access to a suitable (micro)computer. But is it wise to wait until night-fall before asking what comets are visible and where they are in the sky? It would seem much more sensible to spend an hour or two each month plotting the paths over a 'long' period. Use of a planisphere and a lunation table will provide enough information to determine at what time during the night a particular comet is visible, and thus enable us to plan observing sessions several weeks in advance.

It is very important to take care in plotting the paths of comets, especially when we are observing a comet for the first time. We cannot tell beforehand what it will look like, and it may be necessary to use a relatively high magnification to identify it. Each position given in the ephemeris should be marked on the chart with a small cross, interpolating the values of α and δ from the coordinate grid. It is generally enough to mark the position every 5 days, or even sometimes every 10 days if the comet's apparent motion is slow or uniform. The points are joined, preferably with a suitable French curve or flexible curve, to give a smooth path. An irregular line with breaks in it or sharp changes in curvature is likely to give rise to an error of several minutes of arc if the change in α and δ over several days is large or non-uniform. Any error in positioning a cross is soon detected when the points are joined. The appropriate date (month and day) should be marked alongside each cross. Remember that the positions are given for 00:00 UT. It is generally sufficient to subdivide the segments of the curve into equal parts to obtain intermediate dates. If the motion of the comet is very rapid, it is best to mark the position at 00:00 UT every two days, or even every day in extreme cases (as with Comet IRAS–Araki–Alcock).

Examination of the path allows us to identify the various objects that will appear close to the comet: nebulae and galaxies, which must not be confused with it; and neighbouring stars that may be used to determine its position accurately. Anyone who is interested in estimating the comet's magnitude can also check for standard sequences that may be used as photometric references. If we have the *AAVSO Atlas* and the comet is brighter than magnitude 10, no other reference material will be required. We will always be able to find stars in the atlas that are close enough to the comet's path for the magnitude to be estimated. The visual magnitude of these stars may then be checked later in the SAOC. It is not a good idea to know them in advance, which would make our estimates less objective. If, on the other hand, the comet is fainter than magnitude 9.5, calibrated sequences showing stars below the limiting magnitude of the SAOC will have to be used. The only visual-magnitude, calibrated sequences are those available for variable stars. Observers should therefore contact the variable-star associations in order to obtain the required information. [Observers can also consider using the standard North Polar Sequence, which is described in Sect. 15.8. – Trans.]

Here is one practical hint about plotting cometary paths: if two comets are observable at the same time in the same area of the sky, and fall on the same chart in whichever atlas you are using, do not plot both paths on the same chart. Use two separate copies of the chart instead. Ensure that each sheet is marked with the name of the comet whose path is shown. Remember that telescopic images are inverted and so it is necessary to turn the charts upside down when observing at the telescope, and that work has to be carried out in the dark. Confusing one comet with another may lead to unpleasant surprises: there might be apparent sudden, unexpected changes, or even the disappearance of an object that was observed a few nights previously. Such a situation could have arisen frequently in recent years, when comets visible in amateur-sized instruments were in the same area of sky. Examples are: P/Encke and P/Crommelin in 1984 February; and P/Boethin and P/Wirtanen in 1986 May. P/Giacobini–Zinner and P/Halley were also close to one another in 1985 September, but at very different magnitudes (7 and 12 respectively). An example of a plotted cometary path is shown in Fig. 8.9.

Once we have carried out all this preparatory work, we only have to select the charts with the plotted paths, and perhaps charts with comparison sequences for magnitude estimates, and we are ready to tackle observation of comets.

8.3.5 *Locating comets*

The first question is whether the comet is visible with our equipment. Table 8.1 gives an idea of the faintest comets that one can hope to detect with given apertures.

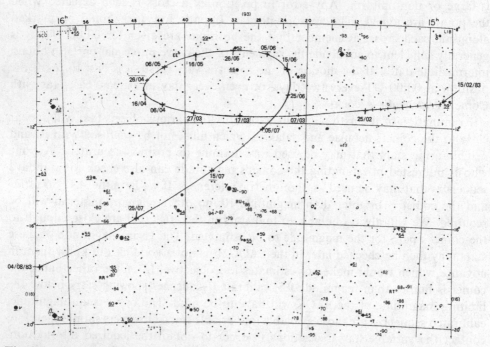

Fig. 8.9. *The apparent path of a comet as plotted on an atlas. This shows the movement of Comet P/Kopff at its 1982–83 apparition, plotted on a chart from the AAVSO Atlas. The ephemeris positions have been plotted every 10 days from 1983 February 15 to August 4. The portion between 1983 February 15 and 25 has been divided into daily segments by simple linear interpolation. At the retrograde points (1983 April and 1983 June), it is preferable to use a non-linear interpolation. (By kind permission of the AAVSO and Sky Publishing Corporation.)*

This is valid for comets whose existence and positions are known. The values given are based on purely empirical results obtained by experienced observers under good observing conditions. Because comets are generally diffuse objects, their light being spread over an area that is generally several arc-minutes across, it is easy to see that it is impossible to detect comets that are the same magnitude as the faintest visible stars. (We shall describe the limits of detectability of comets in the section concerning magnitude estimates.) With the cometary magnitudes in Table 8.1 we have assumed that the comets are 100 % diffuse, i.e., that they have no central condensation. A comet that was stellar in appearance would not, in principle, be any different from a star of equal magnitude, seen under the same conditions.

We shall discuss various factors connected with the choice of observing site in greater detail in the section dealing with comet-hunting. Readers may like to refer to that section now if they require specific information. In any case it will be obvious that searching for comets is difficult if the transparency is affected by mist or moonlight. Stray light (particularly artificial light) also interferes with dark

Table 8.1. *Limiting stellar and cometary magnitudes for given apertures*

Aperture (mm)	Limiting stellar magnitude	Limiting magnitude (comet 100 % diffuse)
50	11.0–11.5	9.5
80	12.0–12.5	10.5
100	13.0–13.5	11.0
150	14.0–14.5	12.0
200	14.5–15.0	12.5
250	15.5	13.5
400	16.5–17.0	14.5

adaptation. All these problems are difficult to reconcile with the wish to obtain accurate observational data, particularly magnitude estimates.

Finding a comet with a reflector or any other instrument is very similar to locating any of the extended objects in the sky, except for the fact that comets move from day to day. Observers will call on their own particular past experience of general observation. Those who have telescopes on an equatorial mounting with setting circles will simply set the right-ascension circle on a bright star, and then simply swing the telescope to the coordinates of the comet. This applies if the comet is sufficiently bright for it to be seen immediately in the eyepiece. It is generally preferable to locate a bright star in the immediate vicinity and to finish setting by using the patterns of faint stars, close to the comet's path, that are shown on the chart. The advantages of an equatorial mounting and graduated circles are fully appreciated when several comets are observable on the same night. Moving from one to the other is very easy. Under such circumstances, the observer should organise the work in order of increasing right ascension, starting, shortly after evening twilight, with objects that have the smallest elongations east of the Sun. In practice, even if graduated circles are used, locating a comet generally ends with checking neighbouring stars to ensure that the correct object has been found.

There is sometimes a problem in correlating the stars observed with those plotted on charts, either because there are difficulties in relating the scales of the charts to that of the field observed in the eyepiece, or because large numbers of stars may be seen that do not appear on the charts, drowning out the brighter reference stars. It is easy to overcome these problems with a few tricks. It is, of course, essential for the telescope to be fitted with a small finder, such as half a pair of binoculars, that is intermediate in size between the eye and the telescope itself, and which will give a fairly wide field whilst showing the brighter reference stars. It is also worth having a series of diaphragms that can be placed in front of the telescope's aperture, thus reducing the number of stars visible that are not plotted on the charts. First attempts at finding a comet should be made with a low magnification, close to the minimum effective magnification $(D/7)$, to give the widest possible field and the maximum illumination. We can use a slightly higher magnification ($D/5$ or $D/6$) to increase the contrast. (It is possible to show, by geometrical optics, that all the light collected by a telescope leaves the eyepiece in a beam with a circular cross-section, which is known as the exit pupil and has a diameter d, such that $M = D/d$, where M is the magnification and D is the diameter of the objective. Obviously d decreases as M increases. It is essential that d should be equal to, or smaller than, the pupil in the observer's eye, otherwise some of the light collected by the objective will be unable to enter the eye and will be lost. Taking the diameter of the pupil of a dark-adapted eye as being 7 mm, the lowest magnification that may be used (known as the minimum effective magnification), when the telescope's exit pupil is equal to the pupil of the eye, is given by $M = D/7$, when D is expressed in millimetres.)

We therefore advise beginners to determine very carefully the fields given by the principal eyepieces. This may be done in three ways:

- by determining the amount of time required for a star to cross a diameter of the field from West to East (the instrument being clamped). The diameter of the field in arc-minutes equals $T \times 15 \times \cos\delta$, where T is the time in minutes, and δ is the declination of the star observed;
- by locating two stars, at diametrically opposite edges of the field; their separation may then be determined from an atlas; the ideal choices for this are stars in the Pleiades;
- by using a graduated (and calibrated) micrometer.

Everyone knows that a comet may be distinguished from stars by its fuzzy appearance, which is generally accompanied by a certain degree of central condensation. In reality, however, things are not quite so clear-cut. Observers are often unable to detect a faint comet, either because it is so diffuse that, with too powerful an instrument, the nebulosity extends over the whole of the field and is thus impossible to detect; or for the opposite reason, because the coma is very tiny and faint and surrounds a star-like central condensation. An example of a strongly condensed comet was Comet Wilson, 1986 I, in 1986 August–September. The tiny coma, only about one arc-minute across, could be detected only at high magnifications. At the opposite extreme, in 1983 June, Comet Sugano–Saigusa–Fujikawa, 1983 V, was a vast, diffuse patch in 7×50 binoculars (the coma being detectable out to about 30 arc-minutes). Being without a central condensation, it was vaguely detectable over a diameter of about 10 arc-seconds in a 260-mm, f/6 Newtonian telescope, and completely undetectable in the 1-m, f/16 telescope at the Pic du Midi. The nucleus was fainter than magnitude 18, and eventually this comet was only visible in wide-field instruments.

Comets also exist that are only visible with fairly high magnifications, around the optimum magnification for resolving objects ($D/2$). This applies, for example, to Comet P/Schwassmann-Wachmann I, which always remains beyond the orbit of Jupiter, at more than 5 AU from the Sun. Normally at about magnitude 18, this comet undergoes sudden, unpredictable outbursts in brightness of about 5–8 magnitudes. The coma that it produces is either very small or very diffuse. It is therefore essential to use the magnification giving optimum resolution to obtain an acceptable contrast, even for simply determining its position relative to the faintest

Table 8.2. *Suggested magnifications for cometary observation*

Magnification	Exit pupil (mm)	Application
30× to 35×	6	Detection of easy objects Magnitude estimates of the same objects General drawing
60× to 80×	3	Detection and magnitude estimation of fainter or more condensed objects Drawing, when conditions preclude the use of high magnifications (poor seeing)
100× to 120×	2	Detection of objects at the telescopic limit Drawing of detailed features under good conditions
200× to 250×	1	Highly detailed drawing of the inner coma under perfect conditions

stars that are visible. Obviously the path of this comet has to be plotted on very detailed charts (from *Atlas Stellarum*, for example).

The use of several different magnifications is therefore essential in observing comets. Table 8.2 indicates what type of magnification is advisable for cometary observation with a classic, 200-mm Newtonian telescope (f/4 to f/8). This information may be extrapolated to other sizes, if reference is made to the size of the exit pupil. Schmidt–Cassegrain telescopes have longer focal lengths (and generally operate at f/10). It is also possible to use slightly higher magnifications for high-resolution work (such as drawings).

The identification of comets is sometimes aided by two factors:

- their tails; in such cases the objects are easy and there can hardly be any doubt over their identification;
- their motion with respect to the stars; this is often apparent only after 15 or 20 minutes observing or with high magnifications (or both).

Although perhaps what we are about to say does not really help the beginner, it is worth bearing in mind that the celebrated comet-hunter Leslie Peltier said that the light from comets was different from that from other nebulosities in the sky. He was undoubtedly talking about the different 'quality' to the light from comets, which is a property that it is difficult to describe. To avoid the various pitfalls posed by cometary observation, an amateur's best course is to observe as many comets as possible. Their varied, and constantly changing appearance will be remembered, and will make the observer more aware of what may be expected.

Fig. 8.10. *Levy–Rudenko, 1984t, photographed on 1985 February 17 when near the galaxies M81 and M82 in Ursa Major. Exposure 7 minutes, beginning 22:56 UT; hypersensitized TP 2415 film; 200/300-mm Schmidt telescope, f/1.5. The comet is indicated by the arrow. Photograph by M. Jäger.*

Fig. 8.11. *Comet Sorrells, 1986n, photographed on 1986 December 5. Exposure 6 minutes, beginning 18:50 UT; hypersensitized TP 2415 film; 200/300-mm Schmidt telescope, f/1.5. Photograph by M. Jäger.*

8.3.6 Making visual observations

To make best use of time spent at the telescope in observing comets, we would recommend that the various points to be noted should be undertaken in a specific order. Let us assume that the telescope is ready, and that we have become properly dark adapted – which can be achieved by using the minimum light, preferably red. We have located the comet, and it is definitely there. Now what should we do?

First, we should already have noted down:

- the date and time (in UT);
- the name or the designation of the object being observed;
- the equipment being used: type, diameter, magnification used for finding;
- details of the observing conditions: transparency, turbulence, humidity, wind, Moon, etc.; perhaps even the temperature and pressure if we feel that this information may be of use.

Once this has been done, it is quite a good idea to give a brief description of the comet. We can record what may be seen at low magnifications:

- Whether the coma is circular, and symmetrical, or not; whether it is parabolic, pear-shaped, fan-shaped, etc.; is the edge more sharply defined on one side?

- Whether there is a distinct central condensation or not; is it stellar in appearance, or is the coma uniform?
- The rough shape of the tail; whether it is distinct or not; is it visible in the finder or binoculars?; perhaps with an approximate indication of any possible particular noticeable structure (such as streamers), using the standard abbreviations: n = north, s = south, f = following (east), p = preceding (west).

Some general remarks about the visibility of the comet should also be given: is it easy to see, visible with averted vision, or at the very limit? The lowest magnification should be used to make an estimate of the degree of condensation (DC) in the coma. This value is somewhat subjective, and is very rough, but it does give some idea of the overall appearance of the object being observed. The degree of condensation is, rather like the diameter of the coma, something that does not give any physical information about the comet. Such details are, however, sometimes useful to anyone examining your observations in deciding on their validity by comparison with other reports that have been submitted. If you describe the object as being very condensed when all the other observers call it diffuse, you may have made a mistake, or may not have observed all the coma. Figure 8.12 gives a schematic representation of the heads of comets with various values of DC. This scale runs from 0 to 9: 0 for a completely diffuse coma, 9 for an object that appears stellar.

Taking the two 'extreme' comets mentioned in the previous section, Comet Wilson had a condensation of about 7, whereas Comet Sugano–Saigusa–Fujikawa had a value of 1 in 7× 50 binoculars.

There is no point is taking hours to decide the degree of condensation. With a little practice, and reference to comets observed previously, it only takes a few seconds to decide what sort of comet we are looking at. Do not forget to note down the instrument and magnification with which the DC has been estimated.

Bearing in mind what we said in Sect. 8.3.2 about the changes in the appearance of cometary heads, the reader will not be surprised to learn that the value of DC varies throughout the period around perihelion. Any more or less ordinary comet may well follow the following sequence:

- discovery (recovery) DC = 4 to 6
- development of the coma DC = 4 → 3 → 2
- perihelion DC = 4 to 7
- shortly after perihelion DC = 3 → 4
- recession DC = 2

Finally, the observer may record the presence of various objects near the comet: clusters (such as M10 in Ophiuchus), an NGC galaxy, individual stars, etc., and also if a particular star causes problems in observing the comet, or even appears merged with it. Indications of motion are more or less useless: if you have found the comet it is because it was near the predicted position. Positions measured visually are not of much interest, unless the comet has only been discovered in the last 24 or 48 hours. Under those circumstances, it may be worthwhile rapidly reporting an approximate position to confirm the discovery and to help other observers to locate

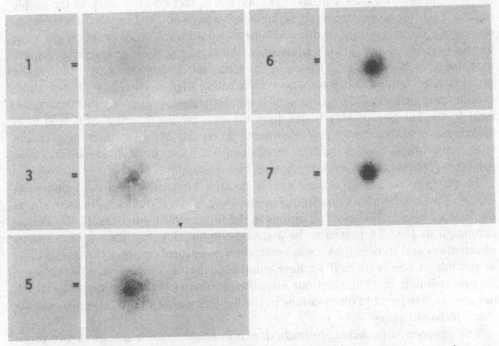

Fig. 8.12. *The degree of concentration (DC) of a cometary coma. This is a schematic diagram showing the various appearances of a simplified coma observed at low magnifications. The extremes are rare or difficult to depict: DC = 0, 100% diffuse comet (without any trace of condensation); DC = 9, very compact, brilliant coma, stellar in appearance or with perfectly sharp edges against the background sky. (After J. E. Bortle,* I.H.W. Amateur Observer's Manual for Scientific Comet Studies, *JPL Publication 83-16, Part I.)*

the object. But such a position is unlikely to be of any use in calculating the orbital elements. It appears to be difficult to obtain a position that has an accuracy of better than 1 arc-minute unless you use a method such as photographic astrometry (*see* Chap. 17).

Later operations are:

- making an initial drawing with a low magnification to show the limits of the coma, the orientation of the tail and its size;
- estimating the integrated visual magnitude;
- detailed drawing.

8.4 Drawing comets

The contribution of amateurs to this field of cometary work represents a continuation of the classic observational astronomy of the 19th century. Modern professionals do not make drawings. They devote their time to spectrophotometry and to other detailed studies at all sorts of wavelengths, but in fact do little work in the visible region. Their telescopes are rarely suitable for visual work, and the use of a telescope several metres in aperture to draw a comet would draw a lot of protests, given the hourly cost of operating such equipment. Photography allows large-scale changes in comets to be followed. However, its technical developments have not really improved its limited ability to cope with a wide range of contrasts. It is possible that developments in CCD detectors will not only allow greater amplification of low light levels, but also provide better discrimination of neighbouring areas with little difference in surface brightness. Computer processing has given interesting results with some bright comets (such as P/Halley), but there are many problems to be solved before this rather tricky technique becomes routine.

Neglecting visual observations merely because of possible future developments would appear to be a very doubtful procedure and would certainly mean an irreparable loss for many cometary apparitions in the foreseeable future. The Halley Watch campaign in 1985–86 proved to be a strong stimulus for the different branches of observation and showed that it was amateurs who ensured continuity in certain classical fields of observation. If we have equipment that is comparable with that used by professionals in 1910, then our visual work obviously has unique potential for helping to interpret old observations in the light of the astrophysical measurements made in recent years.

The observer who decides to make drawings of comets will encounter two problems: 1) that of detecting sufficient detail to warrant making a drawing; 2) that of representing what is seen, and only what is seen, on paper.

Just as there are certain observers who specialize in drawing planetary surfaces and who are able to produce highly detailed drawings of Mars or Jupiter with apertures of less than 200 mm, so there are cometary specialists who observe mainly with modest apertures. Could it be that some people have a certain aptitude, or (perhaps more likely) a certain type of eyesight, that allows them to appreciate fine detail on a bright disk and others to detect low contrast features in nebulosity?

8.4.1 Preliminary techniques

The examination of results obtained by more experienced workers (Fig. 8.13) is not particularly encouraging: the amount of detail shown in their drawings appears to be far greater than what beginners see in their own telescopes. There are two ways of becoming adept at drawing comets and these may be carried out together:

- developing a specific method;
- educating one's eyesight by *regular* observation.

Amateurs generally begin by spending months or years gaining experience by observing fixed objects (galaxies, nebulae, etc.). Repeated observations enable them to discover more and more details in these objects. It is always possible to check the validity of the features that are drawn by referring to photographs taken with amateur instruments. Many amateurs think that all that is needed to detect faint objects or faint detail is a powerful telescope. This is a bit like a violinist who thinks that all he needs to become a virtuoso is a Stradivarius! When one begins observing

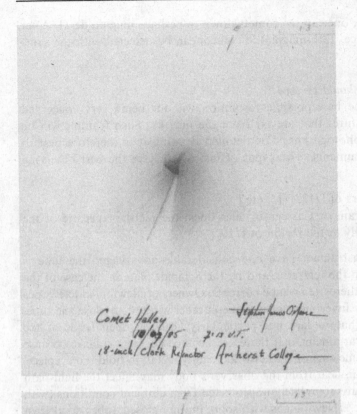

Fig. 8.13. *Drawings of Comet Halley made at* 1^h22^m *interval on 1985 October 7 by Stephen J. O'Meara, using the 450-mm Clarke refractor at Amherst College. These drawings show the changes in appearance that occur in an active comet over a short time-scale. They also show the distinct superiority of large refractors over reflecting telescopes for high-resolution work. Photographic imagery is far inferior. Time of the first drawing: 07:13 UT, second drawing: 08:35 UT.*

with a 250-mm telescope, one can see comets down to perhaps magnitude 11. After two years' regular practice, 13th-magnitude comets can be detected with the same telescope.

8.4.1.1 What instrument should be used?

Our aim is to obtain the best possible resolution without being very concerned about detecting faint features that are far from the nucleus. Such features will be far better recorded on photographs. For detailed drawing it is therefore best to choose longer-focus instruments. Two types of instrument give the best observing conditions:

- classical refractors of f/12, f/15, etc.;
- Schmidt–Cassegrain or Cassegrain telescopes; the relative aperture of the former is generally in the region of f/10.

These two types of instrument have one considerable advantage: the tube is closed (by the objective in the refractor, and by the Schmidt plate in the case of the Schmidt–Cassegrain), so there are no tube currents. Owners of Newtonian telescopes rarely consider the heavy investment of a plane-parallel window to close the tube. The reader will realise that we are not particularly keen on open tubes of the Serrurier-truss type. The argument that there are no tube currents is false, because there always has to be some sort of framework, usually metal, to hold the eyepiece, the spider, etc. In addition, heat from the observer's body may affect the light-path directly, and the mirrors are completely unprotected from ambient conditions (with the resulting humidity, etc.). Newtonians with solid, but not closed, tubes do at least have the advantage of offering a reasonable amount of protection for the primary mirror, right down at the bottom of the tube. It is important to begin by eliminating all sources of turbulence that are within our control; atmospheric turbulence is something that we cannot control, short of finding the best possible site. We are stressing the importance of these points, because they are far more important than one might at first think; turbulence is as damaging to the observation of extended objects as it is to planets.

Observers should ensure that the optical elements in their instruments are properly collimated. Faulty alignment causes spikes on stars and false rays and jets around the central condensation of comets!

What focal length and diameter?   The largest possible! Even refractors of 100 mm diameter, with focal ratios of f/15 are very useful instruments, which are better for making drawings than the classical 115-mm, 900-mm focal-length instruments. It appears desirable to have a focal length greater than 1 m, so that the images are sufficiently large at the prime focus. Bright, active comets may show marvellous structure through a telescope with an aperture of 150 mm. A 200-mm, f/15 refractor is a dream that few amateurs can achieve, unless they have access to instruments at an observatory. Finally, the ordinary 200-mm, f/6 Newtonian that many amateurs build for themselves is an excellent, low-cost compromise. Such an instrument may be used for a lot of different types of work. When discussing magnitude estimates we shall see that Newtonians are definitely to be preferred.

8.4.1.2 Drawing the star field

The aim of drawing comets is to depict the orientation of the jets, rays and streamers accurately. In order to determine the orientations, we need some fixed reference points that can serve to indicate directions. We might consider using a graticule in the focussing mount, with a rotating index that indicates the orientation of one of the lines in the focal plane of the eyepiece. Such a technique is used for double stars, but is not particularly suitable for cometary work, where the details that we need to measure are diffuse. The glass on which the line is engraved also introduces a certain amount of absorption. In addition, it is not always easy to set the line on the photometric centre of a comet, or to measure the change in direction of a curved streamer. This method also has the disadvantage of doing away with concrete information in the form of a drawing, which means that we have no way of subsequently checking the position angles that were measured.

The only way of determining the position angles and the lengths of the features accurately is to draw the star field around the central condensation; measurements may then be made from the drawing after it has been appropriately oriented.

By way of gaining experience in drawing star fields, we suggest that observers should make some sketches of a few open clusters, such as M36, M37, and M38 in Auriga, M44 in Cancer, the Pleiades M45, etc. The method is simple:

- start by drawing two bright stars that are almost diametrically opposite one another in the field of view;
- next draw, in decreasing order of magnitude, stars that form lines or specific right-angle, isosceles, or equilateral triangles with the two bright stars;
- gradually plot the fainter stars, paying attention to the proportions of triangles and alignments as much as possible.

It may help in preparing a drawing of a star field to use an eyepiece with a rectangular grid. It is enough to let one star trail on an east–west line to orient the field. With a little practice, drawing star fields becomes easier, faster and more accurate. Being able to draw the main star patterns visible around objects is a very useful asset. Remember that in finding a field drawn at the eyepiece in an atlas, the most useful stars for identification purposes are faint stars near bright ones. It often happens that there are similar patterns of bright stars not far from one another. The best method of identifying them is then to look for a faint star that is close to one of the bright ones.

In drawing comets we should make two drawings of the star field:

- a drawing at a low magnification (minimum effective magnification or slightly greater), which will show the most important stars and the overall shape of the comet (the coma and the tail);
- a drawing at a high magnification which shows fainter stars and those closest to the comet, as well as all the cometary details visible.

We should always bear in mind that the field may have to be identified in an atlas at a later date. For this to be possible, the drawing made at a low magnification should show stars that are easily identifiable. If there are no SAOC stars close to the comet, the brightest stars in the field should be located with respect to some specific reference points. In some cases it may be a good idea to make a drawing with the finder. Above all, do not commit the fatal error of omitting drawings of the star field when the comet is bright. This is a classic error. Immediately the comet is located, the observer starts to draw all the detail, recording just the stars in the immediate vicinity of the central condensation. But because the orientation has been inadequately recorded, finding the field depicted in an atlas becomes a very difficult task. A 20-arc-minute field only measures 10 mm in diameter on *Atlas Stellarum* (5 mm on the *Atlas Falkauer*), and then you only need to fall foul of a star with a considerable colour index (a very red star), which is visible to the eye, but is absent from the atlas, for everything to be thrown into confusion. You only need to try it to be convinced.

One might think that it was easiest to draw the comet directly onto the chart in the atlas. This dodge can work with very extended comets that may be observed with the naked eye, or with binoculars, where we can draw the outlines of the head and of the tail on a small- or medium-scale chart. Such a preliminary drawing, whether made in the observing book or on a chart, enables the diameter of the coma, and the length and orientation of the tail to be measured. But this does not solve the problem of the more detailed drawing, which should be made at a larger scale. Even the charts in *Atlas Stellarum* have too small a scale (2 arc-minutes per mm), because the aim of the drawing is primarily to show the details that exist in an area no more than about 1–2 arc-minutes across.

The observer should always remember to indicate the approximate direction of north with an arrow. Use of an equatorial mounting makes this easier, because north is always in the same direction with respect to the tube of the instrument. With an altazimuth mount, it is not very difficult to orient the field by letting the stars trail; they leave the field on the western side. Anyone using diagonals or an odd number of reflections should ensure that two directions, north and east for example, are always indicated on their drawings. In so far as it is possible, it is best to keep the optical train simple, preferably with just the main optical elements and an eyepiece. This minimizes unnecessary absorption. Nebular filters may prove useful for observers who work under skies affected by light-pollution. We will not pass judgement on these filters, but surely the value of observing comets under such conditions is rather doubtful?

There is no sure recipe for success. The sensible use of different magnifications, together with the observer's experience, may sometimes give surprising results, even under far-from-ideal conditions.

Instruments on a driven equatorial mounting make observing much easier, especially when high magnifications are used. Here again, much depends on patience and experience. There are even some advantages to using an undriven telescope, because the gradual drift of the comet across the field moves the fuzzy spot of light and the various internal details across different parts of the retina. The image does not stay in the same spot. Deliberate use of averted vision shows that certain parts of the eye are better than others at detecting low luminosities or in resolving power.

The main thing is to be able to see something and to draw it. We have not yet spoken of drawing the comet in the right position. This is the last thing to be done.

We have taken our time over drawing the star field. Once the photometric centre of the comet has been drawn, we must work a bit quicker.

8.4.2 Tricks of the trade in drawing comets

It is hardly necessary to state that drawings are made with pencil (grade HB, for example) on suitable paper. The photometric centre of the comet should be positioned very carefully. All the main features will start from this small point. The length and direction of these features should be carefully drawn in relation to nearby stars. We can now understand why the drawing will be of no value if the comet moves against the background stars to any considerable extent whilst the details are being drawn. The observer starts by sketching the most obvious features: small bright jets, for example, and irregularities in the central condensation. The area all round the centre should be very carefully examined. It may appear elongated or even double. The widest streamers are simply indicated with shading, paying attention to areas of different density. Evening out the areas of different contrast may be finished later when the drawing is tidied up; time should not be wasted on it during observation.

The drawing is made in a somewhat staccato fashion:

- 15 seconds at the eyepiece, taking particular note of one feature;
- this is immediately drawn on the paper, which is illuminated with a very faint light;
- a slightly longer time is spent at the eyepiece (adaptation time plus further careful examination of the same feature);
- the drawing of the feature is finished;
- back to the eyepiece to find another feature, etc., moving the eye frequently to different parts of the field.

If you are doubtful about some feature, change the magnification: higher if the detail appears linear and fairly strong in contrast, so as to eliminate some of the fuzzy background; lower if you want to confirm the way a jet fans out, etc. First impressions are often the best, and drawing should be done spontaneously, not in a laboured fashion.

The nuclei of very active comets (new comets) sometimes break up in a spectacular fashion, as happened with Comet West in 1976 March (Fig. 8.14). Occasionally, when the different pieces have sufficiently large sizes and lifetimes, they may take on the appearance of independent comets. It is very important to record the positions of the various secondary condensations as accurately as possible. Analysis of their motion is practically the only direct way of estimating cometary masses. Observers should, however, always be careful of stars that appear close to the photometric centre or that are surrounded by the general cometary nebulosity. They may create the illusion that the nucleus is double or even that of a straight jet. The same type of illusion that is encountered in the study of double stars and of fine detail on planetary surfaces: the eye tends to join two neighbouring spots, linking them with 'bridges'. Change the eyepiece!

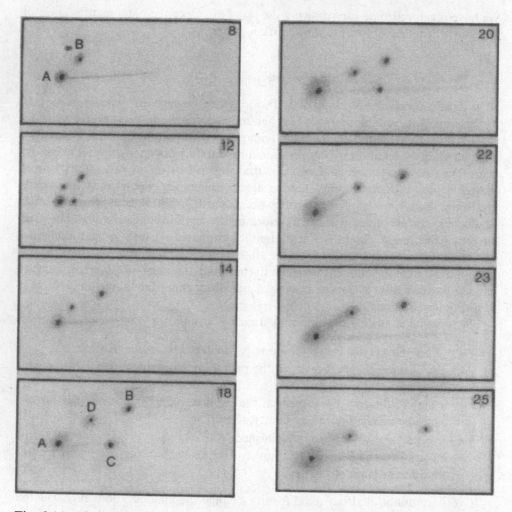

Fig. 8.14. *The break-up of the nucleus of Comet West (1976 VI) in 1976 March as shown in drawings made by John Bortle with a 310-mm reflector (magnification 146×), between 1976 March 8 and 25. The distance between the two components A (the main nucleus) and B was about 4 arc-seconds on March 8, and reached 18 arc-seconds on March 25. Perihelion was on 1976 February 25. Fragmentation began about 12 days before perihelion. According to Z. Sekanina, the first event was undoubtedly separation of component D; it is possible that B split off from D. C was the last component, and separated from the main nucleus about 10 days after perihelion; it had a very short lifetime (about 3 weeks). Each of the two pieces B and D showed its own plasma tail. At maximum brightness both of these fragments reached a magnitude of about 5–6. After such an event, obviously Comet West will never be the same again. (After* Sky and Telescope, *1976 June, p. 387. By kind permission of Sky Publishing Corporation.)*

Some comets also show several levels of brightness, with rays curving in opposite directions. The first look reveals one ray. The next tends to show a different one. Everything should nevertheless be noted down. Not all the features are necessarily visible at the same time. A good test is to make a second drawing with a different magnification. If the second drawing is made truly objectively, we shall probably be surprised to find that the two drawings are similar, despite everything appearing different to begin with.

It is absolutely essential to draw objectively, without being biassed by any large-scale structure that may have been observed. Various magnifications allow us to see details with differing contrast, and at various distances from the photometric centre: dense jets close to the nucleus with a high magnification, diffuse, larger streamers at lower magnification. Everyone tries their own 'tricks' for getting the best results. Some are:

- tapping the telescope to make it vibrate: diffuse areas are easier to see when they move;
- relaxing for a couple of minutes by just looking at the sky. In computer terminology, this is equivalent to 'rebooting'. It is not a good idea to spend too long at the eyepiece. One ends up seeing what one wants to see. The eye is an imperfect detector, which rapidly gets tired but which has the important advantage of being able to react to small stimuli; it is important to treat it properly.

Beginners should not start by making drawings at high magnification. The simple reason is that they will not see very much. It is preferable to gain experience first with lower magnifications. Conditions are also less demanding, because the movement of the comet is less perceptible. The degree of complexity in cometary drawings made by beginners and advanced workers is probably fairly constant. The former have difficulty in drawing the few details that they do see, whereas the latter see so many details that their experience is constantly put to the test and they have to work quickly.

At roughly the magnification that gives the best resolving power, the drawing should be finished in about 10 minutes, say 15 at the most, otherwise one runs the risk of making mistakes in the orientations because of the comet's motion in that time, and may also spoil the drawing by too many erasures. After 15 minutes it is best to begin another drawing, again first drawing the surrounding star field. The time should be noted to within 5 minutes: this should be the time of the mid-point (± 5 minutes) of the period spent drawing the cometary features, counting from the time the photometric centre was placed.

There is one important point for all those who observe in a group: Quiet Please! Everyone should make their own drawing, without commenting aloud, and without looking at anyone else's drawing. If observers influence one another, they rob themselves of a great deal of satisfaction when they do come to compare their drawings and find that they have shown the same features in the same positions.

Figure 8.15 shows the sequence of steps in making a drawing, from the initial location of field stars to the final, quick, penciling-in of areas of different density.

Beginners tend to produce drawings that are too small and illegible. To ensure

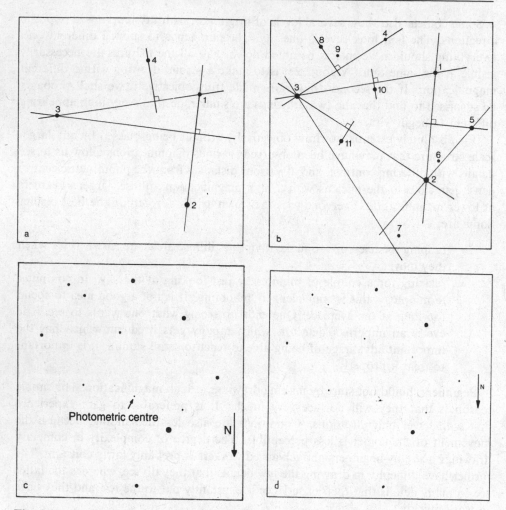

Fig. 8.15. *Making a cometary drawing.*

a: The brightest stars are plotted in decreasing order of magnitude (stars 1, 2, 3 and 4) starting with the stars farthest from one another. The chosen star patterns should encompass the position of the comet.

b: Fainter stars are plotted relative to the patterns of brighter stars. Plot all the stars detected in the immediate area of the comet's photometric centre.

c: Once the star field has been plotted, indicate the approximate direction of north (and also east if the field has been reversed by using a diagonal). Indicate the photometric centre of the comet as carefully as possible. The time will be the nominal time of the observation (T) less 5 minutes.

d: First draw the cometary features with the highest contrast. Diffuse areas are indicated by shading of varying density. Sketching of cometary features should stop at T + 5 minutes.

For subsequent stages, see Fig. 8.18.

that the scale is sufficiently large, it may be an idea to use a standard blank for each magnification that is used. The scale might be 10 mm per arc-minute, for example, which would give quite a good angular resolution (6 arc-seconds per mm). The exact scale should always be checked later with an atlas, as we shall describe in the next section.

Once finished, further comments should be added to the drawing to supplement the details recorded when beginning the wide-field drawing. We might note down verbal descriptions of the features recorded: the type of condensation ('very small and bright, diameter less than 10 arc-seconds, etc.'), curvature of the jets (for example: 'initially towards the north, then curves round towards the west'), details of areas of marked contrast as well as darker areas (the 'shadow of the nucleus') in the antisolar direction, etc. Such information is very useful when we come to work up the final drawing.

On some rare and debatable occasions, some observers have reported scintillation of the 'nucleus'. Don't forget to check whether the stars are scintillating in phase!

With active comets it is very instructive to make several drawings at intervals of 1–1.5 hours. The average lifetime of jets is a few hours, apart from certain persistent features. Some comets appear very different over a period of as little as 2 hours. The appearance also varies considerably from one day to the next. This is not confined to just new comets, but may also apply to periodic comets that are still rich in volatile material (P/Kopff, for example).

8.4.3 Producing a final drawing

The final drawing, worthy of being sent to a central organisation, is prepared from three sets of information:

- the sketch that we have just finished;
- calibration details: scale and orientation;
- a 'clean' drawing, which should be a faithful reproduction of what we saw.

8.4.3.1 Providing a scale

Before preparing the final drawing, we need to calibrate the original sketch. We can start by tackling the problem of identifying the star field in an atlas. Our first concern is to determine the position of the comet at the time of observation. Section 8.9.1 gives a formula for calculating coordinates by non-linear interpolation. Those who have access to powerful methods of calculation may even calculate the coordinates by applying Kepler's laws to the orbital elements to obtain the comet's position at the precise time of observation. As we have already said, a linear interpolation between two dates given in the ephemeris is not adequate unless the motion of the comet is linear and uniform over the period concerned. Assuming uniform motion may cause nearly as many problems in identifying the field as neglecting to draw the field stars. It is also worth mentioning that the α and δ coordinates of the comet must be expressed in the same epoch as that used for the charts in the atlas. Once the position has been calculated, we can start to compare our drawing with the atlas. The approximate indication of north that we included in our observation

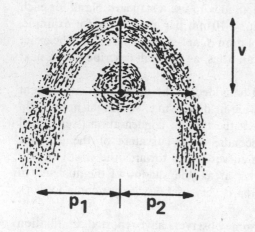

Fig. 8.16. *The semi-latus rectum, and the features that need to be measured in a cometary coma. The envelope found in the coma of a comet is not necessarily symmetrical. Note down the distances p_1 and p_2 across the photometric centre, at right angles to the theoretical direction of the tail. In addition, make a note of the distance between the photometric centre and the top of the envelope (the vertex distance, v). (After F. L. Whipple,* I.H.W. Amateur Observer's Manual.*)*

helps to orient the sketch, which generally has to be turned upside-down, because the usual orientation of telescopic fields is with north at the bottom.

When the field has been located in the atlas, we have to choose two moderately bright stars, fairly widely spaced on each side of the comet, that appear on both the chart and on our drawing. We measure the distance between the two stars on the atlas with a rule of some sort. It is not very difficult to estimate the distance to a tenth of a millimetre. A graduated loupe (or a linen tester) would allow us to get even greater accuracy. Knowing the scale of the atlas, the distance between the two stars may be converted into arc-minutes. With *Atlas Stellarum* the distance in millimetres may be simply multiplied by 2 to obtain the separation in arc-minutes: 4.5 mm gives 9 arc-minutes. With *Atlas Falkauer* we have to multiply by 4, i.e., 3 mm becomes 12 arc-minutes. We have now scaled our drawing, because we know the angular separation between the two stars.

Measurement of the size of the coma and of jets and other features is made with reference to the distance sepatating the two stars on our drawing: we can make the measurements in millimetres and use the simple rule of three to obtain the arc-minutes. The accuracy required is about 0.1 arc-minute for small and average-sized comets. It seems pointless trying to obtain such an accuracy for very extended and diffuse comets that are larger than 15–20 arc-minutes, the limits of which are difficult to detect. We can be content with an accuracy of about 1 arc-minute, or even 5 arc-minutes if the coma is more than 1° in diameter (as with Comet IRAS–Araki–Alcock). In any case, the size of the coma is obtained only for general information. In principle, the diameter of the coma is measured through the 'nucleus', perpendicular to the direction of the tail. This is not necessarily where the coma is widest; Fig. 8.16 shows how the measurement should be made. Some astronomers are interested in the half-width of the envelope (the semi-latus rectum), which is used in analysing the evolution of cometary haloes. This serves as a reminder that any asymmetry should be noted, whether it occurs along the axis of the tail or perpendicular to it.

The length of the jets is measured from the photometric centre using the same method of scaling that we used for the coma. While we have a calculator handy, it is a good idea to calculate the distance that represents 1 arc-minute: $1' = x$ mm. This should be shown on the final drawing in the form of a bar (as in Figs. 8.5 and 8.18).

The length of the tail is measured in the same way, provided it is not too long. If it exceeds $10°$, its length will have to be determined by calculation, because of the distorting effect of the projection used in the atlas. As we have already indicated, under these circumstances, it is possible to record the extent of the tail directly on a small- or medium-scale atlas (such as the *AAVSO Atlas*, *Becvár*, *Norton's*, etc.). The observer should carefully measure the coordinates (α_1 and δ_1) of the head (the photometric centre) and of the farthest visible extent of the tail (α_2 and δ_2). The length L in degrees is then calculated as follows:

$$\cos L = \sin \delta_1 \sin \delta_2 + \cos \delta_1 \cos \delta_2 \cos(\alpha_1 - \alpha_2).$$

The angles are all expressed in degrees; the right ascensions α_1 and α_2 should be converted into degrees, given that one hour of RA equals $15°$. A relative accuracy of 10% is sufficient: $\pm 0.1°$ for a length of $1°$; $\pm 1°$ beyond $10°$ in length.

8.4.3.2 Measurement of position angles

We shall use the standard abbreviation 'PA' for position angle. Determination of PA is made on the rough draft; drawing lines on it to show the direction of North, extending the line of the jets, etc., from which we can then measure the PA. For practical reasons we avoid marking the sketch of the comet itself as otherwise it would be completely lost in a thicket of lines. We continue to use the same two stars that were used to determine the scales. Lay a protractor on the atlas, centred on one of them. We then only have to measure, to an accuracy of a degree, the angle between north and the line joining the two stars. Alignment of the $0°$ graduation on the protractor should be carried out very carefully by use of the coordinate grids. With photographic atlases (such as Vehrenberg's) considerable care should be taken to ensure that the transparent grids are perfectly aligned with the centre and the north–south meridian on the chart before attempting to measure any position angles. We then transpose the angle onto our rough sketch, and indicate the direction of north. It is not a bad idea to check that the direction of north that we have measured does correspond to the approximate direction that we noted down during our observation.

We now only have to measure the PA of the cometary features shown on the sketch. Prolong the direction of the jets or other linear features by a fine line until it cuts the north–south line. Now the protractor: make sure the angle measured with respect to north is in the right quadrant. Remember that PA is measured from north and increases through east (Fig. 8.8). The angles are rounded to the nearest degree. The error in such measurements is generally about $\pm 3°$.

It is not always easy to measure the PA of wide jets; take the central axis and record the angles tangent to the outer limits when the jet is very broad. Two measurements are made for curved jets:

- the PA at which it leaves the photometric centre;

- the angle indicating the direction of the end of the jet. This direction does not pass through the photometric centre. We might note, for example: 'jet at PA 45°, curved towards PA 120°, length 3 arc-minutes'. In this case, the length represents the distance between the photometric centre and the visible end of the jet.

Distortions may sometimes be visible in the gas tail. The PA of the various changes of direction should be measured, as well as the distances at which they appear, taken from the photometric centre (Fig. 8.17).

This part of the process of producing a final drawing is fairly long. What has to be done is quite simple, however. Any amateur who is used to working with an atlas will find no difficulty and will complete that part of the work in a few minutes. Once the information recorded during the observation has been finished, if the drawing is clear and contains all the necessary reference stars, there is no reason to be held up by these measurements. The accuracy of the numerical results is strongly dependent on the care that has been taken in drawing the star field, particularly as regards the position angles.

When the size and the orientation of the jets, etc. have been determined, these may be entered on the detailed forms issued by most of the coordinating bodies (International Comet Quarterly, etc.). It would be a shame to stop there. We have finished most of the work. We need not deny ourselves the satisfaction of preparing a record that summarizes all that we have been able to see of the comet.

8.4.4 *Making the final drawing*

Preparing a final drawing does not have much to do with astronomy. It simply involves another technique that we can learn by following specific instructions. The aim is to provide a record that matches all that we have observed, and that will be immediately comprehensible to anyone examining it. So this is not the time to make an 'artistic' drawing, embellished with various flourishes, or touched up so that the jets appear as beautifully symmetrical fountains.

The final drawing is done on tracing paper, not merely for the sake of appearances, but because this is the only way in which the scale and orientation shown on our rough draft may be properly reproduced. It would be absurd to make another drawing from our original sketch: we would introduce more errors into the lengths and the position angles of the jets, rays, etc. The operations are carried out in exactly the same order as they were originally at the telescope. We start by orientating the sheet of tracing paper with north at bottom, east at right, as in the telescope. The directions should be marked by two arrows in a corner of the drawing. These arrows should be sufficiently long for anyone analysing the drawing to be able to extend the arrows so that any PA may be measured. At the top we should give:

- name and designation of the comet;
- date and time of the observation in UT; the time of the middle of the period in which the cometary structure was drawn, to within 5 minutes;
- the type and aperture of the instrument (in mm), the magnification(s) used.

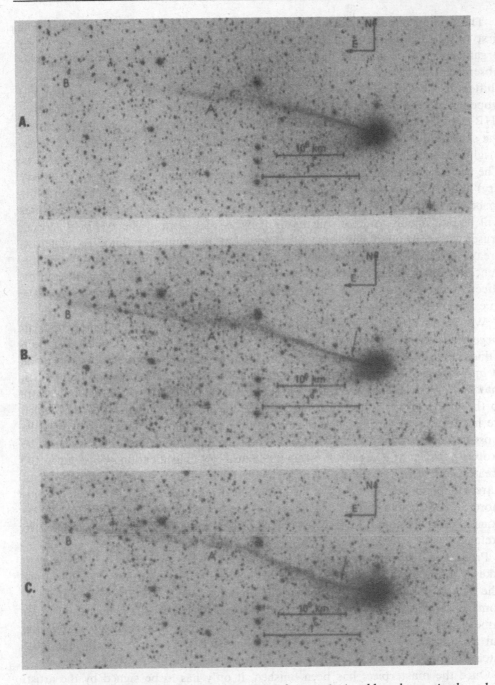

Fig. 8.17. *Comet Bradfield, 1979 X, on 1980 February 6. A sudden change in the solar-wind flux caused an elegant curve in the gas tail. The part of the tail marked A–B retained practically the same orientation. Times of mid-exposure for the photographs, top to bottom are: 02:32:30 UT; 02:48:00 UT; 03:00:00 UT (Photographs: Joint Observatory for Cometary Research, NASA-Goddard Space Flight Center, and New Mexico Institute of Mining and Technology.)*

The tracing paper is laid over the rough draft, with north properly aligned. The first thing to be done is to trace the positions of the stars. This will allow any organisation to which the drawing is sent to check that the correct object was observed. It also has another advantage in that if the tracing paper is accidentally shifted, it is easy to reposition it correctly. The stars can be marked with a fine felt-tipped pen, keeping their relative brightness. The rest is done with a fairly soft pencil (HB–2B). Indicate the photometric centre of the comet. Then follow approximately the same order in which the details were originally recorded: first the small, dense, bright jets near the nucleus, then using light shading, the more diffuse jets and rays. The various diffuse patches are obtained by using a blotting-paper 'stump' or a (very dry) finger. We can hardly expect to reproduce the actual densities. For the drawing to be understandable, we have to exaggerate the contrasts, otherwise the features will be as difficult to detect on the drawing as they were at the eyepiece! Use an eraser with caution, it may leave oily marks that may ruin everything. To lighten an area, a razor blade or scalpel may be used with care and at the very end of drawing. Penciling over scratches left by a razor blade may be disastrous. A blade is most effective when used on very dense areas (black in the drawing). Everyone will have to experiment for themselves.

We would advise beginners to make a general drawing with a moderate enlargement, and to be content with this initially. When some experience has been gained, then something more complicated can be tried: combining drawings made at more than one magnification. This gives a record that as yet no photograph can match, showing all the variations in density from the brilliant photometric centre to the limits of the coma. The first part of the process is identical to that which we have described so far. The tracing-paper copy is laid out and copied from the most detailed drawing made with the highest magnification. The second part of the process consists of using the drawing made at a low magnification, which shows the greatest extent of the coma and possibly the beginning of the tail. The network of stars will again help to lay down the extensions to the luminous jets and the various more diffuse structures that extend several arc-minutes from the central condensation. Figures 8.5 and 8.18 show examples of cometary drawings that combine two sketches.

Producing such an image means working fast at the telescope. It is essential to take no more than 20 minutes for both sketches, at high and low magnifications. The accuracy required is less for distant features, which may be poorly defined. Some observers prefer to start with the strongest eyepiece and then to finish the same sketch by adding the diffuse features. Everyone has their own way of working, but as long as they ensure that the scale is correct.... Don't forget that the most interesting features are those in the immediate vicinity of the nucleus.

Once the masterpiece has been finished, it only has to be signed by the artist. Don't forget to show a small bar indicating the scale (1 arc-minute = X mm). The astronomers who analyse your drawings will be delighted if you also indicate the direction of the Sun. You don't get that by guess-work, but by calculation. The position angle x of the Sun is obtained as follows:

$$\tan \theta = [\cos \delta_\odot \sin(\alpha_\odot - \alpha_C)]/[\cos \delta_C \sin \delta_\odot - \sin \delta_C \cos \delta_\odot \cos(\alpha_\odot - \alpha_C)],$$

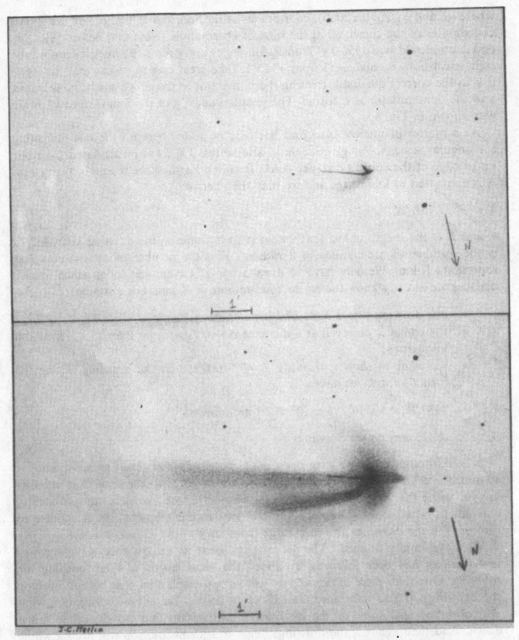

Fig. 8.18. *Comet P/Giacobini–Zinner, 1984e (1985 XII), on 1985 August 28 at 03:15 UT (time of the mid-point of the drawing), made with a 400-mm reflector.*
Top: *Bright structure near the nucleus observed at high magnifications (254× and 407×). The direction of the Sun is towards the right (the east). The narrow gas tail does not appear to originate in the centre of the nucleus but rather from a sort of luminous segment centred on the nucleus.*
Bottom: *The drawing is finished by including diffuse patches (streamers and tail) observed at a lower magnification (81×).*
(Drawing by J. C. Merlin. Taken from l'Astronomie.)

375

where $\alpha_\odot$ and $\delta_\odot$ are the 2000.0 coordinates of the Sun, and α_C and δ_C are the 2000.0 coordinates of the comet, all at the time of observation. [Note that before 1992, the epoch employed was 1950.0 – Trans.] All angles are given in degrees, including the right ascensions $\alpha_\odot$ and α_C (1 hour = 15°). Take great care to ensure that the angle θ is in the correct quadrant, treating the numerator in the above equation as a sine, and the denominator as a cosine. The position angle θ of the Sun is derived in the way shown in Fig. 8.8.

As a matter of interest, a second bar may be added beneath the one indicating the angular scale, showing the scale in kilometres. This is also calculated from our knowledge of the comet's distance Δ AU from the Earth at the time of observation: Δ is converted to kilometres and we may then derive:

$$Y = 3437.75L/\Delta,$$

where L is the length in km that we want to indicate on the drawing and 3437.75 is the number of arc-minutes in a radian. Y is the number of arc-minutes that represents L km. We only have to draw a bar Y arc-minutes long, which is easy because we already know the scale: 1 arc-minute = X mm. For example:

- our drawing is to a scale of 1 arc-minute = 10 mm;
- the comet is observed at a distance $\Delta = 0.5$ AU from Earth, i.e., 75 million kilometres;
- we want to show a distance $L = 50\,000$ km on the drawing. Using the formula for L, we have:

  $$3437.75 \times 5 \times 10^4 / 75 \times 10^6 = 2.3 \text{ arc-minutes},$$

  i.e. 23 mm on the drawing.

N.B. Although attempts have been made to standardize scales (1 arc-minute = 15 mm, for example), there is no need to worry about this. The scale that interests anyone using the drawings is primarily the kilometre scale, which is not particularly convenient when one is actually 'on the job' at the eyepiece. It is simpler to photograph the drawings and to enlarge them later to the appropriate scale.

We have finally finished. The 30 minutes spent at the eyepiece penciling in a few features has been followed by about the same length of time finishing the drawing. One final piece of advice: do not let too much time pass before finishing the drawings. If you wait more than 24 or 48 hours, you will have forgotten what you observed and some of the qualitative information that could not be noted down. And then again, if the drawings begin to accumulate, you will need a whole week-end to sort them out. A bit of organisation is required.

8.4.5 *Making use of the drawings*

An individual amateur does not generally have sufficient records available to be able to draw any conclusions about the rotation of the nucleus, its orientation in space or even the positions of active areas. This sort of information is deduced from statistical analyses that draw on hundreds of position-angle measurements of the jets. The reader who would like to know more about the results obtained from these methods should refer to the remarkable study by Fred Whipple of Comet P/Schwassmann-Wachmann 1 (*see* Bibliography).

If we want to contribute to such work, then we should send our drawings to competent organisations that are able to handle records from all over the world (such as International Comet Quarterly, or the temporary organisations set up to cover certain specific comets, etc.). Before starting to submit regular observations, we should check their accuracy by comparing them either with published observations or with those of colleagues who have observed independently (Fig. 8.19). Like magnitude estimates, drawings by one observer become of greater interest the more there are. It is easier to understand the changes that have taken place in a particular comet if the artist's techniques are familiar.

8.5 Estimating the total magnitude

The estimation of the brightness of a fuzzy patch is a very old problem, which has been the subject of considerable controversy. Again we shall find that the use of a standard procedure, accurate references, and a little practice will enable amateurs to produce usable results. Apart from a few exceptions, which are more because of individual interest than part of their work, professional astronomers do not make estimates of the visual magnitude of comets. Amateurs provide nearly all the data that describe the overall behaviour of comets. Availability of telescopes with apertures larger than 400 mm even enables some observers to detect comets down to 15th magnitude. These are the amateurs who are slowly pushing the limits of cometary observations farther and farther out into the Solar System, both in their detection and in following their activity visually. Photography, on the other hand, is generally limited to the star-like central condensation.

Although photoelectric photometry enables an accuracy of better than one hundredth of a magnitude to be obtained on stars, the same does not apply to extended objects. The differences between measurements made with a photometer and visual estimates are always in the same sense: the brightness is greater when estimated by the eye. This tends to make one think that the measurements do not apply to everything seen by the eye, either because of the size of the coma, or because of the selective nature of the spectral regions covered by particular emulsions or filters (or both) that are used in photometry.

The visual light-curves and the absolute magnitude of comets may be correlated with data derived from other spectral regions, such as the production rate of the hydroxyl radical, OH, obtained from radio observations, for example. Radio astronomers are particularly interested in our estimates. They have the big advantage of being quick. If they are sufficiently reliable, they may be enough to prompt someone to turn a major radio telescope onto a comet. In particular, the information obtained from visual magnitudes enables comets to be classified.

Contrary to drawing, which is more satisfactory if one can use a 'large' telescope (over 150 mm in aperture), estimating the magnitudes of comets is easier with small instruments. They should, however, be suitable for what we have in mind.

377

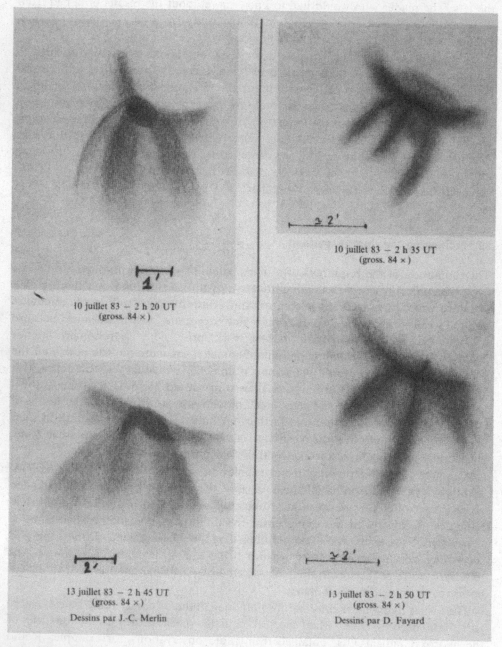

10 juillet 83 − 2 h 20 UT
(gross. 84 ×)

10 juillet 83 − 2 h 35 UT
(gross. 84 ×)

13 juillet 83 − 2 h 45 UT
(gross. 84 ×)

Dessins par J.-C. Merlin

13 juillet 83 − 2 h 50 UT
(gross. 84 ×)

Dessins par D. Fayard

Fig. 8.19. *Comet P/Tempel 2, 1982d (1983 XI). Drawings made independently by two observers using the 600-mm reflector at the Pic du Midi on 1983 July 10 and 13. Although each observer has his own way of rendering on paper what he sees through the eyepiece, the structures shown are perfectly consistent between the observers. It should be noted that these drawings are the only records obtained in visible light that clearly show the dust tail of P/Tempel 2 during its 1982–3 apparition. (Drawings by J. C. Merlin and D. Fayard, originally reproduced in l'Astronomie.)*

8.5.1 History, and basic principles

The first attempt to standardize techniques of estimating the brightness of comets was introduced by Pogson in 1856. He suggested comparing the brightness of the nucleus with neighbouring stars. It soon became apparent that this method led to considerable discrepancies between observers when a comet had an extended coma. Depending on the instrument used, the image of the comet appeared more or less condensed, so observers were not all measuring the same thing.

Towards the end of the 19th century, Holetschek partially solved the problem by suggesting the use of the minimum magnification, specifically so that the image of the comet would be as small as possible, and more easily compared with stars. The principle of defocussing the eyepiece was introduced around 1910–20 by G. Van Biesbroeck (Yerkes Observatory), the aim being to spread out the light from the stars over a diameter equal to that of the comet being observed. Accurate definition of the conditions under which defocussing should be used were standardized more recently by Bobrovnikoff (in 1941), when analysing observations made on P/Halley in 1910. Variant methods have since been proposed by Sidgwick (1949), Beyer (1952) and Morris (1979). We shall describe these techniques in Sect. 8.5.3.1.

From all these various methods, we may extract three basic principles for estimating the total visual magnitudes of comets:

- use the smallest possible instrument;
- defocus the star images;
- compare the comet with stars whose visual magnitudes are known.

The first principle is precisely the advice given by Holetschek. The best instruments are binoculars. If the comet is too faint for binoculars, use a short-focus instrument. The focal ratio should be about f/3 to f/6: classical Newtonian telescopes are perfectly suitable. Refractors are not advised, because their focal ratios are usually f/10, f/15 or even more. Some Schmidt–Cassegrain telescopes may be fitted with focal reducers, which give an effective focal ratio of f/5. The negative lens in the focal reducer does, however, introduce some absorption. The magnification used should be close to the minimum effective magnification ($D/7$–$D/6$).

There is no question about the principle of defocussing (or making extrafocal estimates). It is essential to compare like with like: fuzzy patch with fuzzy patch. 'Guesstimates' made without defocussing are of no value whatsoever. Observers who work this way are quite happy checking to see if their 'estimates' agree with the predicted values. Statistical analysis of their observations would be damning.

The third principle takes care of the problem of comparison magnitudes. The only reliable sources for visual magnitudes are stars. The magnitude of a comet should not be estimated by comparing it with a globular cluster, a galaxy or the integrated brightness of an open cluster, even if the latter appears diffuse to the naked eye or in a small telescope. It just is not possible to guarantee the magnitudes of non-stellar objects that are given in the catalogues. The magnitudes vary from one catalogue to another, and again, they would have had to be measured in a spectral region agreeing with the visual range. In any case, it is difficult to see how comparisons may be made between a comet and any neighbouring clusters or galaxies, which are of different shapes and sizes. A purely qualitative estimate ('comet brighter than M13', for example) is only of interest when an observer having no experience of making extrafocal estimates actually discovers a comet. If you aspire to become one of the lucky few whose names are perpetuated in the sky, don't wait for that fateful day when it would be very useful to give a realistic estimate of the magnitude of 'your' comet. Many reference books and atlases give the visual magnitudes of certain stars. You probably already have in your own library material that can be used for making extrafocal estimates.

8.5.2 Precautions to be taken

8.5.2.1 Predicted magnitudes

Comet ephemerides usually include predicted magnitudes. Two types of magnitude may be given:

- m_1: the apparent total visual magnitude of the coma;
- m_2: the magnitude of the nucleus.

There are two sources from which these predictions may have been obtained. First, proper observations made at earlier returns of previously observed, periodic comets. Second, estimates of the behaviour of new comets, whether periodic or not. The photometric parameters of a comet: the absolute magnitude H_1, and the activity parameter, or measure of the sensitivity of the magnitude to heliocentric distance, n, are defined on p. 347.

As we have already indicated, m_2 magnitudes are, as far as we are concerned, of very general interest only. They are given with the ephemerides of faint, distant comets, for which significant m_1 values may no longer be derived, because activity is generally more or less non-existent beyond 3 or 4 AU. When a comet is stellar in appearance, determination of the absolute magnitude simply comes down to using the equation:

$$H_1 = m_2 - 5 \log \Delta - 5 \log r,$$

and which applies for any time of observation. The value of m_2 is known (it is observed), and Δ and r are given in the ephemeris. We can use this equation to obtain the approximate magnitude for comets that are observed at considerable distances from the Sun, and which are assumed to be inactive. For example:

- On 1982 October 16, when Comet P/Halley was detected by Jewitt and Danielson with the 5-m telescope at Mount Palomar (with a CCD), its magnitude was estimated as $m_2 = 24.2$; it was at a geocentric distance $\Delta = 10.9$ AU, and at $r = 11.0$ AU from the Sun. From this we deduce that the absolute magnitude H_1 of the nucleus was 13.8. According to Michel Festou, the true absolute magnitude of the nucleus should have been closer to 15; it is therefore possible that a certain amount of activity was already occurring in P/Halley in 1982.
- Comet P/Schwassmann-Wachmann 1, which always remains beyond the orbit of Jupiter, was photographed in quiescence in 1983 April by Cruikshank

and Brown, with the NASA 3-m infrared telescope on Hawaii: $m_2 = 17.6$, $\Delta = 5.42$ AU, $r = 6.30$ AU. We obtain an absolute magnitude H_1 of 9.9.

Corrected to 1 AU from the Earth and 1 AU from the Sun, the nucleus of P/Schwassmann-Wachmann 1 is therefore a much larger object than the nucleus of P/Halley. Most short-period comets have absolute nuclear magnitudes fainter than 13. P/Schwassmann-Wachmann 1 is an exceptional object that we shall discuss later.

We may think of m_2 magnitudes as being of interest only to astronomers who photograph comets far from the Sun, using powerful telescopes. We shall return to this question later in Sect. 8.7.5. The magnitudes that primarily interest us are therefore m_1 magnitudes. They mainly involve the head. We are not trying to estimate the magnitudes of tails, which are too diffuse and the visibility of which depends on too many different factors.

For periodic comets where we are observing the 'n-plus-one-th' return, the ephemerides may be accompanied by realistic predicted magnitudes, based on analysis of earlier visual observations. As we might expect there are certain limits to such an extrapolation. First of all, it is essential that the comet should really have been observed visually. There are many very-short-period comets whose earlier returns have all been unfavourable for various reasons: an apparition that was mainly confined to the southern hemisphere, which is not as well covered by amateurs as the northern sky; a small elongation; perihelion at the time of conjunction with the Sun, etc. This is why Comet P/Churyumov–Gerasimenko, which completes its orbit in less than 7 years, was not observed visually until the favourable return of 1982, which was its third known apparition, it having been discovered in 1969. The data collected by amateurs in 1982 were crucial in assessing the level of activity of this object.

We might therefore assume that with a well-observed periodic comet that has had one or more returns we would be able to obtain accurate predicted magnitudes. This it true in two cases:

- Old comets where the dust component has been entirely lost and which have essentially no further resources, or surprises to spring on us – for example P/Encke, which has been observed at 53 apparitions.
- Comets that have a particular type of regular behaviour, such as an outburst a specific number of days before or after perihelion, which may therefore be easily predicted – as with P/d'Arrest, which has been observed at 14 apparitions (Fig. 8.20).

The number of comets with known light-curves is very small. There are also a large number of 'quiescent' comets, where for many decades the brightness has simply followed their variation in heliocentric distance, never departing very much from a value that may be expressed as an average absolute magnitude plus a parameter expressing the amount of activity – in short, they show low activity. And then, one day, we hear that an amateur who has been searching for comets has accidentally come across a known object that is far brighter than predicted. Or else a professional astronomer photographs (no less accidentally), a comet far from the Sun when it is

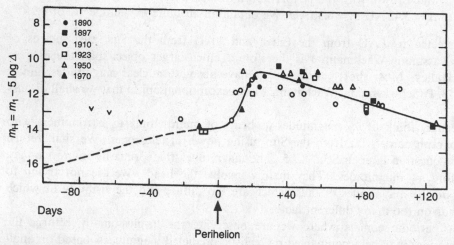

Fig. 8.20. *Light-curve of Comet P/d'Arrest from observations collected over six apparitions. Observations made at each return are identified with separate symbols. The magnitudes have been reduced to a constant distance from the Earth (1 AU): they are heliocentric magnitudes* m_H. *Comet P/d'Arrest regularly shows a jump in brightness a few days after perihelion passage. (After J. E. Bortle,* Sky and Telescope, *1976 July.)*

receding into the depths of the Solar System, and which shows an unexpected degree of activity at a great heliocentric distance. It is worth remembering that comets are objects whose behaviour always warrants close surveillance.

To have some idea of the order of magnitude of these effects, consider the photometric parameters of Comet P/Halley at its return in 1910:

- before perihelion: $H_1 = 5.47$; $n = 4.44$;
- after perihelion: $H_1 = 4.94$; $n = 3.07$.

It was therefore brighter after perihelion than before. It takes longer to fade than it does to brighten. This is one of the reasons why we can follow P/Halley for a long time after perihelion.

Prediction of the magnitude of new comets is far more difficult. Many will remember Comet Kohoutek 1973f (1973 XII), which had been described in many newspapers as the 'Comet of the Century'. It was a typical new comet in Oort's sense. It was discovered from Hamburg Observatory in 1973 March. It was then close to the orbit of Jupiter ($r \approx 5\,\mathrm{AU}$) at a magnitude of about 16. It was therefore relatively bright and a promising object. Perihelion passage would not take place until 1973 December, which left sufficient time for a lot of experiments to be prepared and for many observers to be mobilised ... as well as the general public. The predicted magnitude was calculated using the traditional equation with $n = 4$ (*see* p. 347). This suggested that the comet would attain magnitude -5 at perihelion and shortly afterwards, at the beginning of 1974, become an easy object for anyone to see in the evening sky without any optical aid. Although Comet Kohoutek was thoroughly studied, it was a great disappointment for most of us. It was observable with the naked eye for a few weeks in 1974 January, but hardly as well as P/Halley in 1986. Study of its brightness showed that its absolute magnitude was around 5.5–6 and that *n* was in fact about 2.5! The nucleus was probably relatively large, but with such an activity factor, 1973 XII was certainly not a super-comet. Spectroscopic studies showed that the dust component was minimal, either because the nucleus was poor in dust, or because the degassing did not release the trapped dust. It is undoubtedly the ejection of dust particles that makes some comets so spectacular (Arend–Roland, West, etc.), but such behaviour is unpredictable. The nucleus of 1973 XII was probably not covered in a layer of protective dust like P/Halley. Even at a great distance, the solar radiation was just acting on bare layers of ice. The nucleus had hardly any source of material left after perihelion.

Predicted magnitudes are therefore strongly affected by the magnitude at the time of discovery. If we also take into account the fact that many comets are discovered when they undergo outbursts of activity, we can easily understand that extrapolating magnitudes may cause disappointment when the brightness returns to normal. This problem applies to both new comets and short-period ones.

There are numerous examples that show the value of visual observations, whether these are made at the time of discovery or as routine checks on known comets. An accurate visual estimate is always more valuable than an estimate taken from a photograph. With rare exceptions (such as when the comet is stellar in appearance, or when the observer is particularly experienced), integrated magnitude estimates made from photographs are always pessimistic, or even wildly inaccurate. The differences from visual estimates are between 2 and 5 magnitudes, depending on the appearance of the comet and the extent of its head. If a predicted magnitude for a diffuse comet is based on an estimate from a photograph, then you can cheerfully assume that the comet is 2–3 magnitudes brighter visually. If a magnitude of 15 is suggested, the comet may be between magnitude 11 and 13. The moral is: don't take a photograph, use an eyepiece!

Finally, we would give two pieces of advice to observers:

- Systematically try to find all comets with predicted total magnitudes of more than, or equal to, 15. It is also probably worth being interested in all comets with m_2 magnitudes of about 16–17. Any development of a coma will mean that these objects become visible in apertures of 200–400 mm with m_1 magnitudes of about 11–15. Repeated visual observations would enable us to know more about certain objects that have been neglected in the past: new comets that show sudden outbursts, or periodic comets that have not been observed visually at earlier apparitions. We should also mention that for certain comets negative observations (i.e., comet fainter than magnitude …) are of undoubted interest: these are comets that show outbursts such as P/Schwassmann-Wachmann 1. In such cases, the magnitude of the faintest stars that are visible should be noted.

- Do not let yourself be influenced by predictions. Both as they are approaching, and as they are receding, comets may show random variations from one day to the next, sometimes even within a few hours. Don't forget that the photometric parameters characterise the overall behaviour of a comet throughout an apparition. The intrinsic fluctuations in brightness will be deduced from a detailed analysis of all the observations obtained. They may then be related to the activity detected around the nucleus (i.e., by visual observation).

8.5.2.2 *Magnitude reference material*

We now come to a question that must always concern observers. The collection of reliable material that will enable us to estimate the magnitude of comets is a long-term affair. First, though, we must distinguish between an atlas and a catalogue. We have already spoken about some of the atlases available:

- plotted atlases: *Norton's, Coeli, Borealis*, etc., *Atlas 2000.0, AAVSO, Brun*, etc.;
- photographic atlases: *Falkauer, Stellarum, Papadopoulos, Palomar*, etc.

Among these atlases, two give specific details of visual or photovisual magnitudes alongside certain stars: the *AAVSO Variable Star Atlas* (described in detail in Sect. 8.3.4.2) and Brun's *Atlas Photometrique*. These are the ones of most interest to us. Brun's *Atlas Photometrique* was compiled by the founder of the AFOEV [Association Francaise d'Observateurs d'Etoiles Variables = French Association of Variable-Star Observers – Trans.], about 1910. It is an accurate source of stellar magnitudes down to 6.50 (from the Revised Harvard Photometry), and slightly less accurate for some fainter stars (Harvard Durchmusterung). This atlas was republished (in photocopied form) a few years ago by the SAF. It covers the sky from the North Pole ($+90°$) to declination $-30°$ and shows all the stars down to about magnitude 7.5, with explicit and systematic information about visual magnitudes. The scale is 10 mm to 1 degree: the coordinate grid for 1900 is drawn on the charts. The Brun *Atlas* would be a good basic reference that could be acquired before becoming involved in searching for sources suitable for fainter comets.

Photographic atlases are unusable for estimating cometary magnitudes for two reasons:

- they do not explicitly mention any magnitudes (which is obvious, because they are simply a collection of photographs);
- they were generally taken in a different spectral region than the visual, either in the blue (m_{pg}) as with the two Vehrenberg atlases and the Palomar blue survey, or in the red, as with the red version of the Palomar survey.

Even the *Papadopoulos Atlas*, which was designed to show photovisual (m_{pv}) magnitudes, cannot be used, because there are no photometric sequences that may be used for secondary reduction of any fields that interest us. It is, in any case, not obvious that any such manipulation would be very reliable on prints – it is normally carried out on original negatives.

A catalogue is usually a forbidding work that contains mainly pages and pages of numerical data. Some atlases such as Becvár and Tirion's *Sky Atlas 2000.0* are accompanied by their respective catalogues: *Atlas Coeli Catalogue* with Becvár and *Sky Catalog 2000.0* with Tirion. These volumes (which one buys separately) give numerical data for the majority of the stars plotted in the atlases. In particular,

Sky Catalog 2000.0 gives the visual or V magnitudes (the latter in the UBV system) for stars down to magnitude 8. These magnitudes are reliable. If we want a larger list, containing fainter stars, we could refer to the *Smithsonian Astrophysical Observatory Catalogue* (SAOC) which gives information on about 260 000 stars down to approximately magnitude 9.5. This catalogue is a compilation of older catalogues (such as the AGK2, Cape Catalogue, and Yale Catalogue). The visual magnitudes given for stars north of −30° were determined by measuring the diameters of stars on plates (taken in the second half of the 19th century). The differences between these values and magnitudes obtained by photometry or densitometry are sometimes as much as 0.4–0.8 magnitudes. For lack of anything better, observers can use the SAO magnitudes, but it is generally preferable to search for more recent sources, even if the stars are slightly farther away from the comet. The magnitudes given on the *AAVSO Atlas* (to tenths of a magnitude with the decimal point omitted, as on all sequences) are revisions of older estimates. They should be used for preference. These revisions have often been made as a result of work on variable-star comparison sequences.

This brings us to the best source. Amateurs can obtain variable-star sequences from the specialist groups (AFOEV, AAVSO, BAA Variable Star Section, etc.); some examples of typical sequences are given in the chapter on variable stars. Around 2000 stars are regularly monitored by variable-star observers. Possessing all the existing sequences is an ideal that few will be able to contemplate, but one which we may well try to attain. These are the only charts that give visual magnitudes beyond 13, sometimes as low as 16 or even 18. There is one complication in that sometimes errors are found in variable-star comparison sequences, either when the zero-point is in error, or when one of the comparisons itself proves to be variable. The required correction generally consists of a simple revision upwards or downwards, but because of the way in which comets move, this may have a disastrous effect on cometary light-curves. When a comet is moving rapidly, we are practically forced to use different comparison stars each night. So some parts of a comet's light curve may be affected by faulty comparison-star magnitudes. In every case, observers must scrupulously note down which stars in which sequence have been used for their estimates, so that corrections may be applied if the magnitudes of the comparisons are recalibrated. It is perhaps preferable to use the same sequence for as long as possible, even if, day by day, it becomes farther away from the comet, rather than changing sequence too frequently. After using the same sequence for several days, identification of the comparison stars becomes a very rapid process, and the gain in time is appreciable. Pay attention to differences in transparency between the fields that are being observed (the comet's and the comparison stars'). Some observers take this principle to the extreme and always use the North Polar Sequence, whatever the position of the comet. Stars in the area around the North Celestial Pole have been accurately calibrated down to very faint magnitudes. The sequence has the further advantage of being visible all the year round. However, using this method does imply that, whenever an estimate is made, careful attention must be given to the differences in transparency between the region round the Pole and that containing the comet. In practice, this method has one major advantage; because only one sequence is involved, observers rapidly find the comparisons that they require for their estimates. When observers submit their final reports, they must take care to indicate the sources of their comparison-star magnitudes. In addition, they should carefully retain all their original estimates, as with variable-star observation.

Among other reference sources that show stars down to a sufficiently low magnitude, we should mention the UBV sequences that exist for certain open clusters, and the fields around certain galaxies that have been the site of supernovae. Only V magnitudes should be used. [Another source of visual magnitudes are the charts that have been issued by groups searching for supernovae, such as the UK Nova/Supernova Search Programme – Trans.]. Brun and Vehrenberg's Selected Areas are unusable in their original form because they were calibrated in m_{pg} (B magnitudes). Some Selected Areas were measured in V magnitudes in the 1970s (Landolt, etc.). The use of V magnitudes gives, in principle, results that are very close to visual magnitudes and allow us to attain the required accuracy of ± 0.2 magnitude.

The bibliography gives a list of photometric references recommended by the *International Comet Quarterly* as being suitable for visual, integrated-magnitude estimates. They are listed in order of preference.

8.5.2.3 Observing conditions

Estimates of cometary magnitudes are strongly affected by the prevailing transparency. As we have already indicated, observational reports should include the diameter of the coma that was detected. This will give an indication of how much of the comet was seen and estimated. Magnitude estimates should not be made when various sources of stray light are present, such as street lighting, mist, and moonlight.

Proper dark adaptation is essential. This is why we have suggested that observation should begin with a general drawing at low magnification (Sect. 8.3.6), allowing time for the farthest extensions of the coma to become visible.

8.5.3 Extrafocal methods

8.5.3.1 Techniques

The three different methods are shown schematically in Fig. 8.21. They are:

The Bobrovnikoff method (Fig. 8.21a) consists of defocussing the stars until they are similar in size to the coma, or slightly smaller than it. The estimate is made by comparing fuzzy stars with the *fuzzy comet* [i.e., both are defocussed – Trans.]. This is the oldest method, and the one that is most used. Defocussing enables the effect of the central condensation in comets to be reduced by spreading the light uniformly into a disk. This is also the method that is instinctively used when the comet is close to a standardized sequence of stars.

The Sidgwick method (Fig. 8.21b) was invented to try to overcome the problems posed by comets that are very faint, or very diffuse (or both), which tend to disappear before the stars are sufficiently defocussed. It consists of defocussing until the stars are as extended as *the sharply-focussed coma*. It is therefore necessary to remember the appearance of the coma. Estimates

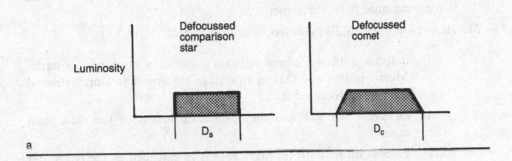

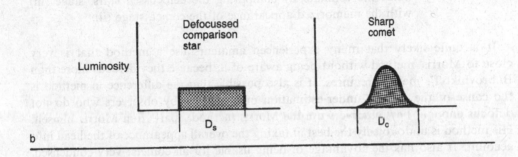

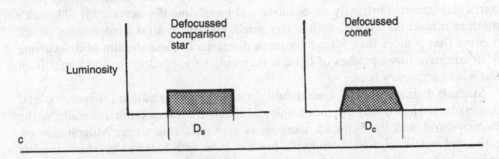

Fig. 8.21. *Schematic representation of the extrafocal methods used to estimate the magnitudes of comets.*
a: The Bobrovnikoff method $(D_S < D_C)$
b: The Sidgwick method $(D_S = D_C)$
c: The Morris method $(D_S = D_C)$
See the text for details. (After J. N. Marcus, Comet News Service.*)*

387

are made by comparing fuzzy stars with the sharp coma, or rather with the appearance as memorized. This method is best applied to faint comets (close to the instrumental limit) or comets that have no marked central condensation. It is sometimes preferred when the comparison-star sequence is some distance from the comet.

The Morris method   (Fig. 8.21c) consists of several stages:

(i)   defocus until the comet appears *uniform* in brightness; Charles Morris points out that at this stage the comet is more extended than the defocussed stars;

(ii)   *memorize* the size and the surface brightness of the defocussed comet;

(iii)   defocus more, until the stars appear as extended as the comet did in stage (i). More defocussing is required than in Bobrovnikoff's method: instead of $D_S < D_C$ we want $D_S = D_C$;

(iv)   make the estimate by comparing the defocussed stars, stage (iii) with the memorized appearance of the comet, stage (ii).

It is quite likely that many experienced amateurs use a method that is very close to Morris' method without being aware of it, because they defocus more than Bobrovnikoff's method requires. It is also possible that the difference in method is the cause of the chronic under-estimation of brightness by observers who do not defocus enough. Few observers use the Morris method, apart from Morris himself. His method is undoubtedly the best at taking the overall appearance of the head into account. It also has the advantage of being usable for all comets: very condensed, very diffuse, or a mixture of the two.

In any case, it is extremely important that the method used should be properly recorded *for each estimate*. One method may be completely unsatisfactory for a particular comet. Obviously an estimate will be of doubtful accuracy if Sidgwick's method is used on a comet with a very small, brilliant, central condensation within a coma that is more than 5 arc-minutes in diameter. Because the aim of defocussing is to compare fuzzy patches of light, it may only be applied to the total light from the whole cometary head.

Statistical analysis of a considerable number of observations, observers, and comets have shown that the mean square difference between estimates made by the Bobrovnikoff and the Sidgwick methods is about 0.1 magnitude. Magnitudes are estimated better with Bobrovnikoff's method than with Sidgwick's, which appears logical, because defocussing loses the outer regions of the coma, which account for only a small part of the total magnitude.

It is pointless trying to imagine what complications and disadvantages may apply to one or other of the methods: practice is the best way of finding out their individual peculiarities. We do not really recommend the use of a graduated reticle to ensure that the coma and the diameter of the defocussed stars are the same. This only leads to further absorption of light, especially from the coma, or to undesirable reflections.

8.5.3.2 *Practical advice on using extrafocal methods*

The term 'extrafocal' is subject to confusion, in that it indicates 'out of focus' without specifying whether we have moved the eyepiece away from the tube of the instrument or in towards it. There seems to be no basic reason to prefer one method to the other. Certain observers use both indifferently, without noticing any significant difference.

The estimate itself should be made by the Argelander method, which is used for variable stars. From the available comparison stars we choose one that, when defocussed, appears slightly fainter than the comet, and one (defocussed) that appears slightly brighter. The reader should refer to Chap. 14 for the details of the precise method used to determine the brightness of an object (a star or comet) from the known magnitudes of two stars. As with variable stars, beginners should start by finding stars whose magnitudes differ considerably from that of the comet. The unknown magnitude will therefore lie somewhere within a range of perhaps ± 1 magnitude. This will give a rough idea of the comet's brightness and major errors will be avoided. The observer then looks for stars that are somewhat closer in brightness, within ± 0.5 magnitude, for example. When there is some doubt about the order of decreasing magnitude in which the stars and comet occur, it may help to make quick use of a method suggested by Beyer. This method consists of extreme defocussing – until the comet disappears. Note which stars disappear just before, and just after, the comet itself. The estimate itself is then made more carefully by means of one of the methods described in the last section, using the stars that have been chosen. Under no circumstances should Beyer's method be used for the estimate itself. If we defocus too much, all we are left with is the central condensation, and we will omit the contribution made by the outer regions, which may amount to as much as 0.3 magnitude of the total magnitude. This is why the magnitude is often under-estimated. This method becomes worse the more extended the comet. We encounter the same sort of problems as in estimating magnitudes visually from a photograph.

Defocussing is not an observer's normal reaction. This is why we emphasize again that it is essential to use an instrument that is not only suitable for this sort of observation but also for the comets that are being observed. It is just as bad to use a 200-mm reflector for a 6th-magnitude comet as for one with a coma 10 arc-minutes in diameter. Use binoculars as often as possible. It is highly instructive to make estimates on the same occasion with several instruments of different sizes and types. Simple 7×50 binoculars are generally sufficient for comets down to magnitudes 9–9.5 when conditions are good; 11×80 lower the limit by at least a magnitude, sometimes even more if the comet is reasonably condensed. The more powerful the instrument, the more it will have to be defocussed, and the more sources of error there will be.

When defocussed, the image of a star in a reflector shows the silhouette of the secondary mirror, which appears as a dark circular spot at the centre of the luminous disk. The greater the amount of defocussing and the brighter the stars, the greater the effect. It seems debatable whether this may truly be compared with the image of a comet, where the distribution of light is completely different. Using a refractor of the same aperture instead of the Newtonian reflector is not a good idea if the focal length of the refractor is more than 1 m. This will again involve too much defocussing and large images at the focal plane.

The error most frequently made by beginners is insufficient defocussing. The stars are not spread out sufficiently and appear too bright. The brightness of the comet is therefore under-estimated by several tenths of a magnitude, and sometimes even by several magnitudes, when inexperience combines with other factors (such as instrument, magnification, transparency, etc.). An error in defocussing by a factor of 2 in the diameters produces an error of 1.5 magnitudes in the estimates!

Even so, there are times when one is forced to use high magnifications or, at least, the magnification for optimum resolution $D/2$ (mm), when the head is undetectable with the minimum effective magnification. This was the case with Comet Wilson 19811 in 1985 September, and occurs with Comet P/Schwassmann-Wachmann 1 when it is active. Again, one of the recognized methods has to be chosen. The Bobrovnikoff method appears most reliable for extended comets.

Two other problems crop up: the colour of the reference stars, and the distance between the comet and the sequence being used. There are only two ways of overcoming the inaccuracy resulting from the colours of the reference stars:

- Eliminate stars that are too red; use only stars that are of spectral classes B and A, for example, which are bluish-white; this information is given for all the stars in the *SAO Catalogue*.

- In the absence of this information, make several estimates with various combinations of comparison stars: e.g. five estimates with 3 or 4 stars. Those estimates that differ by more than 0.2 or 0.3 magnitude are rejected, a note being made of the stars that appear to have doubtful magnitudes.

It is never very scientific to be content with a single magnitude estimate involving just a couple of stars. The average of several estimates made one after the other within a fairly short time (less than 10 minutes) is far more reliable. It is better not to average estimates taken over a longer period, such as an hour, because of the variations that may occur in the brightness of comets, particularly active comets.

We now come to the almost legendary problem of estimating the brightness when the comet and the comparison sequence are not visible in the field at the same time. Although such a practice would be regarded as scandalous when dealing with variable stars, it appears from experience that it may be applied to comets. Their diffuse appearance seems to be remembered better than point sources like stars.

The procedure to be used is as follows:

a Identify the comet and go through the standard routine to determine the limits of the coma and the best instrument to be used for making the magnitude estimate. The observer is advised to pay considerable attention – perhaps using the finder – to the position of the comet with respect to nearby stars so that it may be found quickly;

b Defocus the cometary field to gain an idea of the appearance and the diameter of the stars when defocussed; this step may be used to determine which of the methods just described can best be used for making the estimate;

c Move the instrument onto the selected comparison sequence. Identify the comparison stars;

d Select suitable stars (using the Beyer technique if necessary);

e Defocus, using the same magnification as used in step *b*.

In most cases, and especially when a sequence is being used for the first time, the initial 'jump' has to be made from scratch, and it usually finishes at step *c*. We have practically forgotten the appearance of the coma whilst we have been wandering about searching for the comparison stars. After having properly identified the comparisons as well as any obvious alignments that would help identification, we have to go back to the comet. We have to start again with steps *a*, *b* and *c*, but this time they will be quicker because we have become familiar with the field. Steps *d* and *e* may then be applied, and we can write down our first estimate. We return to step *a*, perhaps noticing a small difference between our first estimate and the appearance of the comet. Under such circumstances, a proper estimate of the brightness requires at least two complete cycles, not counting the initial, identification trip. If we try a third time, we usually find that the result is the same as the second estimate. Each time we get to stage *e*, there is no objection to making estimates with several pairs of stars.

All the various steps that we have described would, on average, take about 10 minutes, or a bit less for an experienced observer who has an instrument that is easy to handle. Once again, we see that binoculars are ideal. When a more powerful instrument is required, it is a good idea to look for comparison sequences that may be found easily from the field containing the comet: fields with the same right ascension; with the same declination; or linked by specific patterns of stars that may be followed with the finder or even directly with the telescope and the magnification used for defocussing. Readers will appreciate the advantage here of frequently using the North Polar Sequence. Each observer will have to establish the best method for a particular occasion, which will depend upon the equipment (telescopes with English mounting cannot reach the Pole, for example), the available reference material and experience.

Some observers ask how they can gain experience of extrafocal estimates before becoming 'operational'. It would seem that the only objects that can be used to learn about comets are – comets. The remarks about the use of non-stellar objects that were made at the beginning of Sect. 8.5.1 remain valid: we just do not have reliable visual magnitudes for galaxies, nebulae and clusters. On average, three or four comets are accessible with normal, amateur instruments each year, down to about magnitude 10. On certain occasions, as many as five to eight comets have been observed in a single night, at all sorts of elongations, both east and west of the Sun, and at opposition. Amateurs who want to start estimating magnitudes can make use of the slack periods that occur by collecting their reference material (atlases, sequences, etc.) – and by reading this book!

8.5.4 Reducing the estimates

When we have finished a session of magnitude estimates, we have a set of raw data that looks something like this:

estimate 1:	a, 3, C, 7, b
estimate 2:	a, 4, C, 2, d
estimate 3:	e, 4, C, 1, d
estimate 4:	a, 3, C, 1, d
estimate 5:	e, 3, C, 2, f

where C is the magnitude of the comet (which we need to derive), and a, b, d, e, and f are visual magnitudes of comparison stars. The intermediate figures are Argelander steps. The example just given is for Comet Wilson 1986l, on 1986 October 8 at 21:30 UT. The observations were made with a 150-mm reflector with 25× magnification. The stars used were part of the comparison sequence for the variable star, R Del, which was about 2° away from the comet.

Note that we need not know the magnitudes of the stars beforehand, just that they may be used and have been calibrated. It is as bad to know the magnitudes of the stars in advance as it is to worry about the predicted magnitude of the comet. On the other hand, we cannot just assess the magnitude of the comet by guess-work, taking a quick glance though the eyepiece and then writing down '$m_1 = \ldots$'. The eye is not a photometer, and can never be one, even after years of practice – time has nothing to do with it! The eye is, instead, uniquely able to place stars in order of brightness, and to determine which object is the brighter, even if the difference in magnitude is very small. Which is quite an achievement.

The R Del sequence gives us the magnitudes of the stars: a = 10.1; b = 11.6; d = 11.0; e = 9.8; f = 10.8. (The letter 'c' is avoided, being used for 'comet'.)

We can then calculate the magnitudes for each estimate of the comet, C:

estimate 1:	10.55
estimate 2:	10.7
estimate 3:	10.76
estimate 4:	10.78
estimate 5:	10.4

The average of these five values is 10.64, which must be rounded to 10.6, because we cannot expect to obtain better accuracy than 0.1 magnitude.

Note that this value is practically identical with the result of the first estimate, which was made with the pair of stars with the greatest difference in magnitude (a 1.5-magnitude range). It is interesting to see that a careful estimate using two stars that are very different still produces a valid result. The estimates all fall within ±0.2 magnitude of the mean; we could hardly hope for better. So we can confidently give the result as:

$m_1 = 10.6$; 1986 October 8, 21:30 UT, using 150-mm Refl., 25×.

The time should be converted into hundredths of a day (giving an accuracy of a quarter of an hour): 1986 October 08.90.

When the reduction gives discordant results, there is justification in rejecting aberrant estimates. If the preceding estimates had given:

10.55
10.7
10.6
10.78
10.08

we would be justified in considering the fifth estimate as doubtful, and that there is an error in the magnitude of either star e or f. In fact, the second reduction was made assuming that e = 9.0 instead of 9.8. Note, however, that the average of this second set of estimates is 10.54, which rounds to 10.5. This is still close to the value obtained previously because we have made five estimates. The result would have been disastrous if we had only made a single estimate with stars e and f!

In this example, the deduced magnitude may be used as it is, without having to apply any correction for extinction, because the comet and the sequence were close to one another in the sky (within 2°). It would be pointless making an estimate if one or other of the fields were covered by cloud! When the two fields are some distance from one another and, in particular, when they are at very different altitudes, it is essential to make a correction for extinction. Under such circumstances we should take the precaution of determining the naked-eye or telescopic limiting magnitude in each of the fields whilst making the estimates. The *AAVSO Atlas* indicates the magnitudes of stars down to 6.5–7.0 fairly systematically. It is easy enough to check very quickly to see if a particular star of magnitude 6.2 or 6.4 is visible near the comet, and then to do the same near the comparison sequence. If, for example, the comet is nearer to the horizon and a nearby star of magnitude 5.9 is just at the naked-eye limit, whereas a 6.2-magnitude star may be seen near the sequence, then the correction for extinction is 0.3 magnitude. Taking our earlier result, that would mean that the comet was in fact brighter, being 10.6 − 0.3, i.e., 10.3. This is the magnitude that would have been estimated if the comet and the sequence had been at the same altitude. Take care not to apply the correction with the wrong sign!

This is all we need to do. The final magnitude reported to any coordinating group would be $m_1 = 10.6$ in the first case (comet close to the sequence), or $m_1 = 10.3$ after correction for extinction (if necessary). We should not make any other adjustments. Later reduction is for those who analyse the observations. Above all, make no attempt to adjust your figures: for example, over a period of several days you might obtain magnitudes of 11.2, 11.0, 10.8, 10.6, 10.8, 10.9, ...,. The '10.6' might be an outburst and the '10.8, 10.9' the beginning of a decline. Such variations may be correlated with results from other observers. We hardly need to reiterate that you should not be influenced by the predicted magnitude, whether for any particular day, or by its variation from day to day. You will be agreeably surprised when you find that your mini-outbursts are confirmed by other observers.

8.5.5 *Reporting observations*

If we want to send our observations to an astronomical association or a competent central organisation, it is essential to obey certain rules designed to ensure that the results may be properly used. Figure 8.22 shows a sample of a form prepared by the *International Comet Quarterly*, the group that acts as a central clearing-house for observations made all over the world. We are not giving this as an example of the sort of report that should be submitted but as the *only* form that should be used, in the interests of international standardization. It summarizes all the numerical data that have been obtained by reducing both magnitude estimates and drawings (Sect. 8.4.3).

The form should be completed as follows:

- one sheet per comet and observer: give the full name and address, and the comet's designation (provisional or final);
- one line of data per observation, i.e., at a particular time, with a particular instrument and magnification.

Each line contains the following information:

- year + month + day + decimals of a day (UT) in hundredths of a day;
- the method used to estimate the magnitude: B = Bobrovnikoff, S = Sidgwick, M = Morris (see ICQ for individual instances);
- the estimated total magnitude (m_1), to a tenth of a magnitude;
- the photometric reference used (see the ICQ code);
- equipment: type R = refractor, L = reflector, B = binoculars, E = naked eye (see ICQ for individual instances); diameter (in cm), focal ratio and magnification;
- the diameter of the coma, to a one-tenth of a degree when this is feasible (*see* Sect. 8.4.3.1);
- DC = degree of condensation (0 to 9), (*see* Sect. 8.3.6);
- tail length: in degrees to a hundredth of a degree when this is feasible (*see* p. 371);
- PA: position angle of the tail in degrees, to the nearest degree (*see* p. 371 and Fig. 8.8).

There is some space for brief notes:

- presence of the Moon (is that possible?);
- comet very low on the horizon;
- tail or jet curved towards PA ...

More detailed comments should be entered on a separate sheet, with an indication of the time of observation to which they relate. Drawings are welcome, of course; see the examples in Figs. 8.5, 8.18, etc.

If you have observed several jets, write down the details of each on successive lines, repeating the information about the date and the equipment. Each of the additional lines will contain just a length in the column headed 'Tail Length' and an entry under 'PA'. Because lengths are given in degrees, for smaller structures (less

COMET OBSERVING REPORT FORM. Send completed sheets to INTERNATIONAL COMET QUARTERLY, c/o C. S. Morris, Prospect Hill Road, Harvard, MA 01451, U.S.A.: or c/o D. W. E. Green, Smithsonian Astrophysical Observatory, 60 Garden Street, Cambridge, MA 02138, U.S.A. Please convert observing times to decimals of a day in Universal Time (e.g., Aug. 3, 18:00 UT = Aug. 3, 3.75 UT). Xerox or photocopy this sheet for more copies. (This form supersedes all previous forms. 1983 May 31.)

PLEASE PRINT OR TYPE ONLY. Use only one sheet per comet.

Name and designation of comet: COMET _____ 19 ____

Observer _____

Address _____

Date (U.T.)*	M.M.*	Total* Magn.	Ref.*	Instr.* Aperture	Instr.* Type	f/*	Power*	Coma Dia.	D.C.	Tail Length	P.A.	Remarks

NOTE: Drawings and additional comments or remarks should be included on separate sheets of paper. To be eligible for publication in the ICQ, columns marked with an asterisk (*) must be filled in.

Fig. 8.22. The International Comet Quarterly form for reporting visual observations of comets: magnitude estimates, position angle, length of tail and jets, etc. See the text for details of its use.

than 5 arc-minutes long), it is suggested that the length in arc-minutes and tenths of an arc-minute should be entered under 'Remarks' at the end of the line. And so on for each of the 15 jets that you measured on your wonderful drawing!

Some special symbols are used to indicate when the data are of moderate accuracy and for various additional comments. The way in which these are used should be obtained directly from the ICQ. An asterisk at the head of a column indicates that the information *must* be included if the observation is to be considered for publication, and that if the details are missing the corresponding magnitude may be rejected. Naturally this does not apply to the lines with details of jets.

As its name indicates, the ICQ is published once every three months. If you want to submit observations directly, it is sensible to forward reports at regular intervals matching editorial dates. Quite apart from the matter of publishing observations, adherence to a standard form will encourage everyone to note down useful information whilst making actual observations.

8.5.6 An examination of possible problems

Before considering how the estimates of m_1 are used, we should ask ourselves why there is so much difference between the estimates of one observer and another. We shall discuss the main causes of the scatter now.

8.5.6.1 Mutual influence by observers

We would certainly not suggest to beginners that they judge their work by the estimates made by so-called 'experienced' observers. Anyone can make a mistake, or have inaccurate reference material, and there are also so-called 'experienced' observers who would 'tweak' their estimates to please a friend. But objectivity and confidence in oneself are, respectively, a duty and a right. A detailed statistical analysis can establish personal equations, errors in procedure, correlations with predictions, and correlations between the curves obtained by different observers. Examination of the numerical data given in the reports enables the worst errors to be eliminated, such as under-estimation of the diameter of the coma because the instrument was not suitable for the comet, or because of unfavourable conditions, a degree of condensation that does not accord with other observations, etc.

Figure 8.23 gives a typical example of a series of observations that shows the influence of one observer on another; where 'any resemblance to a known comet or observers is purely coincidental'. The estimates made by different observers are identified by different symbols. Most observers recorded the continuous decline in the comet's brightness, whereas two observers, X and Y, saw the comet brighten at the same time. In the first place, it is certainly possible to show − by examining least-square differences, for example − that observer X did not see what he should have seen. Again, it is obvious that the estimates by Y 'gravitate' round those of X: the estimates reflect those of X, being shifted by 24 or 48 hours. Obviously frequent contact between X and Y has been detrimental to the objectivity of Y's results. Any attempt to determine the comet's photometric behaviour from all the available data, and without individual analysis, would show the comet's magnitude as having been completely stable over the period concerned.

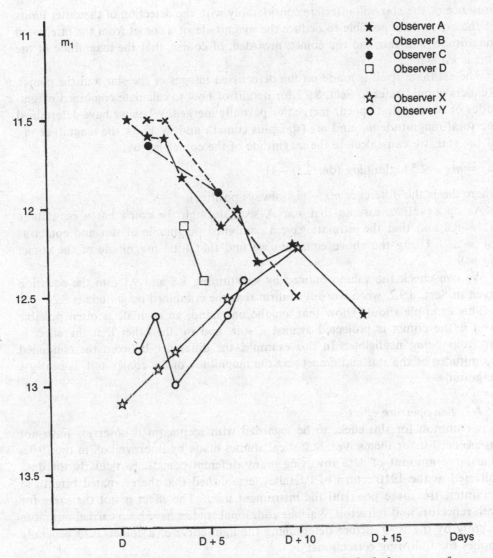

Fig. 8.23. *The effects of mutual bias between observers on the light-curve of a comet. The observations of Y 'orbit' around those of X. They fail to agree with those of observers A, B, C and D, who saw the comet fade regularly. (Diagram by J. C. Merlin.)*

8.5.6.2 *Stars seen through the coma*

It often happens that a comet, whose magnitude we want to estimate, is seen in projection against a background star that is sufficiently bright to affect the estimate. If the difference in magnitude between the star and the comet is greater than 2.5 magnitudes, the integrated magnitude of the star and the comet differs by less than 0.1 magnitude from the magnitude of the brighter of the two, the star or the comet. If the comet is significantly fainter than the star – i.e., by more than the figure just mentioned – the magnitude estimate is seriously compromised. In many cases, the presence of the star will interfere considerably with the detection of the outer limits of the coma. It is possible to deduce the magnitude of a comet from the integrated magnitude of a star and the comet, provided, of course, that the magnitude of the star is known.

The estimate itself is made on the defocussed images of the star and the comet. Readers should refer to Sect. 8.9.2 for details of how to calculate combined magnitudes of objects that appear merged or partially merged. When we have determined the total magnitude m_1, and m_T (star plus comet), and we know the magnitude m_A of the star, we can calculate the magnitude of the comet, m_C by:

$$m_C = m_A - 2.5 \log[\text{antilog } (dm/2.5) - 1],$$

where dm is the difference $m_A - m_T$ (always positive).

As an example, assume that star A seen through the coma has a magnitude $m_A = 8.8$ and that the estimate gave a combined magnitude of star and comet as $m_T = 8.2$. Using the above equation, we find that total magnitude of the comet $m_C = 9.1$.

We can check the value obtained by substituting 8.8 and 9.1 into the equation given in Sect. 8.9.2, when we will confirm that the combined magnitude is 8.2.

This example should show that, on the one hand, an estimate is often possible even if the comet is projected against a star, and on the other that the effect is far from being negligible. In this example, the difference between the combined magnitudes of the star and comet and the magnitude of the comet itself is nearly 1 magnitude.

8.5.6.3 *The aperture effect*

It is common for this effect to be regarded with scepticism if observers have not experienced it for themselves. Statistical studies made by Bobrovnikoff in the 1940s of a large amount of data involving many different comets, particularly the data collected at the 1910 return of P/Halley, established that the estimated brightness is fainter, the more powerful the instrument used. The effect is not the same for both reflectors and refractors. Various additional studies have been carried out more recently by the ICQ. Before determining the light-curve of a comet, ICQ generally applies the following corrections:

- reflectors: corrected m_1 = measured $m_1 - 0.019(D - 6.78)$
- refractors: corrected m_1 = measured $m_1 - 0.066(D - 6.78)$

where D is the diameter of the objective measured in cm, and 6.78 is the so-called standard aperture; it is, in fact, the average value of the apertures used for observations of P/Halley in 1910.

In other words, if we use an instrument with an aperture that is between 40 and 80 mm, our estimates are subject to practically no correction. An estimate of m_1 with a 200-mm refractor gives, on average, a brightness 0.9 magnitude fainter than an estimate with an aperture of 70 mm (e.g., with binoculars). A 200-mm refractor only introduces a difference of 0.3 magnitude over that given by binoculars.

We do not propose to discuss in detail all the aspects of this phenomenon. For a given comet, experience shows that a single observer, using 50-mm binoculars and a 200-mm reflector, often makes estimates that differ by between 0.2 and 0.5 magnitude, the fainter estimate being made with the larger instrument. The most widely accepted explanation suggests that this effect has a physiological cause arising in the eye. It should therefore be strongly dependent on the observer. When there are sufficient observations of a particular apparition of a comet, it is interesting to see if the coefficients 0.019 and 0.066 are truly constant. In fact, it seems that the effect varies from one comet to another, and that the coefficients also differ from one observer to another. The most important point to remember is that large refractors (above 150 mm in aperture), are definitely not recommended.

As we have said in Sect. 8.5.4, observers should not apply any corrections to their own estimates. They should leave it to the central organisation to decide, by appropriate statistical analysis, whether the volume of observations received warrants standardization of the raw data. The results obtained for P/Halley in 1985–6 were sufficiently numerous for all manner of observational effects to be determined. Preliminary analysis of data obtained before perihelion produced an anomalous situation. The raw data showed a large scatter, and lay within a band about 2 magnitudes wide. When corrected for aperture, the estimates showed even more scatter! This result did not call the aperture effect into question, because it had a far more human cause, which was the influence of recognised, experienced observers on others. The experienced observers generally used instruments that were appropriate for making magnitude estimates, and for the brightness of the object being observed. The less-experienced observers allowed themselves to be biassed, and tended to give the same magnitude, regardless of the instrument being used. The use of the average corrections that we have just mentioned only served to increase the dispersion.

At the end of 1986, ICQ carried out a more complete analysis of the brightness of P/Halley on both branches of the curve (before and after perihelion). This analysis, which included more than 1000 estimates made by various observers, did not reveal any significant aperture effect. The reason is simple: the observers were very well-disciplined and always used suitable instruments – i.e., the smallest possible!

8.5.6.4 *The effect of magnification*

The magnification used certainly has an effect on the estimated magnitude. If the magnification is exceptionally high, the brightness will be underestimated: either we lose the outer region or we are led to use too great a degree of defocussing. Estimating brightness is very difficult when the comet has many features close to the nucleus that are broad or of high contrast, or both. The structure of the coma of P/Halley in 1985–6 apparently had far more effect on estimates of m_1 than the aperture effect. This suggests that we should slightly modify the magnification suggested by Holetschek (Sect. 8.5.1), to improve contrast: use a magnification of $D/6$ to $D/5$ rather than the minimum effective magnification ($D/7$).

With small comets, where the diameters are less than 1 arc-minute, we find the opposite effect, especially when a star-like central condensation is visible at low magnifications. If we use the smallest magnification we are likely to believe that we are not defocussing enough. The total magnitude will be under-estimated, because we are ignoring the nebulosity, which is mainly visible with averted vision. This phenomenon was analysed several decades ago, and concerns the behaviour of the eye when dark-adapted. During the day, the resolution of the eye is about 1 arc-minute, or even less for the detection of high-contrast, linear objects (such as an electrical filament). With night vision, the resolution appears to be closer to 20 arc-minutes with the naked eye, which means that a diffuse patch will appear more or less as a point to averted vision until it subtends more than 20 arc-minutes in diameter. Applied as it stands to 'telescopic' observations, this finding means that observation of a coma 45 arc-seconds across with a magnification of 25 times (which gives an apparent diameter of about 19 arc-minutes), is absolutely identical to the observation of a star with averted vision – a fully focussed star, of course. It remains for us to determine whether a 150-mm telescope with 25× magnification will suffice to ensure that the outer regions make a proper contribution. Experiment shows that, with the same 150-mm telescope, the brightness estimated with a higher magnification of 50–75× gives a magnitude 0.2–0.5 times brighter than an estimate using the minimum effective magnification. Holetschek's rule does not apply here.

We earnestly hope that from time to time you will have the chance to observe comets such as Comet Wilson (1986l), so that you are able to practice making magnitude estimates in borderline cases.

8.5.6.5 The Delta effect

It has been suggested that there is an effect that is dependent on the Earth-comet distance (Δ). When a comet comes particularly close to the Earth, the diameter of the coma becomes so large that estimates of brightness become exceptionally difficult. All the effects that may cause under-estimation of the brightness are found and accentuated:

- the aperture effect;
- the magnification effect;
- the observer's experience;
- observing conditions, etc.

Joseph N. Marcus has analysed the light-curves of several comets that came within 0.2 AU of the Earth, in particular:

- P/Halley 1910 II (perigee at 0.153 AU on 1910 May 20);
- Suzuki–Saigusa–Mori 1975 X (0.104 AU on 1975 October 31);
- Bradfield 1979 X (0.198 AU on 1980 January 25).

He found that when comets passed the Earth at such distances, with a relatively steady level of activity (n approximately constant), their heliocentric magnitudes (m_H) were distinctly fainter. The closer the comet, the fainter the magnitude estimates! This phenomenon may be explained as a neurological and physiological effect arising in the eye. The eye has difficulty in perceiving extreme gradients in brightness. If the distance Δ is halved, the diameter D of the coma as seen visually does not double, and the eye sees the diameter as less than $2D$, which causes the magnitude to be under-estimated. Figure 8.24a shows this effect on the light-curve of Comet Suzuki–Saigusa–Mori 1975 X. It illustrates the change with time of the heliocentric magnitude ($m_H = m_1 - 5\log\Delta$), the drop in magnitude observed around perigee, and the subsequent increase. The classical methods of determining the parameters H_1 and n (*see* p. 347 and Sect. 8.5.7) lead to results that are of little significance:

$H_1 = 9.22$
$n = 0.92$ (!)
$R^2 = 0.05$ (the correlation coefficient).

Marcus calculated the photometric parameters from equations of the type:

$m_1 = H_1 + 2.5k \log \Delta + 2.5n \log r.$

He found: $H_1 = 8.87, n = 2.37, k = 1.5$ (instead of 2 as assumed in the classical formula). The correlation coefficient is 0.844.

Figure 8.24b shows the behaviour of 1975 X reinterpreted in terms of the Delta effect. The line marked 'true magnitude' (H) shows the heliocentric magnitude $H = 8.87 + 5.93 \log r$. Instead of varying as $m_1 - 5 \log \Delta$, the observed heliocentric magnitude varies as $m_1 - 2.5k \log \Delta$, or $m_1 - 3.75 \log \Delta$. The apparent heliocentric magnitude H' is shown by the dotted line, a curve that represents the observations quite well. The difference $H' - H$ is a function $f(\Delta) = 1.25 \log \Delta$; this is the Delta effect itself (Fig. 8.24c).

It is possible to determine such clear effects only in comets that pass very close to the Earth (within 0.5 AU). It is tempting to use Joseph Marcus' explanation to account for the various effects that are encountered. It is quite possible that the aperture effect and the magnification effect are, as Marcus believes, merely the Delta effect in different guises. If the magnification is doubled, we encounter just the same type of detection problem: the eye does not see a doubling of the apparent diameter of the coma (which would be the true diameter times the magnification). [It should, however, be noted that many workers remain unconvinced of the Delta effect's reality. – Trans.]

8.5.6.6 Naked-eye comets

When the coma becomes very large, beyond about 20 arc-minutes, extreme defocussing is required, and this becomes a problem. We are often confronted with naked-eye comets: they are a magnificent sight, but magnitude estimates are chaotic. How can we defocus the naked eye? If anyone is short-sighted that might be useful; if not, we shall have to defocus it artificially, using spectacle lenses. Even if it means using one of those dreadful 10-mm aperture finders, it is desirable to find some way of fixing the amount of defocussing, and to limit the size of the field of view with a 'proper' diaphragm. The eye is far too imperfect for us to expect an accuracy of better than 0.5 magnitude in the brightness of diffuse patches of light, seen without any optical aid. There is also the problem of locating comparison stars, which may be on the opposite side of the sky. As for estimating the brightness of super-comets that are visible in twilight or even in daylight, at magnitude -10 – well, we wish you the best of luck!

The analysis of data collected for comets where perigee is very close (and data are numerous because such comets are spectacular), shows a marked increase in the scatter in the estimates. This is a direct result of the large variation in the diameter of the coma as seen by various observers. The amount of defocussing undoubtedly

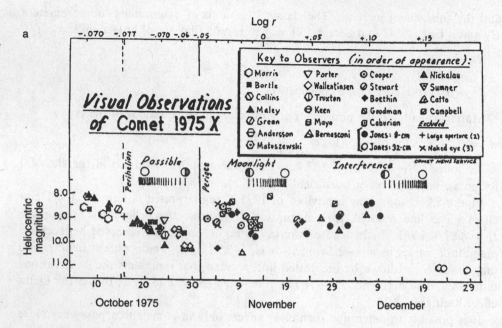

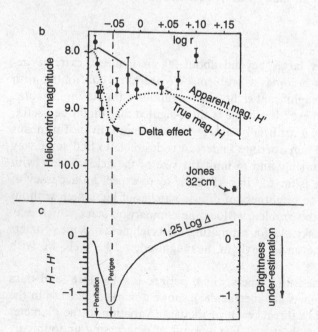

Fig. 8.24. Detection of a Delta effect in the light-curve of Comet Suzuki–Saigusa–Mori, 1975 X.

a: A graph of heliocentric magnitude (apparent magnitude corrected to a constant distance of 1 AU from Earth) versus time. The various observers have been identified with different symbols. The periods when the Moon may have affected observing conditions are indicated, together with the times of perihelion and of perigee.

b: The observations made by Jones in late December 1975 appear to be strongly under-estimated: the comet was very diffuse and a very large instrument was used.

c: See text.

(After J. N. Marcus, Comet News Service, 82-1.)

varies over a range of at least 2 to 1. With bright comets ($m_1 < 4$), where the coma is both large and highly condensed, it is possible that defocussing would be too great if one tried to apply the criteria shown in Fig. 8.21 to the letter. The external regions of the coma at more than 15–20 arc-minutes from the nucleus make, in many cases, a very small contribution to the total brightness (less than 0.2 magnitude). Sidgwick's method should be avoided in such cases. Is struggling to give visual magnitudes for extremely bright comets really justified, when such objects may be examined in a far more detailed way over the whole spectral range?

It should also be noted that it is not easy to construct a curve showing significant activity in a comet that is mainly visible only because it passes close to Earth. The range in r – the Sun–comet distance – is too small. The magnitudes obtained only allow one to determine the absolute magnitude to within an order of magnitude by assuming $n = 4$.

8.5.7 *Light-curves*

Estimates of total visual magnitude are used to plot the light-curve of a comet to determine both the average photometric parameters throughout the period of activity, and to establish any departures from the average level of activity. These differences, if they prove to be significant, are thought to represent outbursts of activity by the nucleus, and may be correlated with photographic or visual images that show the structure near the nucleus, such as the appearance of jets, splitting of the nucleus, reactivation of the plasma tail, etc.

8.5.7.1 *Preparing a light-curve*

We must have sufficient estimates (at least 20 if they are well spread over the whole period of activity, i.e., over a wide range in r), to be able to plot a proper light-curve. The first step is to plot the heliocentric magnitudes in chronological order. It is advisable to identify each observer's points on this graph. For the analysis to have any validity in the statistical sense, we really need at least ten estimates by each observer. This will enable us to carry out individual least-squares analysis. The aim of using heliocentric magnitudes (m_H) is to show the changes in brightness that occurred, assuming a constant distance (1 AU) between Earth and the comet, and given by:

$$m_H = m_1 - 5 \log \Delta,$$

where Δ is the actual Earth–comet distance at the time the estimate of m_1 was made. If the apertures used vary considerably (from 50–250 mm, say), it may be necessary to correct the values of m_1 by using the equations given on p. 398. We will then have:

$$m_H = \text{corrected } m_1 - 5 \log \Delta.$$

The chronological plot of m_1 (Fig. 8.25) illustrates the two branches of the light-curve:

- the ascending branch, which is assumed to peak at perihelion; and
- the descending branch that follows.

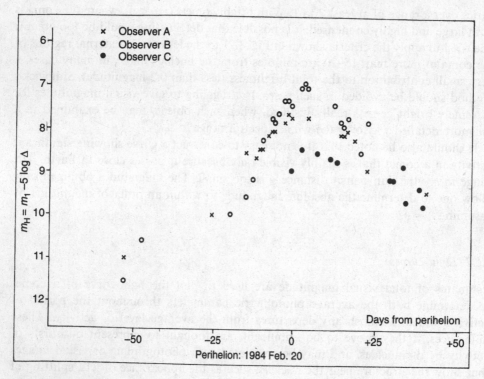

Fig. 8.25. *Evolution with time of the heliocentric magnitude of Comet P/Crommelin, 1983n (1984 IV). The estimates by various observers have been indicated by different symbols. (Diagram by J. C. Merlin.)*

Experience shows that the peak of the curve rarely occurs at the time of perihelion passage, which is where a plot helps to discriminate between the two sets of points. In addition, the curve is very rarely symmetrical about the maximum, so two distinct sets of photometric parameters have to be established. We now calculate the following value for each observation:

$$X = 2.5 \log r,$$

where r is the Sun–comet distance given by the ephemeris.

Calculation of H_1 and n for each branch of the curve is made by using a standard least-squares analysis (or linear regression), with equations of the type:

$$m_H = H_1 + nX.$$

We require the ordinate (or intercept) H_1, and the slope n of a straight line through the points. Readers interested in carrying out this sort of analysis will find the linear regression formulae in Sect. 8.9.3. It is easy to program a microcomputer or scientific calculator to carry out the calculation.

8.5.7.2 *Results of the calculation*

The example shown in Fig. 8.25 is that for Comet P/Crommelin, observed between 1983 December and 1984 March. We have 51 observations, spread over 93 days, 53 before perihelion passage and 40 after. The data come from three observers, who shall remain anonymous. It has been possible to obtain relatively significant results from these estimates. The individual observations have been identified by three different symbols on the graph. Maximum occurs about 3 days after perihelion passage (at time T). The parameters of the ascending branch of the curve therefore include observations made from 53 days before perihelion passage to 3 days after it.

The instruments ranged from 70-mm binoculars to a 350-mm Schmidt–Cassegrain. The formulae for aperture effect given on p. 398 have been applied. The corrections range from 0 for the 70-mm binoculars to 0.4 magnitude for the 350-mm telescope. Calculation of the least-squares fit gives the following:

- ascending branch, between $T - 53^d$ and $T + 3^d$:
  28 points
  $H_1 = 10.27$
  $n = 6.68$
  correlation coefficient: approximately 0.76 %;

- descending branch, from $T + 4^d$ to $T + 40^d$:
  23 points
  $H_1 = 9.49$
  $n = 4.99$
  correlation coefficient: approximately 0.67 %.

The values of H_1 and n calculated for the two sections of the curve are shown in Fig. 8.26, which is a plot of $m_H = f(\log r)$.

8.5.7.3 *Analysis of the results*

The scatter is relatively significant, because the observations lie within a band that is between 1 and 2 magnitudes in width. There are various reasons for this scatter:

- atmospheric conditions: with poor transparency on occasions, and the comet low on the horizon;
- the structure of the comet: diffuse and relatively large, thus being even more sensitive to variations in transparency;
- the variation in brightness is not a linear function of $\log r$: outbursts (at $T - 25$) and plateaux (particularly one between $T - 3$ and $T + 10$) may be identified;
- the equipment used was not always ideal (an f/10 telescope, for example);
- in some cases the method appears to have been a conventional variable-star method rather than a cometary version (was defocussing used?);
- some estimates are approximate, being rounded to the nearest half-magnitude. (There are an abnormal number of estimates of m_1 ending in 0 or 5.)

Despite this long list of criticisms, it does nevertheless seem that the results are reasonably valid. The parameters calculated are very close to the values obtained

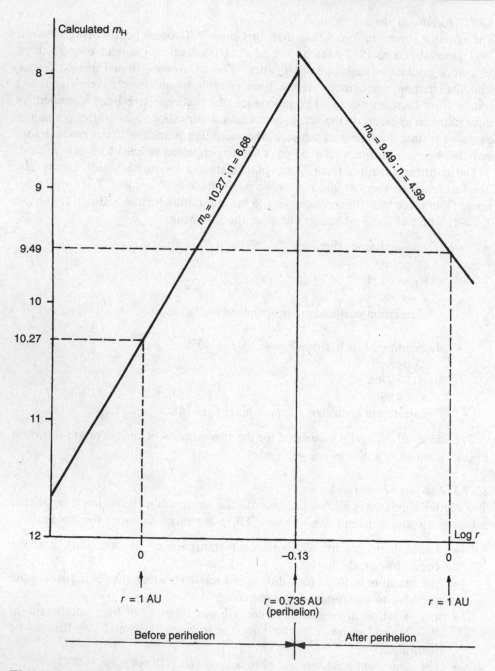

Fig. 8.26. *Mean heliocentric magnitude of Comet P/Crommelin. The straight lines have been calculated from the data shown in Fig. 8.25. (Diagram by J. C. Merlin.)*

from all the estimates submitted during the world-wide observing campaign in 1983–4. As regards the comet, we can make the following points. Its absolute magnitude, around 10, it that of a moderately large object. (For P/Halley $H_1 = 4$ approximately.) The slope n (the activity index) is relatively high for the period before perihelion, showing a rapid rise in activity, which is quite typical of short-period comets. (In this case $P = 28$ years.) The correlation coefficient is not outstanding. The best-fit line is strongly influenced by the three points around $T - 50^d$. The correlation is more significant for the period $T - 25^d$ to $T + 3^d$, with a considerably greater slope: $n = 15$ approximately (!). It would appear that the brightness first rose slowly. Then the comet undoubtedly underwent an outburst about 3 weeks before perihelion, or at least showed considerably increased activity. Around perihelion the brightness was more or less stable until about $T + 10^d$, or slightly later. This period also coincides with the greatest scatter (about 1.5 magnitudes) among the observations received. It is therefore not surprising that the correlation coefficient for the descending branch is so poor (0.67 %). The decline in brightness appears to be slower than the rise: $n = 5$ approximately. It is difficult to say whether the comet underwent an outburst around perihelion: $H_1 = 10.3$ to 9.5. In this sort of calculation the two parameters H_1 and n are very closely linked. Elimination of a few points well away from the mean would doubtless allow far more significant values to be obtained.

8.5.7.4 *Analysis of the* $O - C$ *differences*

This sort of analysis should not stop here. We have already mentioned the importance of having enough observations for each observer (at least ten) to be able to determine personal equations. Figure 8.27 shows an analysis of the differences between the observations shown in Fig. 8.25 and the regression lines shown in Fig. 8.26. We give histograms for the $O - C$ values (observed minus calculated). The observed values are the m_1 values corrected for aperture. The calculated values are: calculated $m_1 = H_1 + 5 \log \Delta + 2.5n \log r$, the values of H_1 and n, calculated previously, being used in the standard equation.

We are lucky in that the values show distinct effects. Observer A's differences are centred on 0, i.e., on the mean. The histogram has a Gaussian distribution (a bell-shaped curve), and the dispersion (2σ) is ± 0.6, which is fairly acceptable, given the irregular nature of the comet's behaviour. Observer B's histogram is not Gaussian. It is essentially flat, with a dispersion ($2\sigma = 1.2$ magnitudes). The differences are not primarily the result of statistical scatter, but rather because of problems with the methods used. The mean of the $O - C$ values for observer C is 0.4, so the magnitude has been more or less systematically under-estimated. The cause of this may be found in the observational reports, which indicate fairly poor transparency and smaller coma diameters than those reported by the other two observers. The histogram is fairly symmetrical even though we have only 11 observations. The problem is less severe than for observer B.

When the histograms are Gaussian, it is possible to consider correcting the observations of each observer by the average $O - C$ value. This is a personal equation, which differs from one observer to another. We also find that for a single observer the personal equation varies between one comet and another. For each comet investigated, the shape of the histograms allows us to define, more or less

Observer A:
Number of observations: 19

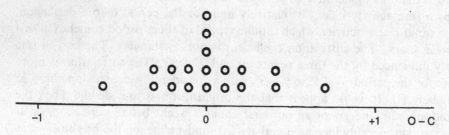

Observer B:
Number of observations: 21

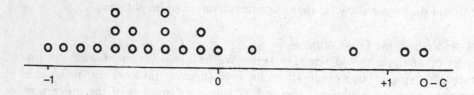

Observer C:
Number of observations: 11

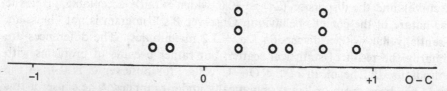

Fig. 8.27. *Personal equations of various observers during observations of Comet P/Crommelin, between 1983 December and 1984 March. These are individual histograms of the* $O - C$ *differences between the observations shown in Fig. 8.25 and the mean lines shown in Fig. 8.26. (Diagram by J. C. Merlin.)*

objectively, a second selection criterion that enables stray points to be eliminated. The curve thus obtained is smooth, and any outbursts appear far more obvious. In the case considered here, this leads to our eliminating the estimates by observer B. Unfortunately not much remains to be analyzed – only about 30 observations in all.

When a comet is active, bright, visible for a long time, and therefore very well-observed, we might expect the correction given by the $O - C$ values to be very informative. Experience shows that the problem actually becomes rather complicated. It is certainly desirable to analyse the differences found for each portion of the curve, before and after perihelion, independently. Despite this, the histograms obtained are more difficult to analyse. Variations in the appearance of the comet produce additional sources of scatter. Analysis of the differences should therefore take the degree of concentration and the extent of the coma into account. Such a study has now been carried out for P/Halley to extract the main features from the thousands of magnitude estimates obtained in 1985–6.

Another phenomenon may cause bias to appear in the mean light-curve. This happens when one observer provides many more observations than others, perhaps 100 observations in 1 month, whereas most observers only obtain that number in 6 months. The conclusions from any analysis of the light-curves of comets are therefore subject to a number of subtle effects. Some idea of their significance may be obtained by examination of the remarks on the observing conditions that were made by the observers on their report forms.

8.5.8 Monitoring comets

We have now covered in detail all the hazards that afflict the techniques used to estimate magnitudes. It will not take long for us to find this out for ourselves. Although it is possible to obtain fairly reliable predictions of the magnitude of certain variable stars that have recognized periods, any prediction of the magnitude of a comet is inevitably subject to error. Comets undoubtedly age, because the loss of material is irreversible. Some attempts have been made to measure the amount of aging and to estimate how many more apparitions of a periodic comet may occur. Such studies have been attempted for P/Encke and P/Halley. When we learn that conclusions have been drawn on the assumption of aging at the rate of 0.1 magnitude per apparition, we may well have doubts, especially when we consider that, even in the most favourable cases, the estimated magnitudes have a margin of error of ± 0.5 magnitude! Trying to determine such effects requires analysis of observations over several centuries – but who may tell what were the photometric standards used in Newton's day? What about extrafocal methods, which were standardized only 50 years ago? Our successors will be able to determine the evolution of periodic comets only from our current contributions.

It would be very sad if only posterity were able to show that amateurs provide useful information. New comets appear regularly, capturing our attention and enhancing the value of our observations. Amateurs can tackle the problem of magnitude estimates in several different ways:

- by observing any object within the grasp of their telescopes whenever the opportunity presents itself, with the aim of obtaining as much detail as possible in the light-curve;
- by ensuring that 'everything that occurs' is monitored.

Short-period comets deserve our attention. Some of them will be the targets of future space-probes. Several short-period comets have shown sudden, large-amplitude outbursts:

- P/Holmes, which was discovered because of a 4-magnitude outburst in 1892;
- P/Tuttle–Giacobini–Kresák: this jumped from magnitude 14 to 4 in one week at the end of 1983 May. The magnitude declined smoothly down to 14–15 on 1973 July 4, then recovered to 6 two days later;
- P/Takamizawa: discovered at magnitude 10 on 1984 July 30; photographs taken before discovery show the comet at magnitude 16 on 1984 July 6 and magnitude 12 on July 8. A photograph taken by T. Seki on 1984 July 26 showed the comet at magnitude 6.5!

The origin of these events is still poorly understood. Given the sudden, random nature of these outbursts, some astronomers think that they may arise from the collision of the nucleus with a large meteoroid. Apart from these spectacular events, some 'quiescent and feeble' periodic comets show very different light-curves from one apparition to another, as with:

- P/Borelly in 1980–1,
- P/Boethin and P/Wirtanen in 1986, etc.

Figure 8.28 shows the light-curves of P/Grigg–Skjellerup for two successive apparitions, in 1977 (Fig. 8.28a) and in 1982 (Fig. 8.28b). During the 1977 apparition, the brightness of this comet continued to rise for 3 weeks after perihelion. The curve appeared symmetrical with respect to the maximum (the decline had the same value of n as the rise). In 1982, the behaviour of P/Grigg–Skjellerup was completely different: a rapid rise during the 3 weeks before perihelion, then a slow decline ($n < 3$), rather like Comet P/Kopff. Yet examination of observations obtained at previous apparitions, before 1977, show a sudden decrease in magnitude after perihelion! Although this is a relatively faint ($H_1 = 12$) and old comet (its period is 5.1 years), it never seems to behave the same at two successive apparitions.

New periodic comets, in particular, should be monitored. For example:

- P/Hartley–IRAS 1983v: discovered independently at magnitude 15 by M. Hartley (with the Anglo–Australian Schmidt) and by the IRAS satellite. Three weeks after discovery is was found accidentally at magnitude 11 by David Levy, an American amateur who was searching for comets with his 400-mm reflector;
- P/Machholz 1986e: discovered 2° from M31 in 1986 May. This comet, which has a very short period (5.24 years), is more or less unique because of its high inclination to the ecliptic (60°) and its very small perihelion distance (0.13 AU). According to the Minor Planet Center, perturbations by Jupiter

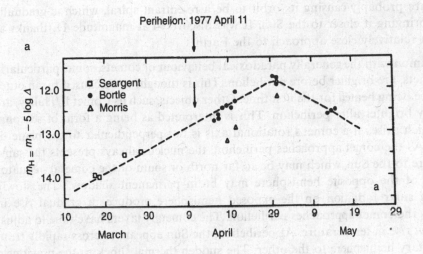

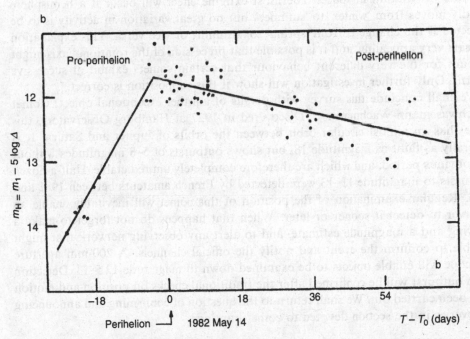

Fig. 8.28. *Light-curve of P/Grigg–Skjellerup (in heliocentric magnitudes) at its returns in 1977 and 1982.*

411

are probably causing its orbit to be a re-entrant spiral, which is gradually bringing it closer to the Sun. It was discovered at magnitude 11, thanks to a relatively close approach to the Earth.

In common with the generally paradoxical behaviour of comets, some, particularly 'new' comets, are brighter before perihelion. This is thought to be caused by an outer layer of ice being heated for the first time. Other objects, such as Comet P/Halley, are noticeably brighter after perihelion. This is interpreted as being a form of seasonal effect, which arises if a comet's rotational axis is not perpendicular to the plane of its orbit. As the comet approaches perihelion, the nucleus always presents the same hemisphere to the Sun, which may be so far north or south of the cometary equator that part of the opposite hemisphere may be in permanent shadow. The slowly increasing solar radiation on the exposed hemisphere produces a gradual rise in activity as the comet approaches perihelion. The cometary layers have time to adjust to the slow rise in temperature. At perihelion, the Sun appears to cross rapidly from one cometary hemisphere to the other. The sudden thermal shock to the previously unexposed side is believed to fracture the comet's outer, partially insulating layer, and thus produce a burst of activity. The effect depends strongly on the relative orientation of the cometary rotational axis to the plane of the orbit, and on its absolute orientation in space. The most extreme effect will occur if a hemisphere rapidly moves from 'winter' to 'summer', but no great variation in activity is to be expected if the change is from 'spring' to 'autumn', or vice versa. This explanation appears very appealing, and it is possible that precession of the rotational axis might account for the very different behaviour that certain comets exhibit at successive returns. Only further investigation will show if this supposition is correct.

We shall conclude this survey with details of another exceptional object: Comet P/Schwassmann-Wachmann 1. Discovered in 1927 at Hamburg Observatory, this comet has an almost circular orbit between the orbits of Jupiter and Saturn. It is generally as faint as magnitude 18, but shows outbursts of 5–8 magnitudes without any obvious period, and which are therefore completely unpredictable. Half-a-dozen outbursts to magnitude 11–13 were detected by French amateurs between 1981 and 1986. Regular examination of the position of this comet will inevitably enable an observer to detect it sooner or later. When that happens do not forget to make a drawing and a magnitude estimate, and to alert any observing network that might be able to confirm the event and notify the official channels. A 200-mm aperture telescope will enable objects to be examined down to magnitude 12.5–13. Detection of an outburst will be confirmed after the traditional checks on position and motion have been carried out. We shall return to the question of confirming and announcing discoveries in the section devoted to comet-hunting.

8.6 Photoelectric photometry of comets

8.6.1 Cometary photometry

Photoelectric photometry of stars is described in Chap. 19. Any amateur who wants to apply this technique to comets should first start by becoming thoroughly familiar with the methods used for stars. Reference should be made to the chapter

Table 8.3. *Filters used in cometary photometry (after I.H.W.s Amateur Observer's Manual)*

Central wavelength (nm)	Bandwidth (nm)	Chemical species	Source
365.0	8.0	Continuum	Dust
387.5	3.9	CN	Gas
406.0	7.3	C_3	Gas
426.0	6.5	CO^+/N_2	Gas
485.6	8.5	Continuum	Dust
511.4	9.0	C_3	Gas
700.0	17.5	H_2O	Gas
719.5	15.0	Continuum	Dust

in question for details of equipment. Then the question of measuring the brightness of an extended object must be tackled.

The main contribution that amateurs may make in this field rests, in part, on the modest size of their instruments. Amateurs' telescopes have fairly short focal lengths, which give small images at prime focus and so allow a large part of the coma of a comet to be examined. Short-focus camera lenses (with focal lengths of less than 600-mm) are particularly interesting in this respect. The aim of making measurements is to determine the brightness of the head and of the central concentration at specific wavelengths and to a high degree of accuracy (± 0.001 magnitude). The required filters are given in Table 8.3, taken from the I.H.W. *Amateur Observer's Manual.* Traditional UBV filters do not allow the various dust and gas components to be distinguished.

Cometary photoelectric techniques are similar to those of stellar photometry. Measurements are made by making the following measurements in succession:

- comparison star,
- comet,
- sky background,

this sequence being repeated several times.

The sky background is measured more than 1° from the comet to eliminate any possible contamination by extensions of the coma, but not more than 5° away (to avoid variations in transparency).

The following information should be recorded together with the measurements:

- name of the observer;
- date and time (UT), for the beginning and end of each measurement;
- the part of the comet that was measured, i.e., the exact position of the diaphragm (with a drawing!);
- the instrument: type, aperture;

- type of photometer;
- diameter of the diaphragm in arc-seconds;
- filter used: with a transmission curve in the case of a non-standard filter;
- measurements, in magnitudes or in MKS units (W/m^2): dark current, comparison star (with identification), sky background, comet;
- comments: any particular conditions that might affect the reliability of the results, etc.

8.6.2 Occultations

The observation of the disappearance or the dimming of a star behind any part of a comet (the nucleus, coma or tail), may give very useful information about the density of the material in the comet. It is interesting to make repeated measurements of the brightness of a star as the comet passes in front of it. Such a series of measurements should start and end with measurements of the star completely free of any effects caused by the comet.

In organising an occultation campaign, it is, of course, essential to know the motion of the comet accurately. Careful research with a photographic atlas or, more frequently, through the tabular information of an astrometric catalogue, will enable one to select stars likely to be involved in interesting events. Both the measurements and the times at which they are made must be known accurately, so that the parts of the comet affecting the brightness of the star may be determined exactly. The timing technique is similar to that used in detecting occultations of stars by minor planets or in monitoring the mutual phenomena of Jupiter's satellites: use a tape recorder to simultaneously record time signals and the occultation information (see Chap. 9, 'Occultations').

Amateurs who want to participate in occultation campaigns should get in touch with professional workers in the field. The latter will tell them the best procedures to follow, and provide full information about which comets to study and which stars to select.

8.7 Cometary photography

We shall not discuss the purely technical side of photography, which is the subject of numerous specialized works. It is obvious that it is not possible to obtain high-quality cometary photographs (Fig. 8.29) unless certain fundamentals of astronomical photography have been met:

- equipment that is suitable for the object being studied;
- a drive that is mechanically, electronically and electrically perfect;
- appropriate films and filters;
- a suitable development technique;
- proper reduction of the photographs.

Cometary photography usually involves an additional problem, which is that of regularly shifting the telescope during the exposure to follow the comet's apparent

Fig. 8.29. *Comet Levy–Rudenko, 1984t (1984 XXIII), on 1985 January 25 at 03:41 UT. Taken with a 200/300-mm Schmidt. (Photograph by M. Jäger.)*

motion. We need to pay particular attention to guiding techniques. Depending on the particular aim in mind, which may be wide-field or high-resolution imagery, astrometry, or just discovery, different degrees of accuracy are obviously required. Those who simply want to form a collection of photographs of fine comets are often content to guide on the stars for exposures of a few minutes. The images are obtained with wide-field lenses having focal lengths of less than 60 mm, and guiding is by means of a separate telescope (usually a refractor) on the same mount. Provided the head of the comet is positioned off-centre in the frame, 28-mm lenses enable the development of cometary tails up to 50–70° long to be recorded. It should, however, be noted that the finest comets are those that are close to either the Earth or the Sun (or both). Small elongations do severely restrict the use of fast lenses (f/1 to f/2) because of the amount of twilight that is present. When perigee is very close (< 0.5 AU) the comet's proper motion during the exposures must be

415

taken into account. A good example of a comet that passed very close to the Earth was IRAS–Araki–Alcock 1983 VII in 1983 May (perigee at 0.03 AU). In a 150-mm telescope, its apparent motion was quite perceptible in just a few seconds. It was necessary to guide on the comet, even for exposures of a few minutes with lenses of 50-mm focal length.

When it comes to the technical data for each exposure, the observer should carefully record full details of the equipment used for the photograph, and those of the guide telescope (diameter, focal length, magnification). Other details should include the type of film, hypersensitization (if any), and filters used. It is a good idea to acquire the habit from the very start of recording the time of the beginning of the exposure to the nearest second – use a watch set from a suitably accurate time-signal – as well as the length of the exposure. Other useful information includes details of the method of guiding and general observing conditions.

8.7.1 Guiding techniques

The techniques used in cometary photography are identical with those employed for the photography of faint objects that are described later (*see* p. 301), except that here the proper motion of comets complicates the issue.

8.7.1.1 Guiding on the comet

This is the most obvious method. Figure 8.30 shows the different relative positions of the image of the comet and the graticule:

- Guiding on the central condensation: if the condensation is sufficiently prominent, guiding may be with it centred on the cross-hairs (Fig. 8.30a), or on a pointer (Fig. 8.30b), which avoids some of the problems caused by the lines of a graticule.

- Guiding tangent to the coma (Fig. 8.30c): the cross-hairs are arranged so that they are tangent to the edge of the coma on the sunward side. The direction of motion is taken as bisecting the opposite quadrant. The edges of the coma need to be sharply defined, and care has to be taken over the orientation.

- Guiding centred on the coma (Fig. 8.30d): the cross-hairs are aligned so as to divide the coma into equal parts. This is the last resort when the central condensation or the coma do not have a sufficiently high contrast, but it is the least accurate method.

Offset guiding (with a secondary eyepiece), is not suitable because we must guide on the comet itself. In addition, as techniques of guiding on the comet are not very accurate, they are recommended only when the photographic instrument and the guide telescope are different, and the focal length of the former is much less than that of the latter – which is true for a small telephoto lens being used with a refractor as a guiding telescope.

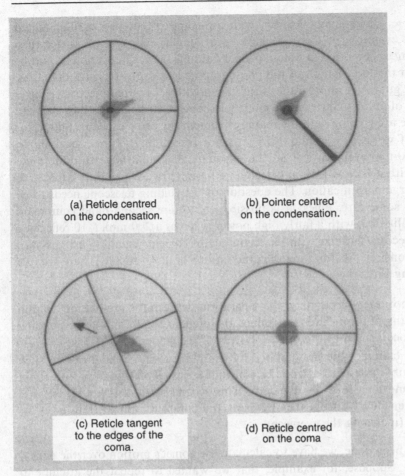

Fig. 8.30. *Techniques for guiding on the head of a comet. (After I.H.W.* Amateur Observer's Handbook.*)*

8.7.1.2 Guiding using a star

Unlike the previous method, this one is accurate, because it is easy to align a graticule very precisely with the diffraction image of a star. In addition, this is the only method that can be used when a comet is too faint to be detectable in the guiding telescope. It is, however, complicated by the fact that the photographic instrument has to be continuously displaced with respect to the star in order to compensate for the comet's proper motion. The rate of this displacement has to be calculated in advance for the data given in the ephemeris, and one must take care to apply it in the correct direction!

This method may use any of the three classical methods of guiding on stars: using a separate guide telescope, offset guiding, or with a beam-splitter. Whichever method is used, the basic technique is the same: we locate a star near the comet that is sufficiently bright to be used as a guide star, and shift the telescope during the exposure so that the star appears to move in the direction opposite to the comet's actual motion and at the same rate. If we want the guiding to be accurate, we shall have to use a high magnification, with an eyepiece that has an illuminated graticule. Most eyepieces of this type have focal lengths equal to, or longer than 10 mm. Experience shows that to obtain proper guiding, the magnification should be at least equal to the focal length of the photographic lens expressed in centimetres. So, in principle, guiding a photograph being taken with a 2-metre objective requires a magnification of at least 200× on the guide telescope. A 10-mm cross-hair eyepiece is sufficient if we are guiding with the main instrument (by using a beam-splitter, for example). But if we use an 80-mm, f/15 refractor for guiding (with a focal length of 1.2 m), the 10-mm eyepiece will not be adequate. We shall have to double the magnification with a Barlow lens, although an 80-mm refractor is hardly capable of taking such a large magnification. The question of the guiding telescope needs to be considered with some care, unless it is possible to acquire second-hand equipment of moderate quality, but with a fairly high power (such as a 200-mm, f/10 reflector).

We need to become familiar with the technique of cometary guiding using a star. The first stage consists of calibrating the graduations of the graticule. There are two methods of doing this:

- If we know (very accurately) the separation between the graduations in mm and tenths of a mm, we can deduce the angular separation corresponding to the focal length of the optical system of the guider.
- We can calibrate the graduations by using a known celestial object, such as a double star with essentially constant separation. (Binaries with rapid orbital motion are to be avoided.) This second method is better, because it is taken directly from the sky using the complete optical train used for guiding (including the eyepiece).

Once this has been done, we have to calculate the comet's motion over the period of the exposure. Suppose, for example, that we wanted to repeat the attempt to photograph P/Halley made by P. Martinez and E. Laffont using the 600-mm reflector at the Pic du Midi on 1985 February 18. [Note that coordinates for epoch 2000.0 would be used for observations after the beginning of 1992. – Trans.] P/Halley was then 4.5 AU from the Earth and 4.8 AU from the Sun. The comet was identified from a one-hour exposure taken between 20:19 and 21:19 UT. (Photographs taken the next night are shown in Fig. 8.34.) The 1950.0 coordinates for the beginning and end of the exposure were be calculated by non-linear interpolation (see Sect. 8.9.1):

- 20:19 UT: $\alpha = 5^h\,00^m\,54.82^s$; $\delta = +13°\,13'\,03.16''$;
- 21:19 UT: $\alpha = 5^h\,00^m\,53.40^s$; $\delta = +13°\,13'\,08.39''$.

The comet's motion in one hour was therefore:

$\Delta\alpha = -1.42^s$ (towards the west)

$\Delta\delta = +5.2''$ (towards the north).

Converting $\Delta\alpha$ into arc-seconds, we have:

$\Delta\alpha'' = 15\Delta\alpha\cos\Delta\delta$, or $\Delta\alpha'' = -20.7''$.

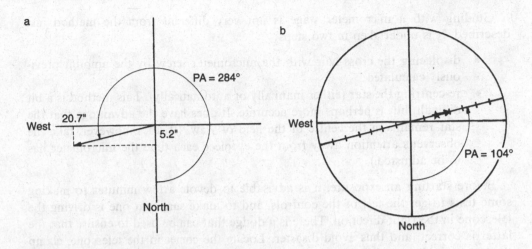

Fig. 8.31. *Procedure for using a star to guide on a comet (see text).*
a: The rate and direction of the comet's motion are: 5.2 arc-seconds towards the north,
20.7 arc-seconds towards the west.
b: The graticule is oriented in the calculated direction. During the exposure the tele-
scope is driven to move the star towards PA 104° (i.e., opposite to the motion of the
comet) at a rate of about 1 arc-second every 3 minutes.

Figure 8.31a shows the direction in which the comet was moving: towards a position angle of 284° in the northwestern quadrant, by 21.3 arc-seconds per hour ($\sqrt{\Delta\alpha^2 + \Delta\delta^2}$). During the exposure, the guide star would therefore have to be displaced by 21.3 arc-seconds in the opposite direction to the comet, i.e., towards a position angle of 104°.

The first step would be to align the graduated scale in the eyepiece in the appropriate direction, 104° to 284°. Second: decide on the frequency of corrections. In this example, the guiding eyepiece was a 12-mm Meade graduated every 40 μm. Because the guider had a focal length of 1906 mm, each graduation therefore corresponds to an angle of 4.3 arc-seconds, which the comet would cover in 732 seconds of time. The observer would therefore control the telescope drive so that the guide star moved, as evenly as possible, from one graduation to the next in 732 seconds. This requires an assistant (or a programmable timer) to give an audible signal every 732 seconds. To increase the accuracy, the audible signal may be more frequent, occurring twice or four times as often; the observer then has to identify half- and quarter-graduations. The assistant should keep track of the signals, so that at any time the observer can be told which graduation should be close to the star. Experience shows that when the exposure is long, seeing sometimes becomes blurred for a short time and, if the comet's motion is fast, one ends up not knowing which graduation the star was last near, nor which one it should be near!

It is not advisable to take turns at guiding; one observer should guide throughout the exposure. If one does take turns, the corrections made during the first few minutes after the exchange are not very reliable – it has even been known for the first correction to be in the wrong direction!

Guiding with a micrometer stage is not very different from the method just described. It is undertaken in two steps:

- displacing the cross-hair with the micrometer screw by the amount previously calculated;
- re-centring the star (either manually or automatically). This method is a bit difficult, but is perhaps more accurate. It does have the advantage that the star remains in the centre of the field of view. (It does, however, take the observer's attention away from the eyepiece each time the micrometer has to be adjusted.)

Before starting an exposure, it is advisable to devote a few minutes to making some tests to get the feel of the controls, and to make sure that one is driving the telescope in the right direction. There is a dodge that can be used to ensure that the latter is correct, and thus avoid disaster. Locate the comet in the telescope, clamp the latter, and bring the star to the cross-hair. If, while looking through the eyepiece, the telescope is pushed as if one were trying to move it in the same direction as the comet's motion on the sky – 'upwards', for example, if Δ is increasing, and 'to the right' if α is decreasing – even a large reflector like the 600-mm at the Pic du Midi will flex sufficiently for the star to move against the graticule. This is enough to show if it moves in more or less the direction that one expects it to follow during guiding.

Those who are perhaps not convinced of the necessity for all this when photographing faint comets may be interested in details of the first attempts made to detect P/Halley with the 600-mm reflector at Pic du Midi in 1984 December. The comet was moving at 40 arc-seconds per hour. If we assume that the diameter of the coma was about 5 arc-seconds, we find that it would only take a few minutes (between 5 and 10) for the comet to move by more than the radius of the image on the emulsion. This means that with a conventional exposure (and no special guiding) the grains in the emulsion would capture light from the comet for rather less than 10 minutes. Under those circumstances, the comet, at a magnitude of around 20, would not be detectable.

Guiding on a star may also be used for bright comets. Indeed, we recommend that it should be used whenever possible in preference to guiding on the comet. It is easier to keep a cross-hair on a point source than it is on a fuzzy patch. There is one instance when this type of guiding is not recommended. This is when the comet is very close to the horizon. Differential refraction changes the apparent rates at which objects rise and set, and leads to errors that are unacceptably large in long exposures.

8.7.1.3 *Electronic help*

The method of cometary guiding on a star that we have just described is less accurate than tracking a star itself, either (in the case when a graduated eyepiece is used), because the star has to be shifted along a specific line, rather than remaining fixed; or (when a micrometer stage is used), because the micrometer head cannot be adjusted without disturbing the observer.

It is possible to avoid these disadvantages by building a special photographic

Fig. 8.32. *The device for moving the eyepiece during cometary guiding, built by Patrick Martinez, mounted on a 150-mm aperture reflector. The cross-hair eyepiece can be seen mounted on a movable carrier, below which is the stepper motor that moves the carrier. The whole assembly is mounted on a plate that may be turned through 360° at the focal plane.*

baseplate, where the film-carrier is moved at regular intervals by a stepper motor. This motor is driven by an electronic timer, which sends an impulse at specific, programmable intervals. Naturally the whole assembly that carries the movable parts should be adjustable in the focal plane so that the axis of motion of the photographic film-carrier may be made to coincide with the direction of the comet's motion. Such a design has been described by the British amateur Ron Arbour (Arbour, 1985). The observer simply guides on the star as if on a fixed object; the motor drive compensates for the comet's movement. The accuracy obtained is as great as that for a fixed object, and the method of operation is almost as simple.

Another solution is to keep the photographic film stationary at the focus of the telescope, but to use the same sort of stage to move the guiding eyepiece (Fig. 8.32). The observer re-centres the guide star on the cross-hairs each time the motor moves by a single step, but as this is very small, the operation is no more complicated, nor any less accurate, than ordinary guiding on a star. Two observers have independently built and described this sort of arrangement: Edgar Everhart in the U.S.A. (Everhart, 1987), and Patrick Martinez in France (Martinez, 1987b).

8.7.2 *Wide-field photography*

Wide-field photography is an area where amateurs with small Schmidt cameras may obtain spectacular images of both gas and dust tails (Fig. 8.33). Because these two types of tail emit at different wavelengths, we can consider using selective filters to isolate individual features. Table 8.3 (p. 413) gives the list of filters used for optical photometry of comets, for both studies of the head and of the tails. The lines from gaseous C_2, C_3 and CN correspond to what is seen visually in tails.

The gas tail may be recorded with the usual Kodak 103-a0 emulsion (which is blue-sensitive). Better definition is obtained by using hypersensitized Kodak TP 2415, combined with a cut-off filter (opaque beyond 500 nm) to eliminate the red. Those who do not have the standard photometric filters can combine Wratten 2B and 47A filters. A red-cut (blue) filter passing wavelengths > 500 nm will allow the dust tail to be isolated (or a Wratten 21, etc.).

Colour emulsions (either hypersensitized or cooled to minimize reciprocity failure) have the advantage of being able to record everything in a single picture, when conditions are such that selective photographs cannot be obtained (such as when the comet is low on the horizon). Darkroom techniques will subsequently enable the different characteristic emissions to be shown selectively on different prints.

The main benefit of wide-field photography appears particularly clearly in worldwide observing campaigns. The number of amateurs involved, and their flexibility – generally having easily portable equipment – enables a comet to be monitored throughout the day, at longitudes where there are no professional observers. Those who are unable to afford Schmidt cameras use medium-focal-length telephoto lenses. Probably telephoto lenses of 200–300 mm in focal length offer the best compromise, in that they are able to record tails and tail-streamers (*see* Fig. 8.37) over a field of several degrees with quite good resolution, without requiring exceptionally accurate guiding. But there is nothing to stop anyone trying to do even better.

8.7.3 *High-resolution photography*

This type of photography tends, of course, to compete with visual observation and drawing of comets. As regards the focal lengths used, high-resolution photography is at the other extreme from wide-field photography. Cassegrain or Schmidt–Cassegrain reflectors (with focal lengths of more than 2 m) and, of course, large refractors, are the most useful. The drive should be accurate, smooth and free from backlash. Exposures times need to be quite short (generally less than a minute) so that just the area immediately around the nucleus is recorded.

Despite the passage of time, it seems that conventional photography is unlikely to give a better performance than the eye in detecting differences in brightness within the head of a comet. Careful methods have to be applied to exaggerate photographic contrasts to give a result comparable with that attained by an artist. In view of the criticism sometimes levelled at visual results, it is fair to point out that image-processing techniques are not totally objective. The choice of the final image does not depend on completely objective factors; it is governed by the eye of whoever is controlling the process. The future probably lies with CCD systems,

Fig. 8.33. *(Above and following pages) – A spectacular disconnection event in the gas tail of Comet P/Halley on 1986 April 11, photographed by G. Klaus from Namibia. Exposures were 10 minutes on hypersensitized TP 2415, with a 200/300-mm Schmidt, f/1.5, field diameter = 10°.*
a: 1986 April 11, beginning of the exposure 21:20 UT; the gas tail is very turbulent.

which will allow exposure times to be reduced considerably. We shall have to wait a few years yet before these systems attain a resolution that is acceptable for use with long focal lengths of 1.5–2 m. It will then not be long before amateurs turn en masse to CCDs (*see* Chap. 20). The difficulty that we shall then encounter will mainly concern software for interpreting the digitized images. The programs will have to be effective enough to eliminate all subjective judgements in the choice of the final image.

The golden age of visual observation has not yet come to an end. It is possible

Fig. 8.33. *b: 1986 April 12, beginning of the exposure 02:10 UT; the gas tail is completely detached from the coma.*

that in a few years, a time will arrive when CCDs dramatically confirm what is seen by the eye. If use of CCDs becomes commonplace, and cheap, then, and only then, will visual observation start to become a thing of the past.

8.7.4 Astrometry

The most important role that amateurs have to play in cometary photography probably lies in measuring positions to obtain improved orbital elements. The IAU constantly wants positions of astrometric quality (with an accuracy of better than 5 arc-seconds). With a new comet, many astronomers are unable to carry out any work if the position is not known to an accuracy of better than one minute of arc. Given

Fig. 8.33. *c: 1986 April 13, beginning of the exposure 23:15 UT; a new gas tail has formed. The dust tail forms a very wide, diffuse, fan. (see also Fig. 8.45)*

that ephemerides are only ever extrapolations from measured positions, it often happens that the position of a comet is not known to less than 5 arc-minutes, and sometimes the uncertainty may be as much as 30 arc-minutes if the extrapolation is over several weeks. A single, recent astrometric position (reported within 24 hours of the photograph being taken), allows the orbital elements to be corrected. The more up-to-date the information provided by people undertaking astrometry, the greater the value of their work for the rest of the astronomical community.

The technique of measuring positions is described in Chap. 17. When applied to comets, it requires extreme care to be taken in locating the photometric centre on the photographs.

Exposure times should be as short as possible, so that the photometric centre is recorded. If at all possible, the equipment should be driven to track the stars, rather than the comet, because the latter procedure is always more risky. There will, however, be no way of avoiding it if the comet is very faint. In so far as possible, measurements should be made using stars that are comparable in brightness with that recorded for the central condensation. If tracking has been on the comet, then two sets of measurements have to be made, one for each end of the trail left by every reference star. The average of the x and y values for each end of the trail should be used in the final reduction, and the time taken is that of the middle of the exposure.

Although use of a focal length of more than a metre is advisable, there is no truly ideal instrument for astrometry, the only proviso being that there should be enough catalogued stars within the field. Trials with telephoto lenses with focal lengths of 200–300 mm have shown that it is perfectly possible to obtain positional accuracies of better than 5 arc-seconds. The fact of being able to use more reference stars (more than ten) more than compensates for the small scale of the image. Schmidt cameras of the sort available to amateurs give excellent results for the same reason, helped by the extremely small, sharp images. Newtonian telescopes are quite usable, provided only the centre of the field is used. (Images become very distorted beyond about 20 arc-minutes.) Their major disadvantage lies in the small size of the field that may be photographed, which is less than one degree square, and which limits the number of catalogued stars that are available. Astrometric positions have been measured on several occasions on cometary photographs taken by amateurs using the 600-mm reflector at the Pic du Midi. Despite the very distorted shape of the stellar images more than 5 arc-minutes from the centre of the field – the mirror has a ratio of f/3.5 – the measurements have had residuals of less than 1 arc-second. The degree of experience of the person making the measurements also has a considerable influence on the accuracy of the results.

8.7.5 Recovery of periodic comets

For several decades, large Schmidt telescopes were the preferred instruments for recovering periodic comets, far from the Sun, as they began their next return. The calculation of predicted positions, taking account of planetary perturbations, has become more and more accurate. It is relatively easy to detect a moving object from a Schmidt photograph, if we have a reference photograph and a blink microscope (see Chap. 16). In the last few years, the situation has changed somewhat. The organisations controlling the financing of the large Schmidt telescopes have certainly decided that they are better employed in photometry, in extragalactic work, or in sky surveys, than they are in detecting a single faint comet on a plate 300 mm square. The majority of rediscoveries are now made with Newtonian or Cassegrain telescopes, sometimes equipped with CCDs, but the work is not undertaken systematically. The professional teams that are interested in this field are well-known:

- Jim Gibson at Palomar (1.6-m reflector + CCD);
- Everhart and Briggs at the Chamberlain Station of the University of Denver, Colorado (400-mm reflector);

- the MacCrosky–Shao–Schwartz team at Oak Ridge (Harvard) (1.5-m reflector);
- Gehrels and Scotti at Kitt Peak (910-mm with the SPACEWATCH camera);
- Gilmore and Kilmartin at Mount John, New Zealand (610- and 250-mm reflectors);

and a few others that function more spasmodically (the astrograph at ESO in Chile, etc.). To this list we can add Tsutomo Seki, the Japanese amateur who observes with a 600-mm reflector, and who has contributed to the rediscovery of four periodic comets between 1983 and 1986. Just to set the record straight, we should also mention that two periodic comets have been recovered visually in recent years:

- P/Faye (1984h), observed visually by C. S. Morris on 1984 June 23, using a 250-mm reflector, scarcely two weeks after photographs taken by Gibson. Morris' observations essentially came as confirmation of Gibson's;
- P/Honda–Mrkos–Pajdusáková (1985c), which was recovered visually by three Australian amateurs (Clark, Pierce and Athanassiou) in 1985 April with a 400-mm reflector. The elongation was too small for the comet to be photographed at its 1985 apparition.

When specific coordinated observing campaigns are envisaged, efforts are always made to recover the chosen comets at greater distances and fainter magnitudes. Recoveries are the preserve of large reflectors (with apertures larger than 1.5 m) fitted with CCD detectors. The record in this respect is held by Comet P/Halley, which was recovered 1600 days before perihelion in 1986 at magnitude 24.2. Most other recoveries are made when the comets are between magnitudes 18 and 20.

What part can the amateur hope to play in the face of such magnitudes? The fact that some comets have still been recovered visually in recent years should be some encouragement. Second, the reader will have noticed from our list of the teams engaged in this work that some of the instruments are relatively modest (250-, 400- and 600-mm reflectors). It is, however, a bit risky jumping into this type of work with just a 250-mm reflector. Any comets recovered would have to be brighter than about magnitude 18, which severely limits the number of objects yet to be recovered that are available for detection. The chances are far better with a 400-mm telescope, which can photograph 19th-magnitude comets with exposures of an hour or so. The number of amateurs who have even larger telescopes drops dramatically, but the availability (currently) of certain 'professional' telescopes (such as the 600-mm at the Pic du Midi) should not be forgotten (Figs 8.34 and 8.35).

The method of determining the expected track that has been described earlier (p. 350) must be meticulously applied. The motion of the comet should be plotted on a very detailed atlas, using *Atlas Stellarum* at the very least for identifying the field and, if possible, the *Palomar Sky Survey* (POSS) for accurate identification of the position and access to far more reference stars. A 400-mm reflector will record stars of magnitude 20–21 with an hour's exposure, which is practically the limit of the first edition of the POSS.

There comes the problem of finding a reliable ephemeris. It is not really possible to count on the conventional sources available to amateurs. Provisional orbital

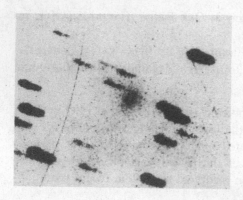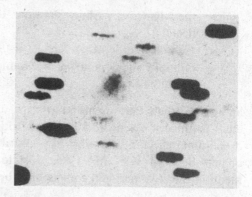

Fig. 8.34. *The first amateur photographs of Comet Halley taken in Europe at its 1985–6 return, obtained on 1985 February 19 by Patrick Martinez and Eric Laffont when the comet was still at magnitude 19–20. The telescope used was the 600-mm reflector at the Pic du Midi, under the scheme in which it is made available to amateurs. The film was hypersensitized Kodak TP 2415. Guiding was calculated to follow the theoretical motion of the comet. Because each exposure lasted an hour, the stars left trails 20 arc-seconds long. Note the motion of the comet between the two exposures (left, 19:26 to 20:26 UT; right, 21:35 to 22:35 UT).*

elements that incorporate allowance for planetary perturbations to which a comet has been subjected since its last apparition may be obtained from at least two organisations:

- the IAU Minor Planet Center;
- the Bureau des Longitudes in Paris.

These orbital elements are osculating elements calculated for the next perihelion passage and, in principle, are usable as such. Their direct application assumes that planetary perturbations may be ignored between the time of detection and the coming perihelion. As detection is only rarely attempted more than a year in advance, when the comet is generally well inside the orbit of Jupiter ($r < 5$ AU), the effect of perturbations is usually very small. Observers who prefer to play safe can request ephemerides directly from the appropriate organisations. The question of the expected magnitude should not be neglected: we are mainly interested in nuclear magnitudes m_2.

Since 1987 the *International Comet Quarterly* has published ephemerides calculated by Syuichi Nakano for all comets whose return is expected in the current year, down to m_2 magnitudes of 25–26. This is obviously an invaluable publication for anyone who wants to try to recover a periodic comet, and it is easily available to amateurs.

Once the motion of the comet being sought is known, then the procedure becomes one of implementing the guiding techniques described earlier (Sect. 8.7.1.2). Obviously the exposure should be duplicated, at the very least, so that the comet can be confirmed by its motion. We must beware of faint minor planets: there are hundreds

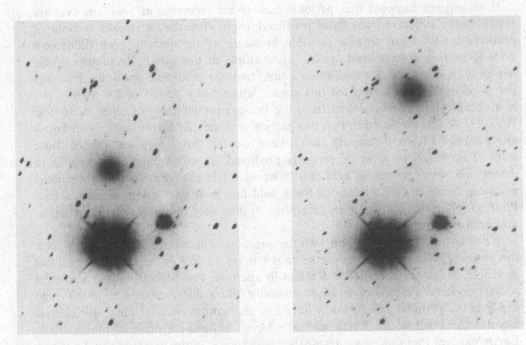

Fig. 8.35. *Appulse of Comet P/Tsuchinshan 1, 1984p (1985 I) and the star 37 LMi (the brightest in the field) on 1985 February 19 and 20. Exposures of one hour, using the 600-mm reflector at the Pic du Midi. Film: hypersensitized Kodak TP 2415 (Photographs by P. Martinez).*
Left: *1985 February 19, beginning 02:09 UT*
Right: *1985 February 20, beginning 01:44 UT*
North at top, east to left.

of them below magnitude 15. Some may have an apparent motion that, in a single observing session, is very similar to that of a comet. The recovery of a periodic comet is of greater value if it is confirmed by further photographs obtained over a period of 24 or 48 hours. It is absolutely essential to have adequate information to hand in attempting to carry out reductions of this type of observation. The *Minor Planet Circulars* are indispensable. Assistance from groups that have calculation facilities should not be neglected, especially in ensuring that you do not inundate the IAU with false alarms, which will only damage your reputation.

If you do succeed in recovering a comet, it is not sufficient just to report the fact. If you really want to convince the IAU Central Bureau, you must include accurate astrometric positions with the message. Only calculation of the orbital elements will show whether you have really detected the expected object. Any estimates of magnitude will be of value; ensure that you state whether they are m_1 or m_2 magnitudes, and remember to give any useful information about the appearance of the object. The way in which such messages are coded may be found in Sect. 8.10.

It sometimes happens that periodic comets are recovered at positions that are considerably different from those predicted, even when the latter take account of perturbations. Apart from a possible break-up of the nucleus, such differences arise from non-gravitational forces. An example of this was P/Kohoutek, 1986k, which was recovered 5 arc-minutes away from the predicted position. This was the third apparition known for this comet, which has a period of 6.6 years – and is a completely different object from the famous (or infamous) Comet Kohoutek, 1973 XII. It was the first time that this particular comet had shown such marked non-gravitational effects. We should always bear in mind that comets are objects whose level of activity can never be properly predicted. It would have been possible to detect this comet within the field of a Newtonian telescope centred on the predicted position, because such a telescope has a field that is at least 30 arc-minutes across. Photographers should examine their photographs very carefully before coming to any final conclusion.

We may finish this discussion with the question: where and when should we look for periodic comets? If an expected comet is on its way in towards the Sun, and is visible in instruments of 400–600 mm in aperture, obviously it will be recovered sooner or later. Comets of this sort generally follow direct orbits (i.e., in the same sense as the planets) and normally have low inclinations ($< 30°$). They are detected beyond the orbit of the Earth, and often even beyond the orbit of Mars, so the Earth 'catches them up'. As a result they will appear in the morning sky. So if you want to recover them early, without waiting for them to move into the evening or night-time sky, you know what you have to do!

Some periodic comets are lost after an apparition (for example, P/Lexell), generally because their motion has been perturbed by a close approach to Jupiter. Recovery of these comets is much more difficult, and more closely resembles the methods used in systematic searching for comets, as described in the next section.

8.8 Searching for comets

The uninformed reader may think that we have now come to the most chancy aspect of the study of comets. It is true that the wide range of orbits and the unpredictable behaviour that may be found among these objects have a significant effect on the detectability of any comet.

Some years ago, the Czech astronomer Lubor Kresák made a realistic analysis of the proportion of comets that we are able to observe. One of his conclusions was that the number of comets that we observe is practically the optimum number for ensuring the existence of cometary science:

- if there were only one-tenth as many comets, the objects would be regarded as very exotic objects, far too rare to prompt special observational and research campaigns;
- if there were ten times as many comets, interest in comets would possibly have already waned, and there might be little enthusiasm for devoting time to systematic study, with a sky continually full of 'Comet West-like' objects!

To continue this rather anthropocentric argument, examination of recent discoveries has confirmed the self-evident fact that more comets must exist than those we discover. The numerous different statistical studies that have been made on the frequency of discoveries and the factors that influence them, all conclude that a large number of bright comets go unseen. It is almost certain that some comets that become brighter than magnitude 10 actually cross the sky without being detected.

Apart from the desire to see one's name perpetuated in the sky, early detection of new comets has an obvious scientific significance. Few comets have been observed at great distances from the Sun. We have little information about the way in which activity begins in new comets. It would also be extremely interesting to detect short-period comets at greater distances in the Solar System – well beyond the minor-planet belt – to find out more about their distribution and the influence of perturbations by the giant planets, Jupiter and Saturn in particular. There is work for everyone, for both visual observers and astrophotographers.

Before describing search techniques, we must briefly review some of the high points in the history of cometary discoveries. This will enable us to set modern amateurs in the proper context in terms of what amounts to an international competition.

We must begin by exhorting observers to take the utmost care in their work. Every year brings its crop of comets lost for lack of follow-up observations or accurate ones. Every month brings its crop of false alarms; of 'Kodak comets' as the photographers call them, that result from emulsion defects that are wrongly interpreted; of rediscoveries of various globular clusters; or of comets that everyone has known about for weeks. It may be essential to develop a proper technique, but it is no less important to have a good knowledge of the sky, and to keep oneself informed about comets that are visible or that are expected. Learning photographic techniques and the reduction of photographs by conducting a programme of comet-hunting is just as foolhardy as beginning visual comet-hunting without knowing anything about the objects in the Messier catalogue. You will be forgiven for the first false alarm, on the grounds that you are a beginner. With the next one, no one will take you as being a truly serious observer. If you think you have discovered a comet, before alerting the IAU or even an observatory, and especially if you are unknown in the astronomical community, get in touch with one of the amateur groups. Others can then check and confirm your discovery. A group has more weight that a single individual when it comes to alerting the scientific community. Don't forget that even in the case of known, respected observers, the IAU always insists on having confirmation by someone else. It is better to be a co-discoverer than to be regarded as a phoney. Observation of known comets is excellent training for comet-hunting because it forces you to gain access to suitable reference material and to keep in touch with what is happening.

8.8.1 *The development of systematic comet searches*

The first observers who set about ensuring complete coverage of the sky were the court astronomers to the Emperor of China, more than 3000 years ago. Several observers each observed part of the sky throughout the night. For the period between the 3rd century B.C. and the beginning of the 17th century A.D., Chinese archives are considered to be complete as regards comets down to about magnitude 2. The techniques developed by the Chinese astronomers subsequently spread to Korea and Japan. It was only from about the 15th century, at the time of the Renaissance, that European astronomy emerged from limbo.

From then until the first half of the 18th century, the names of several European observers are linked with comets. Tycho Brahe discovered two comets, in 1582 and 1590. Johannes Kepler discovered one in 1618. In France, La Hire discovered three comets between 1678 and 1702. An important event occurred in 1680 with the first telescopic discovery of a comet by Gottfried Kirch. The rate of discoveries started to accelerate after about 1740. Observers began to set new records, for example, the Dutch astronomer Klinkenberg discovered at least five comets between 1743 and 1750. The development of observational astronomy in Europe is shown by the geographical distribution of discoverers during the period between 1550 and 1750. Of 45 comets catalogued, the figures may be broken down to give:

30 discoveries in Europe (exactly two thirds)
6 China (13 %)
5 South America (11 %)
3 South Africa (7 %) } (18 %)
1 rest of Asia (India).

Less than 20 % of these discoveries were made in the Southern Hemisphere. If we extrapolate from the numbers for the Northern Hemisphere (37 comets), we can assume that about 30 comets were missed because of this difference. The situation is far from being evenly balanced today, because the Southern Hemisphere still has less instruments (and astronomers). For the record, we may also note that a significant proportion of the cometary discoveries from outside Europe, were made by people, originating in Europe, who were working in what were then colonies.

The situation changed after 1750, with the arrival of two French observers, Charles Messier in 1759, and Pierre-François-André Méchain in 1781. With Méchain's help, Messier prepared a catalogue of non-stellar objects, containing all the 'nebulae without stars' that might be confused with comets. Méchain was not just an assiduous and efficient observer. He was also a natural philosopher of world-wide renown, who devoted much time to geodesy (measuring the arc of the meridian, for example), and who calculated the orbital elements of numerous comets. Messier observed some 40 comets. He has been credited with the discovery of 16 of these, ranging from P/Halley, which he found independently of Palitzsch in 1759 January, to the first comet in 1801, which he discovered 24 hours after Pons. Only 12 comets are named after him, one of which he shares with Méchain. Five other objects appear to have been discovered by Messier (apart from the 16 just mentioned), but it has not been possible to arrive at any consistent orbital elements for them. Seven comets are named after Méchain. Among comets with established orbital elements, five others may be identified as having been discovered independently by Méchain.

Others became famous at around the same time: Bouvard and Montaigne in France, Olbers in Germany, Vishnievski in Russia, etc. The roll of honour has not recorded some of the most active comet hunters, however, such as the Cassinis,

Maraldi, de La Nux, etc. The principle of giving the name of the discoverer to a comet was begun, apparently, by Messier in 1759. The rule was strictly applied, because in most cases the initial discoverer was the only person honoured. It would, incidentally, be unforgivable for us not to mention the work of Caroline Herschel, William's sister. She discovered eight comets in the last 15 years of the 18th century.

In the face of the considerable efforts being made in Europe, the number of comets discovered from the Southern Hemisphere was more or less negligible. It is possible that all the various political and military troubles at the end of the 18th century caused most of the colonial officials to return home.

After the disappearance of Messier and Méchain, new records were set by the most famous comet-hunter of them all, Jean-Louis Pons. The last comet to have Messier's name was in 1798 and Méchain's last two were in 1799. Pons discovered his first comet in 1801 July, using a refractor that he had built himself. This comet was independently discovered the next day by Messier, Méchain and Bouvard. While he was active, Pons made 37 discoveries, detecting 80% of the comets recorded in the first 27 years of the 19th century. Of humble origins, J.-L. Pons was given the job as caretaker at the Observatory in Marseille in 1789. He was promoted to assistant astronomer in 1813 and later moved to Italy, where he was director of the Observatory at Firenze (Florence) from 1825 until his death in 1831. Official recognition of his worth was doubtless a result of the changing climate of ideas at the end of the 18th century. Contrary to what may be generally believed, Pons was by no means alone in making systematic searches at that period. He had serious competitors in Blanpain, Bouvard, Gambart, and Olbers, among others.

After 1820, new European observers became involved. We may mention one amateur astronomer in particular: Wilhelm von Biela. This captain in the Austro–Hungarian army discovered three comets. His name is particularly remembered for the periodic comet P/Biela, the nucleus of which split in two in during the 1845–6 apparition. It is also the parent comet of the 1872 November meteor shower (the Andromedids or Bielids).

Every period seems to be notable for its own set of exceptional observers and also by the way in which the centre of interest moved from one country to another. A few French observers were active at the end of the 19th and beginning of the 20th centuries (Coggia, Borelly, Giacobini, etc.), as well as several Germans, such as Tempel, but the main centre of activity turned more and more to the U.S.A. Five names repeatedly occur in connection with the discovery of several tens of comets over a period of about 50 years: three amateurs (Barnard, Brooks, and Swift), and two professionals (Perrine and Tuttle). The first three were responsible for about 50 discoveries. Legend has it that Barnard only became a professional after having discovered ten comets. Barnard himself is supposed to have said that he built his house from the prize-money that he won every time he found a new comet. The end of his career, at the Lick Observatory, marked a new turning point in the history of systematic searching, because it saw the first photographic cometary discovery in 1892 October (P/Barnard 3).

In the Southern Hemisphere, several amateurs divided discoveries between them during that same period. In South Africa there were Forbes, Reid, and Skjellerup;

in Australia, Tebbutt; and in New Zealand, Grigg. There were also a few individual discoveries in South America.

The U.S.A. was again very well-represented with the appearance of Leslie Peltier, whom Harlow Shapley described as the greatest non-professional astronomer of the time. Alongside his career as a variable-star observer, Peltier discovered ten comets between 1925 and 1954. His period of activity overlaps with the modern period, when two new nationalities make their appearance: the Czechs and the Japanese. A systematic visual programme was organized in 1945 by the Skalatne Pleso Observatory (in the High Tatra Mountains in Slovakia), mainly using 100-m binoculars (i.e., 25×100). The team led by Anton Mrkos recorded numerous successes over about 15 years. Around 20 comets were discovered by Mrkos, Kresák, Becvár and by the female astronomers Pajdusáková and Vozárová.

Modern amateurs will be more familiar with later developments. The first intimation of a new stage came in 1940 when Sigeki Okabayasi and Minoru Honda both detected the first 'Japanese comet'. Honda discovered 12 comets between 1940 and 1968 before turning to the photographic discovery of novae. This change of emphasis was perhaps prompted by the intense competition from his compatriots. The great period for Japan came mainly between 1960 and 1976 when 19 observers were involved in the discovery of 27 comets, three of which were photographic. A drop in the numbers of Japanese discoveries may be seen from about 1970–1. There are several probable causes. It is highly probable that the most important cause was the explosive expansion of Japanese industry, leading to serious degradation of observing conditions over Japan. T. Seki has had to move his observing site, and now concentrates on astrometric work with a larger telescope (a 600-mm reflector). The second factor that has stolen some discoveries from the Japanese is William Bradfield and Co. By systematically searching the sky observable from Australia, Bill Bradfield found 12 comets visually between 1972 and 1984 [and two more subsequently (1987 XXIX = 1987s and 1988 XXIII = 1989c) – Trans.], many before they became visible to northern observers. The Southern Hemisphere also has other successful observers, such as J. C. Bennett in South Africa, and R. Austin in New Zealand, whose discoveries include Comet Austin 1982g (*see* Fig. 8.46).

Discoveries from countries in Europe and North America have shown a resurgence in recent decades, but mainly in the English-speaking countries. G. Alcock in the U.K., R. Meier in Canada, D. Machholz and D. Levy in the U.S.A., and many other observers as well. The U.S.S.R. has also recently been involved in visual discoveries through K. Cernis, a student at the University of Vilnius. It is perhaps arguable whether Cernis is an amateur, because he was training to become a professional and observed with the university's 480-mm telescope. [The opposite perhaps applies to Robert McNaught in Australia, who was professionally engaged in artificial-satellite work, and had the use of other professional instruments, but who discovered Comet McNaught, $1987b_1$, with a 85-mm camera lens. – Trans.]

The amount of dedication shown by amateur astronomers for systematic programmes is probably a corollary of the amount of activity shown by professional astronomers in similar fields. If photographic work on minor planets or supernovae is being carried out, sooner or later an unexpected comet will be discovered. So we should not be surprised that the Schmidt telescopes used in this sort of programme discover a large number of comets, in particular the small Schmidt (460-mm) at Mount Palomar, used by Carolyn and Eugene Shoemaker, and the 1.22-m Schmidt at Siding Spring, used by Malcolm Hartley and Kenneth Russell.

Less than a century after the first photographic discovery, space technology now plays a notable part in cometary discovery. Twenty-two comets have been discovered by three artificial satellites (IRAS, SOLWIND and SMM, the Solar Maximum Mission), one of which was of magnitude 6 (1983 VII). IRAS (Infrared Astronomical Satellite) was designed to map the sky at infrared wavelengths. The scientists who designed it thought that it would be ideal for discovering Earth-approaching minor planets. Although detection of minor planets was very effective – notably the astonishing 3200 Phaeton – the satellite was even more successful in discovering comets, Fig. 8.36. SOLWIND was the name of a coronagraph mounted on the U.S. Army's P 78-1 satellite. Before its wanton destruction in a 'Star Wars' experiment, this satellite monitored the solar corona. Six sun-grazing comets were discovered when the observations were analyzed, some months and even years after they had been made. According to calculations by Brian Marsden of the Harvard-Smithsonian Center for Astrophysics, these comets (Fig. 8.40) all appear to be members of the Kreutz group, which is a family of comets that have retrograde orbits, with inclinations of about 140°, perihelion distances of less than 0.01 AU, and periods of several centuries. They are all probably parts of a single body that broke up in a complex manner at repeated approaches to the Sun. None of the comets discovered by SOLWIND were observed from Earth, for two reasons: they were simply too faint to be detected more than a few hours before perihelion; and they were destroyed at perihelion passage. Several of these comets could perhaps have been discovered with a coronagraph from the ground. It is very probable that many other, similar objects go unseen. After the destruction of the SOLWIND experiment, ten more probable Kreutz comets were discovered by a similar experiment on the Solar Maximum Mission satellite before it decayed in late 1989.

As far as France is concerned (where comet searching may be said to have started), although it may be a country rich in amateur activity, sadly it is one of the poorest where comet-hunting is concerned. The last comet discovery in France (1985p) was made by Jacqueline Ciffréo who, during the holidays, works at a professional observatory, as part of the team using the 900-mm Schmidt at Centre d'Etudes et de Recherches Géodynamiques et Astronomiques (CERGA) near Grasse. Periodic comet P/Ciffréo was discovered on 1985 November 8 on a plate of Comet Halley. For the last discovery by a French amateur we have to go back to 1970 May 21, when Emilio Ortiz was the co-discoverer – from an aircraft – of 1970 VI White–Ortiz–Bolelli. Earlier amateur discoveries have been practically forgotten. They were:

- periodic comet Herschel–Rigollet recovered by Roger Rigollet (Lagny) in 1939 July;
- two comets discovered by Ferdinand Quénisset (Juvisy) in 1893 and 1911.

We may also mention the opportunity lost by Michel Verdenet (Bourbon-Lancy) who detected Comet Kohler 1977 XIV almost two days before the official discoverer, but who did not report his observation in time. One other comet has been discovered

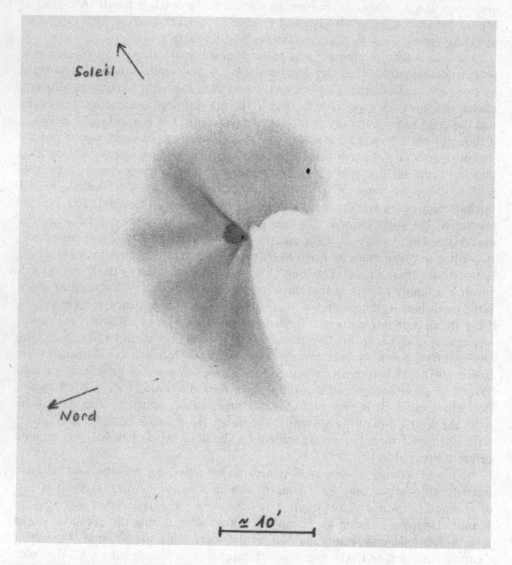

Fig. 8.36. *Comet* IRAS–Araki–Alcock, *1983d (1983 VII), drawn through a 150-mm reflector (magnification 75×). 1983 May 11, 22:20 UT (mid-point of the drawing). (Drawing by J. C. Merlin; from* l'Astronomie.*)*

recently in France by professional astronomers, and this was Heck–Sause 1972 VIII, found with Liège University's 600-mm Schmidt installed at the Observatoire de Haute Provence. Otherwise, nothing since Borrelly and Schaumasse's work at the Marseille Observatory (before 1920)!

This gives some idea of how the fortunes of a particular country may change. We hope that describing the techniques of systematic comet hunting will inspire others to try to emulate Pons, Messier and Méchain and revitalize cometary work in France.

8.8.2 A typical comet-hunter

The one extremely effective way of never discovering a comet, according to Lewis Swift, is to stay in bed! As soon as you turn a telescope on the sky, the probability of finding a comet jumps from zero to a value that is not completely negligible, even in full daylight. If you observe often, or regularly, the probability increases considerably. Whatever sort of instrument is used, there are two ways of discovering comets:

- by accident when one is observing some other astronomical object;
- by design when a specific method is used.

These two basic categories are considerably modified in actual practice. There are observers who are not primarily interested in comets, but who tell themselves that if they observe frequently, even for pleasure, they may be lucky enough to discover one. Others are carrying out programmes devoted to other types of object, but who use techniques that are similar to those employed in systematic comet-hunting. They are therefore always alert to the possibility of discovering a comet. Nowadays, there are no professional teams exclusively engaged in searching for comets, as there was in Czechoslovakia in 1940–60. The relative rarity of these objects is probably inadequate to prompt institutions to allot the appropriate funds for this sort of work. It is also possible that, more or less subconsciously, professionals believe that there are enough effective amateurs to discover the most interesting objects. We have to look among amateurs to find the main characteristics of a typical comet-hunter. We shall examine the results obtained by the different techniques (visual, photographic, etc.), and at different times (the early and modern periods), in more detail in Sect. 8.8.7.

When we examine the list of comets discovered in recent years, especially comets brighter than magnitude 13, i.e., those that are, in principle, discoverable by amateurs, we find that the names of the discoverers may be divided into two groups:

- those that appear once;
- those that appear with a certain regularity.

The first category naturally includes accidental discoveries. It is possible to define fairly accurately the types of work in which amateurs have most chance of discovering a comet accidentally. Variable-star observers undoubtedly have the greatest opportunity. They regularly monitor tens or hundreds of star fields scattered more or less haphazardly over the sky. Whether examining charted fields or in moving from star to star as they position their telescopes, they are ideally placed to detect any new object, particularly visually. Using circles to set a telescope will obviously cause amateurs to lose 90 % of their chance of making an accidental discovery. Variable-star observers who conscientiously follow specific sets of stars also benefit from the experience of the fields that they acquire. Any additional object is immediately identified in a field that has been observed time and time again. A new category of variable-star observers has appeared recently, involving people who search for extragalactic supernovae (*see* Chap. 15). Whether photographic or visual methods are used, the regular examination of fields around galaxies also leads to memorization of star fields, and the acquisition of proper reference material. Supernova hunters also have a trump-card in that they are interested in areas of the sky that are the bane of the comet-hunter's existence. Constellations such as Virgo, Coma Berenices, Leo, Ursa Major, Canes Venatici, etc. are simply torture for visual comet-hunting. More time is spent identifying objects in the NGC than in searching for comets. Although amateurs rarely undertake searches for minor planets, astrometry of these objects is another way of capturing comets. Is this not how many of the Japanese make their discoveries? Finally, we can see that any observational programme that causes us to cover large areas of the sky may give rise to the accidental discovery of a comet. We should get the same sort of results if we are observing galaxies, nebulae or clusters, provided the number of hours is sufficiently high.

A simple fact emerges, which is particularly realised by those who have already discovered 'some' comets: searches are far more effective if we have a thorough knowledge of the sky. This is where we encounter the second group of discoverers, those whose names recur with a certain 'regularity'. We all know the names of Bradfield, Meier, Cernis and Machholtz. These are all experienced observers, who have accumulated years of practice in observing, and who learned to identify dozens of non-stellar, Messier and NGC objects before beginning comet-hunting. During their sessions of visual sweeping, their memory enables them to identify the majority of false comets. They do not have to lose time leafing through atlases, except, of course, for some of the fainter objects. Regular observing has enabled them to become fully familiar with their equipment and to record objects at the limit of detection in a methodical manner. These basic principles also apply to photographers. Rapid identification of a new object in a star field assumes a certain degree of familiarity with examining photographs and properly applying good observational practices, such as making duplicate exposures.

How many amateurs have alerted an observing group because they have discovered a comet or some other object on a single photograph taken a week or more previously? If you do take a single photograph, develop and check it immediately.

8.8.3 Site and observing conditions

Anyone living in a town that is within easy reach of open country is lucky. Amateurs forced to live in a large metropolis are at a disadvantage, but they can rest assured that they are not alone. A large proportion of active comet-hunters travel tens of kilometres to reach their observing sites as, for example, do Bradfield and the Californians such as Machholtz. It is becoming rare to find a site free from light pollution. Bradfield has solved the problem by selecting two sites: one northwest of Adelaide for evening sessions, and one northeast for early-morning work.

The site selected should provide suitable conditions, which are the same as those required for observation of nebulae, galaxies, etc. With clear skies, it should be

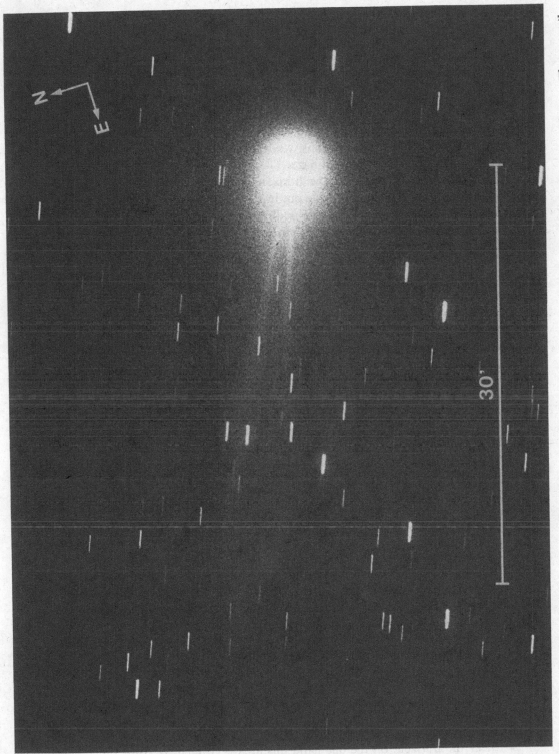

Fig. 8.37. P/Halley on 1985 December 30, exposure 45 minutes beginning 18:15 UT on 103aE film; 200-mm reflector. The gas tail (PA 62°) is accompanied by a tail streamer (at PA 81°). (Photograph by J. P. Damon.)

possible to identify stars down to magnitudes 6.0–6.5 with the naked eye. We need to avoid hilly or other sites that may cause problems with fog and mist, turbulence, or cold winds. We also need to find somewhere quiet, and where we can become dark adapted, so anywhere near a road, with possible car headlights, has to be avoided. Everyone has had their own problems of this sort. The site should also be suitable for the methods being used. We need a clear horizon to the east or west for comet-hunting close to the Sun, or to the north for work during the summer at mid-northern latitudes.

As regards meteorological conditions, we hardly need to say that searching for comets is incompatible with fog and haze. The slightest veil of cirrus cloud affects observation. Beware of mists on the horizon! Aircraft condensation trails are another scourge, and the Moon may be put in the same category. A comet-hunter's calendar is a lunar calendar. For each half of the sky, to west and east of the meridian, the Moon has a deleterious effect for about 12 days each month: from the 4th to the 16th day in the lunation for evening observations, and from the 12th to the 24th day for early-morning observations. So it is possible to observe on both sides of the Sun for eight days around New Moon. Early-morning sessions may continue until day 12 in the lunation, and evening sessions may recommence on day 16 or 17. In adapting the way in which their time is spent, dedicated observers (the true potential discoverers) will only be interrupted for 5 days around Full Moon. You should not laugh at this remark. The percentage of clear nights is an important factor. It is difficult to conduct a really dedicated observing programme from sites where observers have less than 50, clear, Moon-less nights per year. Persistent observers have often indicated that they accumulate 100–200 hours of searching every year in about one hundred (morning or evening) sessions. But it should be remembered that some discoverers, such as George Alcock in the United Kingdom, certainly do not have the benefit of Californian-type skies. Nor should we forget that Pons, Messier, etc. discovered a comet a year, on average, from France, where the weather is by no means ideal.

8.8.4 Photographic search programmes

8.8.4.1 Where should one search?

Before starting a photographic, comet-search programme, we need to look at the probability of having any real success. It is obvious that a single amateur will have difficulty in checking as large an area of sky photographically as can be covered by a visual searcher. There are two ways of increasing the chances of success:

- search elsewhere;
- search for a different class of objects.

The first point means that there are areas of the sky where photographers have little chance of success, and, above all, little chance of being first. These areas are:

- near the horizon: on the one hand this area is covered by visual observers, and on the other, conditions for photography are very poor;

- near the ecliptic: a zone about 30° wide is relatively well-covered by various organisations that are searching for minor planets;
- fields with a large number of galaxies: covered by those searching for extragalactic supernovae.

This does not mean that amateurs have no chance of discovering a comet in any of these areas. An efficient observer, who can analyze plates immediately, may well be able to announce a new object before competitors who take longer to check their photographs.

Searching for long-period comets, or for comets with highly inclined orbits, generally involves the whole sky from the equator to the poles. Lubor Kresák has described a few practical principles concerning comet-hunting closer to the ecliptic. Such comets are strongly influenced by Jupiter, because sooner or later most of them pass less than 1 AU from the giant planet. Incoming comets that have been perturbed in this way are discovered 3 or 4 years after having crossed Jupiter's orbit. During this time, Jupiter has moved between 90° and 120° in longitude. The periodic comets concerned are moving more rapidly and therefore precede the planet in longitude. Kresák therefore advises concentrating one's searching in the 3 months of the year beginning with Jupiter's opposition (when the Sun, Earth and Jupiter are aligned). Searching should begin around the apex of the Earth's orbital motion (90° west of the Sun, and therefore in the morning sky), and should move west, finishing 3 months later in the antisolar direction. We thus monitor ecliptic longitudes greater than that of Jupiter.

The second suggestion (that we should search for other objects) should be taken to mean that we should try to detect fainter comets. As soon as observers eliminate small elongations (less than 45° from the Sun), they reduce their chances of discovering the brightest comets. (The Chinese realised this 3000 years ago.) But if we search for fainter objects, we are tackling a far larger population. Finally, it must be borne in mind that the probability of detecting a comet is not a low as is generally thought.

8.8.4.2 Competition from professionals

Any competition from professional astronomers should really be regarded as an additional incentive. Certainly no one should think, on the basis of the large numbers of comets discovered on the 'Big Schmidt' plates, that amateurs have no chance of discovering comets photographically. Examination of discoveries made in recent decades shows two very well-defined effects.

First, the number of new comets discovered by large Schmidt telescopes (with apertures $\geq 600\,mm$) is less than the total number of discoveries made with smaller instruments and by amateurs. 'Amateur' discoveries are at present almost all visual. The professional discoveries made 'with smaller instruments' concern comets that *are within amateurs' grasp.* These are comets that have magnitudes 12–15, and occasionally slightly less. They are, in 90 % of the cases, too faint to be discovered visually. These comets are definitely well within the range of photographic methods available to amateurs.

The second factor concerns the comets discovered by the large Schmidt telescopes. The average magnitude of these comets is around 17.5, with a range of about ±2 magnitudes. There are always exceptions, such as Comet Hartley–Good, 1985 I, which was discovered with the 1.2-m UK Schmidt at Siding Spring. This is perhaps not a good example, given that the Southern Hemisphere is relatively poorly equipped: a few large instruments, very few moderate-sized telescopes, and few amateurs. There are other examples of comets discovered by large Schmidt telescopes that could have been discovered by amateurs:

- Wilson, 1986l, was recorded on plates taken by the 1.22-m Palomar Schmidt during work on the second Sky Survey;
- P/Ciffréo, 1985p (magnitude 11.5);
- Thiele, 1985m, (magnitude 13), which was also discovered in photographing Comet Halley with the Hamburg Schmidt (800-mm) installed at Calar Alto in Spain. This comet (*see* Fig. 8.44) was photographed by a Japanese amateur, K. Aisawa, with a 200-mm telephoto lens, 24 hours before the professional discovery. Unfortunately for him, he only identified the comet after the official announcement.

When we look at the observational programmes implemented on large photographic telescopes, we find that few of them are used for the study of Solar-System objects (including comets). Many amateurs imagine that Schmidts are used for short exposures. This is wrong. Photometry, or the simple fact of wanting to photograph a field in a particular spectral region, mean that a specific emulsion/filter combination is usually required. This is why exposure times of between 1 and 3 hours are often made with large Schmidt telescopes. If we add up the area of the sky covered by these instruments during a month, we find that very little of the sky is monitored. So the competition from these monsters in the field that is of particular interest to us is not very significant.

Figure 8.38 summarizes discoveries for a period of 20 years between 1962 and 1981. It therefore includes most of the 'Japanese Period'. Three of the SOLWIND comets identified between 1979 and 1981 have not been included. Figure 8.38a shows the 56 discoveries made by amateurs and Fig. 8.38b the 65 professional discoveries. Among the amateur discoveries there are only two photographic ones, both magnitude-14 comets, and both discovered by Nobuhisa Kojima using a 310-mm reflector: P/Kojima 1970 XII and Kojima 1973 II. These are by-products of searching for expected periodic comets (*see* Sect. 8.7.5).

A marked minimum in the number of comets discovered occurs around magnitude 12, dividing the amateur discoveries from the professional, photographic discoveries. When we remember that the number of comets increases by a factor of about 1.6 in moving from one specific magnitude to the next, the minimum becomes a yawning gap. The faintest comet yet discovered by amateurs was the periodic comet Urata–Niijima, 1986o, at magnitude 16. This was, of course, a photographic discovery. The instrument used was a 300-mm reflector, f/5.8, which our Japanese friends enthusiastically use for photographing minor planets and astrometry.

To date there are relatively few comets that have been discovered (or co-discovered) photographically by amateurs:

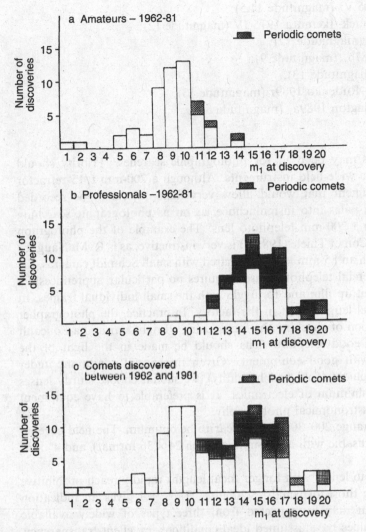

Fig. 8.38. *Histograms of cometary discoveries between 1962 and 1981, in bins one magnitude wide. These histograms include the 121 comets (periodic or non-periodic) that passed perihelion between 1962 and 1981.*

a: Discoveries made by amateurs; the first-magnitude comet is White–Ortiz–Bolelli, 1970 VI; the second-magnitude comet is Pereyra, 1963 V.

b: Discoveries made by professionals: almost half of these comets are periodic.

c: Summing the two preceding histograms shows the dip in discoveries between magnitudes 10 and 15.

(Diagram by J. C. Merlin; data from Kronk, G. W., Comets, a Descriptive Catalog.)

- P/Metcalf 1906 VI (magnitude 11.5);
- P/West–Kohoutek–Ikemura 1975 IV (magnitude 12);
- Sorrells 1986n (magnitude 12);
- McNaught 1987b$_1$ (magnitude 9);.
- Liller 1988a (magnitude 13);
- Okazaki–Levy–Rudenko 1989r (magnitude 13);
- Aarseth–Brewington 1989a$_1$ (magnitude 8.5).

8.8.4.3 Equipment

All types of objectives may be used for photographic searches. Priority should nevertheless be given to wide-field instruments. Although a 200-mm f/15 refractor is a magnificent instrument that would allow very faint objects to be recorded with long exposures, it pales into insignificance, as far as photographic searching is concerned, alongside a 200-mm telephoto lens. The example of the photograph taken by K. Aisawa of Comet Thiele (1985m) is very instructive, as is R. McNaught's discovery of 1987b$_1$ with an 85-mm lens. Compared with small Schmidt cameras, the use of ordinary commercial telephoto lenses requires no particular apprenticeship. It is not necessary to cut up film and to juggle with the small individual frames. In principle, the exact focal length is set in the factory. In practice, the photographer should take the precaution of checking the focus, using a bright star and a Foucault test. The choice of a good telephoto lens should be made in the light of the experience of friends with good equipment. Given the special conditions under which we use this equipment (cold and humidity), we should choose sturdy lenses and bodies, with the minimum of electronics. It is preferable to have equipment specifically devoted to astronomical photography.

Focal lengths in the range 200–300 mm appear to be optimum. The field covered is important: 6° × 9° is usable with a 200-mm lens (on 24 × 36 format), and 4° × 6° with a 300-mm.

The prices of telephoto lenses with longer focal lengths rapidly reach prohibitive proportions, resembling those of 'small' Schmidts. Table 8.4 gives an indication of the performance that may be expected from three types of widely available telephoto lenses. The values have assumed ideal conditions: excellent transparency, perfect guiding, and the use of films with a high signal-to-noise ratio (for example, hypersensitized Kodak TP 2415). The grain structure of Kodak's 103 emulsions frequently leads to confusion when searching for faint patches of light (the 'Kodak comets'). Whatever type of instrument is used, there are always objects right at its limit. Their number increases as the square of the diameter, twice as fast as the number of comets! The maximum exposure times are directly linked to the focal ratio (which governs the rate of darkening by the sky background). The various 'nebular' filters might perhaps increase contrast, but they require an increase in exposure times. The range of cometary magnitudes just given makes allowance for the more or less diffuse appearance of comets.

There is no doubt that the small Schmidt cameras that are within the price-range of amateurs can compete with professional equipment. More and more amateurs are owning Schmidts that are over 100 mm in aperture. The field covered is wider than that of a 200-mm telephoto lens, and the images are far better. A small, 125-mm

Table 8.4. *Limiting magnitudes attainable using small telephoto lenses for photographic searches*

Focal length (mm)	Aperture ratio (F/D)	Usable field	Maximum exposure time (mins)	Limiting magnitude	
				Stellar	Cometary
135	2.8	9° × 14°	15	13	11–12
200	3.5	6° × 9°	20	14	12–13
300	4.0	4° × 6°	25–30	15	13–14

aperture, f/2 Schmidt can record stars down to magnitude 15 with exposures of less than 10 minutes. Schmidts of 200-mm aperture are very formidable instruments (Figs 8.39 and 8.41). Until now few comets have been discovered with these instruments, primarily because of the problems caused by checking the photographs. They show an incredible number of stars and fuzzy objects. A blink microscope is essential.

We shall close this section by mentioning the good old classical Newtonian. The field is usually somewhere between 1 and 1.5 square degrees. This is very small when compared with the field of a Schmidt (25–200 square degrees). It should be remembered that with the same exposure times, two telescopes with the same aperture will reach about the same limiting magnitude (everything else, such as film, guiding and transparency, being equal), whether we are dealing with an f/6 Newtonian or an f/1.5 Schmidt. The greater focal length of the Newtonian (in fact, its focal ratio, f/D) allows longer exposure times to be used, and therefore enables fainter objects to be captured than with a Schmidt of equal aperture. The application of these basic principles to photographic searching is clear: if we have to use a Newtonian, then it is best to make long exposures and probe deeper into the Solar System. Short exposures are fine with a Schmidt, which will attain its limiting magnitude in 15–30 minutes, and cover a wide field. It would be quite ridiculous to attempt to sweep 50 square degrees with a Newtonian: it would need 50 to 100 photographs. When we come to the 50th field, after several months, we will probably hear that a comet has been detected in the first field photographed!

8.8.4.4 Photographic search procedures

Searching for moving objects against a stellar background is made easier if we follow certain specific guidelines. The photographs should be capable of showing us immediately in which direction the objects are moving. Obviously guiding has to be on the stars, because there is no way of predicting how unknown comets are likely to be moving. Once again, there are two conventional ways of solving this problem. The first is to make duplicate exposures. We take two photographs with identical exposures. We should not take one of (say) 10 minutes and one of 30 minutes, because we would run the risk of photographing a faint object on the second plate, and finding that we are no better off than if we had just taken a single

Fig. 8.39. *Comet Austin, 1984i (1984 XIII), on 1984 September 3. Exposure 5 minutes beginning at 02:23 UT; hypersensitized TP 2415 film; 200/300-mm Schmidt telescope, f/1.5. The tail is almost 3° long. An antitail more than 15 arc-minutes long appeared at the beginning of 1984 September. The Earth passed through the comet's orbital plane on 1984 September 13. (Photo by M. Jäger.)*

photograph. Duplicate exposures are a rigorous way of detecting emulsion defects. It is not an absolutely certain way of eliminating the problems created by a ghost image produced by a bright star that is on the edge of the field, or just outside it. The best way of avoiding this problem is by making three exposures: the first two being identical, and the third made by slightly decentring the field in the lens or telescope. Many amateurs are annoyed by having to make several exposures on the same field. We can fully understand this, but we also know that all those that do not are laying themselves open to terrible frustration, because sooner or later they will photograph a moving object. If they take a week to examine their plates, nine times out of ten the object will be lost, especially if we do not even know in which direction the object is moving. Remember that with comets anything is possible, and that movement may be direct or retrograde.

The second method allow us, in principle, to take just a single exposure. It consists of interrupting the exposure for a few minutes, for example two-thirds of the way through the exposure. The sequence might be:

- 10 minutes exposure;
- 3 minutes interruption (with some form of opaque shutter, or by lowering the mirror in an SLR body, if no other method is available), whilst continuing to guide;
- 5 minutes exposure.

The track left by any moving object will be broken into two unequal segments, which will show the direction of movement. Don't make the silly mistake of interrupting the exposure in the middle!

However, we must use this method realistically. It is fine if we are using a focal length longer than 1 m, but not at all effective with very short focal lengths (a few hundred mm). The two parts of the trail are then likely to be confused by the scattering within the film. It would appear to be more sensible in this case to look for movement over more than an hour. The advantage of having several exposures will be decisive if we want to measure the amount of motion, thus allowing other observers to detect it some hours later, or even the next day. Estimation of the daily motion will be more accurate the longer the trail, or the wider the exposures are spaced in time. One of the solutions that might be suggested is to monitor two or three fields per session, arranging the photographs as follows:

- first photograph of field 1
- first photograph of field 2
- first photograph of field 3
- second photograph of field 1
- second photograph of field 2
- second photograph of field 3.

The two photographs of a single area should be separated by at least an hour. A lot will depend on the length of the individual exposures, on the altitude of the fields above the horizon, and on the stability of the meteorological conditions. It is best to work thoroughly and properly, even if it means limiting the number of fields covered.

Whatever region of the sky is monitored, searching is more effective if the observer arranges to sweep a limited area of sky. First of all, it would appear advisable to photograph a given field by always centring the objective on the same reference star. This star, which may also serve as guide star, may be identified on an atlas before beginning the session. It is easy to prepare an overall plan for the photographs by using a small-scale atlas, such as Becvár, Tirion, etc. We can select the fields by drawing a rectangle representing the photographic field onto tracing paper, to the same scale as the atlas. We can then divide up the area of sky into specific, square or rectangular fields along lines of equal declination, with an overlap of 1–2° between neighbouring fields. It is convenient to orient the sides of the frames north-south.

The sort of monthly programme that may be envisaged involves using a 200-mm telephoto lens to sweep an area 25° × 42°. Thirty fields are required. The photographs are oriented north–south, with an overlap of 1° between successive photographs (both north–south and east–west). The exposures are duplicated.

An amateur or a group of amateurs who followed such a monthly programme (which would require 60 photographs), would stand a considerable chance of discovering a new object, whatever area was chosen for monitoring. The work is very similar to what would be needed for making charts. The first series of photographs serve as reference plates for checking subsequent series, with the same field centres, by using a blink comparator (see Chap. 16). The exposure times should be the same throughout the series, or calculated to ensure consistent limiting magnitudes. We may have to convince ourselves of the importance of regular 'mapping' of a zone 25° × 42° in size down to magnitudes 13–14. Don't forget that photography means that we have a record of the exact state of an area of the sky at a particular time. If proper experimental procedures are followed in making the coverage, it has far more value than visual checks of an area, where there can be no objective records. Those who want to check the same area throughout the year will have to choose a circumpolar region. It is, however, always possible to define three or four different areas that are regularly spaced in right ascension – at 8- or 6-hour intervals respectively.

Exposure times Exposure times cannot be left to chance. Let us assume that the resolution of the emulsion being used is approximately 20 μm. A 200-mm telephoto lens would therefore have an angular resolution of about 20 arc-seconds. With a 20-minute exposure most of the comets that we might detect do not show any appreciable size. The comets that might leave trails are close to the Earth ($< 0.5 \, \mathrm{AU}$). If they are faint, they will probably pass unnoticed, because they will only affect the film for a few minutes (the time they take to move by an amount equal to their angular diameter). If they are bright ($m_1 < 10$), they could be discovered quite simply with binoculars. The conclusion is that with wide-field instruments, we need to use the longest possible exposures. The times remain manageable, because the relative aperture limits them to around 30 minutes at most. No filters are used because we are only interested in discovering comets. Provided the focal length is less than 600 mm, the motion of a comet that is more than 1 AU from the Earth remains imperceptible during such an exposure. Obviously the motion is easier to detect the smaller the actual image-size (as with pictures obtained with a Schmidt).

With other instruments such as Newtonians, the first obstacle to overcome is the observer's prejudices. A concrete example of this occurred a few years ago when the American amateur astrophotographer William Sorrells discovered Comet 1986n (magnitude 12) on a 60-minute exposure, 'deep-sky' photograph made with his 400-mm, f/5 reflector. Anyone thinking of using their Newtonian to search for comets would do well not to struggle to obtain exposures that push the equipment to the limits that are imposed by the focal length. As we have seen with the photographs of Comet Halley made with the 600-mm reflector at the Pic du Midi (*see* p. 417), even a comet more than 4 AU from the Earth shows perceptible motion in an exposure lasting an hour. As far as detecting faint comets is concerned, when tracking is on the stars there is no real gain with exposures longer than 30 minutes. The reader will have realised that it is preferable to make two 30-minute exposures than a single hour-long exposure. Even when it comes to recording just stars, there is little gain in the second half-hour.

Do not forget to record the time of the beginning of the exposure to the nearest second. If you detect a comet, you will want to measure its coordinates, so you must aim at securing photographs of astrometric quality.

8.8.4.5 *Overall organisation and checking*

We imagine that there are very few amateurs who are able to carry out all three operations (taking the photographs, developing them, and then checking them), in just a single night. (Some will even begin by hypersensitizing the emulsion.) Many photographers stock-pile the negatives until the next week-end, until the next 'cloudy night', or even until Full Moon. Such a procedure is not compatible with being able to react rapidly in the case of an alert. It is advisable to develop the films and to carry out at least a preliminary check within 24 hours of taking the photographs. The best solution is probably to work in a team and take turns at the telescope and in the darkroom. Our Japanese friends regularly show us what can be done with their numerous discoveries of minor planets and comets. We have already mentioned (p. 441) the case of 1986o, which was discovered at magnitude 16 at the end of 1986 October:

- T. Niijima took three plates 'for minor planets' with his 310-mm reflector;
- T. Urata checked the plates and discovered a moving object. In fact, Urata did not stop there. He sent three astrometric positions to the IAU and, because he had sufficient positions, he even calculated the orbital elements.

Because the field of a Newtonian is quite small, checking is relatively fast. If there is nothing startling in the field, we can go on to examine fainter objects in detail, using a loupe, or by projecting the negative (with either a slide projector or a photographic enlarger). Checking a Schmidt photograph takes longer: the images are very sharp, but the scale is small (about 7 arc-minutes per mm for a focal length of 500 mm) and the number of stars recorded is very high, especially near the Milky Way. A 10-minute exposure with a 200-mm reflector will exceed the limit of *Atlas*

Stellarum. Beyond that again, there are two ways of confirming what objects have been photographed:

- either we have access to the Palomar Sky Survey;
- or we have a second photograph of the same field taken with the same instrument. (But we must have remembered to take one.)

A detailed atlas is required to identify the principal reference stars and to orient the field. This is when we appreciate the fact that all the photographs have been taken with the sides of the field oriented north–south. If we have a second photograph of the field then we shall be happy. A blink microscope enables any change, such as a variation in brightness or position, to be detected immediately. Superimposition of two negatives is only effective for very fast-moving objects. We need to lay the photographs on a ground-glass screen illuminated from below.

If we discover a suspect object, we can immediately see if it is present on the second photograph. If the motion is obvious and if both trails (on the two photographs) show mutually consistent elongations, with an interruption corresponding to the lapse of time between the two exposures, and if the object appears diffuse, then it is probably a comet. We then need to check with the ephemerides to ensure that no known comet is at that position: it could be a newly-discovered comet (either periodic or non-periodic), or an expected periodic comet, which might perhaps be brighter than predicted. If the motion is obvious and the appearance is stellar, then we need to check the ephemerides to see whether a known minor planet is in that position (in the BAA *Handbook*, the IAU *Circulars*, the MPC, etc.). Many faint or distant comets appear essentially stellar in moderately long exposures. Once the direction and rate of the object's motion has been determined, we can try another, longer, exposure guided to follow the movement. We will then have more chance of recording any nebulosity. If there is any doubt about the nature of the object, and if checks show that it definitely is some new moving object, then it should be announced as a 'fast-moving object', without prejudging whether it is a comet or a minor planet. Comets are frequently initially taken to be minor planets and are given minor-planet designations. This happened with 1986o, which was initially designated 1986 UD, and was recognized as a comet only after detailed examination of the plates.

Ideally, the next step is to determine astrometric positions. If the observer does not have the necessary equipment, the positions of the ends of the trail should be estimated as accurately as possible, using an atlas and the coordinate grids. The daily motion of the object should be calculated. The data to be notified to the IAU are described in Sect. 8.10. Before alerting the IAU, it is advisable to contact colleagues who can confirm your discovery.

If a diffuse object is found, but there is no obvious motion, check that there is no photographic fault causing the problem: a plate-flaw, problems with the guiding, the ghost image of a star, etc. There will be serious doubts if the trails on successive plates are contradictory. Check whether there is a galaxy or other nebulous object known at that position and beware of close double stars. This is when one appreciates having three plates!

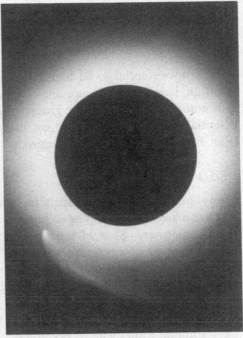

Fig. 8.40. *Comet Ikeya–Seki, 1965f (1965 VIII), photographed on 1965 October 21 close to the solar limb a few hours before its perihelion passage. Left: 02:02 UT; right: 03:27 UT. Photographs taken by F. Moriyama and collaborators using the 100-mm coronagraph at the Norikura Solar Outstation of Tokyo Observatory, Tokyo University. Comet 1965 VIII is a member of the Kreutz group (period about 880 years); it was observed with the naked eye 1° from the solar limb, its magnitude was then about −10!*

8.8.4.6 *A little encouragement*

Searching photographically for comets is tedious work, which requires plenty of patience, attention to minutiae, and discipline. For a single individual it possibly requires the same amount of effort as discovering a comet visually. We find the same sort of constraints:

- either we are sweeping a large area of the sky, using short exposures to detect fairly bright comets down to about magnitude 11–13;
- or we are checking a small area to a greater depth, hoping to detect objects down to about magnitude 14–18.

When compared with visual work, it is obvious that even with very simple photographic equipment, we can record much fainter comets. Although visual work will continue to be successful in finding comets brighter than magnitude 12, we are convinced that photography by amateurs will eventually fill the gap between magnitudes 10–11 and 15–16 (Fig. 8.38).

Experience shows that the effectiveness of photographic searches depends far more on the systematic checking of the photographs than on any other factor (such as elongation, etc.). A comet does not just 'jump out' of a film that is still in the camera or in the refrigerator! We would like to urge amateurs to actually get started on a programme of photographic searching and, if possible, to cooperate with others.

It is not essential to be highly specialized, nor to have complex equipment. In addition, it is easy to inspire one's friends! We can even imagine the foundation of a monitoring network, where groups of observers have divided up the sky, ensuring essentially complete coverage, and using lenses with focal lengths of 200–300 mm, 'Flat-Field' cameras, or small Schmidts. Well, we can always dream!

8.8.5 Visual searches

Many amateurs have tried spending several hours or several evenings zig-zagging across an area of sky, hoping to pick up a comet that they are sure must be awaiting them somewhere. Most of them became discouraged, preferring to concentrate on less haphazard observing, or on careful study of a known object. Although these other means of satisfying the thirst for observation doubtless have their charms, it remains true that they have little chance of affording as much satisfaction as the discovery of a comet.

If we are not successful in our first attempts, it is possibly because we do not have suitable equipment, or because our methods are not appropriate. We shall try to describe the techniques that have been shown to be most effective and which will maximize our chances of success. The various methods of carrying out visual searches largely consist of systematically sweeping a pre-determined area. It is not a question of covering the whole sky every night. Experience shows that even a 90-minute visual observing session is very tiring.

8.8.5.1 Equipment for visual searching

The type of instrument used for visual searches is directly related to the method employed and thus to the area of the sky that is covered. We know that comets are brightest when close to the Sun, so we have more chance of discovering a comet on the western horizon in the evening or on the eastern in the morning. These are the areas where amateurs have most success in visual observation because they are essentially out-of-bounds to photographic work. An important proportion of discoveries are made less than 60° from the Sun. It is useful to know that some comets are even visible in full daylight, close to the solar limb. These are often members of the Kreutz group (see Sect. 8.8.1), which frequently never become visible at night. An amateur who regularly observed the solar limb with a coronagraph would have a definite chance of detecting such objects. Comets are also discovered during eclipses of the Sun, as with Comet 1948 XI, which was observed from the ground and photographed from an R.A.F. aircraft during the total eclipse of 1948 November 1.

Apart from these individual cases, the choice of the area of the night sky to be covered enables us to define the type of mounting that is most suitable:

- an altazimuth mounting for sweeping near the horizon;
- an equatorial mounting for use elsewhere.

Fig. 8.41. *P/Giacobini–Zinner, 1984e (1985 XIII), photographed on 1985 August 12. Exposure 7 minutes beginning 23:24 UT; hypersensitized TP 2415; 200/300-mm Schmidt, f/1.5. (Photograph by M. Jäger.)*

Type of mounting A comet-hunter's equipment should have two qualities: ease of use, and wide field. The movements should be very smooth, in particular the one that corresponds to the direction in which sweeping is carried out, for example, in azimuth for sweeping horizontally. The motion at right-angles should be properly clamped, in order to prevent any unwanted movement. It is essential for the actual mounting to be stable and convenient. Columns are preferable to tripods. The comfort of the observer, who has to move around the mounting while sweeping, should also be borne in mind. Some amateurs have designed observing chairs to which their binoculars are fixed, and others have even made a rotating shelter in which they can sit behind their instruments. Anything goes if the main object is achieved. It is extremely uncomfortable to find oneself bent double throughout a long observation. If, in addition, it is necessary to move around the telescope, then you had best be prepared and book massage sessions! One of the dodges that may be used with a Newtonian telescope on an altazimuth mounting is to place the secondary opposite the declination axis (a design dating back to Nasmyth, and recommended by Antoine Brun, the great French variable-star observer). The eyepiece is mounted in the centre of one of the bearings, so the eye remains at the same height. It would even be possible to observe sitting down – with a chair on castors, it would be even better.

The idea of using an altazimuth mounting should be considered for binoculars. It is very tiring attempting to hold binoculars that are larger than 50-mm in aperture for half an hour. There are commercially available panoramic heads that keep cameras level whilst they are swung horizontally. The most wonderful-looking mountings are not necessarily the best. William Bradfield in Australia has discovered nearly all his comets with a 150-mm refractor built around a salvaged lens: the mounting

453

is a crude altazimuth, based on the design suggested by Couder and Texereau. It is mounted on a column that slides vertically so that the eyepiece may be kept at a convenient height, whatever altitude is being examined. (Similar systems may be found on the altazimuth mountings of certain refractors and commercial tripods.)

For equatorial work, the observer should mainly concentrate on compact designs that allow one to be out of the way of the pier. Refractors on German mountings were extensively used at the beginning of this century. The increase in the numbers of Newtonians, which were cheaper and easier to construct oneself, changes things somewhat after 1945. The fork mounting is less cumbersome than a cradle, and enables sweeping to be carried out in any area of the sky, provided the cell can swing through the fork to allow access to the northern horizon. Everyone will adapt their method of searching to cater for their own equipment. It is not totally impossible to search for comets with an equatorially mounted instrument inside a dome. It merely means using an appropriate method, such as sweeping vertically, for example. The greatest problem with an equatorial mount is the change in position of the eyepiece with changes in hour angle. Some amateurs who are mechanically minded have fitted the eyepiece holder and spider to a rotating top, so that the eyepiece is easily accessible at all times. Short telescopes on fork mountings do not need such a device, if they can be reversed. One should always try to arrange for the eyepiece to be horizontal or downwards, on the eastern side of the mounting when observing before the meridian, and on the western when observing after it.

The optics Conventional refractors with focal ratios of f/8 to f/15 were extremely common at the beginning of this century. Short telescopes became more widespread later, when techniques for making the mirrors and aluminizing them had been mastered. The point should be made that the advantage of Newtonians is in their reduced size; they are not necessarily better in other respects. Whether we use a 150-mm, f/10 refractor, or a 150-mm, f/5 Newtonian, a comet will appear just the same in both if we use the same overall magnification. The darkness of the sky background is the same. The only difference comes in the eyepiece that has to be used (double the focal length in the refractor).

Just as with photography, visual searching requires that a compromise be struck between the area of sky that is covered and the 'depth' to which the hunting is taken, i.e., the magnitude of the faintest comet that can be discovered. Table 8.5 gives details of the performance that may be obtained with different apertures, whether binoculars, refractors or reflectors. We also give the diameter of field that is covered for an exit pupil of 6 mm, as well as the area of sky that can be usefully checked. The limiting magnitude reached for known comets is that obtained under extreme conditions; with thorough searching at high magnification. With comets yet to be discovered, the problem is different. We need to sweep fairly rapidly with the largest possible field. The magnification required is close to the minimum effective magnification, or slightly higher to increase contrast ($D/5$ to $D/6$). We must therefore expect to miss the fainter comets. The immediate reaction to use a larger instrument should be controlled. The number of troublesome objects (galaxies, nebulae, etc.) increases as the square of the diameter, and a beginner will spend a

Table 8.5. *Faintest comets observable with various apertures*

Aperture (mm)	Field		Total magnitude of faintest comets	
	Maximum (°)	Usable (°)	Known comets	Discoveries
50	7.0	5.5	9.5–10.0	8.0
80	5.0	4.0	10.5–11.0	9.0
100	3.5	3.0	11.0–11.5	9.5
150	2.5	2.0	12.0–12.5	10.5
200	2.0	1.6	12.5–13.0	11.0
250	1.5	1.2	13.0–14.0	11.5
400	1.0	0.8	14.0–15.0	12.5

prohibitive amount of time identifying spurious comets. During that time the Earth continues turning, carrying the true comets closer to the horizon.

Theoretical consideration of the number of comets visible indicates that if the aperture of the instrument is doubled, one can expect to detect twice as many comets – not like the troublesome galaxies and nebulae, which become four times as numerous. As a result, the time required to sweep a given area increases as the square of the aperture. This may be expressed by saying that the efficiency of searching decreases by half when the aperture is doubled. We have only half the chance of discovering a comet with a 300-mm telescope that we have with a 150-mm. This conclusion agrees perfectly with the actual discoveries from visual searches. An aperture of 150-mm appears to be a good compromise, capable of searching a large area every evening down to an interesting magnitude (about 11). Amateurs have discovered dozens of comets with instruments of this sort of size – isn't that right, Mr Bradfield? Amateurs who use more powerful instruments have, on average, less success. The record tables (Tables 8.8, 8.9 and 8.10, p. 480) give an idea of the equipment used by the most prolific discoverers. As regards 'large' instruments, we note some equatorials of 400 mm and more in aperture (R. Meier, K. Cernis, D. Levy, etc.).

What about binoculars? Apart from the large field covered (from 3–7° across), binoculars have the considerable advantage that both eyes are used simultaneously. The gain is generally estimated to be 40%, when compared with observations made with one eye. Binoculars with an aperture of 125 mm are as effective as a 150-mm reflector. However, it is always useful to quote some exceptions:

- W. Bradfield used 7×35 binoculars when he discovered Comet 1980 XV (magnitude 6), 22° from the Sun.
- D. Machholtz has, for some time, been using 29×125 binoculars that he made himself from surplus, aerial-camera lenses. He used these to discover P/Machholtz (magnitude 10).

These two results show that it is still possible to discover comets at small elongations with small instruments, and that appropriate binoculars are very effective in detecting relatively faint comets.

Although reality usually lies somewhere between extremes, it must be admitted that we have more chance of being the first to discover a new comet if we use binoculars that are larger than 80 mm in diameter.

The question of magnification is very important. If the magnification is too low, a comet might be confused with a star and thus go unnoticed. We have seen that the eye can detect a fuzzy patch when the apparent size is greater than 20 arc-minutes (real size × magnification). If we assume that most observable comets have an apparent diameter greater than, or equal to, 1 arc-minute, we can deduce that a magnification of 20× is required. The criterion that defines the optimum exit pupil as 5 mm suggests that we should use an aperture of at least 100 mm (5 × 20). So we need not be surprised to learn that 20×100 binoculars are extremely common among discoverers (G. Alcock, the Czechs, the Japanese, etc.).

One last technical comment: eyepieces. Eyepieces of the Erfle type have an actual field of about 60°. This is quite high – indeed, perhaps slightly too high, because it is necessary to move the eye to see the whole of the field. In addition, these eyepieces have significant aberrations at the edge of the field. A Kellner is simpler and not so expensive. It is very suitable for classical Newtonian telescopes (with relative apertures of around f/6). It is also the most common type used on binoculars. In searching for comets with a telescope, a Plössl eyepiece is probably even better, because it produces less internal reflections – we need to beware of non-existent comets!

If you are very familiar with the sky, and have an instrument that meets the criteria described, then all that you have to do now is to adopt a suitable technique.

8.8.5.2 *The azimuth method*

Here we are primarily interested in those areas of the sky where a new comet is most likely to appear:

- on the western horizon, after sunset;
- on the eastern horizon, before sunrise.

We shall also discuss the benefits of monitoring the northern horizon in the summer.

Sweeping horizontally The observing sessions take place outside twilight; start in the evening about $1^h 30^m$ after sunset, or stop about $1^h 30^m$ before sunrise in the morning. There is little point in searching in the twilight, in view of the considerable extinction. We would have considerable difficulty in identifying the field, because so few stars would be visible. It would be most frustrating to see a comet just that moment suspected disappear in the twilight or into the background near the horizon. It is not advisable to sweep less than 10° from the horizon.

Evening sweeping We start by pointing the telescope 30–40° in azimuth farther south than the point on the horizon where the Sun disappeared. The first scan is made horizontally towards the west and north-west, until a point is reached that is symmetrical with respect to the position of sunset. The zig-zag technique consists of making the second scan in the opposite direction to the first (Fig. 8.42a):

- we then move up in azimuth by an amount that is equal to about 80–90 % of the field diameter, so that there is a certain overlap with the previous pass;
- we then scan towards the west and the southwest, until we reach a field vertically above the point where we began the first pass;
- we then move up in altitude again, move back towards the northeast, and so on until we reach the maximum altitude we want to scan.

Figure 8.42b shows, on a chart of the sky, the shape of the area that is swept. The overlap becomes zero shortly after the changes in direction, and between each pass there are vast areas that are not covered at all, because of the effects of the Earth's rotation. The effectiveness of the sweeps is practically halved. If we absolutely insist on sweeping in zig-zags, we could ensure complete coverage by not moving in altitude at the end of each pass as we change direction. Our sweeps would be carried out at a constant height above the horizon, and the Earth's rotation would gradually alter the strip of sky that we are sweeping. This technique has two disadvantages, however:

- the overlap is now too great just after the changes in direction. Half of the area covered on each pass has already been swept in the previous pass; so there is no increase in efficiency;
- we are observing in rather mediocre conditions, always at the same height above the horizon (and therefore with a relatively large extinction).

Two morals may be drawn from this:

- sweeping should be rapid;
- we need to move away from the horizon as quickly as possible.

Bradfield's advice is always to sweep in the same direction – don't zig-zag – for example from southwest to northwest. At the end of each horizontal sweep, move back rapidly to an area above the starting point (i.e., to a constant azimuth, see Fig. 8.42c). It is useful to provide some means of locating this bearing (such as by using either an azimuth circle or a mechanical stop).

Speed of sweeping The amount of shift in altitude should take account of the time required for each pass and of the effects of the Earth's rotation (unless you have been able to identify specific patterns of stars in the first field!). Due west, the vertical movement per minute of time is $15' \times \cos L$, where L is the latitude of the site. For a latitude of 45°, a band 10–11 arc-minutes wide moves below the horizon every minute.

Let us assume that we take 3 minutes to make a single pass through 60°, with a field diameter of 2°, and that we set the vertical overlap between passes as 15 arc-minutes. During 3 minutes, a band about 30 arc-minutes wide will descend towards the band that we have just swept. The offset in altitude should therefore be restricted to $2° - 30' - 15'$, i.e., $1°15'$, to avoid missing a horizontal slice of the sky.

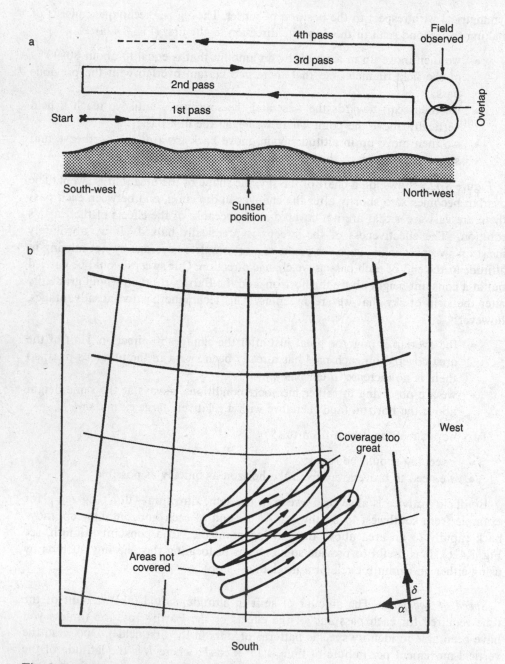

Fig. 8.42. *Horizontal sweeping for visual comet searching.*
a: How it should not be done: the zig-zag method.
b: The effect of using the zig-zag method on part of the sky being observed. There are areas that are not covered and areas that receive too much coverage.

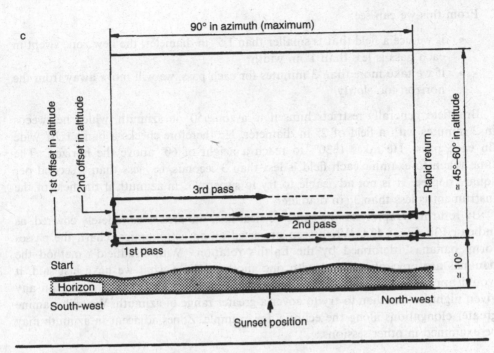

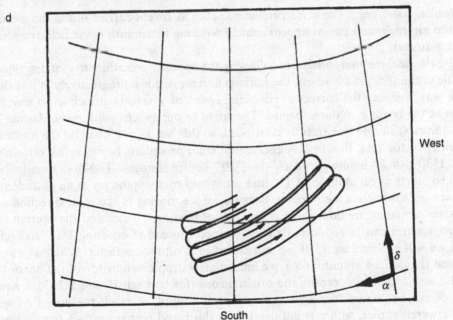

Fig. 8.42. (cont.) *c: How it should be done. After a horizontal pass, make a quick return to a specific vertical line.*
d: The effect of using a proper method on the part of the sky that is being observed. Coverage is complete.

From this we can see:

- if we use a field that is smaller than 1.5° in diameter, the new zone swept in each pass is less than 1° in width;
- if we take more than 3 minutes for each pass, we will move away from the horizon too slowly.

Bradfield generally restricts himself to a zone 50° in azimuth, which he sweeps in 2 minutes with a field of 2° in diameter. He therefore checks a band 1°20′ wide on each pass. He takes 1^h30^m to reach a height of 60° above the horizon. The time taken to examine each field is less than 3 seconds, i.e., less than 1 second per square degree. It is not advisable to try to sweep 90° in azimuth if the field of the instrument is less than 3° in diameter.

By returning rapidly to the same vertical line, the area is completely covered, as indicated in Fig. 8.42d. When seen projected on an astronomical chart, the passes form 'bananas', deformed by the Earth's rotation. With Bradfield's method the bananas are precisely aligned with one another. From what we have just said, it would appear sensible to try to sweep a little farther away from the Sun on any given night, rather than to try to cover a greater range of azimuth. We will examine greater elongations along the ecliptic, for example. Zones adjacent in azimuth may be examined in other sessions.

Morning sweeping   The same principle applies to sweeps carried out in the morning: no zig-zags, and passes approximately 60° long in azimuth if the field is about 2° in diameter.

The classical method consists in following the opposite procedure to that described for the evening. Start 60° *above* the horizon and move down progressively in just the same way, keeping the correct overlap and speed of sweep. In principle we should finish at the horizon as dawn begins. The speed of our sweeps (with passes lasting 2 or 3 minutes) should be carefully controlled, so that we do not 'land' on the horizon too early or too late. Bradfield suggests a different procedure, however. We certainly start 1^h30^m or 2^h before dawn (3^h to 3^h30^m before sunrise). The first zone to be checked starts at an altitude of 15° and is covered by sweeping up to an altitude of 60°, *moving away from the horizon*. Because we are moving in the same direction as the stars' apparent motion, there is less risk of missing a band, and the overlap of 15 arc-minutes can be reduced, or even omitted. Instead of covering 1°15′ on each pass, we will be checking 1°30′, so our coverage in altitude is faster. When we have reached the desired altitude (60°), we immediately drop down to 15° of the horizon. Whilst we have been sweeping the original zone (for between 1^h and 1^h30^m), a new area of sky has risen. We again sweep upwards until we reach the area that we have covered earlier, with a small overlap. A third, and perhaps even a fourth, step consists of sweeping slightly lower, at about 10° from the horizon, which is in any case closer to the Sun, until dawn.

Effectiveness of large instruments close to the horizon   Let us assume that we are using a telescope with a field-diameter of 1° (a 300-mm reflector, for example). To sweep a band 40 arc-minutes in altitude at each pass, with an overlap of 10 arc-minutes, the length of a pass should be kept to 1 minute. Scanning 60° in azimuth means that 1 second is all that can be allowed for checking a field, as against 2–3 seconds with a 150-mm telescope. That being the case, it would appear doubtful whether the full power of the instrument is being used. The apparent field of eyepieces is always about 45–50°. Taking 3 seconds to scan a 1° field means taking 3 minutes to cover each sweep of 60°; during that time the sky rotates 30 arc-minutes when looking due east or west! The width of the band covered in each sweep becomes ridiculous, or else we have to restrict ourselves to a lesser range in azimuth (30°?) and check the neighbouring areas in other sessions. We are certainly far from enjoying optimum conditions for success.

Large instruments (with diameters greater than 250 mm) are very poorly suited to searching near the horizon. We need to be quick. Only very experienced observers who are extremely familiar with the sky can use large diameters. Beginners will encounter shoals of spurious comets, which will slow down the sweeping. They will be completely overtaken by events, caught either by the dawn or else by 'collision' with the horizon.

Sweeping vertically Some observers sweep vertically instead of horizontally (Fig. 8.43). Bradfield's advice remains valid, when suitably modified: at the end of each pass, move back rapidly to the line (parallel to the horizon) where you started, do not make zig-zag sweeps. It is rather doubtful whether vertical sweeping should be used from middle latitudes. The time required to cover 60° in azimuth in 2°-wide vertical strips causes us to run the risk of missing a comet that is close to the horizon at the other limit of the zone being covered. The horizontal method gives priority to checking zones closest to the Sun. This appears to be more sensible. Vertical sweeping is far more suitable for other areas of the sky, particularly near the meridian, where there is no risk of comets setting in the next half-hour. Large Dobsonians may certainly be used for searching for fainter comets, well away from the eastern and western horizons.

Vertical sweeping is also very suitable in two other cases:

- Observation from countries in equatorial regions. Vertical sweeping is then equivalent to equatorial sweeping. Stars set vertically towards the horizon, but – take care(!) they set at 1° every 4 minutes, whereas at a latitude L, they set at 1° in $(4/\cos L)$ minutes, i.e., at about 1° in 5.5 minutes at 45° latitude.
- With a Nasmyth type of mounting (p. 453) there is less moving around the instrument. This is more comfortable during a sweep, particularly because the height of the eyepiece remains constant.

Sweeping the northern horizon At middle latitudes, the northern horizon is of interest in the summer months. The Sun sets in the northwest, so there is a distinct possibility that a comet might appear in the area of sky that lies beneath the celestial pole. (This obviously also applies to the South Celestial Pole during the southern summer.) At 00:00 UT, the Sun is at its lowest point below the northern horizon. The most interesting areas are to the northwest and north (between the end of

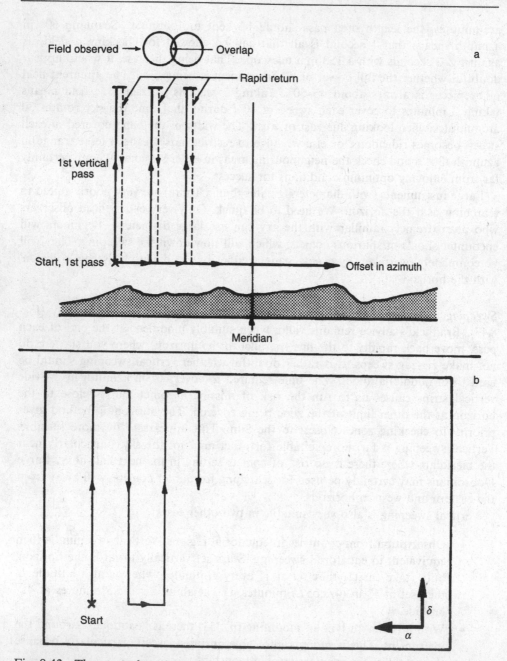

Fig. 8.43. *The vertical method of sweeping for comets. At middle latitudes this method is best for use close to the meridian.*

a: Vertical method used on an altazimuth mount: no zig-zagging, a rapid return to a particular altitude parallel to the horizon.

b: The vertical method as used on an equatorial mount. It is permissible to zig-zag, especially if the mounting is fitted with a drive in right ascension.

462

twilight and 00:00 UT), and to the north and northeast (between 00:00 UT and first light). Freedom of manoeuvre is fairly small, mainly because of the amount of twilight during the short nights between May and July.

The big advantage in discovering a comet beneath the pole is mainly that we have more time to confirm the observation, without the fear of the comet disappearing beneath the horizon.

Sweeping the northern horizon during summer can be carried out in two stages:

- an evening session, sweeping from the west round to the north;
- a morning session, sweeping from the north round to the east.

The horizontal sweeps are centred on the northwest and then on the northeast. The area covered should reach as far as Polaris. The coverage each side of the Sun is better than at other times of the year.

8.8.5.3 *Equatorial method*

The equatorial method does not present the inconveniences found in azimuth methods. Provided our equatorial mounting is accurately oriented, we can sweep along parallels of equal declination. After each pass, we can sweep back to our starting point – i.e., follow a zig-zag path – having shifted the instrument in declination by 90 % of the field diameter. There is no need for us to use a greater overlap if we are sure that the declination axis was properly clamped during the previous sweep in right ascension. An equatorial mounting is even more useful if it has a drive in right ascension. The drive is switched on, and the right-ascension circle is set using a bright star. If we detect any suspect object, we have only to read off the coordinates α and δ on the setting circles, and we can then immediately check with an atlas to see if there is anything at that position.

Sweeping in declination (i.e., at a constant right ascension), along meridians of RA is perfectly feasible using the same method (with a right-ascension drive), and eliminates all the problems with the 'bananas' that arise with sweeping in azimuth (p. 456) and vertically (p. 461).

What region of the sky should be monitored?   All of it, of course! Rolf Meier advises sweeping the sky from one horizon to the other. Searches begin, for example, at the celestial equator, not far from the western horizon if observing in the evening. The chance of discovering a comet is, however, somewhat low. Those who are unable to alter the position of the focussing mount so that it is easy to reach will have to be content with checking half the sky at a time, first from west to the meridian, and then from the meridian over to the eastern horizon after having reversed the mounting. This second area could be the subject of the next (or subsequent) session(s). The shift in declination, either north or south, may be determined having regard to the areas swept previously. Just as with sweeping in azimuth, the observer should allow an overlap of a few degrees from one day to the next, increasing this if the sessions are separated by a longer interval. Observers in the Northern Hemisphere should try to work towards increasing declination, where the transparency is better.

If you want to be the only one checking a particular area, you had best keep away from the ecliptic – although visual observation gains in immediacy when compared with photographic work.

If during preceding evenings (or mornings) you have checked the western (or eastern) horizon using an azimuth-sweeping technique, there is no point in extending the equatorial sweeps very low down towards the western (or eastern) horizon. Covering the sky with a mixture of the two techniques (azimuth and equatorial) is probably the ideal way of ensuring that no comets are missed. A programme that is solely equatorial does not allow the western and eastern horizons to be covered exhaustively: there are unchecked areas that disappear beneath the western horizon (10–11 arc-minutes per minute of time), and areas at specific declinations that appear at the eastern horizon, which have, in principle, already been checked. Some very low comets are missed because of the lengths of the sweeps: sweeping half the sky (from the meridian to one horizon) at a rate of 1° every 2–3 seconds requires between 3 and 5 minutes. The equatorial method has a little more latitude in the speed of sweeping, but not much more, given the lengths of the sweeps (between 6 and 9 hours in right ascension). With 20 passes, the area swept in 1^h30^m with a field of 1° is about 1800 square degrees – say 90° in right ascension by 20° in declination.

Extending evening equatorial sweeps down to the eastern horizon appears to be very desirable for about a week immediately after Full Moon. Any rapid change in the Sun–Earth–comet alignment may cause a comet to move from invisibility to a stage at which it can be detected, in less than 15 days. A comet may thus cross the area of the sky that is normally monitored photographically (along the ecliptic in the morning sky), during the period between Full Moon and Last Quarter. It therefore becomes directly accessible in the middle of the night (on the eastern horizon in the evening) when the Moon moves out of the way.

A similar situation arises in the first 12–13 days of each lunation: a fast comet could cross the evening sky during the 8–10 days around First Quarter. Morning equatorial sweeps should be extended down to the western horizon, about an hour after the Moon has set.

Is there any point in using an equatorial method at the northern horizon? If we follow parallels of equal declination, we skip the most interesting areas lying close to the horizon in the northwest and northeast. It would perhaps be more sensible to sweep in declination, along the meridians of RA, from pole to horizon. In addition, the overlap is always far too great at the pole. The azimuth method undoubtedly causes less problems in this part of the sky.

8.8.5.4 Getting organised

Before each session, anyone searching for comets should carry out a certain amount of preparation: it is pointless rushing blindly into checking an area only to find that it is strewn with nebulous objects that one has never seen before. After having decided on which area of sky should be covered, it is useful to check with a planisphere to see which constellations this includes and which are the brightest stars that may serve as reference points. This is even more important if we are observing close to the horizon, at the end of evening twilight, or before dawn. We can then take a look at an atlas, such as *Atlas Coeli*. The area of sky covered by one chart in this atlas is very similar to the area normally covered in comet-hunting. We can then see what nebulae, galaxies, etc., we can *expect* to encounter as we sweep. If, for example, we miss NGC 2403 in sweeping across Camelopardalis, it shows that our searches are not very effective! We believe that the state of mind conducive to reliable searching is shown, not so much by being able to recognize extraneous objects quickly, as by being able to say 'I have just passed α UMa, so I am just about to reach M81 and M82'.

Many amateurs seem to feel that it is not essential to know the sky well in order to search for comets; but few of these people actually discover anything. It must be admitted that we are more likely to discover unsuspected galaxies than comets.

Our chances of finding a comet are far higher if our observational programme is well-organized. Obviously there is nothing that we can do about the weather. It is naive to think that observing after a long period of bad weather is going to improve one's chances, because naturally the weather has not necessarily been bad elsewhere in the world. On the other hand, the presence of the Moon is a factor that affects observers all round the world. A lot may happen in the evening sky between days 3–4 and 14–15 in a lunation; and the same applies for the corresponding days in the morning sky. In 10 days a moderately fast comet may cover more than 30° and move out of the solar glare. So our motivation to observe as soon as the Moon's interference lessens in the evening or morning should be extremely strong.

The first session in a lunation should therefore be a sweep in azimuth near the Sun. If we find nothing, and provided our technique is thorough – one could say 'infallible' – we can assume that it is not essential to check this area again for a week or so. In the next session we can check a neighbouring area, perhaps slightly farther to the south. We should take care to ensure that the overlap with the previous area is at least 5° in azimuth, or more if the first sweep was made several days earlier. In three evenings it is therefore possible to cover an area of about 180° in azimuth, from north through west to south, up to about 60° altitude. An identical procedure may be applied to the eastern sky, beginning about 3 days before New Moon. If we wait until New Moon before beginning there is every chance that the morning sky will have been covered by others. There are no two ways about it: if you want to be the first to discover something new, you have to be properly organized and start at the appropriate time.

Once we have covered the area around the Sun, perhaps with a 150- or 200-mm telescope, or large binoculars, we can turn to the rest of the sky. Sweeps linking the two areas at low elongations may be carried out with an equatorially-mounted instrument, perhaps a 200- or 250-mm reflector.

The ideal would be to cover the whole sky in a month. Taking the criteria concerning field-diameter and rate of sweeping that were discussed on p. 457 into account, we have to reckon on 15–20 hours to cover the whole of the sky visible from our latitude. A single individual would need some ten clear nights (or evening and morning sessions) each month. Even assuming satisfactory weather conditions, a single observer would probably find this difficult, but a group of observers would have far more chance of being successful. This demands thorough organisation, and a completely fair allocation of the areas to be covered. The validity of this has been confirmed by the Czech observers who worked at Skalatne Pleso (Sect. 8.8.1).

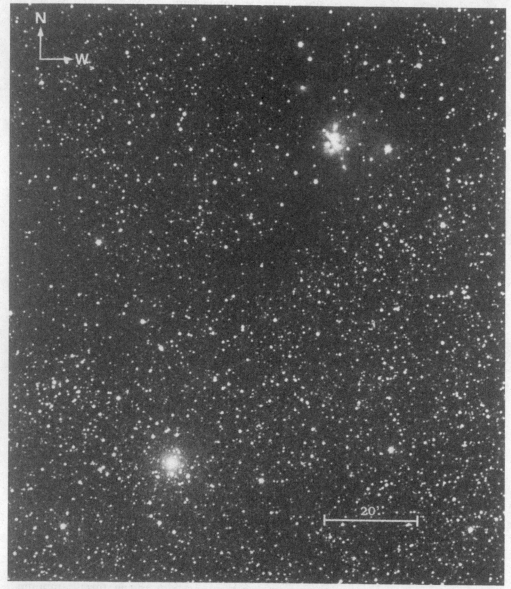

Fig. 8.44. *Comet Thiele, 1985m (1985 XIX), photographed on 1985 October 25 close to NGC 1579 in Perseus. Exposure 7 minutes, beginning 02:59 UT; hypersensitized TP 2415; 200/300-mm Schmidt, f/1.5. (Photograph by M. Jäger.)*

It is important to keep track of all the observing sessions, if only to organize later sessions. If a comet is discovered in a field that you checked some days previously, you should still report your negative observation. It may help to indicate that the comet concerned has undergone an outburst.

Observers should note down the following information:

- date and time of the beginning of the session, and its duration;
- instrumentation used (type, diameter, and magnification);
- method used;
- observing conditions:
  - transparency, either naked-eye or instrumental limiting magnitude;
  - the effects of twilight;
  - phase of the Moon;
- the area covered:
  - limits in azimuth and altitude (or in α and δ);
  - the constellations covered;
  - objects encountered (Messier, NGC, known comets, etc.);
- any interruptions, etc.

If you are successful, totting up the time spent will help to provide statistical information about comet-hunting. The average amount of time required to discover a comet is somewhere around 150–200 hours. These values assume, of course, that the observer has achieved an optimum rate, after several years of apprenticeship.

8.8.6 What should you do if you are successful?

When one encounters a fuzzy patch, the first reactions should be those of a proper observer:

- slightly move the telescope to check that it is not a ghost image of a star (gentle tapping of the tube will usually suffice);
- use another eyepiece to make sure that the diffuse nature of the image does not change: small groups of stars can cause a lot of confusion!

We have said enough about troublesome objects for it to be obvious that the next step consists of checking whether there are any in the observed position. To do this, we must know where we are! This problem does not arise with photographic discoveries, because one is supposed to know the specific coordinates for the centre of the field. With visual observation, it is important to react immediately. It is therefore essential to have a suitable atlas to hand. If we are sweeping equatorially, we can read the setting circles. Even if the motion in RA is not driven, it is still useful to have a declination circle. With searching in azimuth, we can solve the problem by using horizontal coordinates, provided the instrument is fitted with azimuth and altitude circles, and also provided they are correctly set with respect to the horizon and the north–south meridian. We can subsequently transform the readings into hour-angles, and thence into equatorial coordinates, knowing the local sidereal time at the time of observation. If none of these methods are available, then we shall have to use some appropriate technique to obtain the equatorial coordinates of the object observed.

First, keep calm. Don't touch the telescope, and clamp the axis so that you do not lose the field – after having made the checks mentioned at the beginning of this section, of course. The best habit is to look through the finder, and note the characteristic star patterns in the area. It is even advisable to make a sketch of these. If you are still uncertain of the location after having checked an atlas, it is a good idea to scan the area with binoculars to pick up the brighter stars. If this takes some time, don't forget to check whether the ghost – sorry – the object, is still in the telescopic field. If it has completely disappeared, could it have been some atmospheric effect? If you lose the field, there is nothing for it but to sweep again, beginning a few degrees farther north or south.

Once the area has been identified, we need not be worried about going back to the telescope eyepiece. We can then move on to more accurate location, using the same magnification with which the object was discovered. We can check with a detailed atlas (the *AAVSO Atlas* for example). It would be most surprising if there is not at least one SAO star within the telescopic field. From our initial identification through the finder, it should be possible to recognize where we are within a few minutes.

Three cases may arise:

a There is obviously a nebulous object at that position in the atlas. We have discovered a Messier or NGC object. We can also check some of the other publications that list nebulous objects that are detectable in amateur telescopes. Because we did not waste much time in identifying this object, we can carry on sweeping. ('If at first you don't succeed, try, try again ....')

b There is no fixed object at that position, but being well-informed of current events, we know immediately that it is a known comet. If we want to discover a new object, we should not linger over it.

c There is no fixed object there, and neither is there a known comet! If this occurs, then we need to check *a* and *b* again, taking particular care, ensuring that it really is not a known object. The Becvár and Tirion atlases are reliable, even for identifying faint nebulosities or galaxies detected in a 300- or 400-mm reflector. There should be no problem in identifying anything detected with telescopes less than 250 mm in diameter. Various catalogues are also available to amateurs (such as the *Revised NGC*, etc.). The ideal method is probably that of Rolf Meier, who keeps *Atlas Stellarum* handy.

Once all the tests have been carried out – are you really sure that it is a non-instrumental, non-atmospheric object? – you need to act. We should not need to remind you that you should have already:

- recorded date and time (UT);
- recorded instrument and magnification;
- made as accurate a drawing as possible (ensuring that the time is recorded) and, if possible, another drawing under high magnification.

If you have discovered a comet, and do not want to keep the information to yourself (!), you should not only determine its position (to the nearest minute of arc) but also both the speed and direction (PA) of its motion. About 20 minutes are generally required to detect the motion of an 'average' comet, if we are observing with a focal length of about 1 m. This gives us time to make a magnitude estimate and to add some notes about the appearance of the object (see Sects. 8.4 and 8.5.5, etc.). If you are able to take a couple of pictures, that would be perfect. Note down the times of the beginning and end of the exposures to the nearest second, with astrometry in mind. If you are unable to measure the plates yourself, entrust them to a suitably competent group. If you have made the discovery with binoculars, their resolution will not really allow you to detect movement in 20 minutes, unless the object is very fast. It is preferable to check using a telescope; contact a better-equipped friend if necessary.

The observed motion should be converted into daily motion in α and δ. If, for example, the comet is at about $\delta = +20°$, and has, in 20 minutes, moved 0.5 arc-minute towards the east, and 1 arc-minute towards the north, then the daily motion is: 36 arc-minutes east, or $+2.6$ min ($36/15 \cos \delta$) in increasing α, and $+72$ arc-minutes in δ, towards the north.

These estimates of the daily motion should be reported together with the position at the time of discovery so that other observers can recover the object. It will be more difficult to find the object if the estimate of the motion is uncertain or the motion very fast. You are then faced with the horrible dilemma:

- if you report your discovery immediately, the estimate of the motion is inaccurate;
- if you wait until the next day before confirming the motion, there is every chance that someone else will have discovered the same object and announced it before you!

Choose the lesser of the two evils. If your observation has passed test *a, b* and *c* with flying colours, it is absurd to keep the information to yourself. It is quite possible that the sky will be cloudy the next day, so it is important to act quickly, within an hour of making the discovery. Acting quickly does not necessarily mean sending a telegram to the Central Bureau (Sect. 8.10). Those who have sufficient reference material to check their own observations, may perhaps allow themselves to do so, if they are absolutely sure. This will not stop the Central Bureau from awaiting confirmation by another observer, even if you are not merely known, but highly-regarded, or even if you are a professional. The Central Bureau receives dozens of false alarms, which sometimes come, in good faith, from serious observers. On several occasions a provisional designation has been given to a 'Kodak comet', a plate fault, or a reflection that 'moves' from one plate to the next because of a slight shift in the telescope's alignment. The reaction of the Central Bureau is well-illustrated by the discovery of Comet Sorrells, 1986n, towards the end of 1986. On 1986 November 1, Brian Marsden received a telegram from William Sorrells, an amateur from Pleasanton (California), who reported discovering a comet at magnitude 12. The observation appeared to be reliable: detection of the comet on a 1-hour 'deep-sky' exposure, confirmed by several 10-minute exposures, and visual check by the same observer with his 400-mm telescope. Brian Marsden then contacted Guy Hurst in the United Kingdom, asking for confirmation. When confirmation was received (several visual observations and astrometric positions being reported by British amateurs), Brian Marsden officially confirmed the discovery of Comet Sorrells, 1986n.

Contacting the Bureau directly is not necessarily the quickest way, because one must assume that the people running the Bureau will ask for confirmation from selected observers. If your name is completely unknown, your telegram risks having to await accidental, independent discoveries, especially if the details that you send are, to a greater or lesser extent, incomplete. The best solution is to contact observers who will be able to make immediate confirmation (by observing) and who will know how to pass the information on to the IAU. At present the most effective network is the international, European-based one headed by Guy Hurst of *The Astronomer* magazine in the United Kingdom, and which may be contacted by electronic mail, telex, and facsimile, as well as by telephone. Amateurs in other countries who hesitate to contact this network should get in touch with the main observational organisation or individual observers in their own country, who will undoubtedly help them. It is not always easy to know whom to approach at professional observatories. Most of the national or international organisations have sufficient contacts with professionals for the problem of communication to be overcome. [This is certainly the case with the *TA* group, which exchanges an extraordinary volume of data between amateurs and professionals. – Trans.]

We must point out a potential problem occurring with searches made at the eastern or western horizons. The time required to check the discovery and to pass the information on to another observer is generally long enough for the object to become unobservable: either the object has approached the horizon or it has been overtaken by dawn. Don't get too upset if the person you contact is unable confirm the discovery immediately. Anyone you contact by telephone should promise to follow up your call. On your part, you should send *written confirmation* the next day of all the details that you have given over the telephone.

The discovery of a comet is rather like applying for a patent. If you report your discovery to a competent and active organisation, there is nothing to fear. Even if it is not possible to obtain confirmation (because of dreadful meteorological conditions, etc.) and the object is discovered independently from another part of the world, when the organisation that you have contacted receives the announcement of the independent discovery, it will contact the IAU, with details of your discovery's priority. Normally, this is quite effective, provided the provisional designation and the name of the comet have not already been allocated together; it is still necessary to act quickly.

It is just as well to know the various problems that may arise. There is always an element of risk (that of being mistaken) and of adventure in comet-searching. Any observer wanting to start systematic searches should be just as concerned about these questions of credibility, of confirmation of observations, and of (tele)communications as with knowledge of the sky. There will always be some question marks about accidental discoveries made by non-specialists or beginners – but then are they not the persons for whom we have written this book?

8.8.7 *Statistics and factors affecting discoveries*

8.8.7.1 *Prejudice*

A strange form of prejudice appeared some 10 years ago, following the great success of Japanese observers in the 1960s and 1970s. This took the form of imagining that the chance of discovering comets was greater the farther *east* one lived! This belief seemed to ignore the fact that the Earth is round. If a comet is discovered from Japan on any given night, it could have been discovered the night before from points at the same latitude. More or less conscious jealousy of the work of Japanese amateurs is, in fact, merely an outward expression of actual incompetence.

The history of cometary discoveries (as described in Sect. 8.8.1) might well make us think that some irresistible force fiendishly causes the main centre for discoveries to shift continually towards the east. The 'centre of excellence' was in China in antiquity, in France in the 18th century, then in the U.S.A. at the end of the 19th century. It then shifted to Europe at the beginning of this century, moving eastwards in the 1950s, then to Japan, to Australia (Bradfield), then shifting on to California and Canada in the 1980s (Machholtz and Meier). If this fictitious movement only continues observers in Western Europe will again come into their own when it crosses the Atlantic at the end of this century!

8.8.7.2 *The effect of the Earth's motion*

A brief examination of the conditions surrounding discovery of 'amateur' comets shows that two-thirds of them are discovered in the morning sky. Several explanations of this anomaly have been advanced. It is generally recognized that at any given place the meteorological conditions are very often better in the morning than in the evening (there is less atmospheric turbulence). It has also been suggested that the areas close to the Sun are checked more thoroughly in the morning than in the evening: the period of time between ending morning searches and dawn tends to be shorter than that between sunset and the beginning of evening searches. There is some truth in these suggestions, but they are not the whole answer: they do not explain discoveries made more than 30° from the Sun.

Analysis of the motion of comets provides a general explanation that is more convincing. Take the comets that are observed in the evening, at small elongations (say about 45°) east of the Sun. The two types of cometary orbit, direct or retrograde, will give rise to the following situations:

- comets in direct orbits are moving away from the Sun;
- retrograde comets are moving away from the Earth.

In the morning, on the other hand, 45° west of the Sun:

- direct comets are approaching the Sun;
- retrograde comets are approaching the Earth.

As a result, for equal absolute magnitudes, morning comets have a general tendency to increase in brightness, and are therefore more easily detectable. The most favourable conditions are offered by retrograde comets that approach the Earth when they have just passed perihelion. Their brightness remains constant or

Fig. 8.45. *P/Halley close to the galaxy NGC 5128 in Centaurus. 1986 April 15; exposure 7 minutes starting 01:50 UT. Hypersensitized TP 2415; 200/300-mm Schmidt, f/1.5. (Photograph by M. Jäger from Pico de Tiede, Teneriffe.)*

increases significantly if perigee is less than 1 AU. Their increasing elongation means that these objects are easier to discover than comets that are in direct orbits (with decreasing elongations).

This effect would not occur if the Earth were stationary in space, and it would be reversed if the Earth had a retrograde orbit (conditions would be more favourable in the evening). It is the same effect that causes more meteors to be seen in the second half of the night.

Observers therefore know what they have to do: they have twice as much chance of discovering comets if they work in the morning.

8.8.7.3 Elongation

On average, amateurs discover comets much closer to the Sun than professionals. More than 75% of amateur discoveries are made less than 90° from the Sun, the maximum being around 50°. Conditions are most favourable around this elongation. We are scanning a zone where the sky is dark, outside the twilight. We are able to see farther out into the Solar System than when we observe closer to the Sun and we explore a greater extent of ecliptic longitudes than when we observe around the meridian. The very high proportion of comets discovered at these low elongations is undoubtedly one of the assets of the azimuth method (see p. 456). Bradfield discovered his 14 comets less than 60° from the Sun, at elongations of 22° to 57° (the average being about 40°); two-thirds of these comets were discovered in the morning sky. Rolf Meier, who uses the equatorial method (p. 463) discovered his four comets in the evening sky at elongations of 52° to 75° (the average being 67°).

Among the 25% of comets discovered by amateurs more than 90° from the Sun, we find two specific types of object: comets that undergo an outburst, and certain southern comets discovered somewhat late (before Bradfield!).

The discoveries made by professionals are at greater elongations, often around 90° West: these are photographic campaigns taking place in the morning sky. Photographic campaigns also enable professionals to discover comets around opposition (at elongations of 180°). These are usually faint objects detected farther out in the Solar System: comets with large perihelion distances or ones that are captured several months (or years) before perihelion. Such distant detection has sometimes deprived amateurs of some fine discoveries as with Comets Kohoutek (1973 XII), West (1976 VI), etc.

8.8.7.4 Distribution of discoveries and types of comets

Table 8.6 summarizes the distribution of discoveries of new comets between 1900 and 1986. We have distinguished between amateur and professional discoveries by taking the name of the first official discoverer, the first person (or satellite) whose name has been given to the comet. During those 87 years, 376 comets were discovered, i.e., an average of 4.3 per year, both periodic and non-periodic. We have not counted periodic comets discovered before 1900, but then lost, and subsequently rediscovered accidentally in the 20th century (such as P/Herschel–Rigollet, P/Denning–Fujikawa, etc.). Neither have we counted the 'Great Comets', where it is not always easy to decide on the first discoverer.

Of those 376 comets, 133 were discovered by amateurs, i.e., about 35%. It is

Fig. 8.46. *Comet Austin, 1982g (1982 VI), 1982 August 21. Exposure 20 minutes, starting at 20:30 UT. Hypersensitized TP 2415; 300/400-mm, 1000-mm focal length Schmidt telescope. Comet 1982 VI was notable for its very narrow, long tail stretching for several degrees, and for its tail streamers. The coma appeared extremely small and condensed. (Photograph by G. Klaus.)*

Table 8.6. *Distribution of cometary discoveries* *(1900–86)*

	Type of orbit		Total
	Non-periodic	Periodic	
Amateurs	118	15	133
Professionals	160	83	243
Total	278	98	376

interesting to note that between 1759 and 1899, amateurs discovered about 30 % of comets. So the amateur contribution has not declined. There has been a definite increase in the number of amateur observers: the proportion of amateur discoverers rose from 32 % before 1900 to about 40 % today. In addition there have been changes in techniques:

- professionals no longer undertake visual searches;
- amateurs are turning more to equatorial searching, which is better at covering the whole sky;
- amateur photographic discoveries are also increasing.

It should also be noted that amateurs have been co-discoverers of 17 comets first discovered by professionals (two examples are IRAS–Araki–Alcock and P/West–Kohoutek–Ikemura). Amateurs in fact contributed to the discovery of 150 objects (20 of which were of short period).

Of the 376 comets, 96 were short-period (26 %). The difference between the types of comet discovered by amateurs and by professionals is very distinct: one comet in three discovered by professionals is periodic, but this is one in nine for amateurs. This clearly shows the effect of professional programmes aimed at discovering minor planets close to the ecliptic.

Before 1900, about 25 % of the comets discovered by amateurs were periodic. There was no discrepancy, because amateurs and professionals used essentially the same search techniques. It should also be said that the brightest periodic comets were discovered before 1900, as were the comets that passed very close to the Earth. The periodic comets that are being discovered now are, on average, fainter objects or that have perihelion distances farther from the Sun (or both).

The contribution by amateurs is considerably greater when it comes to nonperiodic comets: 118 out of 278, i.e., more than 40 %. This proportion has fluctuated between 30 and 50 % during the 20th century according to the number of new comets that appear and the number of observers looking for them. The high inclinations of these comets make them far more likely to be discovered by amateurs in areas of the sky that are not covered so well by professionals (i.e., at high declinations).

Even though there may be few bright periodic comets still undiscovered, it is fairly obvious that many short-period comets have yet to be detected, because chance has meant that the Earth–Sun–comet configuration has not been favourable at their various apparitions (perihelion having occurred at conjunction). It should also be remembered that the giant planets (particularly Jupiter) regularly inject short-period comets into the inner Solar System by perturbing the orbits of new comets.

When we examine the distribution of inclinations of short-period comets, we discover several effects that may serve to prompt specific observations. We know about 140 short-period comets. The vast majority of these (85 %) have periods less than 20 years; these are 'close' comets that stay within the main planetary system (their semi-major axes being less than 8 AU). Only about half-a-dozen of these comets have inclinations greater than 30°, the others (about 110) being confined to the region around the ecliptic (inclinations of between 0 and 30°). We know only twenty-odd comets with periods between 20 and 200 years, from P/Hartley–IRAS (period 21.5 years) to P/Wilk (period 187 years), including P/Halley of course. The orbital inclinations of these comets are much greater, and five have retrograde orbits ($i > 90°$). At periods over 200 years, the inclinations of the orbits of known comets are far more randomly distributed, and resemble those of 'new' comets as defined by Oort. The reason for the anomalous distribution of comets with periods between 20 and 200 years is a mystery. We do not know whether there are really fewer comets in this group, or whether our knowledge is at fault through lack of systematic coverage away from the ecliptic. Although amateurs are in a good position to detect new comets by searching in azimuth, it would be very useful if more observers were to undertake equatorial searches at greater elongations and at high declinations. This applies both to visual searching and to photographic searches with short-focus lenses (of less than 500 mm focal length). The discovery of Comet P/Machholtz is an exceptional by-product of searches at high declination: its period is 5.25 years and its inclination is about 60°!

8.8.7.5 Visual discoveries

Figure 8.47 shows the number of discoveries between the beginning of the 18th century and 1982. Each point corresponds to the number of comets discovered annually, averaged over 5-year periods. This graph illustrates a number of historical and cometary effects. We immediately see the progressive increase in the annual average during the 18th century, the average stabilizing around one discovery a year about the time of the beginning of Pons' career. The number of observers increased at that period, with advances in equipment, which probably attained a performance comparable to that of present-day 'comet-seekers'.

There seem to be rich and lean periods when more or less comets appear. Some peaks are very marked, such as those at 1818–27, 1843–47, and from 1873 to 1893. We may wonder whether these maxima are the result of the number of comets appearing, or of the degree of activity or number of observers. There is practically only one factor that allows us to assess more or less objectively the true rate of new comets: this is the frequency of Great Comets, the 'monsters' that cannot be missed. Seventeen of these objects are catalogued in the 19th century. The first four Great Comets appeared during Pons' career (1807 to 1823). We can see that the discovery rate was relatively high between 1818 and 1827. After Pons, two Great

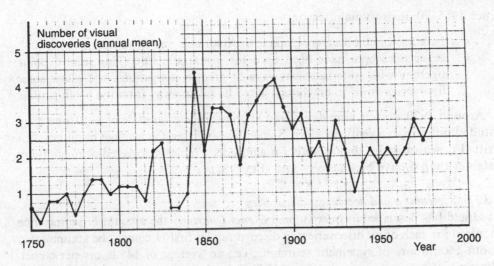

Fig. 8.47. *Variation in the rate of visual discoveries from 1748 to 1982. Each point corresponds to the average annual rate of the number of comets discovered visually, taken over five-year periods. The number of comets discovered this way remains high: there are two or three comets discovered every year by amateurs using small or medium-sized instruments. (Diagram J. C. Merlin; data from* Comets, a descriptive catalog, *G. W. Kronk.)*

Comets appeared in 1830, when the number was a minimum: it is therefore quite legitimate to assume that coverage was poor until about 1840. The 1843–47 peak coincides with the appearance of three Great Comets and with the period of activity of several great observers (de Vico, Brorsen, Mauvais, etc.). We find a similar effect in the 1860s (three Great Comets) and in the 1880s (four Great Comets). The troughs in 1850 and 1870 may perhaps correspond to periods that were poor in comets, but this is by no means certain. The period of Lewis Swift's activity began only after 1875. After 1880, this combined with the activity of Barnard and Brooks to bring the century to an amazing close: 45 comets were discovered between 1877 and 1900 by these three American observers. Forty of these comets are named after a single observer: although Barnard, Brooks and Swift were all observing during the same period, they made few combined discoveries (see p. 480 and Table 8.8 for details of their individual records). The relationship between the number of comets discovered and the number of comet-hunters appears to be quite clear. This is a weighty argument for assuming that many comets go unseen.

Although primarily more systematic, searching was also carried out to a greater depth after 1880: Barnard regularly discovered comets at magnitudes between 9 and 11. The rate of discoveries remained at an average of three to four per year until the end of the 19th century. A rapid fall then occurred with the arrival of photography. The number of comets discovered visually hardly reached three after that, the average stabilizing around two. There has, however, been a definite increase

477

since 1965–70, for two main reasons:

- there are more observers (Japanese, American, and Bradfield!);
- larger telescopes have been used for visual searches. This second point, together with systematic coverage at greater elongations, will allow some discoveries down to magnitude 12 to be taken away from the professionals.

Around 1970, comets were discovered at an average magnitude of 8–10, one magnitude fainter than a century before. Some 15 years later, the average brightness has shifted downwards by about another magnitude. Comets discovered at magnitudes brighter than 6 have become very rare (Fig. 8.38).

8.8.7.6 *Systematic and accidental discoveries*

Bradfield has described in *International Comet Quarterly* the amount of search-time required for each of his discoveries. In discovering his first 11 comets, he accumulated about 1600 hours of systematic searching, i.e., an average of 145 hours per comet. His first discovery was made after 260 hours of sweeping. The extreme periods between cometary discoveries are 9 hours (between the 5th and 6th comets) and 360 hours (between the 6th and 7th). Bradfield's discoveries are exceptional in that none of his 14 were shared with any other discoverer. Only Bradfield's 12th comet is short-period: P/Bradfield 1984a (1983 XIX), with a period of 151 years, discovered at magnitude 11 with a 250-mm reflector. The argument that Bradfield has it all his own way, because he is practically the only observer in the Southern Hemisphere, carries little weight when account is taken of the small elongations at which his discoveries have been made (see p. 473). Bradfield is a very serious and capable observer. It is also instructive to note that the average amount of sweeping that he carries out each month is about 13 hours.

Rolf Meier accumulated just about 200 hours sweeping in discovering four comets over a period of about 10 years. Meier observes much less than Bradfield: his average time spent sweeping per month is less than 2 hours. He has discovered fainter comets: average magnitude about 11 (as against 8.5 for Bradfield). This shows once again that systematic, careful, and patient use of a 'large' equatorial telescope (more than 250 mm in diameter) enables a large number of comets to be discovered.

It would seem that for most discoverers, the hardest part is to find the first one:

- G. Alcock (United Kingdom) accumulated more than 600 hours before finding 1959 IV; he found 1959 VI a week later!
- R. Panther (United Kingdom) discovered 1981 II after about 700 hours sweeping, spread – it is true – over about 40 years!
- D. Machholtz discovered 1978 XII after 1700 hours accumulated in just 3 years 8 months, i.e., an average of 460 hours of observation per year. He observes practically every other night, under fine Californian skies. Average: 38 hours per month!

Alongside the 'great' comet-hunters whose results we shall describe in the next section, there are many accidental discoverers or others whose success has not always been recognized by posterity. This applied to Cernis, who officially discovered

Table 8.7. *Discoveries by 20th-century amateurs (1900–86)*

Number of discoveries	Number of discoverers
1	59
2	12
3	4
4	4
5	6
6	2
7	0
8	0
9	0
10	1
11	0
12	2

two comets, but who was co-discoverer of three other comets after their official announcement. Table 8.7 shows the performance of the 90 amateur discoverers in the 20th century. These 90 individuals share the 150 comets discussed on p. 475: an average of 2.1 discoveries per person. Nearly two-thirds of the discoverers have officially discovered just one comet. If we assume that these are basically accidental discoveries, then we see that the number of dedicated comet-hunters is very low: thirty-odd observers have discovered more than one comet (average: 4.1 discoveries per 'dedicated' observer). The gap between six and ten discoveries even seems to indicate that new comets are shared by an even larger number of observers, whether accidental or not. Honda no longer searches for comets, but he nevertheless discovered 12 'from under the very noses' of his compatriots – and he still has a chance of discovering his 13th during his nova-searches. Given that Bradfield is working in the Southern Hemisphere, should we not conclude from this clumping below six discoveries that at present there is no truly 'great' comet-hunter in the Northern Hemisphere, able to take the lead?

The 150 comets discovered in the 20th century by amateurs resulted from 187 officially-recognized independent discoveries (i.e., the names appeared first, second or third in the official cometary designations). Based on the 59 discoveries that we have taken to be accidental (Table 8.7), we can calculate that accidental discoveries account for more than 30%. This is a very high figure. Whether they observe visually or photographically, amateurs who observe regularly have every chance of discovering a comet. There are comets to be discovered – and places to be won. It's up to you!

Table 8.8. *Principal comet discoverers: 1760 to 1900*

Name	Country	Period of activity	Discoveries Total	Periodic	Remarks[a]
Pons	France/Italy	1801–1827	30	7	1, 2
Brooks	U.S.A.	1883–1911	21	3	1, 3, 4
Barnard	U.S.A.	1881–1892	16	3	1, 4, 5
Swift	U.S.A.	1862–1899	14	5	1, 4, 6
Messier	France	1760–1798	13	1	1, 7
Tempel	France/Italy	1859–1877	13	4	8
Giacobini	France	1896–1907	12	3	1, 9
Winnecke	Germany	1854–1877	11	2	10
Borrelly	France	1873–1912	11	1	1, 11
Méchain	France	1781–1799	9	2	1, 7
Perrine	U.S.A.	1895–1902	9	1	4, 12
Herschel (C.)	Great Britain	1786–1797	7	2	1, 13
de Vico	Italy	1844–1846	6	2	4, 14
Klinkerfues	Germany	1853–1863	6	0	15
Tuttle	U.S.A.	1858–1865	6	4	4, 16
Coggia	France	1870–1890	6	1	1, 11
Gambart	France	1822–1834	5	1	1, 17
Brorsen	Germany	1846–1851	5	2	4, 18
Bruhns	Germany	1853–1864	5	0	19
Donati	Italy	1855–1864	5	0	20
Denning	Great Britain	1881–1894	5	2	21
Schweizer	Russia	1847–1855	4	0	22

[a] *See* Table 8.10 footnote

8.8.7.7 *The roll of honour*

Three tables (8.8, 8.9 and 8.10) show the performance of the most famous comet-hunters, the 59 names that occur at least four times. We have made no distinction between amateurs and professionals. For one thing, the difference is not always easy to determine, and in many cases it is difficult to state categorically whether the observer should be considered one or the other. It seemed more sensible to classify the observers according to whether they worked visually or photographically for most of their career. Table 8.8 almost entirely concerns visual discoveries (up to the beginning of the 20th century). The comments that we have made in Sect. 8.8.1 and subsequent sections, together with the details that are given under 'Remarks', will give the reader some information about the individuals concerned.

The number of discoveries has been taken from the 5th edition of the IAU *Catalogue of Cometary Orbits* (Marsden, 1986), with additions from IAU Circulars up to the end of 1986. The details therefore mainly relate to comets for which

Table 8.9. *Principal comet discoverers: visual discoveries between 1900 and 1986*

Name	Country	Period of activity	Discoveries		Remarks[a]
			Total	Periodic[a]	
Honda	Japan	1940–1968	12	1	1, 23, 24
Mrkos	Czechoslovakia	1947–1984	12	3	1, 25
Bradfield	Australia	1972–1984	12	1	1, 26
Peltier	U.S.A.	1925–1954	10	0	1, 27
Reid	South Africa	1918–1927	6	0	1, 28
Seki	Japan	1961–1970	6	0	1, 29
Metcalf	U.S.A.	1906–1919	5	2	30
Mellish	U.S.A.	1907–1917	5	1	31
Skjellerup	South Africa	1919–1927	5	1	1, 28
Pajdusáková	Czechoslovakia	1946–1953	5	1	1, 25
Alcock	Great Britain	1959–1983	5	0	1, 32
Ikeya	Japan	1925–1937	5	0	33
Fujikawa	Japan	1969–1983	5	1	34
Wilk	Poland	1925–1937	4	1	35
Forbes	South Africa	1928–1932	4	2	1, 28
Sato	Japan	1969–1975	4	0	36
Meier	Canada	1978–1984	4	0	1, 37

[a] *See* Table 8.10 footnote

consistent orbital elements have been derived. It should be emphasized that we are talking about the number of discoveries and not the number of comets. P/Encke was discovered twice by Pons (see Sect. 8.8.1) before it was realized that it was a single object: we therefore have counted these as two discoveries. We have not counted the 'Great Comets', which were discovered independently by a number of observers.

Alongside the name of the observers we have indicated the names of the country (or countries) in which the work was carried out. This is not necessarily the same thing as their nationality: for example, the German Tempel mainly observed in France and Italy. The period of activity is indicated by the dates of the first and last cometary discovery. As well as the total number of discoveries, we give the number of periodic comets that are included in the total. The figures in the 'Remarks' column refer to the notes given after the three tables.

Table 8.8 covers the period prior to 1900. The year 1900 is taken as the cut-off date for the early careers of the observers who span the two centuries (Brooks, Perrine, Giacobini, and Borrelly). Out of the 22 observers, eight were French or worked in France. The U.S.A. comes second. There is just one woman: Caroline Herschel. There is also just one photographic discovery in this Table: Barnard's last discovery (P/Barnard 3).

Table 8.10. *Principal comet discoverers: photographic discoveries between 1900 and 1986*

Name	Country	Period of activity	Discoveries		Remarks[a]
			Total	Periodic[a]	
Harrington	U.S.A.	1949–1955	8	4	38
Shoemaker	U.S.A.	1938–1986	8	3	1, 39
Hartley	Australia	1982–1986	7	5	1, 40
Neujmin	U.S.S.R.	1913–1941	6	5	41
Whipple	U.S.A.	1932–1942	6	1	16
Bester	South Africa	1946–1959	6	0	42
Burnham	U.S.A.	1957–1960	6	1	43
Wild	Switzerland	1957–1960	6	3	44
SOLWIND	U.S.A.	1979–1984	6	0	1, 45
IRAS	U.S.A./GB/NL	1983	6	2	1, 45
du Toit	South Africa	1941–1945	5	3	42
Wirtanen	U.S.A.	1947–1956	5	1	12
Kohoutek	Germany	1969–1975	5	2	46
Gehrels	U.S.A.	1972–1975	5	4	38
Lovas	Hungary	1974–1986	5	2	47
Russell	Australia	1979–1984	5	4	40
Schwassmann	Germany	1927–1930	4	3	46
Wachmann	Germany	1927–1930	4	3	46
Johnson	South Africa	1935–1949	4	1	48
Kowal	U.S.A.	1977–1984	4	4	38

[a] Remarks to Tables 8.8, 8.9 and 8.10

1   *See* Sect. 8.8.1
2   Observatoire de Marseille, then la Marilya and Florence (Italy): 50 OG (!)
3   Phelps then Geneva (New York): 130 and 230 R, 250 OG
4   *See* p. 476
5   Nashville (Tennessee): 100 and 150 OG; then Lick Observatory (California)
6   Rochester (New York): 114 OG
7   Paris: 60 OG
8   Marseille, Venice, Milan and Florence
9   Observatoire de Nice: 460 R
10 Berlin, Bonn, Honigesstein, Karlsruhe, then Strasbourg
11 Marseille
12 Lick Observatory, 51-cm astrograph (Wirtanen)
13 Caroline Herschel, Slough (Buckinghamshire)
14 Osservatorio del Collegio Romano (Rome)

Remarks to Tables 8.8, 8.9 and 8.10—*continued*

15 Sternwarte Heidelberg

16 Harvard College Observatory

17 Marseille

18 Kiel, Altona and Senftenberg: 100 OG

19 Berlin

20 Florence

21 Bristol: 250 R

22 Moscow: 80 OG

23 Seto, Tanokami, then Kurashiki, 100 B, 150 and 380 R

24 *See* p. 478

25 Skalnate Pleso: 100 B

26 Dernancout near Adelaide (South Australia). *See* pp. 453, 454, 456, and 473: 35 B, 150 OG, 250 R

27 Delphos (Ohio): 150 OG

28 Cape of Good Hope: 150 OG (Reid)

29 Kochi: 90 and 150 R, 120 B

30 South Hero (Vermont): 180 OG

31 Madison (Wisconsin): 130 and 150 OG

32 Peterborough. *See* p. 454: 105 B

33 Maisaka: 150 and 200 R

34 Onahara: 160 R

35 Krakow

36 Nishinasuno

37 Ottawa. *See* pp. 454, 463, and 473: 400 R

38 Palomar: 1.22-m Schmidt; Harrington's discoveries took place during the making of the first Palomar Sky Survey (POSS)

39 Palomar: 460-mm Schmidt

40 Siding Spring: 1.22-m Schmidt

41 Crimean Observatory (Simeis)

42 Bloemfontein: Bester's discoveries were with the Ross-Fecker 76-mm camera

43 Lowell Observatory (Arizona): 330 astrometric reflector; the first two comets were discovered visually at Prescott (Arizona), with a 200 R

44 Berne Observatory (Zimmerwald): 400-mm Schmidt

45 From space

46 Hamburg Observatory: Kohoutek's discoveries were with the 0.36-m Schmidt (Bernhardt Schmidt's original Schmidt); Schwassmann and Wachmann worked together, searching for minor planets

47 Konkoly Observatory (Budapest): 600-mm Schmidt

48 Johannesburg, Union Observatory: Franklin-Adams camera (254 mm)

Table 8.9 covers primarily visual discoverers in the 20th century. Apparently two comets were discovered photographically by these 'visual' workers:

- P/Kowal–Mrkos 1984 X, which was not found during the 'Czech period' described in Sect. 8.8.1.;
- P/Metcalf 1906 VI.

We should also note several visual recoveries of periodic comets lost after apparitions observed decades or centuries previously: P/Pons–Brooks, P/Denning–Fujikawa, etc. Again just one woman: Ludmilla Pajdusáková. Japanese visual observers head the list, followed by Americans and South-Africans.

The performances of the various nationalities involved in visual work are even more pronounced when the 90 amateur discoverers are considered as a whole:

26 Japanese,
21 American,
9 South-Africans,
7 Australians,
4 New-Zealanders,
4 Russians,
3 Frenchmen,
2 discoverers in each of Argentina, Germany, Spain and the United Kingdom;
1 discoverer in each of Canada, Denmark, Italy, Norway, the Philipines, Poland, Switzerland and Uruguay.

Table 8.10 shows discoverers primarily using photography. Only Burnham made two visual discoveries as an amateur before, or during, his career at the Lowell Observatory. We have, of course, included two artificial satellites (IRAS and SOLWIND) in this list. The various Comets Shoemaker are shared by Carolyn and Eugene Shoemaker. The Americans, primarily those working at Palomar, head the list.

We have given the place(s) where observations were carried out. Whenever possible we indicate the instrument(s) used:

B = Binoculars; OG = Refractor; R = Reflector. The aperture is given in millimetres, unless indicated to the contrary.

Finally, we may reiterate that the secret of discovering comets is *perseverance*. An old saying, of uncertain origin, says:

> Anyone who discovers one comet is lucky,
> Anyone who discovers two comets is a competent observer,
> Anyone who discovers three comets has understood!

8.9 Mathematical techniques

8.9.1 Non-linear interpolation

We recommend the use of interpolation by the method of finite differences (Vorontsov-Velyaminov, 1969). To obtain sufficient accuracy in the interpolation, it is necessary to have at least four positions from the ephemeris, which will enable the third differences to be calculated.

8.9.1.1 An example using right ascension

Let A_1, A_2, A_3, and A_4 be known right ascensions at times T_1, T_2, T_3, and T_4, respectively, which are at regular intervals, i.e.:

$$T_2 - T_1 = T_3 - T_2 = T_4 - T_3 = 10 \text{ days (for example)}.$$

Let a_1, a_2, and a_3 be the first differences:

$$
\begin{aligned}
a_1 &= A_2 - A_1 = a \\
a_2 &= A_3 - A_2 \\
a_3 &= A_4 - A_3.
\end{aligned}
$$

Let b_1 and b_2 be the second differences:

$$
\begin{aligned}
b_1 &= a_2 - a_1 = b \\
b_2 &= a_3 - a_2,
\end{aligned}
$$

and c the third difference:

$$c = b_2 - b_1.$$

Let Z be a value between 0 and 1 and represent the fraction of the interval $T_2 - T_1$ between the initial time T_1 and the time T, for which the interpolated right ascension is required. We therefore have:

$$Z = (T - T_1)/(T_2 - T_1),$$

the times being calculated in decimals of a day to at least three decimal places (5 minutes = 0.0035 day).

8.9.1.2 The general interpolation formula

The general interpolation formula is given by:

$$A = A_1 + Z\left\{a + \frac{Z-1}{2}\left[b + \frac{Z-2}{3}(c + \ldots)\right]\right\}.$$

Obviously, if a greater accuracy is required, the series may be extended to the nth difference, using $n+1$ positions from the ephemeris. Third differences usually suffice for an accuracy of a tenth of the minute of arc. N.B.: the same procedure, with four reference positions, may be used to interpolate for any time over the whole period T_1 to T_4. Z is calculated from $(T - T_1)/(T_2 - T_1)$ as before; the only difference is that Z is greater than 1, when $T > T_2$. (We may note in passing that conventional linear interpolation only employs first differences: $A = A_1 + Z \times a$.)

To take an example, here are four positions for Comet P/Crommelin 1983n (orbital elements from IAUC 3886):

Date	α (h m)	a	b	c
$T_1 = 1984$ Jan. 01	21 33.90			
$T_2 = 1984$ Jan. 06	21 48.51	+14.61	+1.17	
$T_3 = 1984$ Jan. 11	22 04.29	+15.78	+1.16	−0.01
$T_4 = 1984$ Jan. 16	22 21.23	+16.94		

To calculate A for 1984 Jan. 04 at 18:30 UT, we first convert the date to decimal days: January 04.770 83 UT. We then derive Z:

$$T - T_1 = 3.770\,83; \text{ and } T_2 - T_1 = 5,$$

so $Z = 0.754\,17$.

$$\Delta\alpha = Z \left\{ 14.61 + \frac{Z-1}{2} \left[1.17 + \frac{Z-2}{3}(0.01) \right] \right\} = 10^{m}.91,$$

whence:

$$A = A_1 + \Delta\alpha = 21^{h}33^{m}.90 + 10^{m}.91 = 21^{h}44^{m}.81.$$

Direct calculation using the orbital elements gives the same result, whereas linear interpolation would have given: $A = A_1 + Z \times 14.61$, i.e., $A = 21^{h}44^{m}.92$, a difference of 1.6′ from the ephemeris ($= 0.11 \times 15 \times \cos\delta$).

8.9.2 Calculation of combined magnitudes

8.9.2.1 Calculating the combined magnitude of two stars

We assume that the two stars A and B – A being brighter than B – are sufficiently close to one another to appear combined in a small instrument or at a low magnification. A similar situation applies if a star is seen through the tail of a comet. (In this case the image is assumed to be defocussed.)

Let m_a and m_b be their respective known magnitudes and I_a and I_b their intensities. We need to find the combined magnitude m_t, which corresponds to $I_a + I_b$.

Let the intensity of star B be taken as 1, and I be the ratio of the two intensities: $I = I_a/I_b$. Therefore $I = I_a$, and the combined intensity $I_t = I_a + I_b$, which is given by $I + 1$.

Let Δm be the difference in magnitude between B and A: $\Delta m = m_b - m_a$. According to Pogson's equation, we have:

$$I = \text{antilog}(\Delta m/2.5). \tag{8.1}$$

The difference in magnitude between star B and the combined magnitudes of A and B is: $m_b - m_t = 2.5 \log(I + 1)$, whence:

$$m_t = m_b - 2.5 \log(I + 1). \tag{8.2}$$

Fig. 8.48. *Comet* IRAS–Araki–Alcock, *1983d (1983 VII), the 'surprise' comet, which passed less than 5 million km from the Earth on 1983 May 11. Photographed on 1983 May 11, exposure 10 minutes, beginning 21:45 UT, film 103aE; 400-mm telephoto lens, used at f/5.6. (Photograph by S. Bertorello, AMAS.)*

If we substitute the value of I obtained from (8.1), we obtain the required relationship:

$$m_t = m_b - 2.5 \log[\text{antilog}(\Delta m/2.5) + 1]. \tag{8.3}$$

For example: if $m_a = 8.5$ and $m_b = 9.5$ then $\Delta m = +1$ magnitude, whence; $I = \text{antilog}(\Delta m/2.5) = 2.512$ and: $m_t = 9.5 - 2.5 \log(2.512 + 1)$, therefore: $m_t = 8.14$.

We could equally well take the brighter star as the reference in (8.3) i.e. with $m_a = 9.5$ and $m_b = 8.5$, when we would have: $\Delta m = -1$ magnitude, which would give: $m_t = 8.5 - 2.5 \log(0.398 + 1)$ and thence: $m_t = 8.14$, which is precisely the same result.

The contribution of the fainter star to the combined intensity I_t becomes negligible when the absolute value of Δm is greater than 2.5 magnitudes. If star B is the brighter and $\Delta m = -2.5$ magnitudes, we have $m_t = m_b - 0.095$; so the fainter star contributes less than one-tenth of a magnitude to the combined magnitude.

8.9.2.2 Calculating the magnitude of a star from the magnitude of the other star and the combined magnitude

This is the opposite problem: we know m_t and m_a, and require m_b. In this case we take Δm as the difference in magnitude between m_a and m_t: $\Delta m = m_a - m_t$.

The ratio of intensities $I_t/I_a = \text{antilog}(\Delta m/2.5)$, whence:

$$I_t = I_a \text{antilog}(\Delta m/2.5).\qquad(8.4)$$

We also know that:

$$I_t = I_a + I_b.\qquad(8.5)$$

Combining these two equations, we have:

$$I_a + I_b = I_a \text{ antilog}(\Delta m/2.5),\qquad(8.6)$$

whence: $I_b = I_a[\text{antilog}(\Delta m/2.5) - 1]$.

Expressed in terms of magnitude, this becomes:

$$m_b = m_a - 2.5 \log[\text{antilog}(\Delta m/2.5) - 1].\qquad(8.7)$$

Taking the same values as previously, we have:

$$m_a = 8.5 \; ; \; m_t = 8.14 \; ; \; \Delta m = m_a - m_t = 0.36 \; ; \; m_b = ?$$

Using (8.6), we have: $m_b = 8.5 - 2.5 \log[\text{antilog}(0.36/2.5) - 1]$, whence: $m_b = 9.5$ (rounded value). The true value of m_t is $8.1361\ldots$

Note: if (8.7) is applied to a comet, m_b is the total visual magnitude m_1 that has to be determined. The combined magnitude m_t is the total, defocussed magnitude of star A plus the comet.

8.9.3 Least-squares linear regression

We assume that we have a population of n points x, y, to which we want to fit a linear relationship of the type:

$$y = ax + b.$$

We have to determine the coefficients of a straight line of slope a and intercept b on the y-axis (when $x = 0$, $y = b$). We calculate the following sums:

$$A = \sum_{i=1}^{n} x_i \quad B = \sum_{i=1}^{n} y_i \quad C = \sum_{i=1}^{n} x_i/y_i$$
$$D = \sum_{i=1}^{n} x_i^2 \quad E = \sum_{i=1}^{n} y_i^2$$

The values of a and b are calculated as follows:

$$a = (n \cdot C - A \cdot B)/(n \cdot D - A^2),$$

$$b = (B \cdot D - A \cdot C)/(n \cdot D - A^2).$$

The correlation coefficient R^2 is calculated from:

$$R^2 = (n \cdot C - A \cdot B)^2/(n \cdot D - A^2) \cdot (n \cdot E - B^2),$$

where R^2 lies between 0 and 1.

8.10 IAU Telegram code

The IAU's Central Bureau for Astronomical Telegrams, located at the Smithsonian Astrophysical Observatory, Cambridge, Massachusetts, and directed by Dr Brian Marsden, has defined a method of coding messages that reduces the length and reduces the risks of errors in transmission, and at the same time permits all useful information to be conveyed. Messages may be sent by telegram, by telex or, increasingly, by electronic mail.

Messages mainly deal with discoveries (of novae, comets, etc.). Ordinary mail is generally adequate for other types of information. The Bureau has, however, allowed for coding and transmission of astrometric positions for known objects, and orbital elements for moving objects such as minor planets and comets. This generally applies to objects that have been recently discovered. The coding applies in both directions, both when observers are communicating discoveries to the Bureau, and when the Bureau passes the information to its subscribers. The following details and examples are taken from descriptive information issued by the Bureau.

8.10.1 *Method of coding*

Messages consist of an initial set of 3 alphanumeric groups, followed by a variable number of numerical groups. The latter consist of groups of 5 figures, separated by a space. There are three types of group.

8.10.1.1 *First groups for all telegrams*
DISCOVERER OBJECT OBSERVER AAAAB

Group 1   The name of the discoverer(s) and/or object designation. The latter may be:

- for a comet (year + letter)
- for a minor planet (year + 2 letters)
- for a nova (name of the constellation)
- for a supernova (name of parent galaxy – N = NGC, I = IC, M = Messier – followed by the catalogue number).

Group 2   The type of object: COMET, OBJECT (= minor planet), NOVA, SUPERNOVA, VSTAR (= variable star), etc.

Group 3   The name(s) of the observer(s) and/or the person(s) who carried out the computation of the orbital elements.

Group 4   Equinox, position/ephemeris
AAAA   =   mean equinox (for beginning of year)
  B   =   1 for an approximate position
     =   2 for an accurate position
     =   3 for orbital elements of a Solar-System body
     =   4 for an ephemeris.
Note: when an ephemeris follows the elements, Group 4 is replaced by the word 'EPHEMERIS', and the epoch of the ephemeris is assumed to be the same as that of the elements.

For example, ALCOCK COMET HURST 19501 indicates that Hurst is communicating the discovery of a comet by Alcock, equinox 1950.0, and that the position is approximate.

8.10.1.2 Middle groups for a position

If B = 1 (approximate position), we have:

CDDEE FFFGH IIJJJ LMMNN PQRRS TUUUU VWWXX

If B = 2 (accurate position), we have:

CDDEE FFFGH IIJJK KKKLM MNNPP PQRRS TUUUU VWWXX

Group 5 Date of observation:
  C   =   final digit of year
  D   =   month (01 = January ... 12 = December)
  E   =   day, 01–31 (in UT).

Group 6 Time of observation:
  FFFGH   =   decimals of a day

This group may be omitted entirely for observations of stationary objects.

Group 7 (Letters I to S inclusive)

When B = 1 (approximate position):
  Right ascension   = $II^hJJ^m.J$
  Declination   = $L\ MM°NN'$
  L = sign: L = 1 if declination is negative;
                          L = 2 if declination is positive.
  P = 0

When B = 2 (accurate position):
  Right ascension   = $II^hJJ^mKK^s.KK$
  Declination   = $L\ MM°NN'PP''.P$

For both:
  Q   =   1 for total magnitude m_1 for a comet
      =   2 for nuclear magnitude m_2 for a comet
      =   3 for visual magnitude m_v
      =   4 for photographic magnitude m_{pg}
      =   5 for photovisual magnitude m_{pv}

The last three apply to objects of stellar appearance (including minor planets).

  RR   =   magnitude; if magnitude is negative, add 100
  S   =   appearance, according to the table below;
           or, if the object is not a comet, S = tenth of a magnitude

	Nothing reported about tail	Tail	
		< 1°	> 1°
Stellar appearance	0		
Nothing reported about appearance	1	2	3
Object diffuse, without central condensation	4	5	6
Object diffuse, with central condensation	7	8	9

Group 8
For comets and minor planets, the daily motion:
In right ascension $\quad=\quad$ T UU^{m}UU
In declination $\qquad=\quad$ V WW°XX'
For supernovae, the offset from the nucleus of the parent galaxy:
In right ascension $\quad=\quad$ T UUUU''
In declination $\qquad=\quad$ V WWXX''
T, V = sign: $\qquad\qquad$ 2 = positive (east or north),
$\qquad\qquad\qquad\qquad$ 1 = negative (west or south)

The whole of this group may be omitted if the details are unknown or irrelevant.

8.10.1.3 Middle groups for orbital elements
If B = 3 (orbital elements), we have:
CDDEE FFFGH IIIII JJJJJ KKKKK TTTTT UUUUU

Group 5 Date of perihelion passage
C $\quad=\quad$ final digit of year
DD $\quad=\quad$ month (01 = January ... 12 = December)
EE $\quad=\quad$ day

Group 6 Time:
FFF $\quad=\quad$ time of perihelion passage
$\qquad\qquad\qquad$ in decimals of a day (TT)
G $\quad\quad=\quad$ time interval in days (rounded to nearest
$\qquad\qquad\qquad$ integer) between the first and last positions
$\qquad\qquad\qquad$ used for computation; 0 = 10 days or more
H $\quad\quad=\quad$ number and quality of observations on which
$\qquad\qquad\qquad$ the computation is based, or by which the
$\qquad\qquad\qquad$ elements have been checked, according to the
$\qquad\qquad\qquad$ following table:

	Maximum residuals		
	> 5''	1''–5''	< 1''
Fewer than three accurate positions*	1	2	3
Three accurate positions	4	5	6
More than three accurate positions	7	8	9

* Or other factors that decrease the reliability of the orbit, such as approximate, doubtful or unsatisfactorily distributed positions.

Group 7 Orbital elements
III°II $\qquad=\quad \omega$ (argument of perihelion)
JJJ°JJ $\qquad=\quad \Omega$ (longitude of ascending node)
KKK°KK $\qquad=\quad i$ (inclination).

Group 8   Orbital elements

   T.TTTT   = q (perihelion distance in AU)

   U.UUUU   = e (eccentricity).

The group for e may be omitted if the orbit is parabolic ($e = 1$).

8.10.1.4 *Middle groups for ephemerides*

If $B = 4$ we have:

CDDEE (IIJJJ LMMNN 9TTTT 8UUUU) cddee

Group 5   Date of first position in ephemeris:

   C   =   final digit of year

   D   =   month (01 = January ... 12 = December)

   E   =   day, 01–31 (in UT).

Group 6   Omitted: it is assumed that ephemerides are for 00:00 TT.

Group 7   Position:

   Right ascension   = II^hJJ^mJ

   Declination   = $L\ MM°NN'$

   L = sign: $L = 1$ if declination is negative;

                       $L = 2$ if declination is positive.

Group 8

   T.TTT (preceded by a 9)   = Δ (geocentric distance in AU)

   U.UUU (preceded by an 8)   = r (heliocentric distance in AU).

Groups 7 and 8 (shown above in parentheses) are repeated as often as necessary to complete the ephemeris. Positions are given at regular intervals (10, 5, 2 or 1 days). Group 8 is not always given for each position.

   cddee = date of last position in the ephemeris, in the same form as before.

8.10.1.5 *Final groups for all telegrams*

YYYYY ZZZZZ REMARKS COMMUNICATOR

Group 9   Checksums, remarks, communicator

   YYYYY   =   the last five figures of the sum of all the groups of digits including and following Group 4 (containing the equinox).

   ZZZZZ   =   the last five figures of the sum of all the groups giving the right ascension, declination and magnitude only (or ω, Ω, and i in the case of orbital elements). In practice, this 'second checksum' includes only the items under Group 7 (or involving the letters I to S inclusive).

Note: any digit that is unknown, not significant, or otherwise to be omitted should be replaced by a slash (/); this counts as zero in the checksums. A group of five slashes (/////) should be avoided. If two or more observations are included in one telegram, checksums should follow each one. A checksum of all numerical information is useful even if the telegram is not in code.

Group 10 Any additional remarks, qualifying the observation of the computation. In the case of an accurate position, the location of the observing station should be specified.

Group 11 The name of the communicator.

8.10.1.6 Examples
Readers may like to try decoding the following examples.

1. CLARK COMET CLARK 19501 30610 66/// 20540 13130 01135 2015/ 10002 81068 34805 GILMORE

Translation: Gilmore reports that Clark has discovered and observed a comet as follows:

1973 UT	α_{1950}	δ_{1950}	m_1
June 10.66	$20^h 54^m 0$	$-31°30'$	13

Object diffuse without central condensation, tail $< 1°$. Daily motion: $\Delta\alpha = 1^m 5$, $\Delta\delta = -2'$.

2. BALLY CLAYTON 1968D COMET ROEMER SCHREUR 19502 80827 20246 18513 33623 22222 82157 77090 56515 19502 80827 20872 18513 16823 22225 7//// 48762 25761 CATALINA LPL

Translation: The Lunar and Planetary Laboratory communicates the following observations of Comet BALLY–CLAYTON (1968d), obtained by Roemer and Schreur at the Catalina station:

1968 UT	α_{1950}	δ_{1950}	m_2
Aug. 27.202 46	$18^h 51^m 33^s 36$	$+32°22'22''8$	15
27.208 72	18 51 31.68	$+32$ 22 25.7	

On the first plate the object was diffuse, with central condensation, nothing reported about a tail.

3. 1972f COMET CANDY 19503 20327 72656 25771 15959 12369 09275 75860 54099 EPHEMERIS 20403 00158 14433 91171 80934 00558 14741 01503 15007 90961 80972 03000 15042 20418 49301 64442 CANDY

Translation: Candy reports that he has calculated parabolic elements and an ephemeris for Comet 1972f, as shown below. The elements are based on three accurate observations covering an arc of 5 days, and the residuals are all less than

1″. [Note that because this and the next example were before 1991 December 24, the time is given in ET, not TT (*see* Sect. 8.3.4.1). – Trans.]

$T = 1972$ Mar. 27.726 ET, $\omega = 257\overset{\circ}{.}71$, $\Omega = 159°.59$ 1950.0

$q = 0.9275$ AU, $i = 123.69°$

1972 ET	α_{1950}	δ_{1950}	Δ	r
Apr. 03.0	$00^h 15^m\!.8$	$-44°33'$	1.171	0.934
8.0	00 55.8	-47 41		
13.0	01 50.3	-50 07	0.961	0.972
18.0	03 00.0	-50 42		

4. KOHOUTEK OBJECT AKSNES 19504 11125 00412 11411 90325 81185 00362 11543 00316 11709 00272 11832 00231 11950 00192 12103 90344 81114 00157 12213 11207 69507 84703 APOLLO TYPE ASTEROID MAGNITUDE SEVENTEEN SEKANINA

Translation: Sekanina communicates the following ephemeris by Aksnes for the object discovered by Kohoutek. The object is an Apollo-type asteroid.

1971 ET	α_{1950}	δ_{1950}	Δ	r	Mag.
Nov. 25.0	$00^h 41^m\!.2$	$-14°11'$	0.325	1.185	17
27.0	00 36.2	-15 43			
29.0	00 31.6	-17 09			
Dec. 01.0	00 27.2	-18 32			
03.0	00 23.1	-19 50			
05.0	00 19.2	-21 03	0.344	1.114	
07.0	00 15.7	-22 13			

5. HONDA SERPENS NOVA HONDA 19001 00215 8//// 18257 20238 03053 40764 41548 BRIGHTNESS INCREASING HIROSE

Translation: Hirose reports that Honda has discovered and observed a nova in Serpens, as shown below. The brightness is increasing.

1970 UT	α_{1950}	δ_{1950}	m_v
Feb. 15.8	$18^h 25^m\!.7$	$+2°38'$	5.3

6. N3811 SUPERNOVA ROSINO 19501 90209 11386 24758 0412/ 20005 20003 89982 40264 ASIAGO

Translation: The Asiago Astrophysical Observatory communicates the observation by Rosino of a supernova in NGC 3811 ($\alpha = 11^h 25^m\!.7$, $\delta = +47°58'$, equinox 1950.0), 5″ east and 3″ north of the nucleus. On 1969 Feb. 9 UT the photographic magnitude m_{pg} was 12.

9 Occultations

R. Boninsegna & J. Schwaenen

9.1 Introduction

An occultation is an event involving two bodies, one of which partially or completely hides the other. Generally, however, for an event to be described as an occultation the body that is hidden (the more distant of the two) should have a smaller apparent diameter than the occulting body.

The best-known events are those that involve the Moon, and a star or planet. For example, a series of occultations of the Pleiades by the Moon began in 1987. Strictly speaking, solar 'eclipses' (Fig. 9.1) are occultations, as are the events that occur in 'eclipsing' variables. [Technically, eclipses occur when the *shadow* of a body partially or completely obscures another object, as in a lunar eclipse. – Trans.] When the apparent diameter of the occulting body is smaller than the body occulted, we use the term 'transit'. Transits of the Galilean satellites in front of Jupiter are more frequent events than transits of Venus or Mercury in front of the Sun.

As regards occultations, we have to determine which of the many possible events are visible and are of scientific interest to amateurs. In general, observations are more valuable if the occulted body is a star. Its apparent diameter is so small (< 0.01 arc-second) that timings of disappearance and reappearance are extremely accurate.

One of the most famous examples is the discovery in 1977 March of the rings around Uranus from observations of the occultation of a bright star. Short disappearances that were recorded before and after the occultation that was caused by the disk of Uranus itself enabled the characteristics of the system of rings to be determined. Occultations involving planets and stars (*see* Fig. 9.3) are only of real interest if the planets concerned are remote ones, such as Uranus, Neptune and, in particular, Pluto (Fig. 9.2), which will not be visited by space-probes for many years. It should be noted that most of these events are not observable in amateur telescopes: the brightness of the planet drowns the faint light from the star.

Better known, and far more accessible, are total or grazing occultations of stars by the Moon, which we shall cover in detail in this chapter. The same also applies to the occultations of stars by minor planets. These have only recently been predictable, but are a fascinating study.

Finally, it is possible that by the turn of the century we may be able to predict and observe the occultation of a distant star by another that lies closer to us. Such an event would last several days and would be more akin to variable-star observing. It is possible that the event would not result in a decrease in the combined magnitude of the two stars, but in an increase, because of the gravitational lensing effect that might concentrate the light from the occulted object. But that is another matter, and in any case we shall probably have to await the results from the European

Fig. 9.1. *A total eclipse of the Sun is simply the result of an occultation of our parent star by the Moon. This photograph was taken on 1980 February 16 at Malindi (Kenya) at the focus of a 77-mm aperture, 1-m focal length refractor, using Ektachrome infrared film and an orange filter; exposure 1/8 second. (Photograph by R. Boninsegna.)*

astrometric satellite Hipparcos before any such event may be predicted (Kovalesky, 1982).

Perhaps we should close this short introduction by recalling, for the record, the unique observation made on 1737 May 28 from Greenwich by John Bevis, who, using a refractor with an OG about 70 mm in diameter, was able to observe the occultation of Mercury by Venus, when the two planets were very close to the horizon.

9.2 Determining the time

9.2.1 Timing the instant of the observation

The principle behind all occultation observations is to measure, with an accuracy of 0.5 second or better, the exact times of contact between the occulting body and the one occulted. The methods and the equipment used for this vary from observer to observer; the items most frequently employed are a radio, a tape-recorder and a stopwatch.

The radio receiver picks up time signals from a suitable station: for example (in

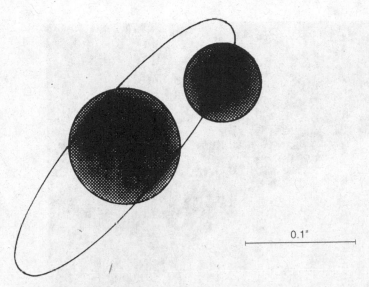

0.1"

Fig. 9.2. *As a result of an alignment that occurs once every 124 years, the Pluto–Charon system began a series of occultations and transits in 1985, continuing until 1990. Some of the events were observable by amateurs equipped with a telescope of at least 300 mm in aperture. The event took the form of a decrease in the total magnitude of the system, over approximately three hours and amounting to about 0.6 magnitude. The orbital period of Charon is 6.39 days.*

Europe) Y3S, Nauen – previously DIZ – at 66.3 m wavelength (4525 kHz) or (in the U.S.A.) WWV, Fort Collins at 2500, 5000, 10 000, 12 500 and 15 000 kHz. These stations broadcast 'pips' every second and a double or longer pulse at the beginning of every minute. The tape-recorder simultaneously records these time signals as well as clicks, electronic tones, or other noises indicating the times of contact, and which may be produced by any convenient means.

[On occasions, observers have been able to arrange for time signals to be retransmitted on a more convenient frequency by a local radio station or radio amateur. In many countries, such as the United Kingdom, however, such a procedure is prohibited under the terms on which licences are issued. It is, in any case, essential to ensure that no delay is introduced into the signals. – Trans.]

9.2.1.1 Measuring a single time

For observations when only a single contact has to be timed, two methods are commonly used. In the first, a stopwatch is started at the third pip after a minute marker. These 3 seconds are very useful for the very simple, and good, reason that one is often surprised by the minute marker. To avoid having too much drift in the stopwatch, it is advisable to choose a time signal only a few minutes before the event. The observer should check carefully whether the stopwatch was started at precisely the same time as the signal, and any difference should be noted. Finally, at the precise instant that the event occurs, the stopwatch is halted.

Fig. 9.3. *The occultation of Saturn by the Moon on 1974 March 3, taken with a 115-mm aperture, 900-mm focal length reflector, exposure 1 s. (Photograph by J. Schwaenen.)*

The second method consists of waiting, halted stopwatch in hand, for the event, and starting the watch as it occurs. A second timing may also be recorded if the stopwatch has a split timing feature. Finally the watch is stopped 3 seconds after a suitable minute marker following the time of observation. All that remains is to subtract the time indicated by the stopwatch.

A 'speaking clock' service may also be used instead of a radio. In the absence of a short-wave receiver or a telephone, it is possible to use, as a last resort, the 'pips' broadcast every hour by many radio stations. The fact that these signals are infrequent, however, means that certain precautions must be taken. The stopwatch should be started at the event and the elapsed time should be noted – without stopping the watch – at several subsequent hour markers so that any errors may be averaged out. In addition, the stopwatch should have been carefully calibrated beforehand to determine its overall drift (which is likely to be fairly considerable over several hours). The results thus obtained may be corrected for reaction time, as described in the section concerning personal equation, before being recorded on the report form. [Note that observations to be submitted to the International Lunar Occultation Centre (ILOC) in Japan are generally recorded as uncorrected times, personal equation being entered separately. If reaction time is subtracted, this must be noted on the report forms. – Trans.]

Table 9.1. *Average reaction times measured with an occultation simulator*

Type of occultation	Average reaction time (s)
Spectacular	0.26 ± 0.03
Very favourable	0.28 ± 0.04
Favourable	0.34 ± 0.05
Marginal	0.74 ± 0.13

9.2.1.2 Measuring several times

When several consecutive times have to be measured, the best method is to record suitable signals and additional time signals simultaneously on a tape recorder. These may then be reduced later (*see* Sect. 9.2.3). Do not forget that, in the case of grazing occultations, recording should start at least 15 minutes before the predicted time, if the ephemerides give only the time of the middle of the event, which may last for more than 20 minutes.

9.2.2 Personal equation

9.2.2.1 The effect of reaction times

One's personal equation or reaction time is a very important factor that must be taken into account in deriving accurate timings. Between the time one detects an event and the instant at which one reacts there are a few tenths of a second. The more experienced the observer, the easier it is to estimate this amount. Everyone has their own personal equation, largely as a result of their own physiology and powers of concentration. Other factors may influence it, however, and observers will find these easier to recognize with experience. (They include fatigue, discomfort, mental state, seeing conditions, telescope aperture, concentration, etc.)

There is a very easy way of determining one's reaction time. With a piece of paper, hide the figures representing seconds on your stopwatch. When you see the minute figure change, press the stop button. You can then read off the time that you required to react in hundredths of a second. This reaction time may vary from about two-tenths of a second to 1 second according to the various factors that we have mentioned.

9.2.2.2 Occultation simulator

In Belgium, J. Bourgois has carried out experiments with an electronic occultation simulator that he built. This has been devised so that when a button is depressed, it starts a stopwatch, and at any time between 4 and 10 seconds later an artificial star disappears or reappears in the eyepiece. The timing is stopped by the observer who depresses a limited-travel push-button. The reaction time is given directly on an illuminated counter. Average reaction times measured with this apparatus are given in Table 9.1.

Another finding from these experiments is that personal equation decreases as children mature into adults. The swiftest reactions appear to be obtained at around 25 years old. Personal equation is then about 0.26 s ±0.03 s. Over 40 years old, there appears to be little change in personal equation.

These values are only valid for good observing conditions: i.e., for an occultation at the dark limb of the Moon, which is not too bright; or for occultations by minor planets when the drop in brightness is at least 1 magnitude. With actual, good-quality observations, average reaction times are 0.40 s ± 0.10 s for favourable occultations, and 0.60 s ± 0.20 s for poor conditions, for both disappearances and reappearances. With poorer observations, these values may reach 0.70 s ± 0.20 s and 1.10 s ± 0.50 s respectively. Such high values for the reaction times may seem surprising, but do indicate that observers have a tendency to under-estimate their reaction times, especially with reappearances of faint stars.

To summarize, we would say that 0.40 s is a good estimate of the average personal equation for a very favourable occultation.

9.2.3 Analysis

9.2.3.1 Deriving times from the magnetic-tape recording

Before beginning to analyze the recording of one's observations, it is useful to determine the exact duration of several minutes as recorded on the tape. This is necessary because the analysis is made at a different temperature from that prevailing when the recording was made, and this nearly always slightly affects the speed of the tape.

The steps to be carried out are as follows: first, count the number of seconds between the preceding minute marker and the audible signal for the event; this will give you the time to the nearest second. Second, obtain the time to a tenth of a second by starting the stopwatch at an earlier second marker and stopping it at the exact instant of the audible signal. If you feel that the measurement is affected by reaction time, this should be taken into account or, better still, the measurement should be repeated. Ideally, the measurement should be repeated several times, and the mean taken. If the personal equation for the event is subtracted from the result, we obtain the time of true first contact. Do the same for any other times of contact that were recorded.

9.2.3.2 Accuracy of the measurements

A point that must be borne in mind is the accuracy of the final timing of the event given the errors introduced by the reaction time and by the method of analysis. This accuracy is weighted as follows: 0.1–0.2 s – excellent; 0.3–0.4 s – good; 0.5–0.7 s – fair; 0.8–1 s – poor. Beyond 1 second the measurement is of doubtful validity. This accuracy is entered on the report forms (*see* Sect. 9.6), and should not be confused with personal equation.

It is also necessary to calculate the geographical position of one's observing site (using a map to a scale of at least 1:25 000). This information should be given to an accuracy of at least 1 second of arc (about 30 m in latitude and 20 m in longitude), together with the height in metres above sea level.

9.3 Occultations of stars by the Moon

9.3.1 General

Occultations of stars by the Moon are astronomical events that (unfortunately) few amateurs observe, although they are of considerable scientific interest. These events may be divided into two distinct types: ordinary total occultations, and grazing occultations.

As we know, the Earth rotates from West to East once in slightly less than 24 hours. Because of this, the stars and Moon appear to rotate around us from East to West, whereas the Moon, because of its own proper motion around the Earth, appears to retrograde with respect to the stars. This causes it to pass in front of various stars lying within a band centred on the ecliptic; we then say that a star is occulted by the Moon.

The star disappears at the eastern limb of the Moon and reappears at the western limb; this sequence of events is known as a total occultation. There is another case when, instead of a star being occulted for some time by the Moon, it appears to graze the lunar limb in the northern or southern polar regions and, because these areas are very rough, the star may disappear and reappear several times behind lunar mountains; this is known as a grazing occultation.

The first observations of occultations were made long ago. In 357 BC Aristotle saw Mars disappear behind the disk of the Moon. Later, in 1497, Copernicus observed an occultation of Aldebaran. In the 18th century, J. J. L. de Lalande realised that observation of the occultation of a single star from different sites could be used to provide information about the difference in longitude between the observers. He mentioned an occultation of Aldebaran observed from Paris and Berlin on 1749 April 6.

Another application of observations of occultations was illustrated when, in 1637, Jeremiah Horrocks watched the Moon cross the Pleiades. He noted that each star, when it was occulted, disappeared instantaneously, and deduced from this that the angular diameter of stars must be very small. This idea was revived in 1908 when P. A. MacMahon suggested that the precise interval of time over which the light from a star declined before disappearing could be used to determine the angular size of the star. Finally, Jacques Cassini, who observed the occultation of the double star Gamma Virginis on 1720 April 21, clearly saw the disappearance take place in two steps.

As for grazing occultations, it would seem that the first observation was made, quite by accident, from Danzig on 1794 March 7 at 19:00 UT. Aldebaran grazed the northern limb of the Moon, disappearing and reappearing twice behind lunar mountains.

9.3.1.1 Requirements for an occultation to occur

In Fig. 9.4, E is the Earth, M the Moon and S the star, assumed to be at infinity. We lay a plane through the centre of the Earth, E, and perpendicular to the direction of the star – this is commonly known as the fundamental plane. The intersection of this plane (which we will call FP), with the equator at N, defines the axis EX, the EZ

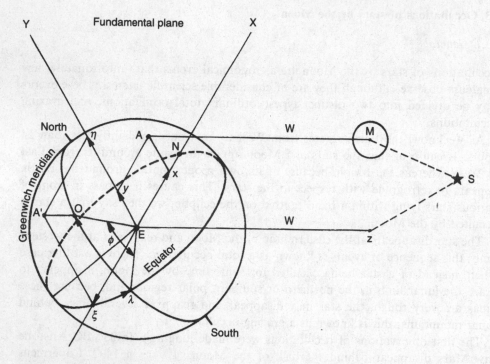

Fig. 9.4. *The projection of the Moon onto the fundamental plane.*

axis being simply the direction W of the star. We take the axis EY as perpendicular to the other two axes.

Given the very short duration of the event, we assume that the plane FP is fixed. The projection of the centre of the Moon (M) onto the plane is A, and the coordinates of this point are x and y, the standard Bessel elements. We then take an observer O on the Earth, at altitude Z, and situated at the point given by the coordinates λ and ϕ. Similarly, we obtain the coordinates ξ and η of the point A' which is the projection, parallel to EZ, of O onto the fundamental plane. Let the distance between the points A and A' on this plane be represented by P, the other parameters x, y, ξ and η being generally expressed in terrestrial equatorial radii. Using this as a unit, the semi-diameter of the Moon is expressed as a constant K, equal to 0.272 495.

It will therefore be seen that a disappearance or reappearance will occur at the times when P = K.

9.3.1.2 Scientific interest

The results of a total occultation give detailed information about the motion of the Moon in longitude, and about its actual position, together with improved knowledge of Ephemeris Time and the fundamental, celestial-coordinate reference system.

The results of grazing occultations enable the motion of the Moon in latitude to be studied in greater detail, in particular the longitude of the nodes of its orbit. They also allow the profile of the lunar limb in the polar regions to be derived, and sometimes enable differences in altitude of just a few tens of metres to be determined. This in turn allows solar eclipses to be analyzed in greater detail, which should enable any possible changes in the Sun's diameter to be determined. Finally, occultations have also been instrumental in the discovery of some very close binaries, with separations of around 0.01 arc-second. Such separations are too small to be observed visually, but are too great to be detected spectroscopically.

9.3.1.3 Instrumentation

A reflector or refractor, a radio receiver or telephone, and a stopwatch reading to one-tenth of a second are basically all that are required for a total occultation, when normally only one time of contact is required. For a grazing occultation, when several timings have to be recorded, it is essential to have a tape recorder on which suitable time signals and audible tones indicating the successive disappearances and reappearances of the star may be recorded simultaneously (*see* Fig. 9.25).

The arrival of video cameras has enabled several amateurs, both in North America and in Europe, to use these for observing occultations. In this method, the video camera is mounted at the prime focus of the telescope, and the image is transmitted directly to a monitor. To record the different times of contact, a visual indication of the time, synchronised with a suitable broadcast time signal, is superimposed on the video image. Because the standard video scan rate is 25 images per second, information may be recorded every 0.04 second (Fig. 9.5).

Since 1977, the Astrometry and Celestial Mechanics Department of the Royal Belgian Observatory has used a television camera of the EBS (Electron Bombardment Silicon) type, mounted on a Cooke–Zeiss equatorial telescope, with an aperture of 450 mm and a focal length of 7 m, for its double-star and occultation observations (Dommanget, 1985). The equipment comprises:

- the camera,
- a monitor,
- a stabilized power supply,
- a time-signal generator, locked onto a signal transmitted by the Time Service,
- a video recorder.

An observation is made by recording a series of images over a period of twenty to thirty seconds around the predicted time of contact. The camera provides 25 double images per second, and individual examination of these images enables a final accuracy of 0.02 second to be achieved. With this procedure, occultations of stars as faint as magnitude 11.5 have been observed at the dark limb under favourable atmospheric conditions (Fig. 9.6).

9.3.2 Periodicity of occultations

Occultations of a star occur in series, and within a series there is an occultation at each conjunction of the Moon and the star, i.e., every 27.3 days. For example, a series involving Spica (α Virginis) was:

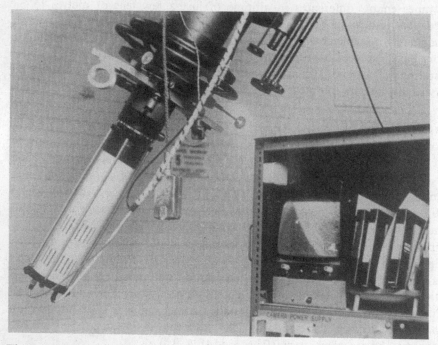

Fig. 9.5. *A T6 CK92A camera, fitted with an EBS tube and a (30×) image intensifier, mounted on the 450-mm aperture, 7-m focal length refractor at the Royal Belgian Observatory. (Photograph by J. Dommanget.)*

1987 April 14	occultation (visible in France)
1987 May 11	occultation
1987 June 7	occultation (visible in France)
1987 July 4	occultation
1987 August 1	occultation
1987 September 24	occultation
1987 October 22	occultation (visible in France)
	etc.

[As implied by this list, successive occultations will not be seen from the same site. The duration of any series is also limited and depends on the ecliptic latitude of the star. – Trans.]

9.3.2.1 Description of the event

The orbital plane of the Moon makes an angle of 5°8′.7 with that of the Earth's orbit, the ecliptic. The intersection of these two planes, called the line of nodes, itself intersects the celestial sphere at two diametrically opposite points known as the ascending and the descending nodes. The line of nodes is not fixed relative to the stars, and it turns slowly in a retrograde direction by 19°21′ per year, taking 18.6 years to complete a single rotation. This motion may be seen from successive values of the longitude of the ascending node:

Fig. 9.6. *Occultation of a double star observed with the video camera at the Royal Belgian Observatory. (Photograph by J. Dommanget.)*

1986 January 1	35°49′
1987 January 1	16°29′
1988 January 1	357°10′
1989 January 1	337°46′
1990 January 1	318°27′
etc.	

Because the nodes move by approximately 19°21′ per year towards the West, while the inclination remains constant, the path followed by the Moon against the celestial sphere at each revolution is slightly different from the previous one.

Figure 9.7 represents the zodiacal zone, with EE′ the ecliptic, and the sinusoidal lines the path of the Moon. Points A and D are the ascending and descending nodes, respectively. The longitude of the ascending node is the distance EA. Let us assume that on 1989 January 1 a star coincided exactly with the ascending node A; i.e., its ecliptic longitude λ was 337°46′ and its ecliptic latitude β was 0°. When the Moon passed the ascending node, an observer on Earth would have seen an occultation of the star.

One year later, the node would have retrograded 19°21′ and therefore been at A′; the Moon then followed the broken line and could not pass in front of the star at A. It is not, however, essential for the Moon to pass exactly over any point such as A for the star to be occulted, because the Earth and the Moon are not points but bodies with finite diameters. Occultations of the star at A would therefore have occurred for several months before 1989 January 1 and ceased only several months

505

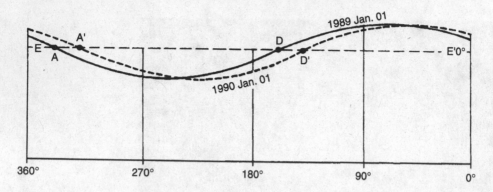

Fig. 9.7. *Apparent path of the Moon along the ecliptic between 1989 January 1 and 1990 January 1.*

after that date. This would give us a total of about 20 occultations over a period of seventeen months. The first occultations in the series would have been visible from the southern polar region, those in the middle from the equatorial region, and those at the end from the northern polar region. This ended the series and for the next 9 years the Moon would pass north of the star. Then it would be the turn of the descending node to pass the star, giving rise to a new series of occultations, but this time first visible in the northern polar regions and ending near the South Pole.

9.3.2.2 The 18.6-year periodicity

The period of 18.6 years therefore includes two alternating series. The same applies to stars that are not exactly on the ecliptic but which are fairly close to it. For stars situated less than 3°56′ from the ecliptic there are two series of occultations every 18.6 years. The duration of each series is, however, longer the farther the star is from the ecliptic:

> 1.4 years for a star at latitude 0°
> 1.5 years for a star at latitude 2°   (north or south)
> 1.8 years for a star at latitude 3°   (north or south)
> 2.2 years for a star at latitude 3°4′   (north or south)

For stars between 3°56′ and 6°21′, there is only one series of occultations every 18.6 years, but this time the greater the ecliptic latitude of the star, the shorter the series:

> 5.9 years for a star at latitude 4°   (north or south)
> 4.9 years for a star at latitude 4°4′   (north or south)
> 3.8 years for a star at latitude 5°2′   (north or south)
> 2.2 years for a star at latitude 6°   (north or south)

Any stars with ecliptic latitudes greater than 6°37′ (north or south) cannot be occulted by the Moon. For a geocentric observer, the maximum latitude that the centre of the Moon's disk may attain is 5°09′. From the Earth's surface, however, it is possible for the Moon to occult stars that are, geocentrically, 1°12′ from its centre, giving a value of 6°21'. In reality, the extreme latitude is 6°37', because we must add the maximum value of the Moon's apparent semi-diameter, 16 arc-minutes. All the stars that may be occulted are to be found in various catalogues; the one that is currently used for calculating occultations is the *XZ Catalog* – which is based on the AGK3, SAO data and the older *Zodiacal Catalog* – because it is the most accurate.

9.3.2.3 *Calculating the periodicity*

To calculate the period between occultations, the star's equatorial coordinates (α and δ) should be transformed into ecliptic coordinates (λ and β) and then the time T is calculated from:

$$T = 1973.87 - 0.051\,665\lambda.$$

To this value of T add or subtract a multiple of 9.3 years (to obtain the desired epoch), this multiple being even if the latitude of the star is positive, and odd if the latitude is negative. The number of years to be added to, or subtracted from T are:

β positive	β negative
0	9.3
18.6	27.9
37.2	46.5
...	...

The corrections a and b are then calculated as follows:

for a: $\cos x = 0.003\,239\,391(\beta + 72'.6)$

for b: $\cos x = 0.003\,239\,391(\beta - 72'.6)$

with a or $b = 0.051\,665x$, where β is expressed in minutes of arc and x in decimal degrees. These corrections are in decimal years, and may be either positive or negative. If there is only a single series of occultations, the correction $-a$ gives the beginning of the series and $+a$ the end. On the other hand, if there are two series of occultations the corrections become:

$-a$ for the beginning of series 1

$-b$ for the end of series 1

$+b$ for the beginning of series 2

$+a$ for the end of series 2.

Taking as an example, a star at position:

$\lambda = 203°14'$, $\beta = -2°05'$.

To find the next series of occultations, we first obtain:

$T = 1973.87 - (0.051\,665\lambda) = 1963.37.$

The latitude is negative, so we add 27.9; which gives:

$1963.37 + 27.9 = 1991.27$

$a = 4.16$

$b = 2.62$

from which we may deduce:

beginning of series 1:	$1991.27 - 4.16 = 1987.11$	
end of series 1:	$1991.27 - 2.62 = 1988.65$	
beginning of series 2:	$1991.27 + 2.62 = 1993.89$	
end of series 2	$1991.27 + 4.16 = 1995.43.$	

There follow details of the occultation series between 1967 and 2051 for six of the brightest stars.

Aldebaran (α Tau): longitude 69°09′, latitude −5°47′

1977.84 to 1981.36
1996.44 to 1999.96
2015.04 to 2018.56
2033.64 to 2037.16

ZC 810 (β Tau): longitude 81°88′, latitude 5°38′

1967.79 to 1971.49
1986.39 to 1990.09
2004.99 to 2008.69
2023.59 to 2027.29
2042.19 to 2045.89

Regulus (α Leo): longitude 149°13′, latitude 0°46′

1969.84 to 1971.25
1979.68 to 1981.09
1988.44 to 1989.85
1998.28 to 1999.69
2007.04 to 2008.45
2016.88 to 2018.29
2025.64 to 2027.05
2035.48 to 2036.89
2044.24 to 2045.65

Spica (α Vir): longitude 203°14′, latitude −2°05′

1968.51 to 1970.06
1975.29 to 1976.84
1987.11 to 1988.65
1993.89 to 1995.43
2005.71 to 2007.26
2012.49 to 2014.04
2024.31 to 2025.86
2031.09 to 2032.64
2042.91 to 2044.46
2049.69 to 2051.24

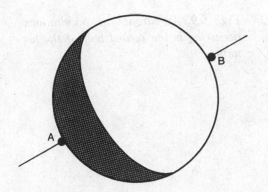

Fig. 9.8. *Appearance of occultations occurring in the first half of the lunation.*

Antares (α Sco): longitude 249°06′, latitude −4°56′

$$
\begin{array}{l}
\text{1967.75 to 1972.85} \\
\text{1986.35 to 1991.45} \\
\text{2004.95 to 2010.05} \\
\text{2023.55 to 2028.65} \\
\text{2042.15 to 2047.25}
\end{array}
$$

Pollux (β Gem): longitude 112°, latitude 6°68′

Because of its proper motion, this star is no longer occulted. The last series of occultations occurred around the year 900 AD.

9.3.3 Total occultations

In total occultations, two types of event (Figs. 9.8 and 9.9) are encountered:

  (i) Between New Moon and Full Moon, disappearances always occur at the dark limb, so observations are easy (A); reappearances are at the bright limb, so observations are almost impossible (B).

 (ii) Between Full Moon and New Moon, disappearances take place at the bright limb, so observations are very difficult (C); reappearances occur at the dark limb, so observation is easy (D).

9.3.3.1 The observations

For an observation to be successful, the two types of event just mentioned need to be considered, as well as the magnitude of the star. If the latter is fairly bright, then observation of a disappearance at C is feasible, whereas if the star is faint, it is practically impossible. Similarly, it is useless to observe reappearances of type B, unless the star is very bright (such as Aldebaran or Spica), but in any case such measurements will always be imprecise.

It is also necessary to know when these events may be observed. Predictions are to be found in various publications, such as the *Handbook* of the British Astronomical Association, the *Observer's Handbook* of the Royal Astronomical Society of Canada,

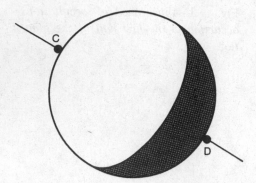

Fig. 9.9. *Appearance of occultations occurring in the second half of the lunation.*

Sky & Telescope, etc. They give the dates and times (in UT) of the times of contact for certain events visible from various stations. These publications also give formulae for calculating the times of events for different sites (the error will not exceed 1 minute of time, if the site involved is less than 300 km from one of the standard stations). Once an observer has become experienced and is obtaining several tens of observations per year, predictions for stars below magnitude 7 (for a specific, properly determined position), will be required. [Until recently, these were provided by the USNO (U.S. Naval Observatory, Washington), but apparently this institution is discontinuing the service – and even the prediction of occultations. At the time of writing, the future of this field remains unclear. Some predictions will probably be provided by IOTA and other organisations. – Trans.] Such predictions primarily give: the date and time of the times of contact, the type of phenomenon (D = disappearance, R = reappearance, and G = grazing), the identification of the star, the magnitude, the position angle (PA), the position of the terminator (CA), etc. (Fig. 9.10).

The cusp angle (CA) is the angle between the closest end of the terminator and the point on the limb where the event will occur. This angle is measured from 0° (North or South) up to 90° (Fig. 9.11). The position angle (PA) is the angle at the centre of the Moon of the point of contact, measured from celestial North through East, i.e., anticlockwise (Fig. 9.12).

Once we have all this information we can make the actual observation. To be certain of the star and to be able to follow it until the moment of disappearance, or to have the correct area in view when waiting for a reappearance, it is sensible to be ready to observe at least a quarter of an hour beforehand. Do not forget that these events occur very rapidly, and that their rapidity often takes beginners by surprise.

Once ready, all we have to do is to wait for the instant of disappearance or reappearance and record it as accurately as possible using one of the methods previously described (*see* Sect. 9.2.1.1). This information is noted on a report form (*see* Sect. 9.6), together with the personal equation, and given a weighting in accordance with the estimated accuracy as described on p. 500.

```
I.O.T.A. / U.S.N.O.    TOTAL OCCULTATION PREDICTIONS FOR 1992          PAGE 53         OPTIONS -- 0,0,2,1          ADDRESS CODE AU148
COORDS., LOCATION, IDENT.-- W  0 56 39.7, +51 25 08.6, +0072M -- READING,BERKS., ENG.-- 250MSP ANDREW ELLIOTT -- STATION CODE SU165

DAY TIME-UT P AC   USNO    O   MAX SP   PCT ELG SN MN   MN CA     PA  VA  WA LONG LAT   A   B   C     DM      SAO      HA     DECL.  RT. ASC.
    H  M  S  D   REF NO V   MAG      SNLT        AL AL   AZ                  LIB LIB  M/O M/O S/K  REF NO   REF NO   O / // O / //  H M  S
                                                                  --------
                                                                  DECEMBER  ****************************************************  DECEMBER
DECEMBER
12/20 11 43/RV 2    1210 96   5.9 A0   89- 142        9  75 47S 244 283 234 -1.2 4.8  0.3 2.0-0.3 +15 1612 097485 -963144 162823  80108.5
12/20 18 29/R  2 X12101 92   9.1      89- 142       10  75 32N 344  23 334 -1.2 4.8 -0.6-1.3-1.9 +17 1742 097495 -950335 165021  80207.9
12/21 07 43/R  2 X12151 82   8.8 A5   89- 141       17  85 64N 312 353 302 -1.2 4.8 -0.4 0.3-1.1 +17 1748 097522 -830612 163948  80333.9
12/22 02 46/R  1 X12203 82   9.0 A0   89- 141       25  95 73N 304 345 293 -1.2 4.8 -0.6 0.5-1.0 +15 1632 097551 -694753 163117  80531.9
12/22 14 23/R  1 X12212 82   8.9      89- 141       27  98 86S 282 323 272 -1.2 4.8 -0.5 1.1-0.8 +15 1634 097554 -665513 162413  80547.1
12/23 02 13/R  2    1228 83   8.4 G0   89- 141       34 107 62N 315 354 304 -1.2 4.8 -0.9-0.1-1.1 +15 1642 097581 -552543 162608  80745.1
12/23 05 18/R  3 X12288 92   8.5 K0   89- 140       35 108 47N 329   8 318 -1.2 4.9 -0.9-0.9-1.3 +15 1643 000000 -544515 162813  80804.5
12/23 37 21/R  3 X12304 91   9.5 A0   88- 140       39 116 53S 250 286 239 -1.3 4.9 -1.0 2.1-0.3 +15 1645 097598 -465121 160409  80839.1
12/23 38 00/R  2 X12305 91   9.2 A    88- 140       39 116 51S 248 283 237 -1.3 4.9 -1.0 2.2-0.3 +15 1645 097599 -464148 160329  80841.3
13/00 10 47/R  2 X12338 81   9.4 F8   88- 140       43 124 73N 304 336 293 -1.3 4.9 -1.1-0.0-0.8 +15 1648 000000 -384613 161415  80950.4
13/00 14 09/R  8 X12348 91  10.0 G0   88- 140       43 125 14S 211 244 200 -1.3 4.9 -1.6 8.4 1.6 +15 1650 000000 -380406 155040  81017.7
13/00 03 17/R  4 X12356 93   8.2 K0   88- 140       42 122 16N   1  34 350 -1.3 4.9 -0.8-5.9-2.5 +15 1651 097627 -404657 162235  81034.2
13/00 28 16/R  6 X12363 92   8.9      88- 140       45 129 12S 209 239 198 -1.3 4.9 ************ +15 1654 097630 -343902 154814  81046.6
X12363 DISTANCE TO SMOOTH-MOON TERMINATOR = 14.9 ARC SEC., DISTANCE TO POSSIBLE 3-KM SUNLIT PEAK =  3.9 ARC SEC.
GRAZE OF X12363 NEARBY -- APPROXIMATE S. LIMIT -- LAT. = 51.106 + 0.079(W. LONG. -  0.944), CA=   1S, SEC. RA=46.612, DEC=14.26.
X12363 DISTANCES TO TERMINATOR=  0.0  0.1  2.7 A.S.,DIST. TO 3-KM SUNLIT PEAK=  0.0  0.0  0.0 A.S. FOR C.A.= -3  1  5, RESP.
13/00 50 49/R  2 X12375 81   9.7 K5   88- 140       48 136 68N 309 336 298 -1.4 4.9 -1.2-0.5-0.8 +15 1655 000000 -290352 160936  81108.0
13/00 44 55/R  2    1235 94   7.4 K0   88- 140       47 134 48N 329 357 318 -1.4 4.9 -1.1-1.4-1.1 +15 1657 097636 -303435 161412  81115.0
13/01 02 43/RA 2 X12384 83   8.3 G5   88- 140       49 140 71N 306 331 294 -1.4 4.9 -1.2-0.4-0.7 +15 1658 097639 -260941 160701  81127.1
13/01 29 04/R  2 X12402 81   9.6 K2   88- 139       51 149 88S 286 305 274 -1.5 4.9 -1.4 0.1-0.5 +15 1661 000000 -194116 155745  81204.0
13/01 49 18/RA 2 X12416 82   9.1      88- 139       52 156 66S 263 279 252 -1.5 5.0 -1.6 0.8-0.1 +15 1663 097654 -144704 154812  81237.4
13/02 14 35/RC 3 X12435 94   7.8 G0   88- 139       54 166 48S 246 255 235 -1.6 5.0 -1.9 1.5 0.3 +15 1667 000000  -83834 153913  81327.0
13/02 48 13/R  2 X12460 82   9.1 K0   88- 139       54 179 68N 309 310 298 -1.6 5.0 -1.2-1.2-0.5 + 0 0000 000000   -2943 155033  81434.7
13/02 53 16/R  2 X12468 82   9.0 F2   88- 139       54 181 63N 315 314 303 -1.6 5.0 -1.1-1.5-0.5 +15 1671 097679    4300 155050  81447.8
13/03 22 40/R  2 X12481 82   9.1      88- 139       54 193 81N 296 288 284 -1.7 5.1 -1.3-1.0-0.2 +15 1673 097682   75628 154112  81525.8
13/04 08 40/R  2 X12517 92   9.0      87- 138       51 210 54S 252 233 240 -1.8 5.1 -1.8 0.0 0.6 +15 1792 097703  190804 152014  81642.5
13/04 28 10/R  2 X12527 81   9.3      87- 138       49 217 66S 264 241 252 -1.8 5.2 -1.5-0.6 0.4 +15 1794 097710  235318 151950  81713.9
13/04 08 05/R  4    1246 95   6.6 K0   87- 138       51 210 12N   5 346 354 -1.8 5.2  0.8-5.4-2.3 +15 1679 097718  184319 154152  81752.7
  1246 DISTANCE TO SMOOTH-MOON TERMINATOR = 16.5 ARC SEC., DISTANCE TO POSSIBLE 3-KM SUNLIT PEAK =  4.6 ARC SEC.
13/05 13 39/R  3 X12563 92   8.7 F0   87- 138       44 230 34S 232 202 220 -1.9 5.3 -2.2 0.6 1.7 +15 1796 097726  345354 150225  81850.4
13/05 10 00/R 18 X12592 92   9.1      87- 138       45 229  6N  11 341 359 -1.9 5.3  1.8-6.2-3.3 +15 1799 097733  334610 152924  81950.6
X12592 DISTANCE TO SMOOTH-MOON TERMINATOR =  4.4 ARC SEC., DISTANCE TO POSSIBLE 3-KM SUNLIT PEAK =  0.0 ARC SEC.
13/06 32 06/R  1 X12629 82   9.1 F2   87- 137       34 250 87S 285 247 273 -2.0 5.4 -0.6-1.6 0.3 +15 1802 097751  540041 150024  82108.1
13/06 44 31/R  2    1257 94   7.5 K0   87- 137-11  32 253 49S 248 209 236 -2.0 5.4 -1.0-1.0 1.1 +15 1805 097762  565954 144719  82135.7
13/06 48 46/R  2 X12652 92   9.1      87- 137-10  31 254 47S 245 207 233 -2.0 5.4 -1.1-0.9 1.2 +15 1806 097765  580108 144548  82145.3
13/06 54 47/R  1 X12657 84   7.4 M6   87- 137 -9  31 255 73N 305 266 293 -2.0 5.4 -0.4-1.9 0.1 +15 1808 097768  592641 150046  82203.7
13/07 10 14/R  3 X12695 92   9.2      86- 137 -7  29 258 14N   4 324 352 -2.0 5.5  0.9-3.7-2.2 +15 1811 097787  625718 150417  82335.6
13/07 23 48/R  3 X12705 92   9.1      86- 137 -5  26 261 16N   2 322 349 -2.0 5.5  0.8-3.4-2.0 +15 1813 097792  661502 150120  82401.4
13/08 56 13/DV 1    1271 81   5.9 A2   86- 135  6  12 278-68S 131  91 118 -2.0 5.7  0.2-1.8-0.0 +14 1899 097843  882051 141356  82815.9
13/20 48 41/R  2 X13668 82   8.5 F2   81- 128        3  74 38S 239 277 223  0.1 6.0  0.4 2.4-0.1 +12 1951 098280-1004844 115726  85917.4
13/21 24 08/R  2 X13705 81  10.0 G5   81- 128        8  81 58N 322   1 306  0.1 6.0 -0.2 0.0-1.1 +12 1953 000000 -922121 121246  90057.5
13/22 23 51/R  2 X13740 71   9.6 F8   80- 127       17  92 76S 277 316 261  0.1 6.0 -0.3 1.3-0.7 +12 1956 000000 -775019 115145  90247.0
13/23 08 30/R  2 X13783 83   8.7 G    80- 127       23 101 43S 244 283 228  0.1 6.0 -0.5 2.7-0.3 +12 1961 098341 -670303 113452  90427.9
13/23 14 24/R  2 X13788 83   8.5 G5   80- 127       24 102 51S 251 290 236  0.1 6.0 -0.5 2.2-0.4 +12 1963 098344 -653628 113554  90436.5
13/23 06 27/M ** X13820 93   8.8      80- 127       23 100  1N  19  58   3  0.1 6.0 ************ +12 1965 098360 -675321 115808  90545.4
GRAZE OF X13820 NEARBY -- APPROXIMATE N. LIMIT -- LAT. = 51.217 - 0.042(W. LONG. -  0.944), CA=   1N, SEC. RA=45.421, DEC= 8.20.
X13820 DISTANCES TO TERMINATOR=  0.0  0.3  6.5 A.S.,DIST. TO 3-KM SUNLIT PEAK=  0.0  0.0  0.0 A.S. FOR C.A.= -3  1  5, RESP.
14/00 26 36/RP 2 X13859 71  10.0 F8   80- 126       34 118 74N 306 341 291  0.0 6.1 -0.9 0.1-0.9 +12 1970 000000 -480757 113758  90706.9
14/00 26 39/RF 2 X13860 71  10.0      80- 126       34 118 74N 307 341 291  0.0 6.1 -0.9 0.1-0.9 +12 1970 000000 -480717 113800  90707.1
14/00 36 13/R  2 X13862 73   8.3 K0   80- 126       36 120 76S 276 310 260  0.0 6.1 -1.0 1.1-0.6 +12 1971 000000 -454532 112813  90711.9
14/01 15 52/R  3    1364 96   6.5 F0   79- 126       40 130 25N 356  25 340 -0.0 6.1 -0.5-3.3-1.8 +12 1979 098400 -362040 113527  90925.4
14/01 52 49/R  2 X13910 72   9.0      79- 126       44 141 63S 263 287 247 -0.1 6.1 -1.5 1.2-0.3 +11 1989 098403 -270657 111005  90928.3
14/01 31 45/R 10 X13919 93   8.9 G5   79- 126       42 135 26N 354  21 338 -0.0 6.1 -0.5-3.1-1.7 +12 1980 000000 -322820 113224  90951.6
14/02 34 45/R  3 X13945 71   9.9 G5   79- 125       48 155 55N 326 342 310 -0.2 6.2 -1.0-1.3-0.9 +11 1995 000000 -170123 111722  91108.2
14/03 38 57/RA 3 X13978 94   8.2 F5   79- 125       49 178 37S 237 239 221 -0.3 6.3 -2.4 2.0 0.8 +11 1996 098437  -11750 104146  91235.5
14/03 20 48/RA19 X13992 92   9.1 F    79- 125       49 171  7N  13  19 357 -0.2 6.3  1.3-7.4-3.2 +11 1998 098442  -55736 111021  91318.2
X13992 DISTANCE TO SMOOTH-MOON TERMINATOR = 14.3 ARC SEC., DISTANCE TO POSSIBLE 3-KM SUNLIT PEAK =  1.7 ARC SEC.
14/03 20 24/RF22 X13993 92   9.6 F    79- 125       49 171  6N  14  20 358 -0.2 6.3  1.5-8.0-3.5 +11 1998 000000  -60258 111021  91318.8
X13993 DISTANCE TO SMOOTH-MOON TERMINATOR = 10.6 ARC SEC., DISTANCE TO POSSIBLE 3-KM SUNLIT PEAK =  0.6 ARC SEC.
14/05 29 01/R  2 X14038 84   8.1 A2   78- 124       44 216 60S 261 239 244 -0.5 6.4 -1.6-0.6 0.6 +10 1965 098462  253438 102404  91528.6
14/06 15 33/R  2 X14075 72   9.6 F5   78- 124       39 230 65N 316 287 300 -0.6 6.5 -0.7-1.9-0.3 +11 2005 000000  364937 102759  91708.9
14/05 55 51/R  5 X14085 91  10.0 F8   78- 124       42 224 15N   6 340 349 -0.5 6.5  0.7-4.0-2.2 +11 2007 000000  314926 103658  91727.7
14/07 22 26/R  4    1384 95   7.4 K0   77- 123 -6  30 246 24S 225 190 209 -0.7 6.6 -2.1 0.3 2.9 +10 1972 098495  530455   94904  91909.7
14/07 36 23/R  2 X14143 72   9.2 F2   78- 123 -4  28 249 74N 307 271 290 -0.7 6.6 -0.4-1.9-0.0 +10 1974 000000  563053  100709  91930.5
14/07 26 52/R 31 X14162 92   9.6 M0   77- 123 -5  30 247  2N  19 343   2 -0.6 6.6  3.0-6.7-6.6 +10 1977 000000  535433  101500  92027.3
X14162 DISTANCE TO SMOOTH-MOON TERMINATOR =  1.4 ARC SEC., DISTANCE TO POSSIBLE 3-KM SUNLIT PEAK =  0.0 ARC SEC.
GRAZE OF X14162 NEARBY -- APPROXIMATE N. LIMIT -- LAT. = 51.850 + 0.734(W. LONG. -  0.944), CA=  -6N, SEC. RA=27.334, DEC= 0.01.
14/08 18 04/R  8 X14177 91   8.5 K2   77- 123  1  22 257 19S 221 182 204 -0.7 6.7 -1.7 0.3 3.7 +10 1980 117614  663320   93501  92102.3
14/23 36 04/RA 2    1469 75   8.0 F5   70- 113       16  99 39S 241 280 222  1.4 6.9 -0.4 2.9-0.3 + 6 2240 118045 -730053   61656  95952.7
14/23 36 52/R  3 X15071 71  10.6 F5   70- 113       16 100 41S 243 282 224  1.4 6.9 -0.4 2.8-0.3 + 6 2240 118046 -724853   61718  95953.2
15/01 04 22/R  2 X15135 61  10.2 K0   69- 113       28 118 82N 300 334 281  1.4 6.9 -0.8 0.4-0.9 + 0 0000 000000 -513437   61423 100243.9
15/01 17 06/R  2 X15139 64   8.7 A3   69- 113       30 121 74S 277 309 257  1.3 6.9 -0.9 1.1-0.7 + 6 2247 118079 -482756   60519 100258.3
```

Fig. 9.10. *An example of occultation predictions prepared by the USNO.*

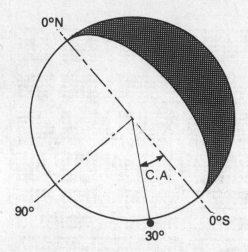

Fig. 9.11. *Definition of cusp angle, CA.*

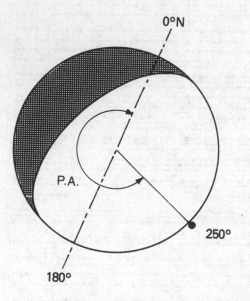

Fig. 9.12. *Definition of position angle, PA.*

9.3.3.2 *Reduction of total occultation measurements*

The results of observations of total occultations of stars by the Moon are used to determine the errors in longitude of the latter. When a large number of occultations have been observed, it is possible to determine, to within several tenths of a second, the difference between Universal Time (UT) and Ephemeris Time (ET). A reduction of the observation may also be made, i.e., the O − C (observed minus calculated) value may be derived.

If the profile of the Moon were truly circular and if the difference between ET and UT were known accurately, we would find O − C. But because the difference between the time scales is only known a posteriori – it cannot be determined in

Table 9.2. *Examples of* O − C *values derived from occultation measurements*

Date	Time (h m s UT)	Mag.	Star	Ph.	O − C(″)
Jan. 01	13:42:14.9	8.6	X13450	R	1.62
Jan. 07	00:50:31.4	8.3	S139562	R	1.75
Jan. 21	11:07:18.05	6.4	S109895	D	1.94
Feb. 01	04:40:19.1	4.5	S119164	D	−1.96
Feb. 08	16:27:47.1	8.6	S186485	R	−1.60
Feb. 21	17:16:42.2	8.4	S76863	D	1.99

advance – and also because the occultation may quite well occur at the bottom of a valley or at the top of a peak, the O − C difference may sometimes attain 1 second of arc and more, as may be seen from the examples of reductions made by ILOC (the International Lunar Occultation Centre, Tokyo) that are shown in Table 9.2. [The full reduction from ILOC actually shows more information than included in this Table. – Trans.]

9.3.4 *Grazing occultations*

These events are rather more complicated than ordinary occultations, because it is necessary not only to have several observers at different observing sites, but also to be prepared to move elsewhere – sometimes several hundred kilometres – if the event is of sufficient importance. Shifting the observing site is essential, because grazes are visible from just a very narrow band on the Earth's surface. For every occultation there is a wide zone where the star simply disappears and reappears, and two zones from which one sees a close appulse, with the Moon passing north or south of the star. It is at the border between these zones (of occultation and appulse) that a star appears to graze the northern or southern polar regions of the Moon.

Because the apparent relief of the lunar surface may vary, according to the altitude of the Moon above the horizon, and the extent of its libration, the width of this band of the Earth's surface may vary between a few hundred metres and about 8 kilometres, depending on whether the occultation occurs at the Moon's North or South Pole, respectively.

An observer in the occultation zone may therefore see the star disappear and reappear several times as it passes behind a peak or across a valley. Within the zone, several observers spaced at intervals along a line perpendicular to the central line will record different times for the same occultation event.

From the positions of the observers, the apparent motion of the Moon, and the observed times, a profile of the polar regions of the Moon may be determined over an arc of several degrees. Because the velocity of the Moon with respect to the stars is approximately 0.5 arc-second per second of time, which is equivalent to about

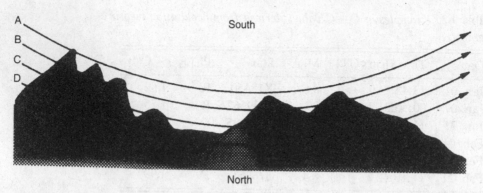

Fig. 9.13. *A fictitious example of a grazing occultation at the southern limb as seen by four observers. The vertical axis represents the lunar relief, without taking the curvature of the lunar surface into account. The apparent paths of the star therefore appear curved.*

1 kilometre on the Moon, timings to an accuracy of one-tenth of a second will resolve relief features on the Moon to about 100 m.

Figure 9.13 shows a fictitious example of the apparent motion of a star behind the limb of the Moon in the south polar region for four observers on the Earth. In this example, observer A, the southernmost in the line, observes a 'miss', or an appulse, i.e., the star just fails to touch the Moon. At B, five fairly short disappearances and reappearances are observed (ten events in all); at C there are three disappearances and three reappearances, and finally the observer at D, the northernmost, sees a single long disappearance – a total occultation.

However, before we get that far there is a large amount of preparation required, which we will now discuss in more detail.

9.3.4.1 Preparation
Each year various individuals or groups – the latter including the International Lunar Occultation Centre (ILOC) in Tokyo, the Computing Section of the British Astronomical Association, and the International Occultation Timing Association (IOTA) – calculate the places where grazing occultations will be visible. These are then plotted on maps of the areas covered (e.g. Fig. 9.14) and help with initial preparations for these events. Such maps appear (usually yearly) in various publications, such as the *Handbook* of the BAA and the *Observer's Handbook* of the RASC, *Sky & Telescope*, and elsewhere.

Generally an experienced, individual amateur or observational group undertakes the organization and planning of grazing-occultation observations. They try to obtain predictions a year in advance and then provide all the information required to prepare for the observing trips.

From the longitude and latitude data given in the predictions (Fig. 9.15) the central line of the event is plotted on a map of the region. This enables the most appropriate area for the observing sites to be determined. The central line is then

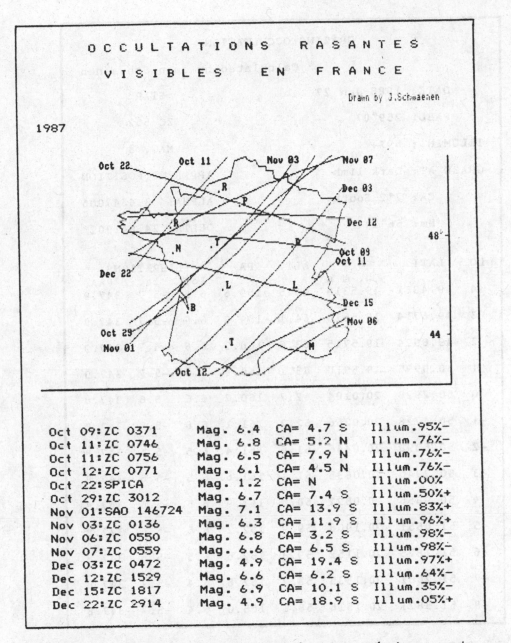

OCCULTATIONS RASANTES

VISIBLES EN FRANCE

Drawn by J.Schwaenen

1987

Oct 22   Oct 11   Nov 03   Nov 07

Dec 03

Dec 12

Oct 09
Oct 11

Dec 22

Dec 15

Nov 06

Oct 29
Nov 01

Oct 12

48

44

Oct 09: ZC 0371	Mag. 6.4	CA= 4.7 S	Illum.95%-
Oct 11: ZC 0746	Mag. 6.8	CA= 5.2 N	Illum.76%-
Oct 11: ZC 0756	Mag. 6.5	CA= 7.9 N	Illum.76%-
Oct 12: ZC 0771	Mag. 6.1	CA= 4.5 N	Illum.76%-
Oct 22: SPICA	Mag. 1.2	CA= N	Illum.00%
Oct 29: ZC 3012	Mag. 6.7	CA= 7.4 S	Illum.50%+
Nov 01: SAO 146724	Mag. 7.1	CA= 13.9 S	Illum.83%+
Nov 03: ZC 0136	Mag. 6.3	CA= 11.9 S	Illum.96%+
Nov 06: ZC 0550	Mag. 6.8	CA= 3.2 S	Illum.98%-
Nov 07: ZC 0559	Mag. 6.6	CA= 6.5 S	Illum.98%-
Dec 03: ZC 0472	Mag. 4.9	CA= 19.4 S	Illum.97%+
Dec 12: ZC 1529	Mag. 6.6	CA= 6.2 S	Illum.64%-
Dec 15: ZC 1817	Mag. 6.9	CA= 10.1 S	Illum.35%-
Dec 22: ZC 2914	Mag. 4.9	CA= 18.9 S	Illum.05%+

Fig. 9.14. *A map showing the visibility limits of grazing occultations occurring over France between 1987 October and December.*

plotted more accurately on a larger-scale (1:25 000) map of the area, using three predicted positions and a quadratic interpolation formula.

Once this more accurate line has been drawn, the values for the lunar profile may be taken from *The Marginal Zone of the Moon* by C.B. Watts. The northern and

```
                    GRAZING OCCULTATION

                        Calculated by Jean Schwaenen

        DATE: 1988 Jan 27                   STAR

        PABL: 259°07                      ZC 552

     ILLUMIN.: 69%+                        MAG. 3

     GRAZE AT: Dark limb             APPARENT POSITION

          CA: 7°2 South              ALPHA:  3.4647006

          Hm: 66°14                  DELTA: 24.041902


 LO    LATI.      U.T      ALT.    PA        LIBRATIONS
                  h. m s                  lo.    la.    PAA
   4   49.4361   19.5319   64.5   159.6    6.6   -5.7   347.9

   3   49.6714   19.5518   64.1   159.9    6.6   -5.7   347.9

   2   49.8924   19.5715   63.6   160.2    6.6   -5.7   347.9

   1   50.0995   19.5910   63.1   160.4    6.6   -5.7   347.9

   0   50.2928   20.0104   62.7   160.7    6.6   -5.6   347.8

  -1   50.4728   20.0256   62.2   161.1    6.6   -5.6   347.8

  -2   50.6398   20.0446   61.6   161.4    6.5   -5.6   347.8

  -3   50.7940   20.0635   61.1   161.6    6.5   -5.6   347.8

  -4   50.9356   20.0822   60.5   161.9    6.5   -5.6   347.8

  -5   51.0651   20.1008   59.9   162.2    6.5   -5.6   347.8

  -6   51.1825   20.1152   59.3   162.5    6.5   -5.6   347.8

  -7   51.2881   20.1335   58.7   162.8    6.4   -5.7   347.8

  -8   51.3821   20.1516   58.1   163.0    6.4   -5.7   347.8
```

Fig. 9.15. *An example of predictions for a grazing occultation as calculated by J. Schwaenen.*

southern limits on either side of the central line may be determined; these may be a few kilometres away from the central line (Fig. 9.16).

A final correction for the altitude of the observing sites has to be made. The standard predictions give data for a line L_0 at sea level, but the actual central line

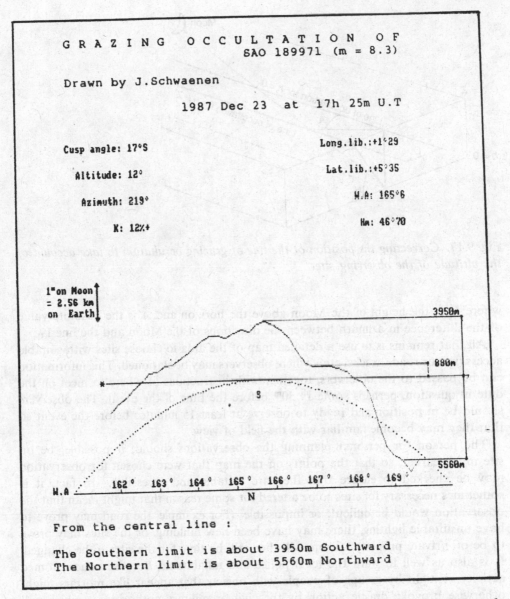

Fig. 9.16. *Profile of the area of the Moon (from* The Marginal Zone of the Moon *by C. B. Watts), where the star will be occulted.*

L_Z for a site at altitude Z will be given by the intersection of a horizontal plane (at the altitude Z) and another plane containing the Moon (Fig. 9.17). (Such a line will generally lie farther south in the Northern Hemisphere.) The distance between the predicted and actual central lines will be given by:

$$D = Z \times \cot(H) \times \sin(A)$$

517

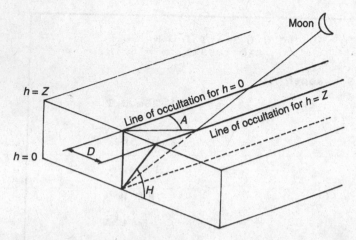

Fig. 9.17. *Correcting the position of the line of grazing occultation to take account of the altitude of the observing site.*

where H is the height of the Moon above the horizon and A is the absolute value of the difference in azimuth between the directions of the Moon and the line L_0.

All that remains is to use a detailed map of the area to choose sites with suitable access (minor roads, tracks, etc.), where observers may be stationed. This information can be passed to the observers, together with instructions of where to meet on the date in question, perhaps some $1^h 30^m$ before the time of the event. The observers should be in position and ready to observe at least 15 minutes before the event so that they may become familiar with the field of view.

The person (or persons) planning the observations should, if possible, be on site much earlier, so that the points on the map that were chosen for observation may be checked to ensure that they are suitable and accessible. In fact, it is sometimes necessary for sites to be altered for some reason that might mean that the observation would be difficult or impossible. (For example, the road may prove to have unsuitable lighting, there may have been new building, or the sites may prove to be on private property.) There will then still be time for the sites to be changed. It is also as well to warn the local police and inhabitants, because unaccustomed activity at night by groups of people with objects that appear like mortars might otherwise provoke drastic action by the civil or military authorities – which will certainly interfere with observations. It is also as well to check with any other local observing groups to ensure that they are not planning to use the same sites. Failure to do so has, on occasion, led to heated words being exchanged between over-enthusiastic observers!

Other factors that should be borne in mind are the size of the instruments to be used and the experience of the observers. These should be used to best advantage. For example the most experienced observers and the largest instruments should be placed where most events are likely to occur. In general, for an occultation at the northern limb, the observers should be spaced about 40–50 m apart over several hundred metres, whereas for an occultation at the southern limb the spacing should be every 100–200 m, but with the line extending over several kilometres. These values strongly depend on the amplitude of the lunar surface relief (Fig. 9.16).

Naturally the number of stations and their spacing depend on the number of observers. They should be located, whenever possible, near suitable reference points (geodetic survey points, minor cross-roads, etc.) that are easily located on the map, so that their exact positions may be determined later.

9.3.4.2 Observations

When the first disappearance is imminent, and throughout the observation, it is important to bear certain points in mind:

(i) identify the minute at the beginning of the recording (for example, after the long, or double, tone of the minute marker, say 'that was the twenty-fourth minute');

(ii) concentrate on what you are doing, and, above all, do not take your eye from the eyepiece throughout the event, because a reappearance may occur just a few seconds, or even a few-tenths of a second, after a disappearance, and vice versa;

(iii) try to react calmly but rapidly at the various events;

(iv) do not add long comments, keep them to a minimum, such as 'three-tenths late', 'gradual disappearance', etc.

Once back home, each observer should analyse their own observations (*see* Sect. 9.2.3). When this has been done, the measurements may be sent to the coordinating person or group, who should compare the different data to check for any possible errors. Finally the appropriate report forms can be completed (*see* Sect. 9.6) for forwarding to the main organisation that collects and analyses the data (*see* Sect. 9.4.7).

9.3.4.3 Reducing a grazing occultation

Anyone keen on mathematics can easily take all the results of a grazing occultation and plot the points where the star encountered the limb of the Moon (both disappearances and reappearances), and obtain a section of the limb profile in that region of the Moon. This may be compared with the predicted profile as will be shown in the following example.

Reduction of an occultation of α Tau A Franco–Belgian group organised a highly successful observation of the occultation of Aldebaran on 1978 April 11. This was observed from La Ferté-Saint Aubin near Orléans in France. No less than 55 events were recorded by eight observers. These events are tabulated in Table 9.3, which gives, for each observer, their position, the nature of the events observed (D = disappearance, R = reappearance, F = flash), the times of the events in UT, the position angle (PA), the Watts angle (WA), and the height, in seconds of arc, of the points of contact above the mean limb. Remember that the Watts angle is the angle at the centre of the Moon between the North Pole and the point concerned, measured eastward.

Table 9.3. *Observations of the occultation of 1978 April 11*

Stations	Ph.	Time (h m s)	PA (°)	WA (°)	Ht (″)
Long. −1° 57′ 18.″1	D	19 17 57.8	174.01	183.29	2.17
Lat. +47° 44′ 48.″4	R	19 22 25.8	181.36	190.64	0.80
Alt. 210 m					
Long. −1° 57′ 18.″1	D	19 17 59.5	174.05	183.33	2.18
Lat. +47° 44′ 43.″1	R	19 19 27.5	176.46	185.74	0.07
Alt. 210 m	D	19 19 29.7	176.52	185.80	0.04
	R	19 19 34.0	176.63	185.91	-0.02
	D	19 19 34.7	176.65	185.93	-0.03
	R	19 22 23.4	181.29	190.57	0.81
Long. −1° 57′ 17.″4	D	19 18 00.3	174.07	183.35	2.18
Lat. +47° 44′ 41.″9	R	19 19 27.3	176.45	185.73	0.09
Alt. 210 m	D	19 19 30.1	176.53	185.81	0.05
	R	19 19 33.8	176.63	185.91	0.00
	D	19 19 35.2	176.67	185.95	-0.01
	R	19 22 22.3	181.26	190.54	0.80
Long. −1° 57′ 36.″0	D	19 18 04.7	174.17	183.45	2.27
Lat. +47° 44′ 20.″9	R	19 18 11.1	174.35	183.63	2.07
Alt. 210 m	D	19 18 16.7	174.50	183.78	1.90
	R	19 18 39.2	175.11	184.39	1.28
	D	19 18 56.8	175.59	184.87	0.87
	R	19 19 13.5	176.05	185.33	0.55
	D	19 19 39.5	176.76	186.04	0.16
Long. −1° 57′ 36.″1	D	19 18 04.8	174.17	183.45	2.28
Lat. +47° 44′ 20.″2	R	19 18 10.7	174.34	183.62	2.09
Alt. 210 m	D	19 18 17.7	174.53	183.81	1.88
	R	19 18 39.3	175.12	184.40	1.29
	D	19 18 57.5	175.61	184.89	0.87
	R	19 19 13.3	176.05	185.33	0.56
	D	19 19 39.5	176.76	186.04	0.16
	R	19 22 22.3	181.24	190.52	1.01
Long. −1° 57′ 36.″2	D	19 18 05.1	174.18	183.46	2.28
Lat. +47° 44′ 19.″5	R	19 18 10.8	174.34	183.62	2.10
Alt. 210 m	D	10 18 17.9	174.53	183.81	1.88
	R	19 18 39.3	175.12	184.40	1.30
	D	19 18 57.8	175.62	184.90	0.87
	F	19 18 59.5	175.67	184.95	0.83
	R	19 19 13.4	176.05	185.33	0.57
	D	19 19 39.7	176.77	186.05	0.17
	R	19 22 25.0	181.31	190.59	1.08
Long. −1° 57′ 37.″2	D	19 18 05.8	174.20	183.48	2.32
Lat. +47° 44′ 14.″3	R	19 18 07.9	174.25	183.53	2.25
Alt. 210 m	D	19 18 18.6	174.55	183.83	1.93
	R	19 18 37.9	175.07	184.35	1.40
	D	19 18 59.7	175.67	184.95	0.90
	R	19 19 10.8	175.97	185.25	0.68
	D	19 19 40.5	176.79	186.07	0.23
	R	19 22 21.3	181.21	190.49	1.05
Long. −1° 57′ 37.″3	D	19 18 06.0	174.20	183.48	2.32
Lat. +47° 44′ 13.″6	R	19 18 07.7	174.25	183.53	2.27
Alt. 210 m	D	19 18 19.3	174.57	183.85	1.92
	R	19 18 38.6	175.09	184.37	1.39
	D	19 19 00.1	175.68	184.96	0.90
	R	19 19 10.9	175.98	185.26	0.69
	D	19 19 40.1	176.78	186.06	0.24
	R	19 22 20.2	181.18	190.46	1.04

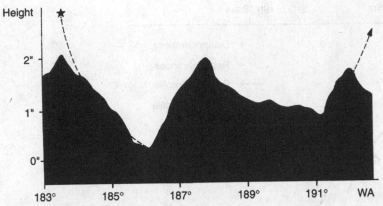

Fig. 9.18. *Predicted profile and apparent path of Aldebaran on 1978 Apr. 4 as seen from one of the stations.*

Figure 9.18 shows the predicted lunar profile from Watts' *The Marginal Zone of the Moon*, together with the apparent path of the star as seen from a single point. The vertical scale is exaggerated in this diagram: 1 arc-second represents 1927 m on the Moon and 3286 m on the Earth's surface. On the horizontal scale 1° is equivalent to 30 334 m.

Figure 9.19 shows, in greater detail, the observed profile compared with the predicted one. We can see that the two profiles are similar in shape, but that the actual profile is slightly lower than that predicted.

The enlargement in Fig. 9.20 shows the discovery of a small valley thanks to the observation of a 'flash'. From this observation, assuming that the duration of the flash was only about 0.5 second, we may deduce that the projected width of the valley is about 420 m, and that its depth is less than 12 m, because the neighbouring stations saw nothing.

9.3.4.4 Conclusions

Jean Meeus was the first to calculate and successfully observe a grazing occultation on 1959 November 20, near Louvain in Belgium. Inspired by him, Belgian observers subsequently organised and attempted 135 different occultation trips between 1959 and 1986. Among these only 34 produced results, although some were exceptionally successful, including those on 1978 April 11 near Orléans in France, on 1981 February 12 in the Netherlands, and on 1984 April 21 near Toulouse (the latter being organised by French observers). Similar successes have been obtained by observing groups in many other countries.

As we can see from this example, failures have been more frequent than success, because, unfortunately, we have no control over the weather. But the real scientific value of these observations means that it is worthwhile persevering and trying to organize groups of observers to travel to observe these disappearances and reappearances, which, when all is said and done, do have an austere beauty.

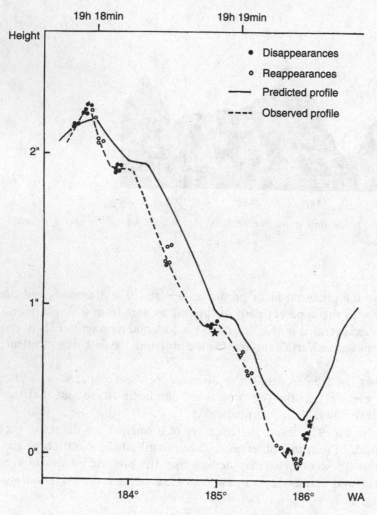

Fig. 9.19. *A comparison of the observed profile (broken line) with the predicted profile (continuous line). The region marked ★ is enlarged in Fig. 9.20.*

9.4 Occultations of stars by minor planets

In 1952, Gordon Taylor, an astronomer at the Royal Greenwich Observatory, and Director of the BAA's Computing Section, undertook the prediction of occultations of minor planets. Initially, the ephemerides of only four minor planets – 1 Ceres, 2 Pallas, 3 Juno and 4 Vesta – were sufficiently accurately known for their positions to be compared with the Yale catalogue. The result of this study was rather meagre: there were very few truly observable events in a year. Although it seems that this type of occultation had been observed accidentally earlier, it was 1958 February 19 before such a predicted event was accurately recorded. That night Per Ake Bjorklund and Svend Aage Müller, near Malmö in Sweden, observed the occultation of the star BD +6° 808 by 3 Juno. Three years later, a second success came with the

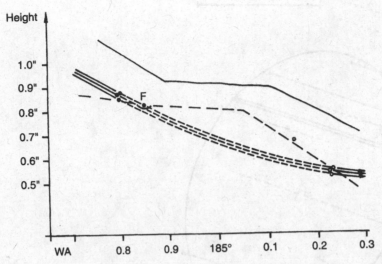

Fig. 9.20. *The discovery of a small valley, thanks to a flash recorded by the observer at station 6.*

occultation of BD −5° 5863 by 2 Pallas, observed from Naini Tal in India with a photoelectric photometer. It was not until the beginning of 1975 that several stations were able to observe the same event simultaneously (433 Eros and κ Geminorum), the subsequent analysis of results providing interesting information. As yet the best-observed occultation by far was the one organized in the U.S.A. on 1983 May 28: 130 stations recorded the occultation of 1 Vulpeculae by 2 Pallas. In Europe, the event of 1983 January 3 when AGK 3 +25° 0989 was occulted by 106 Dione, was observed by 10 stations in the Netherlands, Denmark, and the Federal German Republic (Fig. 9.21). To date, only about 20 of these events have been sufficiently well observed to add to our knowledge of these still-enigmatic objects.

9.4.1 Current and future predictions

Towards the end of the 1970s, improvements in methods of calculation and in our general knowledge of the ephemerides of minor planets made it possible to increase the number of predictions, by including both more minor planets and more stars. G. Taylor in the U.K., L. H. Wasserman and D. W. Dunham in the U.S.A., and E. Goffin in Belgium were primarily responsible for this. On the other hand, the accuracy of these predictions has hardly increased at all: the error in locating the occultation zone is still very large. The average error in the position of a star may be taken as about 0.5 arc-second, but it is very difficult to obtain comparable accuracy in the positions of minor planets, which are subject to a large number of complex perturbations. Quite frequently, their positions are subject to errors of between one and several seconds of arc. The average apparent diameter of a minor planet 150 km across is 0.1 arc-second, so we can understand the considerable uncertainty involved in the new 'sport' of chasing occultations of stars by minor planets.

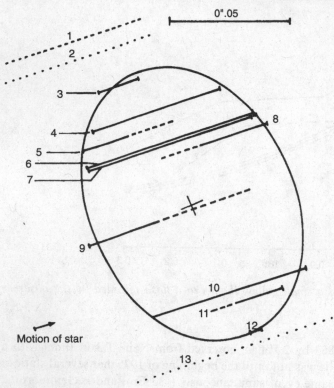

Fig. 9.21. *The results of the occultation of AGK3 +25° 0989 by the minor planet 106 Dione. The continuous lines indicate the duration of the occultation recorded at each of the station. The broken and dotted lines correspond to incomplete or negative (no occultation) observations respectively. (After L. K. Kristensen, 1984.)*

It is possible to reduce this uncertainty a few days before the event by taking, with astrometric telescopes, plates that simultaneously record both the star and the minor planet. The positions may then be measured and compared with the predictions. However, this practice does not allow the uncertainty to be reduced below about 0.3 arc-second. The Bordeaux Observatory uses a completely different technique to improve the data. This involves employing a transit instrument fitted with a suitable photometer, which enables calibrated measurements to be made over an extended period of time. The error is then reduced to 0.1 arc-second or even less. Once the area of visibility of the occultation is accurately known, coordinators are able to warn by telephone those observers in the network who are most favourably placed for observation (*see* Sect. 9.4.7).

From the data provided (we hope!) by the European Hipparcos satellite in the early 1990s, we may reasonably expect better positions for stars and better ephemerides for the minor planets. It is by no means impossible that by the end of the century it will be possible to organize special trips, similar to those made for grazing occultations. Such observations should enable the characteristics of the minor planets to be determined accurately.

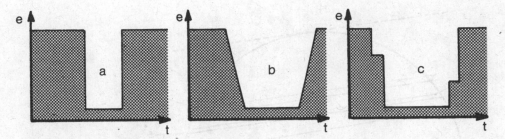

Fig. 9.22. *These three curves represent the three types of decline seen in occultations (the ordinate is brightness and the abscissa time): a) normal; b) gradual: slow minor planet and large-diameter star (or both); and c) double star.*

9.4.2 A description of an event

Before the event, when the minor planet is approaching the star, the motion is generally apparent over a period of a few minutes, which enables the observer to check that the correct star has been located. A little before the event, the minor-planet/star pair becomes difficult to resolve. As the two objects apparently fuse, the total brightness increases. An occultation is indicated by a sudden drop in brightness (the minor planet only being visible), and then, after a few seconds, the overall brightness suddenly returns to maximum. The decline in brightness is all the greater, the fainter the minor planet (*see* Sect. 9.5.1). It is considered possible to observe a drop of 0.5 magnitude under normal conditions (star neither too faint nor too bright, low atmospheric turbulence).

Unfortunately, the observer is most likely to see the two bodies approach one another, fuse, and then separate again, with no actual occultation. Such an appulse is still of interest, however. If an occultation occurs, the event may sometimes take a somewhat unusual form. With a slow minor planet (velocity less than about 10 arc-seconds per hour) and a large-diameter star (over 0.001 arc-second), or both, the expected decline in brightness may be less abrupt and may even last a few seconds in extreme cases. The object being observed appears to shrink, like a toy balloon. If the star is a very close double, the occultation may appear to take place in two steps (Fig. 9.22).

9.4.3 Scientific interest

It is obvious that the observer's task is to measure, as accurately as possible, the times of disappearance and reappearance of the star. The accuracy should be close to 0.1 second, and should not exceed 0.5 second (*see* Sect. 9.2.3.2).

Two or three good observations may provide a lot of information. For example:

- the diameter and the shape of the minor planet, from the duration (the projected chords) of the various occultations (see Fig. 9.23);
- the accuracy of the ephemeris of the minor planet (even with just an appulse);

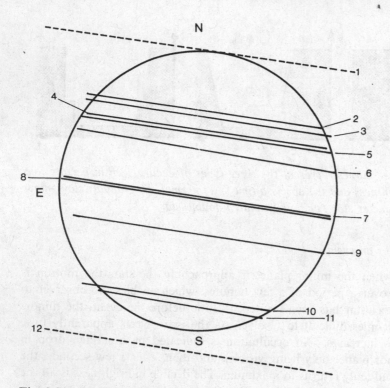

Fig. 9.23. *On 1982 November 22, the occultation of a star by minor planet 93 Minerva allowed the diameter of the latter to be established (170.8 ± 1.4 km). Most of the observations were made in the United States, but others came from Europe. Note that the negative observations at 1 and 12 enabled the number of solutions for Minerva's shape to be reduced (after R. L. Millis et al., 1985).*

- the position and the diameter of the star.

Under certain conditions, and even though a photoelectric observation is obviously better, it is possible to deduce the apparent diameter of stars to within 0.001 arc-second, which is equal to the accuracy given by good speckle-interferometric measurements. In the same way, we may add:

- detection of close double stars;
- the presence of any possible satellites to the minor planet. Although the existence of such companion bodies is still in some doubt, several observations appear to confirm their presence (Fig. 9.24).

9.4.4 Observational equipment

Stars for which predictions are available are generally brighter than 10th magnitude. So that observation will be reasonably easy, it is a good idea to use a telescope that will reach magnitude 11–12, if one wants to follow all the events predicted. Obviously

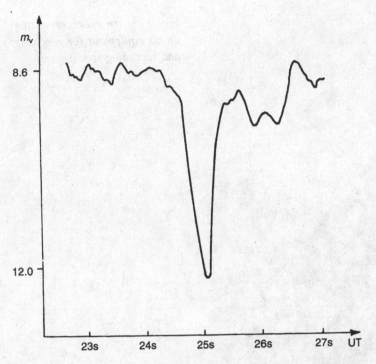

Fig. 9.24. *The light-curve reproduced here has been derived from a video tape recorded on 1982 April 18 at Meudon Observatory. The horizontal scale represents seconds after 20:23 UT, and the vertical scale the brightness. The brief drop in the star's brightness (lasting only 0.6 s), was caused by an object that might well be a satellite of the minor planet 146 Lucina. The occultation caused by the minor planet itself was observed around 1000 km farther south (after J. Arlot et al., 1985).*

a simple pair of binoculars may suffice for events down to about magnitude 8. It is not necessary for the minor planet to be visible, but certainly the star must be, because it is its light that will be hidden. If your mounting has a drive, monitoring, which generally lasts twenty-odd minutes, will be much easier. It is also very useful to have a properly aligned finder with sufficient aperture to reach magnitude 7 easily. Finally, a low-power eyepiece will help you to find the star, and will also allow you to monitor the brightness of stars close to the target star (see the next section).

9.4.5 Observational techniques

A good observation is very valuable, but it is vital for it to be confirmed by another independent station some tens to hundreds of metres away. It is therefore quite a good idea for observers to set up double stations. Isolated individuals should not be discouraged, their observations will retain their value, especially if the event has been particularly well-covered. It is also possible to use photographic equipment, which can play the part of a second observer, as described in the next section.

527

Fig. 9.25. *An observer setting up his equipment for receiving and recording time-signals.*

The first difficulty is to identify the star in question correctly. To ensure this, the European network, for example, prepares detailed observing charts (Fig. 9.26). If the observers are not used to this sort of exercise, it is advisable for them to identify the star several days or, at the very least, a couple of hours before the event. Correct identification of the star may be confirmed by detecting (if possible) the movement of the minor planet. Once the star has been found, the most suitable eyepiece for easy observation should be selected, while still retaining a sufficiently large field for the brightness of neighbouring stars to be monitored. If a disappearance occurs, an elementary precaution is to check to ensure that a cloud has not obscured the stars!

During the period of continuous surveillance, which may last 15–20 minutes, centred on the predicted time of the event, it is essential for the observer to be as comfortable as possible, so maximum concentration may be maintained. It is advisable not to stare at the star, because the tension that this produces is likely to cause the image to wander onto the eye's blind spot (Fig. 9.27), so that one runs the risk of thinking that short disappearances have occurred. It is preferable to scan the field slowly and continuously. This technique is not necessary with binocular observation.

The disappearance, when it occurs, always takes the observer by surprise. Pay

Table 9.4. *Choice of focal length as a function of duration of the event*

Focal length (mm)	300	500	800	1000
Diagonal of field covered with 24×36 format (degrees)	8	5	3	2.5
Time taken for star to cross the field of the camera (minutes)	32	20	12	10
Minimum duration of occultation (in seconds) required to produce a detectable trace 100 µm long	4.6	2.9	1.7	1.4

attention to your reaction time (*see* Sect. 9.2.2): a second is soon gone! It would be useful to make experiments on one's reaction time under various conditions, and to try to restrict the value to less than 1 second. If an occultation is recorded, it is still useful to continue observing for a while, because a second event may occur – one never knows.

As the reader will have realised, absolute honesty and a rigorously scientific attitude are essential.

9.4.6 Photographic assistance

Fit a camera body with a telephoto lens with a focal length of at least 500 mm, and mount it firmly on a tripod. Use a fast film. Place the star at one edge of the field, making allowance for its apparent motion during the following period of time. Because of the rotation of the Earth, the image of a star shifts several minutes of arc per minute of time, as given by $15 \times \cos \delta$, where δ is the declination of the star. An indication of the amount is given in Table 9.4, which is calculated for a declination of $0°$. For another star, divide the values in the last two lines by $\cos \delta$.

This technique is only capable of confirming the authenticity of an event observed visually. So far, only one photograph of this sort has been obtained, and that was by Paul D. Maley (Fig. 9.28).

9.4.7 Organisations handling observations

Whether the observation was positive (an occultation) or negative (an appulse), observers should carefully reduce their observations and accurately determine their geographical positions (*see* Sect. 9.2.3), fill in the appropriate forms and forward them quickly. (An example of a report form and an explanation of how it is completed are given in Sect. 9.6.)

A negative observation may be very important and may fix the limit of the zone of occultation. This was true for stations 1, 2, and 13 in Fig. 9.21, and for stations 1 and 12 in Fig. 9.23.

There are various observational groups that coordinate these observations in different parts of the world. Certain European observations are coordinated by European Asteroidal Occultations Network (EAON), for example, which is based in Belgium; by both the Asteroids and Remote Planets Section and the Lunar Section of the British Astronomical Association; and by the European section of the International Occultation Timing Association (IOTA), the main body of which principally handles occultations occurring in North America. (*See* the Bibliography for addresses of these organisations.) [The majority of reductions of lunar occultations (and some graze predictions) are handled by the International Lunar Occultation Centre (ILOC) in Tokyo. In the absence of details about the availability of predictions with the closure of the USNO prediction service, readers should contact one of the organisations mentioned for further information. – Trans.]

9.4.8 Reducing the observations

In is not possible to describe in detail here the methods of reduction that have to be applied to results of occultations by minor planets, because they are quite involved. The main steps, however, are:

- calculation of the apparent position of the star;
- calculation of the geocentric rectangular coordinates of each observer;
- calculation of the position of the observers on the fundamental Bessel plane at the times of contact;
- determination of the parameters (velocity and position angle of the path) describing the motion of the minor planet on the basis of the chosen ephemeris;
- calculation of the chords and of departures from the ephemeris.

Fig. 9.26. *(Opposite) – An example of the chart given to each observer in the network. At the top of the sheet are details of the event: minor planet, star, date, and time of the event for the mid-point of the track (in this case in Asia), followed by the suggested period of observation for Europe (top right). On the left are given useful details of the minor planet: magnitude, diameter, apparent velocity, horizontal equatorial parallax, and the reference ephemeris used. On the right the data are those for the star: 1950.0 position, visual and photographic magnitudes, and spectrum. The last line gives details of the change in brightness and the expected duration of the event, the elongation of the star from the Sun and Moon, as well as the percentage illumination of the latter. The remainder of the sheet carries three diagrams, one of which shows the predicted path of the shadow across the Earth, the terminator of which is shown by the bold line. The other two sections help to locate the star. The first, 15° square, shows stars down to magnitude 6.5. The second shows stars down to magnitude 10, including the star to be occulted (circled), and the minor planet's path, with a cross indicating its daily position at 00:00 UT.*

156 Xanthippe — PPM 95674

1993 aug 30   2h39.5m U.T.

Minor planet :		Star :	Source cat. PPM
V. mag. = 14.74 Diam. = 126.0 km = 0.05''		α = 6h15m52.946s	δ = +21°06'07.10''
μ = 40.20''/h π = 2.52'' Ref. = MPC19472		V. mag. =	Ph. mag. = 9.20
Δm = 6.4 Max. dur. = 4.5 s		Sun : 63°	Moon : 139° , 96%

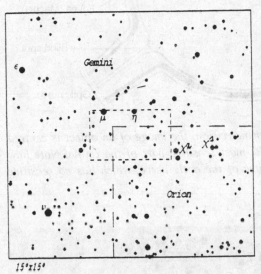

15°x15°

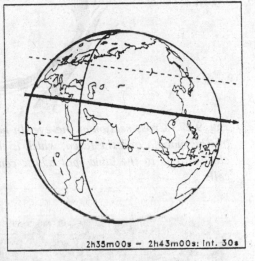

2h35m00s — 2h43m00s: Int. 30s

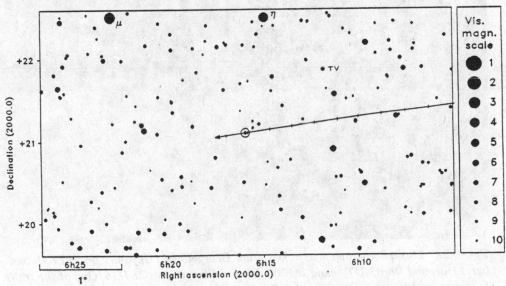

531

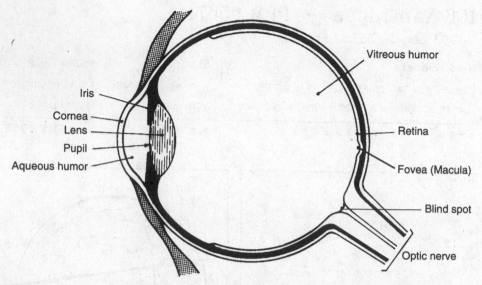

Fig. 9.27. *Cross-section of the eye. In normal vision, the image of an object is centred on the macula and the fovea, which is the most sensitive part of the retina. Note how close this is to the blind spot at the centre of the optic nerve, which has no sensitive cells.*

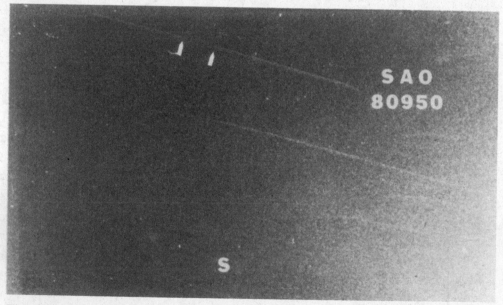

Fig. 9.28. *This photograph was taken from Georgetown (Guyana) on 1979 December 11 around 08:05 UT, using a 1000-mm focal-length lens at f/16. One of the trails is interrupted for about 27 seconds, the result of an occultation of the star SAO 80950 by 9 Metis. (Photograph by P. Maley.)*

If at least three chords are known then an examination of the shape of the minor planet may be undertaken. For calculating the chords, a direct, but low-accuracy method is given in Sect. 9.5.2.

9.5 Calculations

9.5.1 Calculating combined magnitudes

Two commonly asked questions relevant to observations of occultations of stars by minor planets are: 'What is the combined magnitude of two objects that appear as a single diffraction disk?' and: 'What drop in magnitude may be expected when a minor planet occults a star?'

The decline in magnitude is calculated as follows:

$$\Delta m = 2.5 \log(10^x + 1) \tag{9.1}$$

where

$$x = 0.4(m_2 - m_1), \tag{9.2}$$

m_1 = magnitude of the star, and m_2 = magnitude of the minor planet.

The combined magnitude (M) of the two bodies is:

$$M = m_2 - \Delta m. \tag{9.3}$$

(The derivation of this is described in detail in Sect. 8.9.2)

9.5.1.1 Examples

(i)   What was the combined magnitude when the image of 74 Galatea (mag. 11.5) merged with that of the star SAO 145609 (mag. 9.1) on 1987 September 8? Calculate the possible drop in magnitude.

From (9.2), we have:

$$x = 0.96,$$

from (9.1), we have:

$$\Delta m = 2.5,$$

and from (9.3), we obtain:

$$M = 9.0.$$

The results are rounded to the nearest tenth of a magnitude. Obviously the total drop in magnitude is applicable only if the instrument being used is able to reveal the faint minor planet. If, for example, a refractor were being used that only reached magnitude 10.0, the observable drop would be about 1 magnitude. [It may be noted that it is frequently of advantage if the minor planet cannot be detected in the instrument being used. The merging of the two images is then inconspicuous, but an actual occultation results in a complete, and very definite disappearance. – Trans.]

(ii) How much brighter than a star must a minor planet be for the drop in the combined magnitude to be detectable (assuming $\Delta m = 0.5$ mag.)?

For $\Delta m = 0.5$, from (9.1), we have:

$$\log(10^x + 1) = 0.2,$$

whence

$$x = -0.23,$$

and from (9.2), we have:

$$m_2 - m_1 = -0.23/0.4 = -0.6.$$

So the observer can detect the drop in brightness if the minor planet is 0.6 magnitude brighter than the star.

9.5.2 Calculating chords of minor planets

From the detail given on a typical occultation chart (e.g. Fig. 9.26), we may calculate the length of the chords represented by specific durations of occultation. Let:

V be the velocity of the minor planet at the predicted time in km/s,
μ this velocity expressed in arc-seconds/hr,
D the diameter of the minor planet in km, and
d the apparent diameter in arc-seconds,

whence $V = (\mu \times D)/(d \times 3600)$.

The length of a chord (C) is then given by: $C = V \times T$, where T is the duration, in seconds, of the observed occultation, allowance having been made for reaction time.

This calculation is only approximate, because the data given on typical charts are rounded values, and in addition the rotation of the Earth should be taken into account.

9.5.2.1 Example

From the data given in Fig. 9.26, what is the length of the observed chord, if the exact duration of the occultation was 8.7 seconds? We have: $C = 10.02 \times 8.7 = 87.2$.

9.6 Report forms

A typical double-sided form used for reporting observations to ILOC is shown on pages 539–540. Most of the required details are self-explanatory, but some specific items may be mentioned. [This section has been added to amplify the original French edition which reproduced the form without explanation. Details are taken from (ILOC, 1982). – Trans.]

The abbreviations used are:

Telescope types:

    R   refractor;
    N   Newtonian reflector;
    C   Cassegrain (or Schmidt–Cassegrain) reflector;
    O   Other (details should be given).

Mounting:

    E   Equatorial;
    A   Altazimuth.

Drive:

    D   Clock drive;
    M   Manual.

Longitude (accompanied by an indication of whether E or W of the Greenwich meridian), latitude (north or south), and height are self-evident, but it helps in reduction if the geodetic datum is specified, e.g., ED (European Datum), or NAD (North American Datum 1927), etc. If unknown, details of how the positions were derived should be given.

Names of the actual observers and recorders (if any) are given on the reverse of the form, together with any special remarks about specific events.

For each observation, certain specific information is required. The date is given in the standard order: year (last two digits), month, day; followed by the time in Coordinated Universal Time (UTC) *not* Local Standard Time. Only significant decimal fractions of a second should be included. (For visual work, for example, the time should be given to one-tenth of a second.)

Under 'Star Name', 'Ct' stands for 'Catalogue' and should be coded as follows, in descending order of preference:

    S   SAO Star Catalogue,
    R   Zodiacal Catalogue (ZC), by Robertson,
    X   U.S. Naval Observatory reference number,
    F   FK4,
    A   AGK3,
    D   Durchmusterung (BD, CD),
    Other   (to be specified on the reverse).

The declination zone should be stated for catalogues such as the AGK3 or Durchmusterung.

Note that care is required in giving a star number from one of the Durchmusterung listings. Stars occulted by the Moon south of declination $-22°$ always have numbers between 10 000 and 19 999. The leading digit ('1') may be omitted. For zone $-22°$ it should be stated in the comments whether the BD or CD numbers are being used. (As just mentioned, identifications from other catalogues are preferable.)

The codes for Station, Telescope, Observer and Recorder are allocated by ILOC, being given when reductions are returned to observers. Otherwise these columns are left blank. Changes in telescope siting require a new code to be allocated.

The phenomenon (Ph.) is coded as follows:

1	Disappearance at dark limb,
2	Reappearance at dark limb,
3	Disappearance at bright limb or sunlit feature,
4	Reappearance at bright limb or sunlit feature,
5	Disappearance in umbra during lunar eclipse,
6	Reappearance in the umbra during lunar eclipse,
7	Blink,
8	Flash,
9	Miss (i.e., no occultation),
Other	(details to be given in the comments).

The method of timing, MR:

P	Photoelectric,
K	Keytapping,
S	Stopwatch,
E	Eye and ear,
X	Chronograph,
T	Tape recorder,
C	Camera and clock,
V	Television,
Other	(details to be specified).

The method of timekeeping, MT:

R	Radio time signal,
C	Clock (corrected by time signal),
M	Any other medium corrected by time signal,
T	Telephone (speaking clock),
Other	(to be specified).

Personal Equation PE applied (A), chosen from:

S	The following value of personal equation has been subtracted from the observed time.
U	The following value of personal equation is known, but has not been subtracted.
E	Personal equation unknown, but thought to have been eliminated in the method of timing, so is therefore not given.
–	Not known.

The appropriate personal equation is entered in the adjacent columns (B).

The estimated accuracy of the time in seconds (as previously described on p. 500):

Very good 0.1–0.2 s,
Good 0.3–0.4 s,
Fair 0.5–0.7 s,
Poor 0.8–1 s,
Very poor ≥ 1 s (to be specified in the comments).

The value is the weight to be given to the overall observation, not the timing alone.
The certainty, Ce, is coded as:

1 Definite event;
2 Possibly spurious;
3 Probably spurious.

For photoelectric observations the signal-to-noise ratio (S/N) should be entered. The signal (S) is the change as read from the chart or from a digital recorder, and the noise (N), the amplitude of the noise with the star imaged on the photomultiplier (or other device).

Details of a component of a double star, or an unidentified star are entered in the column headed 'X', coded as follows:

W Preceding (W) component,
E Following (E) component,
N North component,
S South component,
B Brighter component,
F Fainter component,
U Unidentified star,
Other (details to be entered in comments).

With unidentified stars it is useful to give approximate values of position angle or cusp angle, and the magnitude, if known.

Sky conditions are entered in two columns, the first, stability, under 'St', and the second, transparency, under 'Tr'. Both are coded:

1 Good,
2 Fair,
3 Poor.

Unusual circumstances are entered under 'C', using the most significant of the following codes if more than one special circumstance applied:

1 Not instantaneous, gradual,
2 Dark limb visible,
3 By averted vision,
4 Star faint,
5 Through thin cloud,
6 Many clouds,
7 Strong wind,
8 In strong twilight.

Temperature in °C. If below −10°C, give in the comments. Columns 54–55 are to be left blank.

Grazing-occultation observations are to be recorded in the order in which they occurred, on consecutive lines of the form. Code 6 should be entered in the column headed 'G' to indicate that one of a sequence of graze events is concerned. In addition, for grazes, the following circumstances should be recorded in this column:

7   Failed to observe event(s) for some reason,
8   Started or resumed observing,
9   Stopped observing temporarily or finally.

The relevant times for grazing events should be entered in columns 1–15, and column 35 (Phenomena) should be left blank. All events should be given in order of their occurrence (i.e., with increasing time).

Columns 74–78 are to be left blank. If the telescope, observer, or recorder code numbers are known these are entered in columns S1–S3, respectively. If unknown, the codes a, b, or c, are to be entered corresponding to the information given at the top of the form. This information is required only once per form if the details remain the same throughout.

On the reverse side, space is available for comments. Columns 56–72 are to be left blank.

Similar information is required for occultations by minor planets, but in such a case the number and name of the minor planet should be given.

OCCULTATION OBSERVATIONS

PLACE NAME _____

ADDRESS _____

TELESCOPES and POSITIONS

	Type	Aperture	Focal length	Mounting	Driving	Longitude ° ' "	Latitude ° ' "	Height m	Geodetic datum
a	R N C₁ O( )	___ cm	___ cm	E A	D M	___._ E / ___._ W	___._ N / ___._ S	___._	_____
b	R N C₁ O( )	___ cm	___ cm	E A	D M	___._ E / ___._ W	___._ N / ___._ S	___._	_____
c	R N C₁ O( )	___ cm	___ cm	E A	D M	___._ E / ___._ W	___._ N / ___._ S	___._	_____

| 1 2 | 3 4 5 6 7 8 | 9 10 11 12 13 | 14 15 16 17 18 19 20 | 21 22 23 24 | 25 26 27 28 29 | 30 | 31 32 33 | 34 35 36 | 37 38 39 | 40 41 42 43 | 44 45 46 47 | 48 49 50 | 51 52 53 | 54 55 | 56 57 58 59 60 | 61 62 63 64 65 66 67 68 69 70 71 72 | 73 74 | 75 76 77 78 | 79 80 81 82 83 |

	Date and Time (UTC)					Star Name		Station	Tel	Obs	Rec	Ph	Mr	WT	PE		Accur	Co	S/N	X	Sky		C Temp		G	T O R
No.	Yr	Mth	Day	Hr	Min	Seconds	Ct	Decl	No.								A	B					str				
1																										
2																										
3																										
4																										
5																										
6																										
7																										
8																										
9																										
10																										
11																										
12																										
13																										
14																										
15																										
16																										
17																										
18																										
19																										
20																										

PLEASE RETURN THIS FORM TO: International Lunar Occultation Centre MORE FORMS REQUIRED?
Geodesy and Geophysics Division
Hydrographic Department   YES / NO
Tsukiji-5, Chuo-ku
Tokyo, 104 Japan

OBSERVERS and RECORDERS

a _____

b ------------------------

c ------------------------

d ------------------------

e ------------------------

f ------------------------

COMMENTS

No.	56	57	58	59	60	61	62	63	64	65	66	67	68	69	70	71	72
1																	
2																	
3																	
4																	
5																	
6																	
7																	
8																	
9																	
10																	
11																	
12																	
13																	
14																	
15																	
16																	
17																	
18																	
19																	
20																	

10 Artificial satellites

R. Futaully & A. Grycan

10.1 General

10.1.1 Historical

On 1957 October 4, the U.S.S.R. placed the first artificial Earth satellite into orbit. Every radio and television news broadcast carried the famous 'beeps' emitted by the two transmitters (at 20 and 40 MHz) that were on board SPUTNIK 1 (Fig. 10.1). This metal sphere was 580 mm in diameter, with four aerials, and had an overall weight of 83.6 kg. Its instrumentation was very simple: a temperature sensor, a meteoroid counter, a probe that measured the intensity of cosmic radiation, and another that checked the level of ultraviolet radiation.

SPUTNIK 1 had a orbit inclined at 65° to the equator, with perigee at 215 km, apogee at 939 km, and an initial period of 96 minutes. This first satellite burnt up in the atmosphere 3 months later. By then SPUTNIK 2, carrying the dog Laika, had been launched on November 3. The space age had begun.

10.1.2 The number of satellites in orbit

At the beginning of 1986 the number of objects in orbit officially reached 6000. This was divided as follows: 51 % of the objects belonged to the United States, 45 % to the Soviet Union, and the remaining 4 % to various other nations.

The radars and surveillance cameras operated by the North American Aerospace Defence Command (NORAD) have recorded 16 000 satellites since 1957, many having fallen back into the atmosphere and broken up. It is estimated that 40 000 other pieces of debris greater than one centimetre across escape detection, either because they are in too high an orbit, or because their radar signals are too weak. Some plastic- or glass-fibre-based materials are very poor reflectors of radio waves at centimetre wavelengths.

10.1.3 The classification of artificial satellites

There are two main categories: payloads and debris.

10.1.3.1 Payloads

A payload is the object that carries one or more experiments, and which is the justification for the whole enormous cost of a launch. The approximately 2800 launches in the first 30 years of the Space Age placed into orbit some 900 Soviet payloads, 500 American payloads, and 150 payloads originating in other countries, which joined somewhat belatedly in the conquest of space.

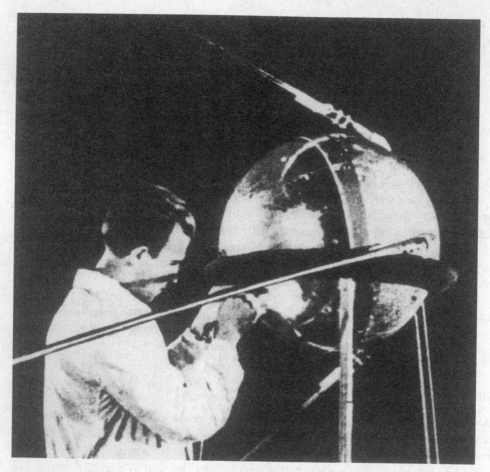

Fig. 10.1. *The first artificial satellite,* SPUTNIK *1.*

The lifetime of electronics equipment is relatively short, and only about 10 % of payloads are still functioning. They are classified according to their mission:

- scientific satellites, technological satellites (materials testing), manned space-craft;
- applications satellites for telecommunications, navigation, meteorology, and for remote-sensing;
- military satellites, for photographic reconnaissance, electronic surveillance, nautical reconnaissance, ballistic-missile early-warning, arms inspection, and the destruction of enemy vehicles in orbit ...

It should be noted that satellites that have the same task generally occupy similar orbits (i.e., they have similar inclinations, periods, etc.).

10.1.3.2 Debris

This category includes:

- Carrier rockets: the last stage of the rocket is injected into orbit along with the payload. It is subject to considerable atmospheric braking, because of its high area/mass ratio, which causes its re-entry (*see* Sect. 10.2.3).

- Support structures: these are the physical supports that carry the payload at the top of the rocket. These two elements are often designed to remain together in orbit.

- Sundry debris: this includes a varied range of items that may be jettisoned during or after the payload is injected into orbit: protective shrouds, fixings, fragments, etc.

We are witnessing pollution of space around the Earth with an increasing quantity of debris, particularly as a result of fragmentation (interception tests, explosion of carrier rockets, self-destruction of secret military satellites, etc.). When the American satellite 1965-82A suddenly exploded, radar catalogued a record 467 fragments. More recently, the ill-fated SOLWIND satellite, with a coronagraph on board which had discovered six remarkable comets, disintegrated into hundreds of fragments when an ASAT (anti-satellite) missile scored a bulls-eye on it as it orbited at an altitude of 555 km.

10.1.4 International designations

International agreements require every country to announce each satellite launch; or, to put it another way, to announce every object that at least reaches Earth orbit. COSPAR (Committee on Space Research) is responsible for giving the official designations. Up to 1962 December, the identification number included the year of launch, followed by a Greek letter indicating the order of launch (SPUTNIK 1 = 1957 α^2). Beginning in 1963, a new system was introduced, with the somewhat cumbersome Greek letters being replaced by numbers. For example, 1986-17A is the designation for the MIR space station (17th launch in 1986). The letter A refers to the payload, and B is the principal carrier rocket if there is just a single payload. When fragmentation occurs, the sequence continues after Z with AA, AB, etc. (The letters I and O are omitted to avoid confusion with the numbers 1 and 0.)

10.2 The orbits of artificial satellites

If the Earth was a perfect sphere, and if there were no other bodies (such as the Sun and the Moon) to cause any perturbations, satellites would follow perfectly regular orbits. In reality, things are completely different. The Earth is not a perfect sphere: it approximates to an ellipsoid that is flattened at the poles. In 1976 the International Astronomical Union adopted the values of 6378.140 km for the equatorial radius (R_e), and 1/298.257 for the degree of flattening (ϵ).

10.2.1 Orbital elements

Five parameters (known as orbital elements) are required to define the size, shape and orientation of an elliptical orbit, and a sixth element gives the angular position of the satellite in that orbit. The various parameters are shown in Fig. 10.2, where:

a = semi-major axis,
e = eccentricity,
i = inclination (relative to Earth's equator),
ω = argument of perigee,
Ω = right ascension of the ascending node, N
M = mean anomaly.

Note that apogee is the point farthest from the Earth, and perigee the closest point.

The Keplerian period is the time T required by the satellite to complete a full orbit of this ellipse. It is given by the equation:

$$a^3/T^2 = GM_\oplus/4\pi^2,$$

where G is the universal gravitational constant, and $M_\oplus$ is the mass of the Earth. In practice:

$$a^3/T^2 = 3.6348 \times 10^7,$$

where a is expressed in kilometres and T in minutes.

The orbital velocity is given by:

$$v = \sqrt{GM_\oplus(2/r - 1/a)},$$

where $GM_\oplus = 3.986 \times 10^5$, r is the distance in kilometres between the centres of the Earth and satellite (also called the radius vector), and v is expressed in kilometres per second.

Naturally, if the orbit is circular, we have:

$$v = \sqrt{GM_\oplus/(R_e + H)},$$

where H is the altitude above the surface of the Earth. For a circular orbit at an altitude of 200 km, the velocity is 7.8 km/s; at 36 000 km it is only 3.1 km/s.

A few additional equations allow us to derive values for some specific parameters of a satellite's orbit.

- The radius vector r_p of perigee is given by:

    $$r_p = a\,(1 - e).$$

- The radius vector r_a of apogee is given by:

    $$r_a = a\,(1 + e).$$

- The semi-major axis equals:

    $$a = (r_p + r_a)/2.$$

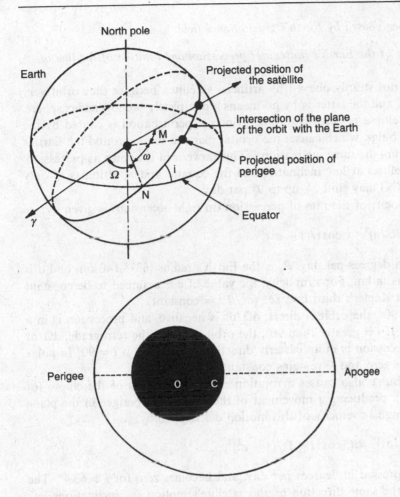

Fig. 10.2. *Orbital elements*

O = centre of the Earth

C = centre of the ellipse

The altitude of perigee or apogee is obtained by simply subtracting the local value for the radius of the Earth from the radius vector r_p or r_a.

The eccentricity may be expressed as:

$$e = 1 - r_p/a, \text{ or } e = r_a/a - 1.$$

For the space station MIR, $e = 0$, and the semi-major axis $a = 6723$ km. Using the equatorial radius given above (6378.14 km), we can deduce that the average altitude H is approximately 345 km (at the equator).

10.2.2 *Perturbations caused by Earth's gravitational field*

10.2.2.1 *The effect of the Earth's flattening: precession and rotation of the line of apsides*

Kepler's laws are not strictly obeyed by artificial satellites, because they orbit very close to the Earth, and the latter is by no means a completely homogeneous sphere. So the Keplerian ellipse is perturbed. The principal perturbation is caused by the Earth's equatorial bulge, which causes the orbital plane to rotate around the Earth's axis, while preserving the same inclination i. This movement is known as precession. In the case of satellites at low inclinations to the equator and at altitudes of only 200 km, the node (N) may shift by up to 9° per day.

The angular velocity of the rate of precession (in right ascension) is given by:

$$d\Omega/dt = -9.964°(R_e/a)^{3.5} \times \cos i/(1 - e^2)^2$$

and is expressed in degrees per day. R_e is the Earth's radius (6378.140 km) and a is the semi-major axis in km. For simplicity, the value of a is assumed to be constant in accordance with Kepler's third law, i.e., $(a^3/T^2 = \text{constant})$.

If i is less than 90°, the orbit is direct, $d\Omega/dt$ is negative, and precession is in a westerly direction. If i is greater than 90°, the orbit is said to be retrograde, $d\Omega/dt$ is positive, and precession is in an easterly direction. Obviously, if $i = 90°$ (a polar orbit), precession is zero and Ω remains constant.

The equatorial bulge also causes a rotation of the major axis of the ellipse (of the line of apsides), producing a movement of the position of perigee in the plane of the orbit. The angular velocity of this motion $d\omega$ is given by:

$$d\omega/dt = 4.982°(R_e/a)^{3.5} \times (5\cos^2 i - 1)/(1 - e^2)^2,$$

where $d\omega/dt$ is expressed in degrees per day, and becomes zero for $i = 63.4°$. The motion occurs in the same direction as the satellite's motion for inclinations less than this value, and in the opposite direction for greater inclinations. An inclination close to 63.4° is chosen when it is necessary to maintain perigee or apogee at a specific latitude in one or other hemisphere.

Again taking the space station MIR as an example, we have: $e = 0$, $i = 51.61°$, and $a = 6723$ km. From this we can deduce that $d\Omega = -5.145°$, and $d\omega = 3.845°$.

Anomalistic period and nodal period The existence of perturbations caused by the Earth's flattening leads us to define two periods that differ slightly from the Keplerian period: the anomalistic period T_a, which is the period between two successive passages of perigee, and is the value given in ephemerides; and the nodal period, T_n – the basic period of interest to observers, required for predicting transits – which is the time between two successive transits of the satellite across the equator, or across a given parallel of latitude. These times of passage are determined accurately from observations. Changes in the nodal period give us information about the various perturbations to which the satellite is subject. They allow us to determine, for example, the density of the upper atmosphere, which is closely linked to solar activity; they also allow us to predict the satellite's lifetime in orbit.

The nodal period, T_n is given by:

$$1/T_n - 1/T_a = d\omega/(1440 \times 360),$$

the periods being expressed measured in minutes. Atmospheric friction is neglected. For MIR, where $T_a = 9.144$ min., $T_n = 91.38$ min.

10.2.2.2 The geoid

The unequal distribution of mass within the Earth has been revealed by both short- and long-period perturbations of the orbital elements of satellites, enabling us to derive a gravitational model of the terrestrial globe. This differs slightly from the shape of an ellipsoid, and from place to place: it is known as the geoid. This is a very valuable tool for fundamental studies of the Earth (plate-tectonics, determining the position of gravity anomalies, etc.) and for certain military operations (refining the trajectories of ballistic missiles). A map of the geoid is shown in Fig. 10.3.

10.2.3 Other perturbations

10.2.3.1 Atmospheric friction

Above 200 km, the altitude at which most artificial satellites orbit, there is still an atmosphere, which, although very thin, does brake satellites, especially as they pass perigee. This causes apogee to decrease, so that the orbit becomes lower and lower, leading eventually to the decay of each satellite. Although the lifetime of satellites will be discussed in Sect. 10.6, we may mention here that a satellite in a circular orbit at 200–250 km will have a lifetime of a few tens of days; between 300 and 500 km, it will amount to a few years; and for a satellite between 500 and 800 km, a few tens to several hundreds of years.

10.2.3.2 The effect of solar activity

The density of the upper atmosphere and its temperature depend partly on the illumination (the day/night effect), but primarily on the degree of solar activity. At a time of major activity, the decay of satellites is accelerated by the increase in atmospheric density. Knowledge of solar activity is therefore of great importance for calculating satellite orbits, and it is regularly measured by radio from Ottawa in Canada, at a wavelength of 10.7 cm.

10.2.3.3 The effect of solar radiation pressure

As well as its effect on the tails of comets, solar radiation pressure is most noticeable with low-density satellites, particularly the balloon type, such as ECHO 1 and 2, PAGEOS, and EXPLORER 9, 19, 24, and 39. The main effect is a change in the eccentricity of the orbit, i.e., of variation in the distance at perigee (which may increase) and at apogee.

10.2.3.4 Effects caused by the Sun and the Moon

Because most satellites orbit at altitudes of less than 1500 km, the perturbations caused by the Moon and the Sun are very weak. On the other hand, with very

Fig. 10.3. *The* GRIM3-L1 *geoid. The contours show variation in height relative to the ellipsoid:* $a = 6378.140 \, km$ *and* $\epsilon = 1/298.257$ *(after C. Reigber and G. Balmino).*

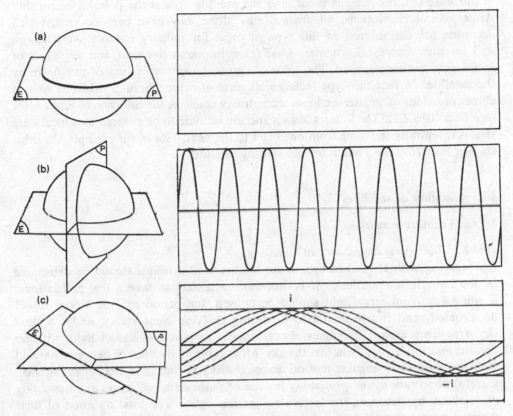

Fig. 10.4. *Orbital planes and ground tracks of some typical orbits.*

elliptical orbits, the changes in the orbit are slow and gradual. In general, we observe a slow increase in the height of perigee, as well as changes in the inclination of the orbital plane.

As luni-solar perturbations are well-known, it is easy to predict the decay of very eccentric satellites a long time in advance: the precise date of re-entry of HEOS 1 was calculated several months in advance.

10.2.4 *The various types of orbit*

Three main types of orbit may be distinguished. In Fig. 10.4, the planes E and P represent, for each of the three cases, the plane of the equator (E) and the plane of the orbit (P), respectively.

For case (a), which is that of an equatorial satellite, the two planes E and P are one and the same, so the projection of the orbit onto a plane map is a straight line. One variant of this type is the geostationary satellite, which is launched into an equatorial (or near-equatorial) orbit at a distance of 36 000 km from Earth, and which has a period of 24^h, thus appearing stationary to an observer on the ground (*see* Sect. 10.7).

For case (b), the orbit is polar, and the satellite crosses the poles at each orbit. After several revolutions, all areas of the globe may have been covered, which accounts for the interest in this type of orbit for military observation satellites, and for meteorological satellites. Case (c) is the most common, and all values of inclination are encountered. The value of i gives the extreme latitudes over-flown by the satellite. In fact this type includes all sorts of orbits: circular orbits as well as elliptical orbits of greater or lesser eccentricity, such as the Sun-synchronous orbit (*see* Sect. 10.3.2.2). The latter allows a specific latitude to be covered at a local time that is essentially the same from one orbit to the next. This is, for example, the orbit chosen for SPOT, the French remote-sensing satellite.

10.3 Visibility of satellites

10.3.1 Visibility conditions

10.3.1.1 The visual appearance of satellites
If observing with the naked eye, about ten artificial satellites should be detectable in the hour following dusk. It is, however, necessary to take a few precautions: a site away from stray light should be chosen, the period of Full Moon should be avoided, and it is essential to be sitting or lying comfortably, under a clear sky, free from clouds. Satellites always appear as point sources of light, with no colour, and move rapidly across the sky. Most orbit at altitudes of between 200 and 1000 km, and their angular motion across the sky is similar to that of an aircraft. Careful observation, and preferably the use of binoculars, will avoid any possibility of mistake, by detecting the latter's navigation lights. The total duration of their passage across the sky does not exceed a few minutes.

10.3.1.2 Satellite eclipses
Satellites are only visible because they reflect sunlight towards the observer. The only exceptions are rare geodetic satellites which have their own energy sources so that, on instruction, they can emit flashes of light.

In the evening, the Earth's shadow rises swiftly in the sky from the eastern horizon (the one opposite the Sun), eventually eclipsing all of the objects, whose progressive disappearance can be readily confirmed. Those satellites that have the lowest orbits (less than 500 km), and which are normally the brightest, are the first to be overtaken by the cone of shadow. This effect limits the useful period of observation after sunset (or before sunrise for morning observations). This duration varies according to the seasons. In winter, the period of visibility does not exceed 1 hour. The situation is distinctly better during the summer, when the shallow angle of illumination can cause the satellite to be illuminated even in the middle of the night, provided the altitude is not too low (Fig. 10.5).

10.3.1.3 The effect of inclination
The inclination i of the orbital plane with respect to the equator is critical, because it determines the regions that can be over-flown. The latitude of the point on the ground directly beneath the satellite (the sub-satellite point) is always between $+i$

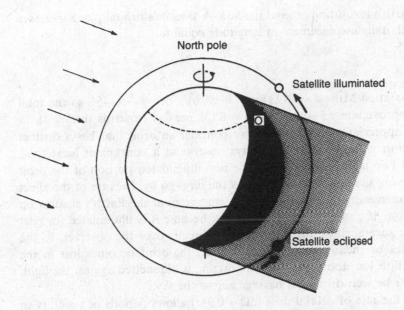

Fig. 10.5. *An indication of why visibility is particularly favourable in summer (for northern observers). The declination of the Sun is strongly positive, and the portion of the orbit above the observer (O) is illuminated even in the middle of the night.*

and $-i$. A satellite in polar orbit ($i = 90°$) is therefore observable from every point on the Earth's surface.

To obtain the maximum benefit from the Earth's rotation (0.463 km/s at the equator), satellites are launched towards the East. [We have: $\cos i = \sin Az \times \cos \phi$, where Az is the azimuth of the launch, measured from North (through East) and ϕ is the latitude of the launch site.] So the inclination of the orbit corresponds to the latitude of the launching site. For Cape Canaveral, the latitude is 28°, so many American satellites are unobservable from much of Europe, because they cross only the tropics and the equatorial zone. The higher latitudes of the Soviet launch sites (Tyuratam-Baikonur, Plesetsk and Kapustin-Yar), on the other hand, mean that the SOYUZ craft, the SALYUT space stations, and also the various objects in the COSMOS series, have excellent visibility. Launched at the rate of two per week, the latter series now accounts for about 1800 objects! An inclination that is less than the latitude of the launch site means an alteration in the trajectory during powered flight (the orbital plane is altered). This requires a costly expenditure of energy, and results in a decrease in the useful payload. This manoeuvre is only justifiable for certain specific missions (such as those that require a geostationary orbit).

10.3.2 The effects of precession and the visibility cycle

The Earth's sidereal period of rotation is $23^h56^m04^s$. The length of the mean solar day (24 hours) introduces an apparent precession of 0.9856° towards the West, caused by the Earth's revolution around the Sun. A satellite's orbital plane therefore undergoes a total, daily displacement in longitude equal to:

$$(-d\Omega + 0.9856°).$$

For the space-station MIR: $a = 6723$ km; $i = 51.61°$; $d\Omega = -5.145°$, so the total orbital drift is approximately $(-d\Omega + 0.98°) = 6.13°$ per day (towards the West).

It is easier to appreciate what occurs if we consider an orbit that has a distinct angle of inclination. Assume that we always observe at a convenient local time, between one and two hours after sunset. The non-illuminated portion of the orbit crosses the observing site earlier and earlier as the days go by, because of the effect of precession. One evening, the orbit begins to emerge from the Earth's shadow on the eastern horizon. The satellite becomes visible, because it is illuminated for part of its path. After several days, the orbit transits directly over the observer, so the satellite culminates overhead. Some ten days later, the orbit becomes lost in the twilight. The satellite, low above the western horizon, is silhouetted against the light, and can no longer be seen during its passage across the sky.

Knowledge of the rate of orbital drift $(d\Omega + 0.98°)$ allows periods of visibility to be predicted a long time in advance, provided there are no significant changes in the orbital parameters. It is possible to determine the dates that are most favourable for culminations overhead and for the orientation of the path across the sky. This is an effective way of retrieving lost satellites.

10.3.2.1 *The visibility cycle*

Depending on the arc observed, it is possible to distinguish between the ascending portion of the orbit (when the satellite is moving from South to North) and the descending portion (when it moves North to South). To find when the same part of an orbit appears under the same conditions, let us take MIR as an example. The daily motion of the orbit is 6.13°. The period required to complete a full revolution (one cycle) is: $360°/(-d\Omega + 0.98°)$, which is equal to 59 days. The orbit is then at an almost identical situation relative to the horizon (the satellite itself is at some unspecified point on its orbit). Only the conditions of illumination have altered, because of seasonal changes.

10.3.2.2 *Sun-synchronous orbits*

If $d\Omega = +0.98°$, the drift in longitude is zero. The orbit crosses the observing site at the same time each day. This very specific orbit is occupied by reconnaissance satellites, remote-sensing satellites, and meteorological satellites.

Two cases may arise. In the first the nodal period T_n is a sub-multiple of one day, for example 90 min ($= 1/16$ day) as shown in Table 10.1. The satellite crosses the observer's parallel of latitude at the same time each day. In the second case there is a slight shift in the longitude of the ground track of the orbit at the time of the satellite's transit the next day, because of the rotation of the Earth. For example, after exactly 26 days, the SPOT satellite has made 369 nodal revolutions, and follows exactly the same track at the same time. Its nodal period is 101.46 minutes.

Table 10.1. *Number of nodal revolutions per day*

Number	T_n = nodal period (m)
17	84.71*
16	90.00
15	96.00
14	102.86

* Theoretical only: the satellite would be orbiting at the surface of the Earth.

10.3.3 Some famous satellites

10.3.3.1 ECHO 1 and ECHO 2 (1960 ι[1] and 1964-04A)

Designed to be passive communications reflectors, these spheres of aluminized mylar were undoubtedly the objects most frequently observed by the general public. Their diameters were 30.5 and 41 m respectively, and this meant that ECHO 1 reached magnitude 0, and ECHO 2 magnitude −1, despite being at an altitude of approximately 1200 km. Their motion across the sky, which was particularly slow because of their considerable altitude, resulted in their being continuously visible for about 20 minutes when they crossed the zenith. These very low-density balloons re-entered the Earth's atmosphere suddenly, on 1968 May 23 for ECHO 1 and on 1969 June 7 for ECHO 2 (Figs. 10.6 and 10.7).

10.3.3.2 PAGEOS (1966-56A)

Identical to ECHO 1, this balloon satellite was launched on 1966 June 24, being placed in a polar orbit ($i = 87°$) at an average altitude of 4200 km. Its brightness was relatively faint, being about magnitude +2.5 at the zenith. The satellite crossed the sky very sedately, and the total time of visibility sometimes exceeded 50 minutes. Its mission, which was to act as a passive target for geodetic triangulation measurements, enabled the form of the Earth to be refined.

On the evening of 1975 July 12, amateurs who regularly followed this satellite were surprised to note several fragments of magnitude +6 to +10 in the immediate vicinity of PAGEOS. The balloon had started to disintegrate, either because the surface had become perished, or, more probably, because it had collided with a piece of debris not recorded by NORAD. A similar event occurred on 1976 January 20 (radar detected 44 new fragments), causing the remnants of PAGEOS to drop to the limit of naked-eye visibility.

10.3.3.3 SKYLAB (1973-27A)

Injected into a circular orbit, 435 km above the Earth on 1973 May 14, this American space station was remarkable for its size (25 m long, maximum diameter 6.6 m). For several months it was accompanied by the S II stage (10 m long, 7 m diameter) of

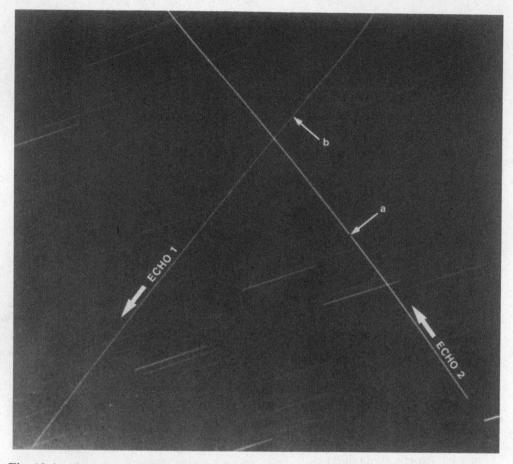

Fig. 10.6. *The satellites* ECHO *1 and 2 photographed on 1966 August 11 as they crossed Cetus in the course of their 26 408th and 12 335th orbit, respectively. The photograph was taken from the Solar Tower at Meudon Observatory, with a Focasport camera (F = 45 mm, f/2.8) using Kodak Plus-X. The breaks at a and b were made at 02:13 and 02:22 UT (photo R. Futaully).*

the Saturn V rocket that launched it. At favourable transits the brightness of these two objects rivalled that of Jupiter. The space station broke up when it re-entered the atmosphere on 1979 July 11, after having caused considerable anxiety because of its mass (75 tonnes) and the wide range of inhabited zones that it regularly crossed ($i = 50°$).

10.3.3.4 SOYUZ, SALYUT *and* MIR

Beginning in April 1971, the Soviets launched several space stations of the SALYUT type (length 15 m, with a maximum diameter of 4 m) from their cosmodrome at Baikonur. The orbit chosen, a circular one at a height of 350 km and with an inclination of 51.6°, meant that they frequently passed over much of North America

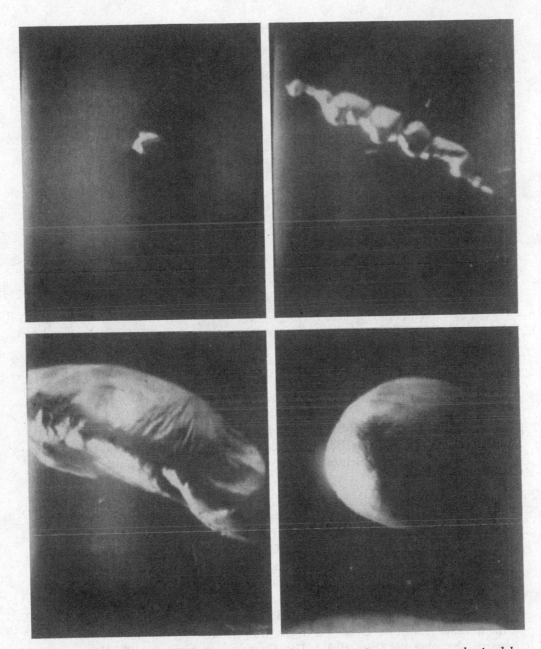

Fig. 10.7. *The inflation of the Echo 2 balloon satellite. The images were obtained by a television camera on board the last stage of the launch rocket (NASA photos).*

and Europe. SALYUT had a magnitude of 0 at culmination. It was sometimes accompanied by either a SOYUZ craft (for the crew) or by a PROGRESS craft (for freight). During their brief period of flight as individual objects, the two latter craft had a magnitude of +1 to +2.

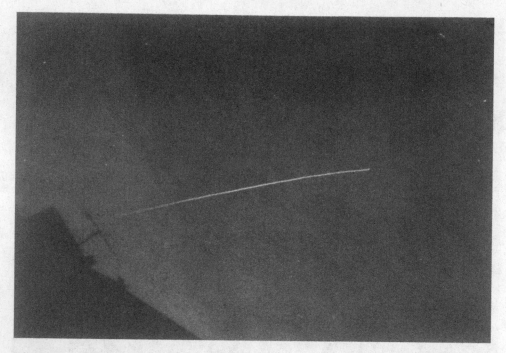

Fig. 10.8. *The space shuttle* COLUMBIA, *photographed on 1983 December 3 over Toulouse. The magnitude was estimated at* −2 *(photo C. Sanchez).*

MIR is a modified form of SALYUT, fitted with six docking ports, which can accept specialized modules. It may lead to the construction of a large orbital complex in years to come, parts of which may be served by a form of space shuttle.

10.3.3.5 SPACE SHUTTLE

The Shuttle's length of 37 m and wing-span of 24 m mean that it is a very conspicuous satellite with a magnitude of −1 or −2 at its height of 300 km. Generally launched with an orbital inclination of 28.5° from Cape Canaveral and with a typical mission lasting 8 days, the Shuttle is impossible to observe from Europe and much of North America. For this it is necessary to await the rare flights that carry the European SPACELAB module, which requires an inclination of 57°. This occurred in 1983, when the Shuttle COLUMBIA (mission STS-9) reached a magnitude of −4, during a memorable transit on December 1. Its wings were fully illuminated and turned towards the observer (Fig. 10.8). The CHALLENGER accident (1986 January 28) considerably reduced opportunities for seeing the Shuttle, because few SPACELAB flights are included in the revised launch schedule.

10.3.3.6 BIG BIRD

This satellite warrants being mentioned, because it consists of a 15-m-long cylinder, 3 m in diameter. Fitted with powerful cameras, this American, photographic reconnaissance satellite has a lifetime of 6 months, and is in a Sun-synchronous orbit

Table 10.2. *Typical magnitudes of various celestial objects*

Standard	Magnitude
Venus	−4.0
Jupiter	−2.0
Sirius	−1.6
Vega	0.0
Polaris	+2.0
Naked-eye limit	+6.0
Limit for 50-mm binoculars	+9.0

between 160 and 270 km, with an inclination of 97°. It streaks overhead, with a relatively constant magnitude of −1. A more developed version, KEYHOLE (or KH11) appeared in 1976. Two satellites of this type are maintained in orbit permanently.

10.4 Satellite photometry

By making visual magnitude estimates, comparing the brightness with those of nearby stars with known magnitudes, it is possible to establish the individual optical characteristics of each satellite. From measurements made at several transits and a brief analysis, we can deduce its size and general shape (a sphere, a cylinder, etc.). Any rotation can also be detected. The magnitudes of some specific objects that may be used to obtain a rough indication of magnitude are given in Table 10.2.

10.4.1 The magnitude of a satellite

The magnitude primarily depends on the surface area S (the cross-section), the coefficient of reflection A (the albedo), the distance, and the phase angle.

- The phase angle is the angle between the Sun and the observer as seen at the satellite. The latter is commonly observed under a phase angle that is close to 90°, which, by analogy with the Moon, would correspond to First (or Last) Quarter (Fig. 10.9). A gain of 2 magnitudes may be expected for a change in phase from $\theta = 90°$ (quarter) to $\theta = 0°$ (full), except in the special case of aluminized balloons of the ECHO type (a reflecting sphere), where the brightness is independent of the phase angle.
- The coefficient of reflection is often very high (the surface is either covered in white paint, or is metallic in nature). The value of A is generally assumed to be 0.8.

Under these conditions, an object with a surface area of 1 m² appears as magnitude +7 at a distance of 1000 km. It is therefore visible with a pair of binoculars. Taking these values, the magnitude m of a satellite is given by:

$$m = -8 - 2.5 \log (S/d^2),$$

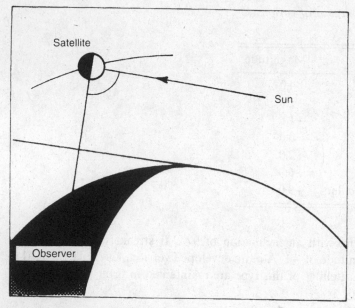

Fig. 10.9. *Phase angle*

where S is the surface area (cross-section) in m^2, and d is the distance to the observer in km. For a spherical satellite: $S = \pi \times R^2$ (where R is the radius of the satellite). If the albedo differs from the value of 0.8 that has been assumed, a correction is required. This correction = 2.5 log $(A/0.8)$.

Apart from the exceptional case of disintegration as a satellite re-enters the atmosphere, satellites are not sufficiently bright to be seen during daylight.

10.4.2 *Photometric behaviour*

A relatively constant brightness indicates that the projected area of the object as seen by the observer is more or less the same. It may be a satellite with a fairly regular shape, spherical, for example, or a satellite that is maintained in the same position (attitude stabilized). The latter applies to most payloads (particularly to space stations, and manned craft).

Factors that contribute to a progressive variation in magnitude throughout a transit are: the phase effect, the initially decreasing and then increasing distance, and by atmospheric extinction – which becomes important if the satellite's apparent distance above the horizon is less than 15°.

Periodic changes in brightness indicate that the object is probably cylindrical and that it is rotating or tumbling. This is often encountered with spent rocket stages, for example (Fig. 10.10). Once the payload has been injected into orbit, the carrier rocket, which also becomes a satellite, starts to tumble. It sometimes presents the whole of its illuminated side to the observer (maximum brightness), or sometimes just the end (minimum brightness). Once one knows the distance, it is

Fig. 10.10. *The* SPUTNIK *3 rocket (1958 δ_1), over Malo-les-Bains in France, at an altitude of 900 km, on 1958 August 11. The trail begins (right) at 22:12:36 UT and enters eclipse at 22:40:06. Altair is visible in the approximate centre of the field, with β and ν Aql on either side (photo P. Neirinck).*

easy to determine the diameter, and then the length as discussed in the previous section. It should be noted that this tumbling motion may occur in a plane that is at right-angles to the line of sight. Under these circumstances the apparent magnitude does not vary.

Some satellites only show brief flashes, and a photographic trail appears as a dotted line. These are rotating objects with sides that are facets of polished material. Sometimes a bright flash, the intensity of which is often under-estimated by the eye, suddenly increases the brightness of certain satellites that are otherwise constant in magnitude. This flash is caused by sunlight being reflected from a flat, polished area that is either metallic or vitreous in nature, and which acts as a mirror. (Generally, any bright peak in the light-curve indicates a specular reflection.) Panels covered by solar cells are obvious candidates. Magnitudes of −3 to −5 have been reported for the Soviet space station MIR when the solar panels (which span 30 m) are favourably placed.

10.4.3 The value of observations

Photometric observations lead to a rapid classification of the different objects that are put into orbit. They give information about the attitude, orientation, and motion of the satellite (such as the axis of rotation), which is sometimes important for interpreting the data returned, and is always useful for predicting reentries.

This information enables the collisional cross-section that the satellite presents to the molecules in the rarefied upper atmosphere to be calculated. This, together with the satellite's mass, is required to calculate the atmospheric density from alterations in the satellite's orbit. If the orbit is circular, the atmospheric density ρ is given by:

$$\rho = (4.8 \times 10^{-5}/a) \times (\Delta T/T) \times (m/S),$$

where ρ = density in kg/m^3, a is the semi-major axis in km, ΔT is the variation in the period, in minutes per revolution, T the period in minutes, m the mass in kg, and S the (average) collision cross-section in m^2.

Conversely, the information offers the possibility of calculating the mass of a satellite, provided its altitude is not too great.

10.4.4 Determining the photometric period

As we have seen previously, a periodic variation in magnitude indicates that the body is rotating. The photometric period is the interval between two maxima. In reality it corresponds to half the period required for one complete rotation, at least for carrier rockets (the problem becomes slightly more complicated with satellites that have reflecting, faceted sides).

A progressive increase in the photometric period is noticed as a satellite gradually enters the denser layers of the atmosphere. This is the result of aerodynamic forces decreasing the rate of tumbling. At high altitudes the decrease is primarily caused by the Earth's magnetic field, which also acts as a brake. Currents induced in the tumbling metallic object create a magnetic field that opposes the terrestrial magnetic field.

Periods that have been measured run from a few seconds to several minutes in the case of carrier rockets. Any venting or sudden rupture of a pressurized tank results in an increase in the rate of rotation. This type of event can also cause a definite change in the orbital elements.

If the photometric period is short, the time taken over 5 or 10 maxima should be determined using a watch or a stopwatch, and the resulting figure divided by the number of cycles. This increases the accuracy slightly. It should be noted that the apparent period measured during a transit does depend, to some extent, on the position of the satellite in the sky with respect to the observer.

The photographic method consists of recording the broken trail of the satellite on photographic film, which is not difficult if the satellite is bright. To measure the rotation, it is necessary to interrupt the exposure briefly (for example with the hand) at specific times, either against time signals or by using a metronome. The ideal, of course, is to use a rotating chopper in front of the objective (as with meteors). This

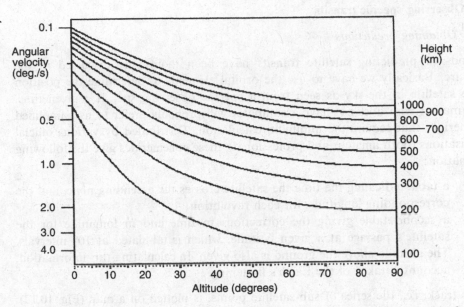

Fig. 10.11. *Angular velocity of a satellite as a function of its altitude above the horizon.*

method also provides a very accurate means of determining the angular velocity (Fig. 10.11).

10.4.5 Is it possible to detect the shape of satellites?

Amateur instruments without automatic tracking mechanisms are unsuitable for high-resolution observations of satellites. The largest objects (such as carrier rockets and space stations), are some 30 m long. Orbiting at a minimum altitude of 200 km, they speed across the sky at the high rate of 2° per minute. Tracking such objects is very difficult if the magnification is greater than 100×, which is the minimum required if there is to be any chance of seeing any actual shape. Professional astronomers use telescopes some 600 mm to 1 m in diameter, tracked by computer. A resolution of 1 arc-second may be obtained by very short photographic exposures (which are required to avoid any trailing), thanks to the use of long focal lengths. This corresponds to 1 m at a distance of 200 km. Recently, CCD cameras and very sophisticated image-processing techniques allow a resolution of less than 1 second of arc with this type of object, and are therefore able to reveal many structural details of satellites in orbit. In certain cases it is even possible to detect the progressive erosion of a satellite's paint (discoloration and darkening) after several months' exposure to ultraviolet and particle radiation.

10.5 Observing specific transits

10.5.1 Obtaining predictions

Methods of predicting satellite transits have been thoroughly discussed in the literature. Basically we have to use the orbital elements to determine the position of the satellite in the sky as seen from a specific observing site, at a given time. Any amateur who does not have a computer will probably prefer to use simplified ephemerides that require the minimum calculation. Distributed by various official organisations (both amateur and professional), these ephemerides give the following information:

- a table indicating the time the satellite crosses the ascending node, and the corresponding longitude, for each revolution;
- a second table giving the corrections in time and in longitude for the satellite's passage at a given latitude, which is tabulated at 10° intervals. The altitude above the ground is also given. In calculating this information, account is taken of the Earth's flattening.

The track, i.e., the series of sub-satellite points, is plotted on a map (Fig. 10.12). By interpolation, the time at which the satellite crosses the observer's parallel of latitude and the time of culmination are obtained. For a low-inclination orbit, it is useful to know the time of meridian transit. The height h in degrees above the horizon is given by:

$$\tan h = [\cos \phi - R/(R + H)]/ \sin \phi$$

where R is 6378 km, H is the height of the satellite (in km), and the angle ψ is measured directly from the chart (Fig. 10.13). Most astronomical books contain the necessary equations to plot the predicted position on a chart of the sky.

Over a single revolution, the ground track's shift in longitude (Fig. 10.12) is mainly caused by the rotation of the Earth, and amounts to 1/4 degree per minute of time (360° in 24 hours). If we take a typical orbital period as 90 minutes, the track is shifted by: $90 \times 1/4° = 22.5°$ towards the west. If the satellite was visible on the first transit, it will often be invisible at the second, being too distant and too low on the horizon.

Over a period of a day, it is necessary to allow for precession of the orbit (Sect. 10.3.2), which may no longer be neglected. The complete equation for the shift in the ground track per (nodal) revolution is given by:

$$\Delta L = \{1/4° + [(-d\Omega + 0.98°)/1440]\} \times T_n,$$

where T_n is expressed in minutes.

For MIR, we have: $\Delta L = 23.23°$. This shift in longitude enables the station to pass over the same points 2 days later, after 31 nodal revolutions: $31 \times 23.23°$ being approximately equal to $2 \times 360°$, enabling another SOYUZ to be launched to join the station.

In the case of the SPOT satellite, the nodal period is 101.46 minutes and the shift ΔL is 25.37°. After 369 nodal revolutions (exactly 26 days), the satellite follows an identical ground-track: the orbit is said to be in resonance with the Earth.

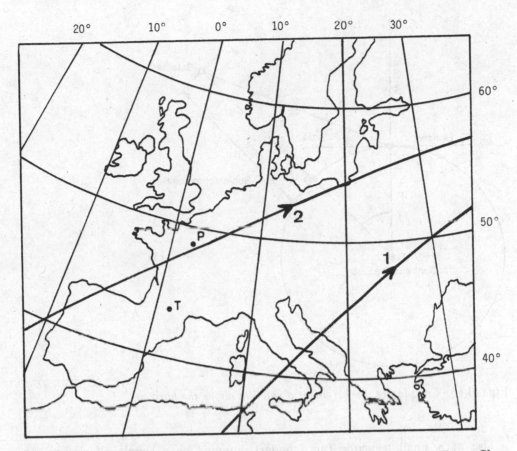

Fig. 10.12. *Ground-tracks of successive transits of a satellite (similar to a Space Shuttle).*

10.5.2 Equipment

Observation of the apparent position of a satellite consists of determining its angular coordinates (generally equatorial ones), at a given time. To do this the observer requires a certain minimum amount of equipment: a sighting instrument, a stopwatch and an atlas of the sky.

10.5.2.1 Optical equipment

Although observation with the naked eye suffices for the brightest objects with magnitudes greater than +4, a pair of binoculars is essential for fainter satellites. Binoculars also have the advantage of restricting the field of view, so the observer is less likely to be 'distracted'. Given the apparent velocity of satellites and their often complex photometric behaviour, the binoculars' field of view should be about 5–6° and the magnification about 8–10×. They may be fixed to a photographic tripod, or else the observer can recline in a garden chair, holding the binoculars with both hands. This is far more comfortable!

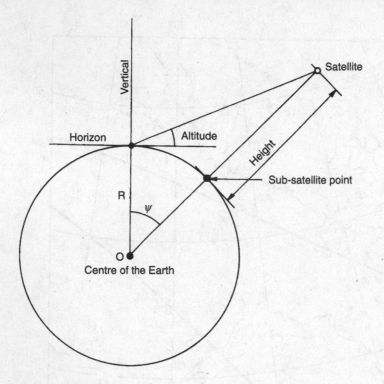

Fig. 10.13. *Calculating the height of a satellite above the horizon.*

Use of a small telescope (or a finder) mounted on a simple altazimuth (or equatorial) mount is to be avoided, because of the inverted field. Even if it is designed to give an erect image, the field needs to be at least 3° in diameter.

10.5.2.2 *Determining the time*

Nowadays, anyone can have a quartz stopwatch that has a theoretical accuracy of 1/100th of a second. Regardless of the quality of the stopwatch, it is not necessary to know its 'rate', which was not the case with old-fashioned, mechanical stopwatches.

It is essential to have access to the speaking clock (whose pips are accurate to 0.05 s) after stopping the stopwatch (*see below*), unless one has a second quartz watch, whose rate is accurately known, and which can serve as a 'standard'. [Other possibilities, of course, include radio time signals, or radio-controlled clocks. – Trans.]

10.5.2.3 *Charts and atlases*

The easiest way for amateurs to determine the position of a satellite is to plot it relative to the stars, using an atlas with the largest possible scale. Unfortunately, many of the atlases that are widely used by amateurs are unsuitable because their scales are too small.

- The *SAO Atlas* was originally produced on transparent overlays for reducing plates taken with the Baker–Nunn cameras, and is available as a set of 152 sheets. It is very accurately plotted, with a scale of $1° = 6.95$ mm. However it does have the disadvantage of not being complete down to its nominal limiting magnitude of +9.

- The atlases produced by Becvár, with a scale of $1° = 20$ mm are very accurate: *Atlas Borealis* (from the North Pole to declination +30°), *Atlas Eclipticalis* (from +30° to −30°), and *Atlas Australis* (from −30° to the South Pole) contain all stars down to magnitude +9.

- The atlases mentioned above are all drawn for epoch 1950.0. At the beginning of the 1980s, atlases using epoch 2000.0 began to appear, the first being Wil Tirion's *Sky Atlas 2000.0*. It contains about 43 000 stars down to magnitude +8 and has a scale of $1° = 7.8$ mm.

- Even more recent is *Uranometria 2000.0*, by Tirion, Rappaport and Lovi, available in two volumes covering the whole sky (Vol.I: North Pole to −6°; Vol.II +6° to South Pole). This has a limiting magnitude of +9.5 and a scale of $1° = 18.5$ mm.

10.5.3 Visual observations

Possessing ephemerides, local predictions, binoculars and an atlas, at last we can begin the joys of observation and ...of verifying the accuracy of the calculated predictions!

10.5.3.1 Techniques

After having checked the star field that the satellite will cross, as well as the direction in which it will travel, we have to wait patiently, stopwatch in hand (and set to zero). It is advisable to begin watching several minutes before the predicted time T and, sometimes, to continue several minutes afterwards.

When the satellite appears, looking like a faint, fast-moving star, it should be followed, in the hope that it will occult a star. If it does, the stopwatch should be started at the instant of the occultation. The stopwatch is then stopped at the exact moment of a subsequent time signal. A typical event occurred when a satellite occulted η Cyg on 1983 December 2. The stopwatch, which was stopped at 17:20 UT, showed 1 min 12 sec; so the time of the event was '17.20 minus $1^m12^s = 17:18:48...$.' The satellites' coordinates were those of η Cyg: $\alpha_{2000.0} = 19^h56^m18^s$ and $\delta_{2000.0} = +35°05'00''$.

There may not be any occultation, in which case an estimate is made of the distance at which the satellite transits between two stars (Fig. 10.14). The stopwatch is started when the satellite passes point S, the appropriate values of α and δ being determined by interpolation.

Apart from the equipment mentioned, it is sometimes very useful to have a small tape recorder, because it can replace a notebook. When multiple satellite transits are being recorded it is also possible to record 'clicks', each of which corresponds to a particular position. Before the tape recorder is stopped it is essential to record a time signal from the speaking clock or radio broadcast. It is then easy to replay

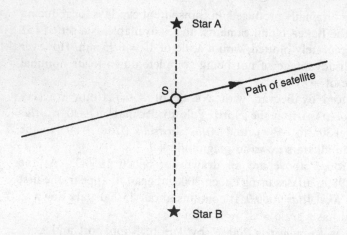

Fig. 10.14. *Measuring the position of a satellite. The satellite passes between stars A and B, and the observer estimates the distance AS – in this case two-fifths of the distance AB.*

the tape, with a stopwatch in hand, and to measure the time intervals and thus determine the various times of observation. However, this method produces a slight loss of accuracy, because of irregularities in the tape recorder's speed.

10.5.3.2 *The accuracy of the measurements*
In the case shown in Fig. 10.14, if the stars A and B are assumed to be 5° apart, it may be estimated that the errors will probably not exceed a quarter of a degree in α and δ and 0.5^s in time. The closer the check stars, the higher the accuracy. When a star is occulted it may attain 1′ in position and 0.1^s in time.

10.5.4 *Photographic observations*

It is always interesting to take photographs of a satellite's passage, but these cannot be expected to give very accurate position measurements, because of the difficulty in obtaining time references along the satellite's track. However, it is easy to determine photometric periods, which are sometimes complex and not detectable visually.

As for photography itself, ordinary, wide-field cameras may be used: a 35-mm body fitted with a 50-mm or 28-mm lens, used at maximum aperture, with a colour or black-and-white film with a minimum speed of 200 or 400 ISO. Exposures of several minutes are required, so the camera must be mounted on a tripod.

10.5.5 *Radio observations*

In 1957–8, many radio hams tuned their receivers to wavelengths of 7.5 and 15 m and picked up the coded transmissions from the first satellites directly. Nowadays the frequencies have changed: many satellites transmit in the 136–137 MHz band. In addition, some satellites (the OSCAR series) have been launched that allow radio hams to contact one another on 144 MHz. No discussion of this field would be complete without mentioning the remarkable results obtained by Geoffrey Perry and the group of pupils at Kettering Grammar School in the United Kingdom, in tracking satellites by radio.

10.6 The life and death of satellites

We should perhaps explain that here we are concerned with the 'physical' lifetime of a satellite, and not with its 'useful' or 'active' lifetime, during which it carries out measurements and transmits information.

10.6.1 *The evolution of orbits*

In Sect. 10.2 we mentioned the various perturbations that alter an orbit's position, gradually deform it, and eventually result in the satellite's destruction when it re-enters the denser layers of the atmosphere. Fig. 10.15 illustrates how an orbit decays: although normally it is elliptical at first, the orbit tends to become circular. It will be seen that the altitude of perigee is relatively stable, but that apogee decreases rapidly. (The four orbits shown are not successive, but represent four different stages of the satellite's overall lifetime.) Subsequently, perigee and apogee (which have become essentially the same) decrease together. Once the orbit has reached the 'critical' distance of around 130 km, the object – whose velocity has increased – rapidly decays into the denser layers of the atmosphere. It heats up and begins to vaporize at an altitude of about 100 km. It then disintegrates completely and the majority of fragments disappear at around a height of 60 km.

10.6.2 *Satellite lifetimes and the prediction of decay*

Although it remains difficult to calculate precisely, the lifetime of a satellite is easier to calculate nowadays than it was at the beginning of the space age, because specialists now have excellent models of the atmosphere. In addition, perturbation theories have been refined.

Let us first express the strength of the force F, opposing the motion of a satellite, that is exerted by the resistance of the medium (in this case the terrestrial atmosphere). We have:

$$F = (C_D S \rho V^2)/2$$

where:

ρ = density of the medium (the atmosphere) in kg/m^3,

S = cross-section of the object at right-angles to the direction of motion, in m^2,

V = velocity of the object with respect to the atmosphere, in m/s,

C_D = aerodynamic coefficient, which is generally 2.2, but often adjusted to fit the observed data with unknown satellites.

Estimating the number of revolutions before a satellite decays implies a prior study of atmospheric friction. It requires a full knowledge of the parameters appropriate

567

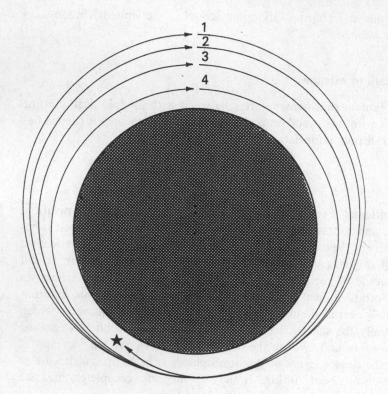

Fig. 10.15. *Evolution of an orbit through atmospheric braking*

to each satellite (C_D, S, m) and of the atmosphere (the density ρ, and the relative velocity V).

After a satellite has been injected into orbit, the variation in the orbital period is determined from observations. It is then possible to calculate the remaining lifetime by applying the equation:

$$L_R = -3/4 \cdot T/\Delta T \cdot e \cdot f(a, e, H)$$

where the parameters (C_D, S, etc.) are implicit in the term ΔT.

The lifetime L_R is obtained as the number of revolutions remaining; f is a complex function involving the scale height H, which describes the way in which the atmospheric density varies with height (*see below*). T, the period, is expressed in minutes per revolution. ΔT is a negative quantity, indicating a decrease in period.

The accuracy of the value calculated for the remaining lifetime is only about 10 % for satellites that regularly undergo atmospheric friction, because of the unpredictable variations in the density as a result of solar activity. Whilst remaining relatively simple, and easy to apply, this formula gives a good 'margin' by allowing the determination of decay several months in advance.

If the orbit is circular, the equation becomes:

$$L_J = -1/960 \cdot T^2/\Delta T \cdot H/a.$$

Table 10.3. *Scale height and density for average solar conditions** ·

Altitude (km)	Scale height (km)	Density (kg/m³)
150	18	2.0×10^{-9}
200	31	2.4×10^{-10}
250	39	5.8×10^{-11}
300	46	1.8×10^{-11}
350	50	6.3×10^{-12}
400	52	2.4×10^{-12}
450	55	9.4×10^{-13}
500	57	3.8×10^{-13}
550	60	1.6×10^{-13}
600	64	7.1×10^{-14}

* (After Jacchia)

L_J is expressed in days for greater convenience; H, the scale height, and a, the semi-major axis, are generally expressed in km.

10.6.2.1 Atmospheric density and scale height

Thanks to satellites, we have a means of measuring the atmospheric density at very different altitudes and times. It was soon realised that the density ρ varied widely, directly linked with solar activity or, more accurately, with the amount of incident ultraviolet radiation, which originates in active regions of the Sun and from the solar surface itself. This ultraviolet radiation heats the Earth's upper atmosphere with a resulting fluctuation in the density ρ with the solar rotation period of 27 days. Above all, however, there is a major variation over a period of 11 years, which corresponds to the solar cycle. The value of ρ may change by a factor of 10 from its average value (*see* Table 10.3) when the Sun is quiet (a decrease in density corresponding to solar minimum) or active (an increase in density corresponding to solar maximum). The effect is particularly marked at 600 km altitude. The scale height H varies to a lesser extent.

Naturally the density (measured at perigee), varies as a function of illumination. This is the day/night (or diurnal) effect. The arrival of the flux of particles (protons and electrons) emanating from the Sun also causes disturbances in the upper atmosphere (geomagnetic activity). Such an event causes an increase in density and therefore an increase in atmospheric friction.

Finally, a semi-annual variation in the density is observed, and this doubtless arises in the lower atmosphere. The maxima occur at the beginning of April and the end of October. The minima are in mid-January and at the end of July, the density ρ varying from a value of unity to about 2 over the course of the cycle.

It will be seen that determining the atmospheric density is a complex problem. This discussion gives some idea of the difficulties encountered in the accurate calculation of the lifetime of satellites.

10.6.3 Satellite re-entries

Accurate knowledge of the shapes, sizes and masses of satellites and the use of very sophisticated decay models enable re-entries to be predicted to a high degree of precision. Sometimes it is possible to manoeuvre the object at the end of its life using its motors – this applied to the SALYUT space station, which had a mass of 20 tonnes – or to modify its attitude, with the aim of either increasing or decreasing the atmospheric braking – we recall the case of SKYLAB in 1979 – or even to cause it to disintegrate before it re-enters. The last technique was used with COSMOS 1402, which carried a nuclear reactor, to increase the dispersion of radioactive debris. Since 1957, around 11 000 artificial satellites (payloads, carrier rockets, and various debris) have re-entered the atmosphere, but less than one hundred falls have been observed visually. It is relatively rare for any unvaporized debris to reach the ground.

10.6.3.1 Conditions and frequency of decays

The frequency of decays is tending to decrease. Nowadays, orbits are often higher than they were in the first decade of the space age, because rockets are more powerful and payloads have become far more advanced, and are therefore more valuable. On average there is one re-entry a day if one includes pieces of debris, which often do not exceed a few centimetres across.

The artificial satellite section at the Royal Aircraft Establishment at Farnborough in the United Kingdom only predicts about ten 'interesting' re-entries (of large objects) every quarter. As for the chances of observing one, these are low. It must be remembered that the first requirement for observation is that the event should take place when it is night-time for the observer. In addition, the event is short-lived: it does not last more than 2 or 3 minutes and begins at a height of less than 100 km. So observation is only possible over a very limited area of the Earth.

But if one is fortunate enough to see a re-entry, do not forget to be a useful witness, and follow the advice given – which is equally valid for a fireball, or any other unusual object. The following details should be noted:

- date, hour and place of observation;
- trajectory: appearance, disappearance (with respect to either terrestrial or stellar reference points);
- culmination: altitude and azimuth;
- duration of the observation;
- brief description of the event (both visual and audible);
- name and address of the observers.

If your observation is to be of use to those who specialize in analyzing such events, it is best to send the information to one of the regional or national observing groups who handle such data, and who will then communicate it (together with other observations) to the specialist centres, such as the BAA, Amateur Satellite Observers, etc. (*see* Sect. 10.9.2).

The magnitude attained by current carrier rockets when they re-enter may reach −6 to −10. The fireball may measure several degrees across and surround numerous incandescent fragments, and may leave behind it a luminous trail that may sometimes reach 10–20° in length.

The very varied colours recorded by observers reflect the chemical composition of the materials used in the object's construction. Maximum brightness occurs at an altitude of about 80 km, when the spectacle may be very impressive (Fig. 10.16).

10.6.3.2 Famous re-entries

The first re-entry observed was that of SPUTNIK 2, which remained attached to its carrier rocket, and ended its life over the Atlantic, near the Caribbean Islands on 1958 April 14. On 1962 September 5, several brilliant objects (magnitude −6) marked the destruction of SPUTNIK 4, but a lump of material weighing 9.5 kg was recovered from a street in Manitowoc in the United States. The first event that anyone succeeded in photographing occurred over the Indian Ocean on 1965 August 24, with the re-entry of the Gemini-5 rocket.

In France we had to wait until 1967 July 18 to see the beautiful re-entry of the COSMOS 169 rocket, which followed the line Rennes–Bourges–Geneva. A similar sight was seen on 1968 September 20, visible from Great Britain and the north of France: this time it was the re-entry of the COSMOS 253 rocket (Fig. 10.17). Less than a year later, on 1969 September 23, French and Belgian observers witnessed the destruction of COSMOS 300, which was an unsuccessful lunar probe that failed to leave Earth orbit, and which was then making its third and final revolution (Fig. 10.18).

But the most closely covered and also one of the most spectacular re-entries was undoubtedly that of SKYLAB, which had an overall weight of 75 tonnes. Excellent scientific cooperation was set up between various countries to try to eliminate the possibility of debris falling on inhabited territory. As it happened the re-entry was imperfectly covered. It occurred on 1979 July 11 at the predicted point, over the Indian Ocean and Western Australia. Among the debris recovered there was an oxygen tank that was 2 m by 1 m, and which weighed 500 kg!

10.6.4 Other phenomena that may be observed

There are numerous luminous events that anyone may witness. They may be summarized here and grouped according to their origin.

10.6.4.1 Events caused by spacecraft operations

On three occasions it was possible to see with the naked eye luminous phenomena associated with the Apollo missions. On 1968 December 21, a few hours after the launch of Apollo 8, observers discovered a 'new' star in Aquila, that expanded to reach nearly 3 degrees across. It was actually caused by the venting of the oxygen and hydrogen that remained in the tanks of the S IV B stage of the Saturn V carrier rocket.

The same type of event was seen under similar conditions on 1969 November 14 on the occasion of the Apollo 12 flight (Fig. 10.19). In both cases the clouds could be seen from Europe, at a distance that varied between 20 000 and 40 000 km. On

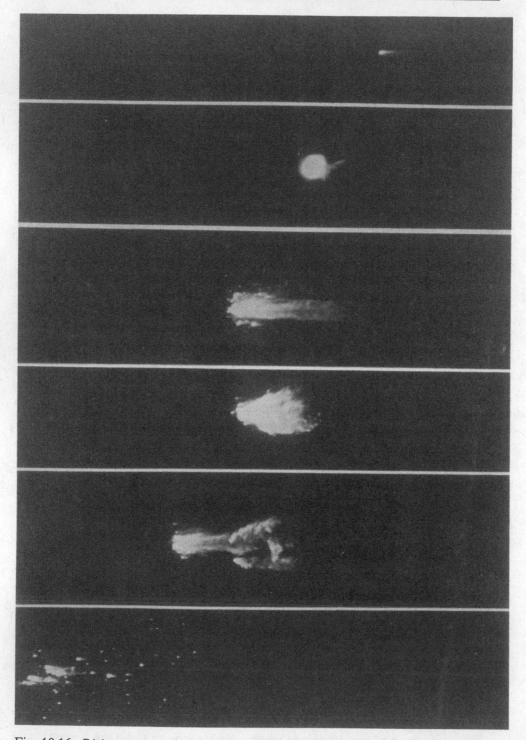

Fig. 10.16. *Disintegration of the Apollo 11 Service Module on 1969 July 24. Between the first and last images, the module dropped from an altitude of 95 km to 55 km (NASA photos).*

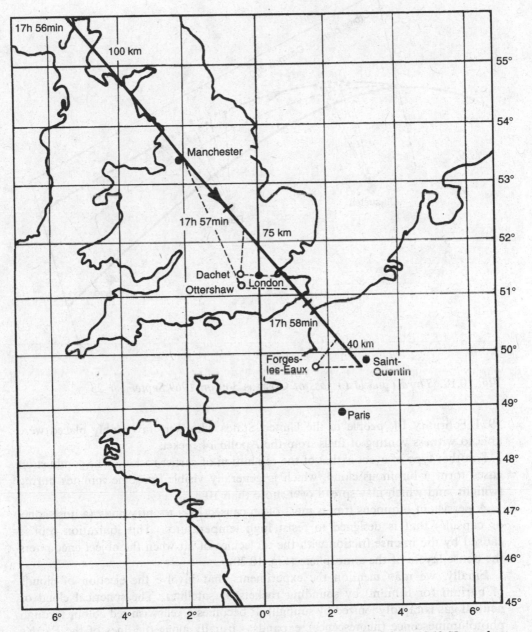

Fig. 10.17. *Trajectory of the* Cosmos-253 *rocket when it re-entered on 1968 November 20. The dashed lines indicate the azimuths of appearance and disappearance for each observing site.*

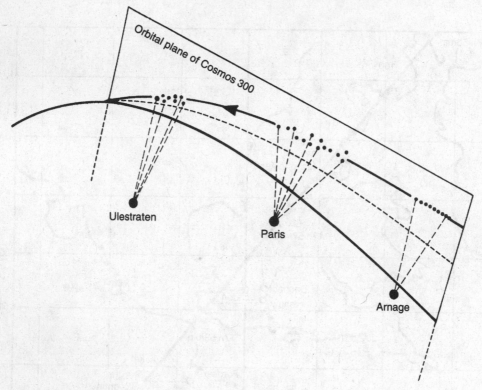

Fig. 10.18. *Third (and last) pass of* Cosmos *300 on 1969 September 23*

1971 February 14, people in the United States who were favourably placed were able to witness venting of fuels from the Apollo 14 rocket.

Another type of event is that of the re-ignition of carrier rockets. The combustion gases form a luminous cloud, which is generally visible for some minutes during twilight, and which may spread over more than 10°!

A persistent luminous trail is most often caused by a re-entry body (a nose-cone, or capsule) that is designed to resist high temperatures. This ionization trail is caused by the intense friction with the molecules of air when the object encounters the dense layers of the atmosphere (Fig. 10.20).

Finally, we may mention the experiments that involve the ejection of clouds of barium (or lithium) by sounding rockets or satellites. The spherical cloud of neutral gas is rapidly ionized by sunlight. It becomes a yellowish-red colour through photoluminescence (fluorescence), expands – usually along the lines of the Earth's magnetic field – and then fades over a period of several minutes. When carried out close to the Earth, these studies are designed to investigate the magnetosphere; farther out, the solar wind. The latter type of experiment is sometimes called an artificial comet.

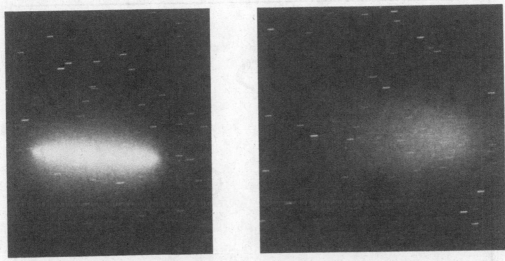

Fig. 10.19. *Fuel dump from* ApOLLO *12 on 1969 November 14 at 20:29 UT left, and 20:31 right (photos W. Seggewiss).*

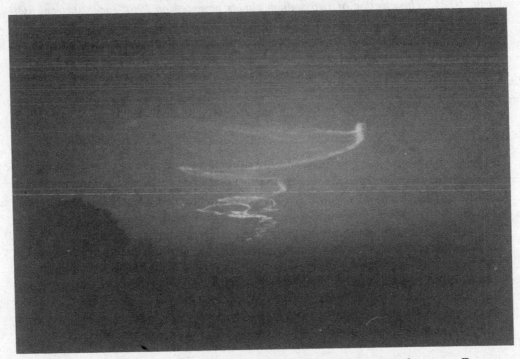

Fig. 10.20. *The launch of a payload from the Landes test centre in southwestern France on 1975 July 16. The ionization trail was rapidly distorted by upper-atmosphere winds (photo H. Le Tallec).*

575

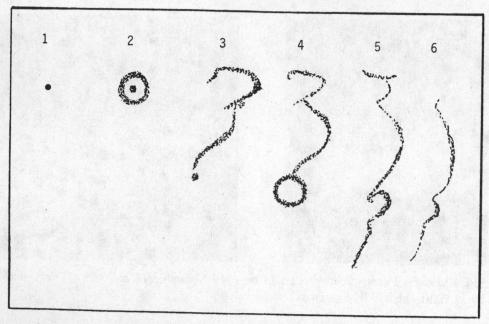

Fig. 10.21. *The ejection of two sodium clouds observed on 1961 June 10, from Sidi-Brahim-El-Guerich (350 km NE of Hammaguir) in Algeria.*
1: 20:10 UT: first ejection, 35° above the south-western horizon,
2: formation of a luminous halo,
3: second ejection, 5° lower, magnitude +1,
4 & 5: distortion of the luminous trail,
6: end of naked-eye visibility at about 20:15 UT.
(Drawing by R. Futaully)

10.6.4.2 *Phenomena of atmospheric origin*

A few lucky people, who happen to be in the vicinity of a launch site, may have witnessed the clouds of sodium that are ejected by certain sounding rockets. Visible during twilight, the clouds are still in sunlight and become luminescent by optical excitation. The sodium clouds are subject to the strong winds found at high altitudes, and are used to study the upper atmosphere by determining its temperature, density, etc. (Fig. 10.21).

Finally, we may mention the balloon experiments, which some people still mistake for flying saucers!

10.7 Observation of geostationary satellites

10.7.1 *General*

Geostationary satellites are a distinct class of object, partly because they are relatively faint, but primarily because of their apparent immobility in the sky. It was the science-fiction writer Arthur C. Clarke – the author of '2001, A Space Odyssey' –

Table 10.4. *Longitude of some geostationary satellites*

Name	Longitude		Longitude
Marecs B-2	26° W	Positions assigned to	34.5° W
Telecom 1-A	8° W	the Intelsat network	31.0° W
Telecom 1-B	5° W	(over the Atlantic)	27.5° W
Meteosat 2	0°		24.5° W
Ecs 2	7° E		21.5° W
Ecs 1	13° E		18.5° W
Arabsat 1-A	19° E		4.0° W
Arabsat 1-B	26° E		1.0° W

who, in 1945, first suggested the idea of placing a satellite at a distance of 36 000 km from the Earth in the plane of the equator, to serve as a telecommunications relay satellite. At that height, it takes 24 hours to complete one orbit, a time that is identical to the Earth's rotation, so that to any observer on the ground it appears fixed in the sky. Syncom 3 (39 kg), launched on 1964 August 19, was the first truly geostationary satellite. Its two predecessors had residual inclinations of 33°, which caused them to swing backwards and forwards across the equator, so they were only geosynchronous (rather than geostationary) satellites. [Geostationary satellites are relatively stable in longitude (Table 10.4). Normally, however, there is a small residual drift in longitude because of irregularities in the geoid, for which compensation has to be made from time to time. – Trans.]

Nowadays there are more than 300 geostationary satellites. They are used for a wide range of missions: meteorology, scientific research, direct broadcasting, early-warning, etc. There has been a spectacular increase in the payloads orbited, with satellites of up to 2 tonnes being placed in this unique orbit.

10.7.2 Injection into geostationary orbit

Shortly after launch, the carrier rocket and the payload follow an orbit with perigee at about 200 km and apogee at around 36 000 km. The thrust required to give a circular orbit is provided by a rocket motor, which is often incorporated into the satellite, which separates from its carrier rocket in transfer orbit. The ignition was first observed in 1963 July, when Syncom 2 was launched. The plume of gas caused by the apogee motor caused the satellite to appear like a short-lived comet.

10.7.3 Brightness of the satellites

Satellites stabilized in three axes that are fitted with solar panels and large parabolic aerials are good targets for amateur astronomers, who will, however, require a telescope at least 150 mm in diameter to detect them easily. A fairly high magnification

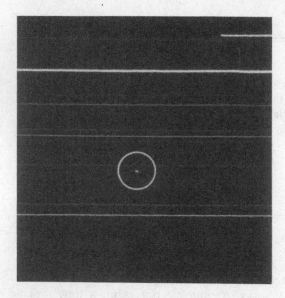

Fig. 10.22. *The geostationary telecommunications satellite* ARAB- SAT *1-B photographed on 1985 October 11, using a 310-mm reflector, exposure 10 minutes on Kodak Tri- X film (photo by A. Grycan and H. Le Tallec).*

is required to pick them out from the stars, which drift by at a rapid rate – 15 arc-minutes per minute of time – (Fig. 10.22).

Here are a few magnitude estimates made at the Jolimont Observatory at Toulouse, with a 310-mm reflector:

INTELSAT-V	+10
ARABSAT	+10.5
TELECOM	+11
MARECS	+11
ECS	+11.5

Because of the phase effect, satellites that are close to the meridian are usually best visible in the middle of the night, when they are aligned with the Sun and the Earth. Satellites that are spinning, and in the form of cylinders covered with very dark solar cells, are very much more difficult targets, especially because they frequently have mesh aerials. An object such as METEOSAT has a magnitude of +14. However, they may gain 4 or even 5 magnitudes at the equinoxes, particularly when the declination of the Sun is similar to the local declination of the satellite. Under these circumstances sunlight is specularly reflected in the direction of the observer from a narrow band that runs the full length of the cylindrical body of the satellite. This is also the period when eclipses may occur, and which are only possible between February 27 and April 12, and between August 31 and October 16. Their durations cannot exceed 1^h12^m. Visually, a distinct decrease in the brightness of the satellite occurs about 1 minute before it disappears completely inside the Earth's shadow (Fig. 10.23).

Accurate position measurements (to an accuracy of 2 seconds of arc) of geostationary satellites can be major contributions to the study of the Earth's internal structure, because the changes in the gravitational field over a period of time that

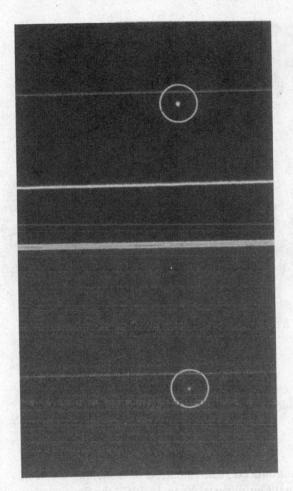

Fig. 10.23. *Eclipse of* Telecom *1-A photographed on 1985 October 10, with a 310-mm reflector. Taken a few minutes apart, these two images show the satellite's progressive entry into the Earth's shadow (photos by A. Grycan and H. Le Tallec).*

they reveal indicate mass-flow beneath the crust. Recently, the international CO-GEOS programme (Campaign for Optical Observation of Geosynchronous Satellites) has been approved, and is supported by the International Coordination of Space Techniques for Geodesy and Geodynamics (CSTG). Amateurs interested in this programme should contact the 'Artificial Satellite Section' of the British Astronomical Association.

10.8 Professional methods of observing

Initially, the techniques used for observing satellites were directly derived from astronomical techniques. They were abandoned in the 1970s, but with the introduction of new detectors (CCDs) in the past 5 years or so, telescopes have again been employed. It is now possible to determine the shape and attitude of satellites accurately from the ground.

Fig. 10.24. *The* ASTRO *2 theodolite on the top of the Solar Tower at Meudon was operational until 1969 (photo R. Futaully)*

10.8.1 Visual work

The U.S.A. and the U.S.S.R. both announced their intention of launching artificial satellites during 1958, the International Geophysical Year. In 1957, the Smithsonian Astrophysical Observatory (SAO) and the U.S.S.R. Academy of Sciences set up networks of voluntary observers: the MOONWATCH and INTERCOSMOS networks.

With the naked eye, it is possible to determine the position of an object that is moving rapidly relative to the stars to an accuracy of 0.5° in position and to 0.5–1.0 s of time. Using a small telescope, however, allows an accuracy of 0.1° in position, and the event may be timed to 0.1 s of time with a stopwatch. This is why MOONWATCH observers were provided with 6×50 monocular telescopes, with a field of 12°. Some countries preferred to use theodolites of the type normally employed for meteorology or geodesy. In France, theodolites (ASTRO 1 and 2) were specially built for satellite observation (Fig. 10.24). Cinetheodolites, which were already in use for aeronautical purposes and by the armed services, were used for more peaceful purposes.

10.8.2 *Photographic techniques*

In 1956, the SAO laid down the characteristics of a typical camera, because photographing satellites against the background of stars would enable satellites' positions to be determined to an accuracy of 2 arc-seconds. This is the arc described by a satellite crossing the zenith, 500 km above a station, in 0.0005 s (and which amounts to about 5 m). The optical system had to be as fast as possible, which meant a large aperture, as well as a focal length of 0.5–1 m, and a field of at least 10°.

The mounting needed to allow different modes of photography:

- fixed: the camera is stationary, and can only photograph bright satellites;
- sidereal: the camera is equatorially mounted; stars therefore appear as points of light, but again only the brightest satellites are photographed;
- tracking: the camera tracks the satellite, making it possible to detect very faint objects (magnitude +14).

Shutters and a timing system enable reference marks to be recorded on the satellite's track (with fixed or sidereal cameras), or on those of the stars (with a tracking camera), the times being determined to an accuracy of 0.001 s or even better.

Among the tracking cameras used in the period 1975–80, we may mention the Soviet VAU cameras (Maksutov design, $D = 500$ mm, $F = 700$ mm); the Baker–Nunn (Super-Schmidt, $D = 500$ mm, $F = 500$ mm) and AFU 75 cameras ($D = 210$ mm, $F = 740$ mm); the German SBG camera (Schmidt, $D = 430$ mm, $F = 790$ mm); the British Hewitt camera (Super-Schmidt, $D = 610$ mm, $F = 610$ mm); and the only French camera, the Antares ($D = 300$ mm, $F = 900$ mm).

10.8.3 *Laser techniques*

This technique became operational in 1964 October, with the first satellite to be fitted with laser reflectors, EXPLORER 22. The principle of laser ranging is measurement of the time taken for the laser pulse to make the return trip between the laser and the (specially equipped) satellite. Although the principle is simple, a laser station consists of a very complex set of equipment, including:

- a transmitter (the laser itself);
- a receiver (telescope and photomultiplier);
- an altazimuth mounting;
- timing equipment;
- an accurate clock (for precise determination of the timing);
- computation equipment (one or more computers), for computing predictions, driving the mounting, reduction of the data, etc.;
- miscellaneous peripheral equipment.

This technique is the one most frequently used nowadays. Between 20 and 30 stations observe half-a-dozen satellites that are fitted with laser retro-reflectors, and work both during daylight and at night. Using this method, the distances are measured to an accuracy of a few centimetres. It should also be worth recalling that

Fig. 10.25. *The third-generation laser telemetry tower at the CERGA Observatory. The receiving telescope, a Cassegrain (diameter 1 m, F = 10 m), carries a TV imaging system (on top), capable of reaching magnitude +14, and the afocal, laser system (on the side); (photo R. Futaully).*

laser ranging is also used with our natural satellite, the Moon, and that it provides fundamental data for the study of the Earth–Moon system (Fig. 10.25).

10.8.4 Radio techniques

Since the 'beep, beep' of the first SPUTNIK, there has been a vast amount of progress in using radio to determine the position of satellites and of ground stations. Several, quite different techniques have been developed to determine their relative positions, using sophisticated analysis of radio signals.

10.8.4.1 Radio interferometry
This method involves measuring the interference pattern set up between signals received by two separate aerials. For about 20 years, NASA has used its MINITRACK network (which had 12 stations) to determine the positions of satellites. The advantage of this system was that it could be employed day and night, even with cloudy skies, provided the satellite had a radio transmitter, generally emitting on a frequency of 130 MHz. This method has now been replaced by the TDRS system, using relay satellites in geostationary orbit.

10.8.4.2 Doppler ranging

The method is simple: it consists of measuring the Doppler effect, in other words the variation in the frequency of a transmitter when the latter is in motion. The Doppler technique has been used with great success by the U.S. Navy with five TRANSIT satellites in polar orbit at 1100 km altitude. Each satellite carries two, very stable transmitters emitting at frequencies of 150 and 400 MHz. The orbit of these satellites is determined to an accuracy of 1 m, 24 hours in advance. With appropriate receiving equipment, it is possible to receive the details of the orbital parameters that are transmitted by the satellites, and thus to determine one's position on the Earth. By the beginning of the 1980s, the TRANSIT system had more than 10 000 users, primarily ships, oil-production platforms, etc.

10.8.4.3 The NAVSTAR system

The principle of this system is also simple. Each NAVSTAR satellite is fitted with a very accurate clock which emits 'pips', these being tagged with an identifier as they are emitted. The receiver (a ship, ground station, etc.), is fitted with a small black box that contains a synchronized clock, and picks up the 'pips' and immediately determines the distance between it and the NAVSTAR satellite. A second measurement using a second NAVSTAR satellite enables the exact geographical position to be determined.

With three satellites, it is possible to determine one's altitude, and a fourth enables the clock error to be corrected. By 1982, six NAVSTAR satellites were in service, transmitting on frequencies of 1575 MHz and 1227 MHz. The positions obtained are accurate to 10 m, but with 'state of the art' equipment, the accuracy is 1 m.

10.8.5 Radar techniques

The radar equipment used for tracking satellites is directly derived from radars used in aeronautical applications. The radio waves are emitted in a very narrow beam that diverges little with distance. If a satellite crosses this beam, it reflects a portion of the waves back towards the aerial. This simultaneously provides both the direction and the distance.

As far as accuracy is concerned, this is generally similar to that obtained by an observer equipped with a pair of 7×50 binoculars (for a satellite orbiting at a height of 1000 km). This accuracy decreases very rapidly as the distance between the radar and the satellite increases. For example, a radar capable of detecting a satellite 1 m in diameter at 1000 km, is only able to detect a satellite at 5000 km if it is at least 25 m across. The angular accuracy remains low, because the width of the radar beam cannot be reduced below about 0.25°. Currently, however, the NORAD network has sophisticated radars that are able to detect satellites 100 mm across at a distance of more than 4000 km.

It is also possible to combine the radar and radio-interferometric techniques. This method has been applied with success in the NAVSPASUR system developed by the U.S. Navy.

10.8.6 *New optical techniques*

Beginning in 1964, and for the subsequent 20 years, the U.S. Air Force developed new optical techniques for tracking satellites at the Cloudcroft Observatory in New Mexico. The work primarily concerned the development of cameras capable of being worthy successors to the famous Baker–Nunn cameras.

The first trials of the GEODSS (Ground-based Electro-Optical Deep-Space Surveillance) system, began in 1982. It consists of five stations, located in New Mexico, Hawaii, South Korea, Diego Garcia in the Indian Ocean, and Portugal. Each station has three domes, each of which holds a telescope. The control and service buildings (like the domes and telescopes) are designed to be dismantled and transported to another site by air.

Each station has two telescopes, 1 m in diameter, 2.20 m in focal length (with a field of 2.1°), as well as an auxiliary telescope, 380 mm in diameter, with a 6° field. Each telescope is fitted with a television camera with a Vidicon tube, 80 mm in diameter, with a SIT (Silicon-Intensified Target) detector. The magnitude attained is between +16 and +18.5 with the 1 m telescopes and +14.5 with the auxiliary telescope.

The use of particularly sophisticated image-enhancement techniques enables the field to be completely analyzed and the position of the satellite determined within 1 or 2 minutes – which is about 100 times faster than with Baker–Nunn cameras. In fact, thanks to the magnitudes that can be reached, it is possible to obtain images of all satellites currently in orbit around the Earth, including those the size of a football at a distance of 36 000 km!

10.9 Observing networks

It is important to recall here the involvement of amateur observers in tracking these space-age satellites.

10.9.1 *Historical*

In 1955, during preparations for the International Geophysical Year, the Smithsonian Astrophysical Observatory (SAO) decided to create a vast network of amateur, artificial-satellite observers, equipped with short, wide-field telescopes (*see* Sect. 10.8.1).

The aim was to gather groups of observers who would observe together. Each observing station, which consisted of some ten instruments, a time-signal receiver, a tape recorder, etc. (Fig. 10.26), was to obtain the positions of all artificial satellites observable visually, and to forward the data to the SAO. This became known as the MOONWATCH network.

It was an unprecedented success. Attracted by the novelty of the programme, a large number of amateurs from all walks of life took part. At the beginning of 1958, the MOONWATCH network had 230 teams, half in the United States, with the others scattered over Canada, Japan, Australia, South America, the Pacific islands, etc.

At the end of the first year of activity, 6816 observations had been made; rising

Fig. 10.26. *The* MOONWATCH *station operated by a high school in Albuquerque (NM)*

to 100 000 after ten years, and to a grand total of 400 000 observations of 6000 artificial satellites when the programme came to an end on 1975 June 30.

In 1957, Paul Muller, who had observed the first SPUTNIK from the Pic du Midi Observatory, set up an observing network at several French observatories (Meudon, Strasbourg, Besançon and Bordeaux). This professional network continued until 1968.

At the same time, under the auspices of the Royal Aircraft Establishment in the U.K., Desmond King-Hele created the Radio and Space Research Station (RSRS) at Slough; largely drawn from amateur observers, this network may be considered as a European MOONWATCH. At the beginning of 1980, about 100 participants received various publications produced by the Appleton Laboratory (the former RSRS): ephemerides and information bulletins (the latter being prepared by a Frenchman, Pierre Neirinck).

The INTERCOSMOS network was the equivalent of MOONWATCH in the U.S.S.R. and the Eastern bloc countries.

10.9.2 *The current situation*

In 1991, the only major network that remains is one based in the United Kingdom, co-ordinated by the Optical Working Group of the Royal Society. Predictions are produced by the Royal Greenwich Observatory (RGO), primarily for its laser-ranging work, but also for priority satellites, as determined by the Working Group. The RGO compiles a data archive of observations, which are available to any interested scientists. The demand for data remains high, and any observations by amateurs are likely to be used. There is an extensive amateur network that extends well beyond the United Kingdom, including observers in France, Belgium, the Netherlands, Germany, Japan, Canada and the United States. The most experienced amateur observers receive predictions directly from the RGO, while others, particularly beginners, obtain details through the Artificial Satellite Section of the British Astronomical Association (Burlington House, Piccadilly, London W1V 9AG). Observations from all observers are sent to the RGO. Certain predictions for bright objects (such as MIR and the SPACE SHUTTLE) are also made available by Astronomy Ireland (P.O. Box 2888, Dublin 1, Ireland), some of which are distributed through the *Early Warning Circular* electronic-mail service provided by *The Astronomer*.

In North America, orbital elements are available on request from Goddard Space Flight Center, Maryland, but users must then derive their own predictions. There is also the Amateur Satellite Observers group (J. Hale, HCR 65, Box 261-B, Kingston, Ark. 72742). Revised orbital elements for more than 200 satellites are made available every few weeks on the CompuServe computer network, together with public-domain and shareware prediction programs. (To access, enter 'GO ASTRONOMY'.)

10.9.3 *The value of amateur observations*

In Sect. 10.8 we saw that professional techniques have become more and more precise, which may tend to discourage some amateurs. In fact, amateur observations are still of value today, especially in the case of very low satellites (with very limited lifetimes). The observations may be used to determine the density of the upper atmosphere, which is a function of solar activity, or to detect changes in orbits (caused by manoeuvres) – provided the observers report their observations to some central organisation.

Measurements of photometric periods also give information about the attitude and motion of the object, provided the measurements are made, if possible, over about 20 maxima. Such observations may appear unappealing to some, but to others they form part of a fascinating hobby. These observers hope that one day they may be privileged to witness the disintegration of a satellite, one of the most spectacular sights offered by any observation of the sky.

Notes, references and bibliography – Volume 1

Chapter 1: The Sun

References

Hale, G. E., 1935, 'A Spectroscope and Spectroheliograph – a Solar Observatory for the Amateur', in *Amateur Telescope Making*, Vol.1, ed. Ingalls, A. G., Scientific American, p. 203

Kitchin, C. R., 1984, *Astrophysical Techniques*, Adam Hilger, Bristol, p. 359.

Quentel, G., 1980, 'Risques oculaires des observations astronomiques', *l'Astronomie*, **94** (Jly–Aug.), pp. 311–17

Paul, H. E., 1953, 'Building a Birefringent Polarizing Monochromator for Solar Prominences' in *Amateur Telescope Making*, Vol.3, ed. Ingalls, A. G., Scientific American, p. 376

Petit, E., 1953, 'The Interference Polarizing Monochromator', *ibid.*, p. 413

Roques, C. E. & J. M., 1961, 'Un coronagraph d'amateur', *l'Astronomie*, **75** (Feb.), pp. 67–72

Bibliography

Beck, R., et al., (in press), *Handbook for Solar Observers*, Willmann-Bell, Richmond, Virginia. Translation from the German of *Handbuch für Sonnenbeobachter*, Berlin, 1982

Bray, R. J., Loughead, R. E. & Durrant, C. J., 1984, *The Solar Granulation*, (2nd edn), Cambridge University Press

Bray, R. J., & Loughead, R. E., 1964, *Sunspots*, Chapman & Hall, London/John Wiley & Sons, New York (reprinted 1979, Dover Publ. Inc., New York)

Durrant, C. J., 1988, *The Atmosphere of the Sun*, Hilger, Bristol

Foukal, P. V., 1990, *Solar Astrophysics*, John Wiley & Sons, New York

Hill, R., 1984, *Handbook for the White Light Observation of Solar Phenomena*, ALPO, Tucson

Noyes, R. W., 1982, *The Sun, our Star*, Harvard University Press

Stix, M., 1991, *The Sun*, Springer-Verlag, Heidelberg (corrected reprint)

Taylor, P. O., 1991, *Observing the Sun*, Cambridge University Press

Veio, F. N., 1991, *The Spectrohelioscope*, publ. by author, Clearlake Park, CA

Wenzel, D. G., 1989, *The Restless Sun*, Smithsonian Institution Press

Zirin, H., 1988, *Astrophysics of the Sun*, Cambridge University Press

Chapter 2: Observing the Sun with a coronagraph

References

Dollfus, A., 1983a, 'Bernard Lyot et l'invention du coronographe et l'étude de la couronne: une cinquantenaire', *l'Astronomie*, **97** March, pp. 107–29

Dollfus, A., 1983b, 'Un pionnier de l'astronomie moderne, Bernard Lyot et le coronographe: la suite de l'oeuvre', *l'Astronomie*, **97** July–August, pp. 315–29

Dollfus, A., 1983c, *Journal of Optics*, **14**, Paris

Françon, M., 1984, *Séparation des Radiations par les Filtres Optiques*, Masson, Paris

Lyot, B., 1931, *l'Astronomie*, **45**, p. 248

Lyot, B., 1932, *l'Astronomie*, **46**, p. 272

Lyot, B., 1937, *l'Astronomie*, **51**, p. 203

Mazereau, P. and Bourge, P., 1985, *A la Poursuite du soleil. La Construction du Coronographe d'Amateur*, Eyrolles, Paris

Roques, C. E. & J. M., 1961, 'Un coronographe d'amateur', *l'Astronomie*, **75** (Feb.), pp. 67–72

Chapter 3: Solar eclipses

References

Dunham, D., *et al.*, 1980, *Science*, **210**, p. 1243

Koutchmy, S., 1983, 'Les eclipses totales et l'environnement solaire', ['Total solar eclipses and the solar environment'], *l'Astronomie*, **97**, pp. 177–85

Meeus, J., Grosjean, E. and Vanderleen, W., 1966, *Canon of Solar Eclipses +1898 – +2510*, Pergamon, Oxford

Zirker, J. B., 1980, *Science*, **210**, p. 1313–19

Bibliography

L'Astronomie, 1986 November, special number 'Eclipses totales solaires' ['Total solar eclipses']

Laffineur, M., 1969, 'L'observation photographique ponderé de la couronne solaire' ['Weighted photographic observations of the solar corona'], *l'Astronomie*, **83**, pp. 337–53

Couderc, P., 1961, *Les Eclipses*, Collection 'Que sais-je?', P.U.F., Paris

Zirker, J.B., 1984, *Total Eclipses of the Sun*, Van Nostrand Reinhold, New York

Chapter 4: The Moon

References

Asaad, A. S. & Mikhail, J. S., 1974, 'Colour contrasts of lunar grounds', *The Moon*, **11**, pp. 273–99

Fitton, L. E., 1975, 'Transient lunar phenomena – a new approach', *J. Brit. Astron. Assoc.*, **85** (6), pp. 511–27

Kopal, Z., 1966, 'Photometry of scattered moonlight', in *An Introduction to the Study of the Moon*, D. Reidel, Dordrecht, pp. 323–36

Mikhail, J. S., & Koval, I. K., 1974, 'Dependence of normal albedo of lunar regions on wavelength', *The Moon*, **11**, pp. 323–6

Mills, A. A., 1980, 'Possible physical processes causing transient lunar events', *J. Brit. Astron. Assoc.*, **90** (3), pp. 219–30

Pieters, C. M., 1978, 'Mare basalts on the front side of the Moon: A summary of spectral reflectance data', in *Proceedings of 9th Lunar and Planetary Science Conference*, ed. Merrill, R. B., Lunar and Planetary Laboratory, pp. 2825–49

Westfall, J.E., 1991, 'The Luna Incognita Project', *Sky & Telescope*, **82**, pp. 556–9 (November)

Bibliography

Guest, J. E. and Greeley, R., 1977, *Geology of the Moon*, Wykeham Publications, London

Kopal, Z., 1966, *An Introduction to the Study of the Moon*, D. Reidel, Dordrecht, Netherlands

Kopal, Z. and Goudas, G. L., 1967, *Measure of the Moon*, D. Reidel, Dordrecht, Netherlands

Link, F., 1969, *Eclipse Phenomena in Astronomy*, Springer-Verlag, New York

Rükl, A., 1976, *The Moon, Mars and Venus*, Hamlyn, London

Rükl, A., 1991, *Atlas of the Moon*, Hamlyn, London

Taylor, S. R., (ed.), 1980, *Lunar Science: A post-Apollo view*, Pergamon, Oxford, and Lunar and Planetary Institute, Houston

Viscardy, G., 1985, *Atlas-Guide Photographique de la Lune*, Association franco-monégasque d'Astronomie

Items concerning Luna Incognita

J. Assoc. Lunar and Plan. Obs., 1982, **29**, (9–10), pp. 180–3

J. Assoc. Lunar and Plan. Obs., 1983, **30**, (3–4), pp. 61–6; 1984, (9–10), pp. 199–205; 1984, (11–12), pp. 221–6

J. Assoc. Lunar and Plan. Obs., 1986, **31**, (5–6), pp. 119–23; 1986, (11–12), pp. 258–60

Sky & Telescope, 1984 Mar., pp. 284–6; 1991 Nov., pp. 556–9

Specialized journals

Journal of Geophysical Research
Proceedings Lunar and Planetary Science Conference
Earth, Moon and Planets
Icarus

Organisations undertaking the study of the Moon

Association of Lunar and Planetary Observers, (Dr John E. Westfall), P.O. Box 16131, San Francisco, California 94116, U.S.A.

British Astronomical Association, Burlington House, Piccadilly, London W1V 9AG

Observations of lunar eclipses should be sent to: *Sky & Telescope*, Sky Publishing Corporation, P.O. Box 9111, Belmont, Mass. 02178-9111, U.S.A.

Chapter 5: Planetary surfaces

References and bibliography

Alexander, A. F. O'D., 1962, *The Planet Saturn*, Faber and Faber, London

Antoniadi, E., 1930, *La Planète Mars*, Hermann et Cie, Paris; English edition: trans. Moore, P., Keith Reid, Shaldon, Devon, 1975.

Boyer, C., 1965, 'Recherches sur la rotation de Vénus', *l'Astronomie*, **79**, p. 223

Briggs, G. A. and Taylor, F. W., 1982, *The Cambridge photographic Atlas of the Planets*, Cambridge University Press, Cambridge

Budine, P. W., 1981, 'Jupiter in 1978–1979', *Strolling Astronomer*, March

Budine, P. W., 1981, 'Jupiter in 1979–1980: rotation periods', *Strolling Astronomer*, July

Budine, P. W., 1983, 'Jupiter in 1980-1981: rotation periods', *Strolling Astronomer*, March

Budine, P. W., 1983, 'Jupiter in 1981-1982: rotation periods', *Strolling Astronomer*, June

Budine, P. W., 1986, 'The 1983 apparition of Jupiter: Rotation periods', *Strolling Astronomer*, January

Camus, J. and Dollfus, A., 1956, 'Les dessins des surfaces planétaires', *l'Astronomie*, **70**, p. 48

Capen, C. and Rhoads, 1978, 'Mars 1975–1976 aphelic apparition, report I', *Strolling Astronomer*, April

Capen, C. and Parker, D., 1979, 'Mars 1975–1976 report II', *Strolling Astronomer*, October

Capen, C. and Parker, D., 1981, 'What is new on Mars – Martian 1979–1980 apparition, report II', *Strolling Astronomer*, July

Dollfus, A., 1953, 'Observation visuelle et photographique des planètes Mercure et Vénus l'observatoire du Pic du Midi', *l'Astronomie*, **67**, p. 61

Dragesco, J., 1972, 'La planète Mars en 1971', *l'Astronomie*, **86**, July-August

Dragesco, J., 1975, 'La planète Mars en 1973', *l'Astronomie*, **89**, January

Dragesco, J., 1978, 'La planète Mars en 1975–1976', *l'Astronomie*, **92**, January

Dragesco, J., 1981, 'La planète Mars en 1977–1978', *l'Astronomie*, **95**, January

Dragesco, J., 1983, 'La planète Mars en 1979–1980', *l'Astronomie*, **97**, March

Dragesco, J., 1986, 'La planète Mars en 1981–1982', *l'Astronomie*, **100**, July-August

Dragesco, J., 1969, 'La vision dans les instruments astronomiques et l'observation physique des surfaces planétaires', *l'Astronomie*, **83**, pp. 355–65, 399–408, 439–47

Dragesco, J., 1970, 'La prochaine opposition de la planète Mars, conseils aux observateurs', *l'Astronomie*, **84**, pp. 119–27

Hunt, G. and Moore, P., 1981, *Jupiter*, Mitchell Beazley Publications, London

Lyot, B., 1943, 'Observations planétaires au Pic du Midi en 1941', *l'Astronomie*, **57**, p. 49

Mackal, P. K., 1982, 'Synoptic report of the 1977–1978 apparition of the planet Jupiter', *Strolling Astronomer*, April

Mackal, P. K., 1982, 'Synoptic report of the 1979–1980 apparition of the planet Jupiter', *Strolling Astronomer*, December

Mackal, P. K., 1985, 'Synoptic report of the 1980–1981 apparition of the planet Jupiter', *Strolling Astronomer*, February

Mackal, P. K., 1985, 'Synoptic report of the 1980–1981 apparition of the planet Jupiter (concluded)', *Strolling Astronomer*, October

McKim, R., 1982, 'Observing Mars in the 1980s', *J. Brit. Astron. Assoc.*, **92** (4), pp. 170–6

McKim, R., 1984, 'The opposition of Mars 1980', *J. Brit. Astron. Assoc.*, **94** (5), pp. 197–210

McKim, R., 1986, *J. Brit. Astron. Assoc.*, **96** (3) p. 166

Martinez, P., 1983, *Astrophotographie – les techniques de l'amateur*, Société d'Astronomie populaire, Toulouse

Néel, R., 1982, 'La planète Jupiter en 1978–1979', *l'Astronomie*, **96**, June

Néel, R., 1983, 'La planète Jupiter en 1979–1980', *l'Astronomie*, **97**, January

Néel, R., 1983, 'La planète Jupiter en 1980–1981', *l'Astronomie*, **97**, September

Néel, R., 1984, 'La planète Jupiter en 1981–1982', *l'Astronomie*, **98**, July–August

Néel, R., 1986, 'La planète Jupiter en 1983', *l'Astronomie*, **100**, March

O'Leary, B., Beatty, J. K. and Chaikin, A., 1990, *The New Solar System*, Cambridge University Press, (3rd edn)

Olivarez, J., 1984, 'On the blue cloud features of Jupiter's NEBs-EZn region', *Strolling Astronomer*, August

Olivarez, J., 1984, 'Some recent 1984 observations of Jupiter', *Strolling Astronomer*, November

Parker, D. C., Capen, C. F. and Beish, J. D., 1983, 'Exploring the martian arctic', *Sky and Telescope*, pp. 341–52

Peek, B. M., 1958, *The Planet Jupiter*, Faber and Faber, London; reprinted with minor corrections, 1981 and also published by Dover, N.Y.

Sidgwick, J. B., 1961, *Amateur Astronomer's Handbook*, Faber and Faber, London, p. 49

Vaucouleurs, G. de, 1945, 'Instruction complémentaires pour l'observation de la planète Mars', *l'Astronomie*, **59**, p. 175

Vaucouleurs, G. de, 1950, *The Planet Mars*, Faber and Faber, London

Chapter 6: Planetary satellites .

References

Arlot, J. *et al.*, 1982, 'Les resultats de la compagne d'observation PHEMU-79 des phénomènes des satellites galiléens de Jupiter en 1979', *Astronomy and Astrophysics*, **111**, pp. 151–70

'Journées PHEMU-85 sur l'observation des phénomènes mutuels des satellites de Jupiter en 1985', *Supplément aux Annales de Physique* (1987), **12**, Colloque n° 1, Supplément n° 1 (1987 February)

Bibliography

Morrison, D., 1982, ed., *Satellites of Jupiter*, University of Arizona Press, Tucson.

Peek, B. M., 1958, *The Planet Jupiter*, Faber and Faber, London, reprinted with minor corrections, 1981 and also published by Dover, N.Y.

Sidgwick, J. B., 1955, *Observational Astronomy for Amateurs*, Faber and Faber, London (and later editions).

Chapter 7: The minor planets

References

Bowell, E., *et al.*, 1979, 'Magnitudes, colors, types and adopted diameters of the asteroids', in *Asteroids*, ed. Gehrels, T., University of Arizona Press

Combes, M. A. and Meeus, J., 1986, 'Nouvelles des Earth-grazers – 6', *l'Astronomie*, **100**, (April), pp. 181–6

Gunther, J. U., 1985, 'Asteroids and amateur astronomers', *Mercury*, **14**, (Jan.–Feb.), p. 9

Townsend, C. and Rogers, J., 1986, 'Favourable near-Earth asteroid approaches: 1986–2000', *J. Brit. Astron. Assoc.*, **96**, p. 106

Bibliography

ALPO, *Minor Planet Bulletin*, quarterly journal of the Minor Planets Section of the Association of Lunar and Planetary Observers, ed. Binzel, R. P., University of Texas, Austin

Binzel, R. P., 1983, 'Photometry of asteroids', in *Solar System Photometry Handbook*, ed. Genet, R. M., IAPPP (International Amateur–Professional Photoelectric Photometry Association), Willmann-Bell, Richmond, Va

Binzel, R. P., *et al.*, 1989, *Asteroids 2*, University of Arizona Press

Burns, J. A., ed., *Icarus, International Journal of Solar System Studies*, Academic Press Inc., Ithaca, N.Y. (A journal containing professional papers on all aspects of Solar–system studies.)

Edberg, S. J., ed., 1983, 'Astrometry' in *International Halley Watch Amateur Observer's Manual for Scientific Comet Studies, Part I, Methods*, NASA JPL

Gehrels, T., 1979, ed., *Asteroids*, University of Arizona Press (new edition: Binzel, 1989)

Lagerkvist, C. I., Rickman, H., *et al.*, 1983 & 1986, *Asteroids, Comets, Meteors*, Vols. I and II, *Proceedings* of a colloquium held at Uppsala, Sweden, 1983 June 20–22 and 1985 June 3–6, University of Uppsala Press

Institute of Theoretical Astronomy of the U.S.S.R. Academy of Sciences, Leningrad, *Ephemerides of Minor Planets*, annual. [Previously distributed by the Minor Planet Center, Cambridge, Mass., but now published (at vastly increased cost) by White Nights Trading Company, Seattle – Trans.]

Marsden, B. G., ed., *IAU. Circulars*, Central Bureau for Astronomical Telegrams, Smithsonian Astrophysical Observatory, Cambridge, Mass.

Minor Planet Center, Cambridge, Mass. *Minor Planet Circulars*

Wetherill, G. W., 1979, 'Apollo objects', *Scientific American*, **240**, p. 54 (March)

Chapter 8: Comets

Bibliography

Arbour, R., 1985, *J. Brit. Astron. Assoc.*, **96**, (1), p. 12

Everhart, E., 1987, *Sky and Telescope*, **73**, (February), pp. 208–12

International Comet Quarterly, Smithsonian Astrophysical Observatory, 60 Garden St., Cambridge, MA 02138, U.S.A.

Martinez, P., 1987a, 'L'utilisation du T 60 pour la redécouverte des cometes periodiques', *l'Astronomie*, **101**, (May).

Martinez, P., 1987b, 'Une platine pour le suive cométaire', *l'Astronomie*, **101**, October, pp. 537–40.

Marsden, B.G., 1986, *Catalogue of Cometary Orbits*, 5th edition, IAU, Cambridge, Massachusetts

Vorontsov-Velyaminov, B. A., 1969, *Astronomical Problems*, Pergamon Press, Oxford

Chapter 9: Occultations

Addresses

Lunar Section, *and* Asteroids and Remote Planets Section, British Astronomical Association, Burlington House, Piccadilly, London W1V 9AG, United Kingdom

European Asteroidal Occultations Network (EAON), Rue de Mariembourg, 35, B–5670 Dourbes, Belgium

International Occultation Timing Association (IOTA), 7006 Megan Lane, Greenbelt, MD 20770–3012, United States of America

International Lunar Occultation Centre (ILOC), Geodesic and Geophysics Division, Hydrographic Department, Tsukiji 5, Chuo-Ku, Tokyo 104, Japan

References

Arlot, J. E., Lecacheux, J., Richardson, C., & Thuillot, W., 1985, *Icarus*, **61**, p. 224

Dommanget, J., 1985, *Bulletin Astronomique de l'Observatoire de Belgique*, **X**, (1)

Kovalesky, J., 1982, 'The satellite Hipparcos', (*Proceedings* Int. Colloq. on the scientific Aspects of the Hipparcos Mission) ESA SP-177, p. 15

Kristensen, L. K., 1984, *Astronomische Nachrichten*, **305**, (4), p. 207

Millis, R. L., Wasserman, L. H., Bowell, E., Franz, O. G., & Nye, R., 1985, *Icarus*, **61**, p. 124

Bibliography

Dunham, D. W., 1985, *Manual for Occultation Observers*, International Occultation Timing Association (IOTA), Topeka, Kansas

Evans, D. S., 1977, *Sky and Telescope*, p. 164

Huxley, A., 1942, *The Art of Seeing*, Chatto & Windus, London

International Lunar Occultation Centre, 1982, *Guide to Lunar Occultation Observations*, Tokyo

Meeus, J., 1979, *Astronomical Formulae for Calculators*, Willmann-Bell, Richmond, Virginia, p. 171

Provenmire, H. R., 1979, *Graze Observer's Handbook*, JSB Enterprises, Indian Harbor Beach, Florida

Chapter 10: Artificial satellites

References

King-Hele, D. G., 1983, *Observing Earth Satellites*, Macmillan, London, and Van Nostrand, New York

Bibliography

Helms, H. L., 1988, 'Listening to Soviet spacecraft', *Sky & Telescope*, pp. 648–649, (June)

King-Hele, D. G., 1981, *R.A.E. Table of Earth Artificial Satellites 1957–1980*, Macmillan, London

King-Hele, D., 1986, 'Observing artificial satellites', *Sky & Telescope*, pp. 457–63, (May)

Maley, P.D., 1986, 'Photographing Earth satellites', *Sky & Telescope*, pp. 563–67, (June)

Mansfield, R. L., 1990, 'Determining Highest Elevation for Viewing Satellites', *Sky & Telescope*, pp. 532–4, (November)

Miles, H. G., 1974, *Artificial Satellite Observing*, Faber and Faber, London

Taylor, G. E., 1986, 'Geostationary satellites', *Sky & Telescope*, p. 557, (June)